INTRODUCTION TO PROCESS SYSTEMS ENGINEERING

INTRODUCTION TO PROCESS SYSTEMS ENGINEERING

Basic Principles and Calculations with MATLAB and Python

IOANNIS K. KOOKOS

Department of Chemical Engineering, University of Patras, Patras, Greece

ELSEVIER

Elsevier
Radarweg 29, PO Box 211, 1000 AE Amsterdam, Netherlands
125 London Wall, London EC2Y 5AS, United Kingdom
50 Hampshire Street, 5th Floor, Cambridge, MA 02139, United States

ISBN: 978-0-443-45444-8

For information on all Elsevier publications visit our website at https://www.elsevier.com/books-and-journals

Publisher: Priscilla Braglia
Senior Acquisitions Editor: Anita Koch
Editorial Project Manager: Sweta Bharti
Production Project Manager: Gomathi Sugumar
Cover Designer: Gopalakrishnan Venkatraman

Typeset by TNQ Tech

Contents

1

Introduction to chemical and process systems engineering

1.1 Definition of chemical and process systems engineering

1.1.1 What is chemical engineering

Chemical engineering is a broad and dynamic discipline that focuses on the transformation of raw materials into valuable products through chemical, physical, and biological processes. It encompasses the design, development, and optimization of equipment, systems, and processes used in the large-scale manufacturing of chemicals, fuels, pharmaceuticals, food products, and advanced materials. At its core, chemical engineering bridges the gap between laboratory discoveries and industrial-scale production. Chemical engineers take innovative processes developed in research settings and convert them into efficient, scalable, safe, and economically viable manufacturing operations. Once implemented, they continue to monitor, refine, and enhance these systems to improve efficiency, safety, sustainability, and product quality.

The foundation of chemical engineering lies in the core sciences and engineering principles—mathematics, physics, and chemistry—while biology is playing an increasingly significant role, especially in fields such as biotechnology, pharmaceuticals, and environmental engineering. Chemical engineers play a central role in a wide array of industries by designing and troubleshooting processes involved in the production of everything from everyday consumer goods to cutting-edge medical devices. Their responsibilities typically include:

- Conducting **research and development** (R&D) to improve existing processes or create new ones.
- Designing **equipment layouts** and production workflows for maximum efficiency and safety.
- Estimating **production costs** and **optimizing** resource use to improve profitability.

Introduction to Process Systems Engineering. https://doi.org/10.1016/B978-0-443-45444-8.01001-0

- **Monitoring process performance** and making data-driven adjustments to enhance productivity.
- **Ensuring compliance** with health, safety, and environmental standards and regulations and **implementing safety protocols** for the handling and processing of hazardous substances.

Chemical engineers are employed across various sectors, with a strong presence in large-scale manufacturing facilities, research laboratories, and pilot plants. Their work environment often includes both indoor and outdoor settings, and they may need to operate in proximity to complex machinery and high-temperature or high-pressure systems. As a result, they are frequently required to wear personal protective equipment, such as hard hats, safety goggles, lab coats, and steel-toe boots.

Beyond technical expertise, **collaboration** and **communication** are critical to a chemical engineer's success. Many projects require close teamwork with professionals from diverse backgrounds, including chemists, mechanical engineers, project managers, and regulatory specialists. Those working in business, operations, or management roles often travel between offices, laboratories, and production sites to oversee project implementation and performance.

Chemical engineering is a versatile and impactful discipline that drives innovation and efficiency in countless sectors. Through their expertise in science and engineering, chemical engineers play a critical role in addressing some of the world's most pressing challenges—from sustainable energy and clean water to advanced healthcare and high-tech manufacturing. Their contributions not only improve the quality of products but also help ensure the responsible use of resources and the protection of human health and the environment.

1.1.2 The origins of chemical engineering

The origins of chemical engineering can be traced back to 19th-century England, a time marked by the rapid expansion of industrial chemistry and the growing demand for key chemicals such as sulfuric acid and soda ash. This period laid the groundwork for what would become the modern discipline of chemical engineering. Two of the most notable milestones in the early chemical industry are the development of the Lead Chamber Process for the production of sulfuric acid—pioneered by John Roebuck in the 1740s—and the Leblanc Process, developed by Nicholas Leblanc in 1791, which enabled the large-scale conversion of common salt into soda ash (sodium carbonate), a vital ingredient in glass, soap, and textile manufacturing. These innovations marked the beginning of chemical manufacturing at industrial scales and signaled a shift from artisanal chemistry to a more systematic and engineering-driven approach.

With the rise of mass production came the increasing complexity of chemical processes. The chemical industry began to require professionals with not only a strong foundation in chemistry but also technical and practical expertise to enhance the efficiency, safety, and environmental sustainability of chemical operations. This need led to the emergence of a new professional identity: the chemical engineer. In response to the growing challenges faced by the chemical industry—ranging from pollution control to economic competitiveness—industrial chemists and manufacturers began to recognize the need for a dedicated professional organization. Between 1879 and 1881, a series of meetings took place among key figures in the field to discuss the formation of a society that would support the advancement of chemical engineering as a distinct discipline. A pivotal moment came in April 1880, during a meeting at Owens College in Manchester, where the name "Society of Chemical Engineers" was proposed for the new organization. This marked a significant conceptual shift, acknowledging chemical engineering not merely as applied chemistry, but as a profession in its own right.

One of the most influential figures in this formative period was George E. Davis (1850–1907), often regarded as the founding father of chemical engineering. Born in Eton and educated at the Slough Mechanics Institute and the Royal School of Mines, Davis had a varied career that deeply influenced the development of the field. Initially working as an analytical chemist, Davis was appointed in 1878 to the Alkali Inspectorate, a regulatory body established under the 1863 Alkali Act—the first piece of environmental legislation aimed at controlling industrial emissions, particularly hydrochloric acid released from alkali plants. In this role, Davis inspected chemical manufacturing facilities across northwest England, gaining extensive firsthand knowledge of industrial processes and the technical challenges associated with scaling chemical production. Davis's work as an inspector convinced him of the need for a more systematic approach to managing chemical processes—one grounded not just in chemistry, but in engineering principles, safety practices, and environmental responsibility. After resigning from the Inspectorate in 1884, he established a private consultancy practice, where he applied his insights to improve plant design, process efficiency, and pollution control.

His most enduring contribution came in 1887, when he published a series of 12 lectures on chemical engineering, laying out for the first time a structured curriculum for the discipline. These lectures were later compiled into a book—"A Handbook of Chemical Engineering"—which became the first formal text on the subject and helped define chemical engineering as a field distinct from both chemistry and mechanical engineering.

The early efforts of Davis and his contemporaries helped formalize the profession and laid the groundwork for the establishment of dedicated institutions such as the Institution of Chemical Engineers (IChemE), founded in 1922 in the UK. Their vision—to apply scientific principles and engineering methods to the design, optimization, and regulation of chemical processes—continues to guide the field today.

1.1.3 Process Systems Engineering

Process Systems Engineering (PSE) is a vital and evolving subdiscipline within chemical engineering that focuses on the systematic analysis, design, optimization, control, intensification, and integration of complex chemical, physical, and biological processes. It adopts a holistic, **systems-level approach** to understanding how individual units operate within larger process networks, with the overarching goal of improving performance in terms of efficiency, safety, sustainability, and profitability.

Emerging in the mid-20th century as chemical processes grew in complexity, PSE evolved to address the limitations of traditional process design and control methods. It integrates principles from chemical engineering, mathematics, operations research, control theory, and computer science to manage and optimize both steady-state and dynamic systems and encompasses the full lifecycle of a process (from conceptual design and simulation to operation, control, and optimization), allowing engineers to make informed decisions at every stage of development and implementation. The core areas of PSE are:

1. Process synthesis, analysis, and design
2. Modeling and simulation
3. Process control and Real–Time Optimization (RTO)
4. Optimization
5. Process intensification and integration
6. Enterprise-Wide and Supply Chain Optimization
7. Safety, Reliability, and Risk Assessment

In recent years, the scope of PSE has expanded to include data analytics, machine learning, and digitalization. These tools enhance predictive modeling, enable soft-sensor development, and support real-time optimization in smart manufacturing environments.

The value of PSE lies in its ability to translate complex industrial challenges into structured mathematical and computational problems. By doing so, it provides engineers with the tools and methodologies to design smarter systems, respond to operational uncertainties, reduce environmental impact, and ultimately drive innovation in process industries.

One of the earliest and most celebrated examples of a systems-level approach to chemical manufacturing is the **Solvay Process**, patented in 1861 by the Belgian industrialist Ernest Solvay. In many respects, Solvay can be regarded as the first Process Systems Engineer. His 1872 implementation of an ammonia-based process for the large-scale production of soda ash (sodium carbonate) represented a fundamental breakthrough in chemical manufacturing. What distinguished Solvay's contribution was not merely the chemistry involved, but his pioneering synthesis of multiple unit operations into a cohesive, continuous process. He successfully integrated gas–liquid contacting, exothermic reaction coupled with cooling, and separation techniques into a tightly coordinated system. To achieve this, Solvay not only conceptualized the overall process flow but also designed novel equipment capable of performing these integrated operations efficiently and continuously at industrial scale.

Solvay's approach exemplifies the foundational principles of PSE: the deliberate integration of chemical and physical operations, the optimization of operating conditions, the strategic use of recycle streams to enhance yields and minimize waste, and the pursuit of process efficiency from both economic and environmental perspectives. His process design inherently considered materials handling, equipment design, and process integration—all hallmarks of a systems engineering mindset. In many ways, the Solvay Process stands as an early archetype of a fully integrated chemical system—one in which the entire process is treated as a unified whole rather than as a series of isolated steps. Solvay's system-wide perspective marked a significant departure from the prevailing practices of the time, laying the groundwork for what would later be formalized as the discipline of PSE.

The enduring influence of Solvay's work has been highlighted by scholars such as Skip Scriven, who, in his influential paper "On the Emergence and Evolution of Chemical Engineering," described Solvay—despite his lack of formal university education—as an embodiment of the modern chemical engineer. More specifically, the nature of Solvay's thinking and the integrated character of his process design clearly align with what we now identify as **Process Systems Engineering**.

This historical perspective supports a compelling assertion: the original chemical engineers were, in essence, process systems engineers. For the first several decades of the profession's development, the core challenges and contributions of chemical engineering were fundamentally systems-oriented, and the profession itself was shaped by pioneers who viewed processes not just as collections of operations, but as integrated, optimized systems—a philosophy that continues to define PSE today.

1.2 The aims of this book

To understand the purpose of this book, consider the following illustrative example: Suppose you are exploring the utilization of carbon dioxide (CO_2) captured from a fossil-fuel power plant to produce methanol. This transformation can be achieved by reacting CO_2 with renewable hydrogen—commonly referred to as green hydrogen, which is produced via water electrolysis powered by renewable electricity sources. A representative process flowsheet, drawn from open literature, is shown in Fig. 1.1. The flowsheet includes several key process specifications. First, the raw material streams (CO_2 and H_2) are compressed to 100 bar and mixed with a recycle stream containing unreacted

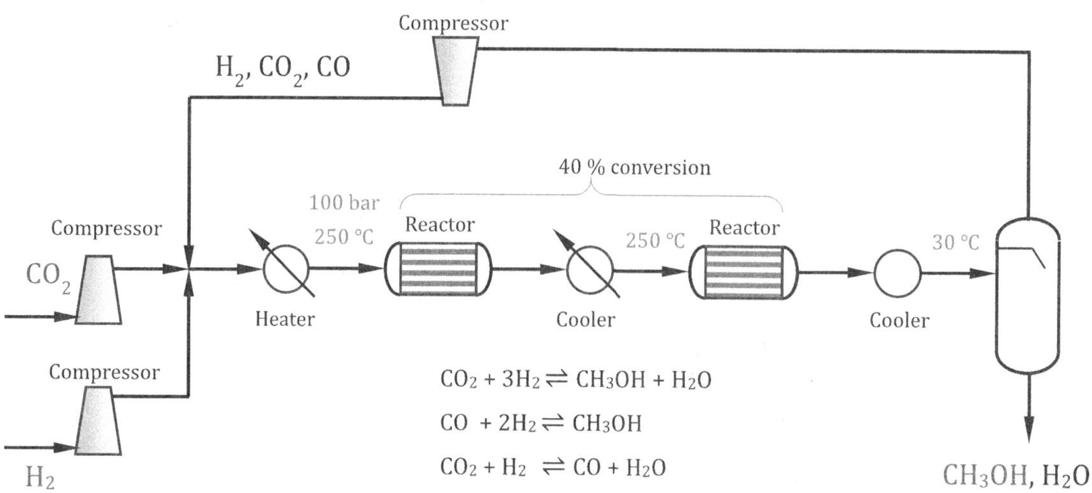

Figure 1.1 A potential process for producing methanol from carbon dioxide and hydrogen.

gases. This combined stream is then heated to 250 °C and introduced into a reactor train consisting of two catalytic reactors with intercooling between them to manage reaction heat and improve conversion. The reactor effluent is subsequently cooled and sent to a flash distillation unit, where methanol and water are condensed and removed as a liquid product. The remaining unreacted gases, primarily CO_2 and H_2, are recycled back to the feed, minimizing waste and improving overall process efficiency.

This example highlights the types of complex, integrated chemical processes that this book aims to address: processes that require systems-level thinking for their design, simulation, analysis, optimization, and control. Throughout the book, you will learn how tools from PSE can be applied to analyze such processes.

To systematically evaluate a potential process flowsheet—such as the CO_2-to-methanol conversion example introduced earlier—it is essential to conduct a detailed process analysis. This analysis involves determining the following key variables for all process streams and units:

- mass or molar flow rates,
- chemical composition,
- temperature, and
- pressure.

In addition to these stream properties, you must also calculate the heat duty (or thermal load) of each processing unit—whether it is a reactor, heat exchanger, compressor, distillation column, or any other piece of equipment.

This initial step, often referred to as process simulation or steady-state modeling, is a foundational task in PSE. It provides the essential data required to:

- Estimate equipment size and specifications.
- Determine utility requirements, including electricity, high- or low-pressure steam, and cooling water.
- Quantify material and energy balances across the system.
- Identify opportunities for heat integration and process optimization.

Ultimately, this comprehensive analysis forms the basis for evaluating both the economic feasibility and environmental performance of the proposed flowsheet. It enables engineers to make informed decisions about process viability, cost-effectiveness, sustainability, and potential for scale-up. Without such quantitative insights, it is not possible to move forward with confident design, optimization, or comparison of alternative process configurations.

In this book, the aim is to present the PSE tools necessary to perform the lower–level tasks of dynamic and steady–state simulation based on the fundamental principles of mass and energy conservation combined with a number of constitutive equations. The development of these constitutive equations will take most of your undergraduate studies and is the core material covered in thermodynamics, transport phenomena, chemical kinetics and reactor design, and separation process principles.

1.3 The contents of this book

This book comprises 12 chapters, not including the one you are currently reading. Chapter 2 is particularly important, as it provides a comprehensive review of the foundational concepts necessary to establish a common language for the chapters that follow. It covers the International System of Units (SI) and discusses the role of uncertainty in reporting the results of engineering calculations. Additionally, it revisits fundamental scientific and engineering concepts, including key process variables such as temperature, pressure, and density, as well as various methods for expressing concentration. The chapter also reviews essential aspects of the ideal gas model, including the internal energy and enthalpy of the ideal gas.

Chapter 3 is arguably the most important chapter in this book, as it introduces the general balance equation, a fundamental framework that underpins the analysis of all types of engineering processes. The chapter then applies this general equation specifically to the balance of total mass, laying the groundwork for more advanced material in later chapters. What sets this chapter apart from traditional treatments is its strong emphasis on process dynamics rather than static or steady-state systems. The concept of accumulation is central to this dynamic perspective, highlighting how material quantities change over time within a system. This dynamic approach provides a more realistic and practical understanding of how processes behave in real-world engineering applications. The chapter concludes with both analytical and numerical methods for solving the resulting balance equations, which often take the form of ordinary differential equations (ODEs). Readers are introduced to techniques for solving these ODEs using computational tools such as MATLAB and Python, equipping them with practical skills essential for modern engineering analysis and simulation.

Chapters 4 and 5 build upon the foundation laid by the general balance equation by applying it to two essential types of balances in engineering systems: the component (or species) mass or mole balance and the energy balance. In Chapter 4, the focus is on component mass balances, which track the behavior of individual chemical species within a system. A key distinction is made between systems with chemical reactions and those without, as the presence of reactions introduces additional complexity due to the generation and consumption of species. Unlike total mass, which is always conserved, individual chemical components are not necessarily conserved in reactive systems. This makes the component balance particularly important—and unique—as it is the only balance involving a nonconserved quantity that is fundamental to chemical and process systems engineering. The chapter also introduces the numerical solution of systems of ODEs, which often arise when modeling time-dependent behavior of multiple interacting species. Techniques and examples using tools such as MATLAB or Python are provided to illustrate practical implementation.

Chapter 5 shifts the focus to energy balances, essential for understanding the thermal behavior and efficiency of chemical processes. The chapter addresses both closed and open systems, incorporating concepts such as enthalpy, heat transfer, and work. Energy balances often result in nonlinear algebraic equations, especially under steady-state conditions or in systems with complex property relationships. This chapter therefore also introduces methods for the numerical solution of nonlinear algebraic equations, providing the computational tools necessary to analyze energy transformations in realistic engineering problems. Together, Chapters 4 and 5 equip readers with the analytical and numerical tools to model and solve complex, dynamic processes involving mass and energy flows, both with and without chemical reactions.

Chapter 6 further advances the discussion of mass balances by focusing on steady-state conditions, which are critical in the design and analysis of continuous processes in chemical and process engineering. The chapter emphasizes the application of material balances across multiple interconnected process units, reflecting the complexity of real-world industrial systems. A systematic approach is introduced to organize and solve these problems efficiently, enabling students and practitioners to handle large-scale process networks with clarity and confidence. The methodology presented in this chapter mirrors the core algorithms used in modern commercial process simulation software, making it highly relevant to practical engineering applications. By mastering this approach, readers gain insight into how simulators function under the hood and how to structure and troubleshoot complex process models. The analysis includes both nonreactive systems, where components are simply transferred between units, and reactive systems, where chemical transformations must be accounted for. This dual focus provides a comprehensive framework for material balance analysis under steady-state assumptions.

Importantly, Chapters 3 through 6 represent the core content of this book, forming a cohesive and rigorous foundation in process modeling. These chapters are presented with a strong emphasis on the use of modern computational tools, particularly MATLAB and Python, which are widely used in both academic research and industrial practice. By integrating theoretical concepts with hands-on computational implementation, the material prepares readers to tackle realistic engineering problems using the tools and methods prevalent in today's process industries.

Chapters 7 and 8 focus on the theoretical foundations required for calculating the thermodynamic properties of both single-component and multicomponent streams, which are essential for accurate process modeling and simulation. These chapters introduce basic thermodynamic models used to estimate a wide range of physical properties, including density, enthalpy, internal energy, and other key state functions. A central focus is placed on the modeling of phase equilibria, particularly vapor-liquid equilibrium (VLE), which is fundamental to the design and analysis of separation processes in chemical engineering. The concepts presented serve as a gateway to the practical application of thermodynamics in solving industrial case studies, especially those involving complex mixtures and separation units. One of the main applications explored is the ideal flash distillation model, which is analyzed in detail. This model is not only foundational in understanding multicomponent separation but also forms the core computational logic used in commercial process simulators. Mastering the ideal flash model enables readers to understand how simulators handle multicomponent streams and lays the groundwork for extending to more complex nonideal systems.

Moreover, Chapters 7 and 8 highlight the importance of understanding the assumptions and limitations of VLE models. This theoretical insight is critical for the effective and responsible use of simulation software, allowing users to recognize when model predictions may deviate from real-world behavior and when more advanced models or experimental data are required. By bridging theoretical thermodynamics with practical simulation tools, Chapters 7 and 8 empower readers to perform robust property calculations and make informed decisions in the modeling, design, and optimization of chemical processes.

Chapter 9 serves as a capstone to the theoretical and computational foundations developed in Chapters 2 through 8 by introducing the application of these principles in the steady-state simulation of large-scale, multi-unit chemical processes—a field known as process flowsheeting. This chapter presents the two most widely used approaches to process flowsheeting: the sequential modular approach and the equation-oriented approach. Each method is explained in detail, highlighting how they differ in terms of computational structure, flexibility, convergence characteristics, and suitability for various types of processes. By understanding these approaches, readers gain valuable insight into how commercial process simulators operate, including how they structure and solve the underlying mathematical models of chemical processes. Key components of process simulators—such as unit operation models, thermodynamic packages, property databases, and solution algorithms—are introduced and discussed. The chapter compares different types of simulator architectures, examining their advantages and limitations in relation to problem complexity, computational efficiency, and user control. To ground the theoretical discussion in practice, a relatively complex case study involving a multi-unit process is presented and analyzed in detail. This example demonstrates how to build a process flowsheet, specify inputs, interpret simulation outputs, and identify performance bottlenecks or optimization opportunities. The case study also reinforces the role of accurate thermodynamic models, mass and energy balances, and numerical methods in developing a reliable simulation. By completing this chapter, readers take a significant step toward the effective and productive use of commercial flowsheeting software. More importantly, they gain a systems-level understanding of chemical process design and analysis, enabling them to model, troubleshoot, and optimize real-world processes with confidence and competence.

Chapters 10 to 13 present material that, while independent from the core topics covered in Chapters 2 to 9, is nonetheless fundamental to a well-rounded understanding of chemical and process systems engineering. These chapters can be studied separately and offer valuable insights into both historical and modern engineering practices.

Chapter 10 focuses on graphical methods for process design, many of which were developed in the early 20th century, prior to the advent of digital computing. Despite their age, these methods remain highly relevant and are still widely used in industry and education—particularly for the preliminary design and conceptual understanding of complex systems such as chemical reactors and equilibrium-based separation units (e.g., distillation columns). These techniques are prized for their simplicity, visual clarity, and accessibility. They allow engineers to estimate key design parameters without relying on computational tools, making them especially useful during the early stages of process development or when only experimental data is available. Moreover, their graphical nature provides intuitive insights into the behavior of systems—insights that are often obscured in purely numerical or algorithmic approaches. Chapter 10 showcases these methods as elegant examples of engineering ingenuity, reflecting how early process engineers developed powerful visual tools to tackle complex design problems with minimal computational resources. Techniques such as the McCabe-Thiele method for binary distillation and Levenspiel plots for reactor design are explored in depth, illustrating how such approaches continue to support rapid analysis and decision-making in modern engineering contexts. By revisiting these classic methods, students and professionals gain a deeper appreciation for the evolution of process engineering tools, as well as practical techniques that remain effective, efficient, and relevant in many real-world applications today.

Chapter 11 introduces the concept and construction of psychrometric diagrams, which are powerful graphical tools used to analyze the thermodynamic behavior of gas–vapor mixtures. The chapter presents a step-by-step approach to developing these diagrams from fundamental thermodynamic principles, offering both theoretical background and practical guidance. Psychrometric diagrams are most commonly used to represent the properties of dry air–water vapor mixtures, with temperature and relative humidity as primary coordinates. Once constructed for a given total pressure (typically atmospheric) and a specified temperature range, these diagrams become highly effective tools for solving complex engineering problems graphically. They are particularly valuable in the design and analysis of air-conditioning systems, HVAC (heating, ventilation, and air conditioning) systems, and cooling towers—systems that are critical to both industrial operations and building environmental control. Engineers use psychrometric diagrams to evaluate processes such as humidification, dehumidification, cooling, heating, and mixing of air streams, allowing for rapid and intuitive assessment of thermodynamic performance without resorting to iterative numerical methods.

Importantly, while the air–water system is the most widely studied and applied, the psychrometric approach is not limited to this specific pair. The methodology can be extended to any mixture consisting of a nearly incondensable carrier gas and a condensable component that exists in both vapor and liquid phases. An example presented in this chapter is the nitrogen–toluene system, which demonstrates the broader applicability of psychrometric principles to chemical process systems beyond environmental and HVAC contexts.

By the end of the chapter, readers will have a solid understanding of how psychrometric diagrams are constructed, how to interpret and use them, and how they can be adapted to a wide range of engineering problems involving gas–vapor thermodynamics.

Chapter 12 addresses the important topic of design under excess degrees of freedom, a common scenario in chemical process modeling and optimization. In such cases, the mathematical model describing the process contains more variables than equations, resulting in a system that is underdetermined. The difference between the number of independent variables and the number of independent equations defines the system's degrees of freedom. When multiple degrees of freedom are available, the process designer is faced with a range of possible design choices, each of which could lead to different operating conditions, costs, or efficiencies. Rather than arbitrarily selecting values for the free variables, Chapter 12 introduces a systematic approach to managing this flexibility by defining a performance criterion—a quantitative measure of process effectiveness that should be either maximized or minimized. Common performance criteria include energy consumption, production rate, profitability, environmental impact, or product purity. Once a performance criterion is established, optimization techniques can be employed to explore the available degrees of freedom and identify the combination of decision variables that yields the optimal process configuration. This transforms an underdetermined system into a well-posed optimization problem, allowing engineers to make informed, data-driven decisions that align with specific design goals.

The chapter discusses both analytical and numerical optimization methods, with examples that illustrate how design decisions—such as flow rates, operating temperatures, and recycle ratios—can be adjusted within a feasible range to achieve superior performance. Tools such as MATLAB or Python are introduced to show how these problems are tackled in modern engineering practice. By the end of Chapter 12, readers will understand not only the theoretical implications of excess degrees of freedom in process design but also how to leverage optimization strategies to convert design flexibility into measurable performance improvements.

Chapter 13 stands apart from the rest of the text in both focus and scope, as it addresses a subject of critical importance to the real-world implementation of chemical processes: the challenge of process scale-up. While previous chapters concentrate on modeling, simulation, and design at a conceptual or pilot level, this chapter shifts attention to the practical task of translating a process developed at laboratory or bench scale into full-scale industrial operation. This transition—from small-scale experiments to large-scale production—is rarely straightforward and presents a unique set of theoretical and engineering challenges that go well beyond the material previously discussed. Process scale-up involves maintaining the desired process performance, safety, efficiency, and product quality while increasing the size and complexity of the system. Factors that may be negligible at lab scale—such as heat and mass transfer limitations, hydrodynamics, equipment constraints, material compatibility, and control strategies—often become dominant and must be addressed explicitly during scale-up. The chapter explores both theoretical principles and practical strategies for successful scale–up. Key topics include dimensional analysis, similarity criteria (geometric, kinematic, dynamic, and thermal), the role of pilot plants, and the use of computational tools and empirical correlations. Real-world case studies may be presented to illustrate common pitfalls and best practices in scaling up reactors, separation units, and integrated processes. Understanding the principles of scale-up is essential not only for chemical engineers working in process development and design, but also for those involved in technology transfer, commercialization, and industrial innovation. Without careful consideration of scale-up issues, even the most well-designed laboratory process can fail during implementation, leading to costly delays, inefficiencies, or safety hazards.

By the end of Chapter 13, readers will have a clear appreciation of the complexity and importance of scale-up, along with practical tools and insights necessary for bridging the gap between theory and industrial application.

Chapter summary

In this chapter, a clear and thoughtful definition of Chemical Engineering and Process Systems Engineering is presented, providing the reader with a solid conceptual foundation for the material that follows. These definitions are accompanied by a brief overview of the historical evolution of the discipline, tracing its development from the early days of industrial chemistry to the modern, multidisciplinary field it is today.

The chapter also highlights the core responsibilities and roles of chemical and process engineers in both traditional and emerging industries. These responsibilities include process design, optimization, control, safety, sustainability, and innovation, spanning sectors such as energy, pharmaceuticals, food processing, materials manufacturing, and environmental engineering. Attention is given to the increasing importance of digital tools, data-driven decision-making, and systems–level thinking in the modern practice of chemical engineering.

Finally, the chapter offers an overview of the structure and content of the book, explaining how each chapter contributes to the development of key competencies in process modeling, simulation, and design. The progression from fundamental principles to advanced applications is discussed, giving readers a roadmap for how the material builds toward a comprehensive understanding of chemical process systems. This orientation ensures that readers understand not only what will be covered, but also why each topic is important and how it fits into the broader scope of engineering practice.

References

Many definitions have been retrieved from the Wikipedia, the free encyclopedia.

Campbell, W.A., 1971. The Chemical Industry. Longman Group Limited, London.

Freshwater, D.C., 1980. George E. Davis, Norman Swindin, and the empirical tradition in chemical engineering. In: E Furter, W. (Ed.), History of Chemical Engineering, Advances in Chemistry Series no. 190. ACS, Washington.

Peppas, N.A., 1989. One Hundred Years of Chemical Engineering. Kluwer Academic Publishers, Dordrecht.

Perkins, J.D., 2003. In: Darton, R.C., Prince, R.G.H., Wood, D.G. (Eds.), Chemical Engineering – the First 100 Years, Chemical Engineering: Visions of the World. Elsevier.

Scriven, L.E., 1991. On the emergence and evolution of chemical engineering. In: Colton, C.K. (Ed.), Perspectives in Chemical Engineering – Research and Education, Advances in Chemical Engineering, vol 16. Academic Press, San Diego.

Stephanopoulos, G., Reklaitis, G.V., 2011. Process systems engineering: from Solvay to modern bio- and nanotechnology. A history of development, successes and prospects for the future. Chem. Eng. Sci. 66, 4272–4306.

2

Introduction and basic definitions

2.1 Measurements in science and the systems of units

Science is concerned with improving our understanding of the systems we study. The first step is to use measuring devices or sensors to measure variables that are related to the state of the environment. Using measurements is also crucial to our everyday life. Watches and clocks are used to measure time, calories are used to express the energy value of foods, speedometers are used to control the traveling speed of our cars, etc. Measurement is crucial to understanding our physical environment, but first let's discuss the physical sciences and the methods of scientific investigation to establish why measurements are so crucial in science.

Science (from the Latin scientia, meaning "knowledge") may be defined as any system of knowledge that is concerned with the physical world and its phenomena and that entails unbiased observations and systematic experimentation. In general, science involves a pursuit of knowledge covering general truths or the operations of fundamental laws. The natural sciences are divided into the biological sciences (sometimes called life sciences), which are concerned with the study of living matter, and the physical sciences, which involve the study of nonliving matter.

Today's scientists come to conclusions in a very well–organized and structured way, and measurements are the basis of scientific investigation. Observations of phenomena lead to questions regarding how or why they happen. These questions are investigated by the scientific method, which can be broken down into the following elements:

1. Making *observations* and *measurements* (quantitative data).
2. Statement of a *hypothesis* (a possible explanation for the observations).

Introduction to Process Systems Engineering. https://doi.org/10.1016/B978-0-443-45444-8.00003-8

3. Performing *experiments* (testing of the hypothesis under controlled conditions to see whether the test results confirm our hypothesis and are reproducible and consistent).
4. Develop a *theory*. If a hypothesis is validated by the experimental tests and predictions that are based on theory are proven correct, then it takes on the status of a theory (a well-tested explanation of observed natural phenomena).
5. Accept the theory as a *law*. If a theory withstands the test of many well-designed experiments, then the theory may be accepted by scientists as a law (a concise statement in words or mathematical equations that describes a fundamental relationship of nature).

It is self-evident that measurements of physical quantities are vital to our understanding of natural physical and chemical phenomena.

2.2 Systems of units

We measure each physical quantity in its own units, by comparison with a **standard**. The **unit** is a unique name we assign to measures of that quantity, for example, the meter (m) for the quantity length. The standard corresponds to exactly 1.0 unit of the quantity. The standards usually have a rather complex definition, as we will see shortly. The standard for length, which corresponds to exactly 1.0 m, is the distance traveled by light in a vacuum during a certain fraction (1/299,792,458) of a second!

We can define a unit and its standard in any way, but it must be sensible and practically useful so that scientists around the world will agree with our selection. Once we have set up a standard, such as the standard for length, we must work out procedures by which any length (ranging from the size of a molecule to the perimeter of our planet) can be expressed in terms of the standard. Rulers, which *approximate* our length standard, give us one such procedure for measuring length. However, many of our comparisons must be indirect, as we cannot use a ruler to measure the radius of an atom or the distance to the Sun.

There are so many physical quantities, and the problem is how to organize them. Fortunately, physical quantities are not all independent. The speed of an object is, for example, the ratio of length to time. Thus, a small number of *independent physical quantities*, such as length, time, and mass, are selected and assigned standards. We then define all other physical quantities in terms of these **base quantities** and their standards (called

base standards). Speed, for example, is defined in terms of the base quantities length and time and their base standards.

Base standards must be both accessible and invariable. If we define the length standard as the length of the foot of a person, then we offer an accessible standard—but our length standard is not invariable (as it varies from person to person). The demand for precision in science and engineering pushes us to sacrifice accessibility for invariability. We then need to make duplicates of the base standards that are accessible to those who need them.

A group of standard units and their combinations is called a **system of units**. Two major systems of units are in use today: the **metric system** and the **U.S. customary units** (emanated from the **British system**). The metric system is used throughout most of the world, and the United States is the only major country that has not gone completely metric. In this text, we will use the metric system exclusively.

2.2.1 The international system of units

In 1971, the 14th General Conference on Weights and Measures (www.bipm.org) selected **seven quantities as base quantities** to form the basis of the **International System of Units**, abbreviated **SI** from its French name (Système international d'unités) and known as the **metric system**. Table 2.1 shows the units for these quantities. Many SI derived units are defined in terms of these base units. For example, the SI unit for force, called the newton (N), is defined in terms of the base units using Newton's law: 1 N is the force necessary to provide a mass of 1 kilogram with an acceleration of 1 meter per second per second.

Table 2.1 SI base units (base quantities and dimensions).

Name	Typical symbol	Name	Symbol	Dimension
Time	t	second	s	T
Length	l, x, r, etc.	meter	m	L
Mass	m	kilogram	kg	M
thermodynamic temperature	T	kelvin	K	Θ
Amount of substance	n	mole	mol	N
Electric current	I, i	ampere	A	I
Luminous intensity	I_V	candela	cd	J

$$1 \text{ newton} = 1 \text{ N} = 1 \text{ kg} \cdot \frac{\text{m}}{\text{s}^2} \tag{2.2.1}$$

where the last collection of unit symbols is read as kilogrammeter per second squared.

To express the very large and very small quantities we often run into, we use scientific notation, which employs powers of 10. In this notation,

$$1\,234\,000\,000 \text{ m} = 1.234 \times 10^9 \text{ m} \tag{2.2.2}$$

and

$$0.000\,000\,789 \text{ m} = 7.89 \times 10^{-7} \text{ m} \tag{2.2.3}$$

Scientific notation on computers sometimes takes on an even briefer look, as in 1.234E9 m and 7.89E–7 m, where E stands for "exponent of 10" (on some calculators E is replaced with an empty space).

The main advantage of the metric system is that it is a decimal (base-10) system while the British system is a duodecimal (base-12) system (12 inches equals a foot!). A base-10 system allows easy conversion to larger and smaller units using a series of metric prefixes to express the multiples of 10. These prefixes are summarized in Table 2.2, but you will only need to be familiar with a few common ones: G, M, k, d, c, m.

Table 2.2 Most used SI prefixes.

Name	Symbol	Value	Pronunciation	Derivation G: Greek, L: Latin
tera (trillion)	T	10^{12}	TERR-uh	G:τέρας, monster
giga (billion)	G	10^9	JIG-a, jig as in jiggle, a as in about	G: γίγας, giant
mega (million)	M	10^6	MEG-a, as in megaphone	G: μέγας, great
kilo (1000)	k	10^3	KIL-o, as in kilowatt	G: χίλια, 1000
Hector (hundred)	h	10^2	HEK-to, heck-toe	G: εκατό, 100
deka (10)	da	10^1	DEK-a, deck plus a as in about	G: δέκα, 10
Deci (one-tenth)	d	10^{-1}	DES-i, as in decimal	L: decimus, tenth
Centi (one-hundredth)	c	10^{-2}	SENT-i, as in sentimental	L: centum, 100
Milli (one-thousandth)	m	10^{-3}	MIL-li, as in military	L: mille, 1000
Micro (one millionth)	μ	10^{-6}	MI-kro, as in microphone	G: μικρός, small
nano (one billionth)	n	10^{-9}	NAN-oh, an as in annual	G: νάνος, dwarf

Physical quantities are organized in a system of **dimensions**, and each of the seven base quantities used in the SI is regarded as having its own dimension. The symbols used to denote the dimension of each base unit in SI are shown in the last column of Table 2.1. All other quantities are **derived quantities**, which may be expressed in terms of base quantities according to the equations of physics, such as Newton's law. The dimensions of the derived quantities are written as products of powers of the dimensions of the base quantities using the equations that relate the derived quantities to the base quantities. The dimension of any quantity Q is written in the form of a dimensional product (the symbols dym{Q} or [Q] are used to denote "the dimensions of the quantity Q")

$$\dim\{Q\} = [Q] = \mathrm{T}^{\alpha}\mathrm{L}^{\beta}\mathrm{M}^{\gamma}\Theta^{\delta}\mathrm{N}^{\varepsilon}\mathrm{I}^{\zeta}\mathrm{J}^{\eta} \qquad (2.2.4)$$

where the exponents α, β, γ, δ, ε, ζ and η which can be small positive or negative integers, or zero, are called the **dimensional exponents**. For the example of force used in Eq. (2.2.1) as the product of mass and acceleration, we have the following

$$\dim\{F\} = [F] = \mathrm{T}^{-2}\,\mathrm{L}^{1}\,\mathrm{M}^{1}\,\Theta^{0}\,\mathrm{N}^{0}\,\mathrm{I}^{0}\,\mathrm{J}^{0} \qquad (2.2.5)$$

There are certain quantities for which the defining equation is such that **all** of the dimensional exponents in the equation for the dimension of Q are zero. Such quantities are simply numbers, and the associated unit is the **unit one**, symbol 1, although this is rarely explicitly written. An important example is that of plane and solid angles, which are treated within the SI as quantities with the unit one. The symbols rad (radian) and sr (steradian) are written explicitly where appropriate to emphasize that, for radians or steradians, the quantity being considered is, or involves, the plane angle or solid angle, respectively. It is noted that in mathematics and some areas of science, rad and sr are omitted. This is a source of confusion and debate. You need to remember that one radian is defined as the angle at the center of a circle in a plane that subtends an arc whose length equals the radius of the circle. The magnitude in radians of a subtended angle φ is equal to the ratio of the arc length (s) to the radius of the circle (r); that is, $\varphi = s/r$ ($[\varphi] = [s]/[r] = \mathrm{L}/\mathrm{L} = 1$). A right angle is exactly $\pi/2$ rad and a complete revolution is 2π rad. This explains the key fact: the **radian is a dimensionless unit** equal to 1 (and the same holds true for the sr).

As we have seen, the derived units are defined as products of powers of the base units. When the numerical factor of this product is one, the derived units are called **coherent derived units**.

The base and coherent derived units of the SI form a coherent set, designated the set of **coherent SI units**. The word "coherent" here means that equations between the numerical values of quantities take *exactly the same form as the equations between the quantities themselves*. Some of the coherent derived units in the SI are given special names and are summarized in Table 2.3.

There are 22 SI units with special names, which, together with the seven base units (see Table 2.1), form the core of the set of SI units. The units with special names that are relevant to this book are summarized in Table 2.3. All other SI units are combinations of some of these 29 units. However, when prefixes are used with SI units, the resulting units are no longer coherent, because the prefix introduces a numerical factor other than one. Prefixes may be used with any of the 29 SI units with special names except for the base unit kilogram. For historical reasons, the kilogram is the only coherent SI unit that includes a prefix in its name and symbol. Names and symbols for decimal multiples and sub-multiples of the unit of mass are formed by attaching prefix names and symbols to the unit name "gram" and the unit symbol "g," respectively. For example, 10^{-6} kg is written as milligram, mg, not as microkilogram, μkg!

The General Conference on Weights and Measures (CGPM) has responsibility for the International System of Units. In 1960, the 11th CGPM formally defined and established the SI and subsequently revised it in response to advances in science and technology. The most recent and perhaps **the most significant revision of the SI** since its establishment was made by the

Table 2.3 SI units with special names and symbols that are relevant to this book.

Derived quantity	Special name of unit	Defining equation	Symbol	Expressed in base units	Expressed in other SI units
Force	newton	$F = ma$	N	kg m s^{-2}	
Pressure/stress	pascal	$P = F/A$	Pa	kg m^{-1} s^{-2}	N/m^2
Energy/work/heat	joule	$W = F\delta x$	J	kg m^2 s^{-2}	N·m
Power	watt	$\dot{P} = W/\delta t$	W	kg m^2 s^{-3}	J/s
Celsius temperature	degree Celsius	–	°C	K	
Plane angle	radian	–	rad		1
Frequency	hertz	–	Hz	s^{-1}	

26th CGPM (2018) and is documented in the ninth edition of the SI Brochure published by The International Bureau of Weights and Measures (BIPM). According to this revision, the definition of the SI units is established in terms of a set of **seven defining constants**! The complete system of units can be derived from the fixed values of these defining constants, expressed in the units of the SI. These seven defining constants are the most fundamental feature of the definition of the entire system of units. These constants were chosen after having been identified as being the best choice, considering that the previous definition of the SI was based on seven base units. It is important to note that **any of the seven base units and the 22 SI units with special names can be constructed directly from the seven defining constants**. In fact, the units of the seven defining constants include both base and derived units. The reason for this significant revision was the following:

> ... to be of any practical use, these units not only have to be defined, but they also have to be realized physically for dissemination. In the case of an artefact, the definition and the realization are equivalent – a path that was pursued by advanced ancient civilizations. Although this is simple and clear, artefacts involve the **risk of loss, damage or change**. The other types of unit definitions are increasingly abstract or idealized. Here, the realizations are separated conceptually from the definitions so that the units can, as a matter of principle, be **realized independently at any place and at any time**.

These seven constants are presented in Table 2.4. The use of a constant to define a unit disconnects definition from realization.

Table 2.4 The seven defining constants of the SI and the seven corresponding units they define.

Constant	Symbol	Numerical value	Unit
Hyperfine transition frequency of Cs 133 atom	$\Delta\nu_{Cs}$	9 192 631 770	Hz
The speed of light in vacuum	c	299 792 458	$m{\cdot}s^{-1}$
The Planck constant	h	$6.626\ 070\ 15 \times 10^{-34}$	J·s
The elementary charge	e	$1.602\ 176\ 634 \times 10^{-19}$	C
The Boltzmann constant	k	$1.380\ 649 \times 10^{-23}$	$J{\cdot}K^{-1}$
The Avogadro constant	N_A	$6.022\ 140\ 76 \times 10^{23}$	mol^{-1}
Luminous efficacy	K_{cd}	683	$lm{\cdot}W^{-1}$

This offers the possibility that completely different or new and superior practical realizations can be developed, as technologies evolve, without the need to change the definition.

Starting with the definition of the SI in terms of fixed numerical values of the defining constants, definitions of each of the seven base units are deduced by using, as appropriate, one or more of these defining constants to give the following set of definitions for the five base units of interest to this book (see the SI Brochure for a complete list).

The second (s):	is defined by taking the fixed numerical value of the cesium (Cs) frequency, $\Delta\nu_{Cs}$, the unperturbed ground-state hyperfine transition frequency of the Cs 133 atom, to be 9 192 631 770 when expressed in the unit Hz, which is equal to s^{-1}. The effect of this definition is that the second is equal to the duration of 9 192 631 770 periods of the radiation corresponding to the transition between the two hyperfine levels of the unperturbed ground state of the 133Cs atom.
The meter (m):	is defined by taking the fixed numerical value of the speed of light in vacuum, *c*, to be 299 792 458 when expressed in the unit $m{\cdot}s^{-1}$, where the second is defined in terms of the caesium frequency $\Delta\nu_{Cs}$. The effect of this definition is that 1 meter is the length of the path traveled by light in vacuum during a time interval with duration of 1/299 792 458 of a second.
The kilogram (kg):	is defined by taking the fixed numerical value of the Planck constant, h, to be 6.626 070 15 $\times$ 10^{-34} when expressed in the unit J·s, which is equal to $kg{\cdot}m^2\ s^{-1}$, where the meter and the second are defined in terms of c and $\Delta\nu_{Cs}$. The effect of this definition is to define the unit $kg{\cdot}m^2\ s^{-1}$. Together with the definitions of the second and the meter this leads to a definition of the unit of mass expressed in terms of the Planck constant h.
The kelvin (K):	is defined by taking the fixed numerical value of the Boltzmann constant, *k*, to be 1.380 649 $\times$ 10^{-23} when expressed in the unit $J{\cdot}K^{-1}$, which is equal to $kg{\cdot}m^2\ s^{-2}\ K^{-1}$, where the kilogram, meter, and second are defined in terms of h, c, and $\Delta\nu_{Cs}$. The effect of this definition is that 1 kelvin is equal to the change of

thermodynamic temperature that results in a change of thermal energy kT by $1.380\ 649 \times 10^{-23}$ J.

The mole (mol) One mole contains exactly $6.022\ 140\ 76 \times 10^{23}$ elementary entities. This number is the fixed numerical value of the Avogadro constant (number), N_A, when expressed in the unit mol^{-1}. The amount of substance (n), of a system is a measure of the number of specified elementary entities. An elementary entity may be an atom, a molecule, an ion, an electron, any other particle, or a specified group of particles. The effect of this definition is that the mole is the amount of substance of a system that contains $6.022\ 140\ 76 \times 10^{23}$ specified elementary entities.

Although these "exact" definitions might look awkward, especially to engineering students, their definition has only theoretical and conceptual value. Anyone who feels uncomfortable with them can still feel free to assume that 1 m is the length of a prototype meter bar and 1 kg is the mass of a prototype cylinder composed of platinum–iridium stored somewhere in a protected environment (see Fig. 2.1).

Figure 2.1 Historic Dutch replicas of metric standards in the collection of Rijksmuseum, Amsterdam: iron meter with case (constructed by Étienne Lenoir, 1799; catalog number NG-2001–16-C-8), copper grave kilogram with case (1798; catalog number NG-2001–16-D-9), copper volume measures (1829; catalog number NG-2001–16-B-28). From Wikipedia https://commons.wikimedia.org/wiki/File:Metric_standards_Rijksmuseum.jpg. (Metric Standards, Rijksmuseum, photo by AKS.9955, licensed under CC BY-SA 4.0)

2.2.2 Accepted non-SI units

It is recognized that some non-SI units are widely used in both industry and academia and are expected to continue to be used for many years. Therefore, the International Committee for Weights and Measures has accepted some non-SI units for use with the SI. The non-SI units that are used in this book but are acceptable by CIPM are summarized in Table 2.5. If these units are used, some advantages of the SI are lost. The SI prefixes can be used with several of these units, but not, for example, with the non-SI units of time (i.e., mmin or Gd should not be used), mass (kt, Mt are used but not mt, or μt) or pressure (mbar is commonly used by kbar is never used).

2.2.3 The correct use of unit symbols and names of SI units

There is now a consensus on the general principles for the "correct" and consistent writing of unit symbols and numbers within the SI system of units. Compliance with these rules and style conventions supports the readability and unambiguity of numerical results expressed in SI units and eliminates the sources of expensive or lethal mistakes. One of the expensive failures that was due to lack of consistent use of units was the failure of the Mars Climate Orbiter in 1999 (a $125 million spacecraft that

Table 2.5 Accepted non-SI units.

Quantity	Name	Symbol	Value in SI unit
Time	minute	min	1 min = 60 s
	hour	h	1 h = 60 min = 3600 s
	day	d	1 d = 24 h = 1440 min = 86 400 s
	year	a or y	1 y = 365 d = 8760 h = 31 536 000 s
Volume	liter	L	1 L = 10^{-3} m^3
Mass	(Metric) ton or tonne	t	1 t = 1000 kg = 1 Mg
pressure	Bar[a]	bar	1 bar = 100 000 Pa
energy	Electronvolt	eV	1 eV = 1.602 176 634 × 10^{-19} J
Angle	Degree	o	1°= (π/180) rad
Area	hectare	ha	1 ha = 10 000 m^2

[a]bar has previously been accepted for use with the SI, but is now no longer accepted for use with the SI. However, it is widely used in the chemical and petrochemical industry and will continue to be used for years to come.

was lost in the atmosphere of the red planet). These missions were the second installment in NASA's long-term program of robotic exploration of Mars initiated in 1996. The investigation committee concluded that "... the Mars Climate Orbiter Mishap Investigation Board has determined that the root cause for the loss of the Mars Climate Orbiter spacecraft was the failure to use metric units in the coding of a ground software file, used in trajectory models ...".

When writing unit symbols and names and expressing the values of quantities, the following rules must be respected.

Rule 1. Unit symbols are printed in upright type regardless of the type used in the surrounding text.

Rule 2. Unit symbols are printed in lower-case letters unless they are derived from a proper name, in which case the first letter is a capital letter. An exception is that either capital L or lower-case l is allowed for the liter, in order to avoid possible confusion between the numeral one (one) and the lower-case letter l (el).

Rule 3. A multiple or submultiple prefix, if used, is part of the unit and precedes the unit symbol without a separator. A prefix is never used in isolation, and compound prefixes are never used.

Rule 4. Unit symbols are mathematical entities and not abbreviations. Therefore, they are not followed by a full stop except at the end of a sentence, and one must neither use the plural nor mix unit symbols and unit names within one expression, since names are not mathematical entities.

Rule 5. In forming products and quotients of unit symbols the normal rules of algebraic multiplication or division apply. Multiplication must be indicated by a space or a half-high (centered) dot (·), since otherwise some prefixes could be misinterpreted as a unit symbol. Division is indicated by a horizontal line, by a solidus (oblique stroke, /) or by negative exponents. When several unit symbols are combined, care should be taken to avoid ambiguities, for example by using brackets or negative exponents. A solidus must not be used more than once in a given expression without brackets to remove ambiguities.

Rule 6. It is not permissible to use abbreviations for unit symbols or unit names, such as sec (for either s or second), sq. mm (for either mm^2 or square millimeter), cc (for either cm^3 or cubic centimeter), or mps (for either m/s or meter per second).

Rule 7. Unit names are normally printed in upright type and they are treated like ordinary nouns. In English, the names of units start with a lower-case letter (even when the symbol for the unit begins with a capital letter), except at the beginning of a sentence or in capitalized material such as a title. In keeping with this rule, the correct spelling of the name of the unit with the symbol °C is "degree Celsius" (the unit degree begins with a lower-case d and the modifier Celsius begins with an upper-case C because it is a proper name).

Rule 8. When the name of a derived unit is formed from the names of individual units by juxtaposition, either a space or a hyphen is used to separate the names of the individual units.

Rule 9. The numerical value always precedes the unit and a space is always used to separate the unit from the number. Thus, the value of the quantity is the product of the number and the unit. The space between the number and the unit is regarded as a multiplication sign (just as a space between units implies multiplication). The only exceptions to this rule are for the unit symbols for degree, minute, and second for plane angle, °, ′ and ″, respectively, for which no space is left between the numerical value and the unit symbol. This rule means that the symbol °C for the degree Celsius is preceded by a space when one expresses values of Celsius temperature.

Rule 10. The symbol used to separate the integral part of a number from its decimal part is called the decimal marker (either the point on the line or the comma on the line). The decimal marker chosen should be that which is customary in the language and context concerned.

Rule 11. If the number is between +1 and −1, then the decimal marker is always preceded by a zero.

Rule 12. For numbers with many digits the digits may be divided into groups of three by a space, in order to facilitate reading. Neither dots nor commas are inserted in the spaces between groups of three. However, when there are only four digits before or after the decimal marker, it is customary not to use a space to isolate a single digit. The practice of grouping digits in this way is a matter of choice; it is not always followed in certain specialized applications such as engineering drawings, financial statements, and scripts to be read by a computer.

Rule 13. When multiplying the value of quantities either a multiplication sign × or brackets should be used, not a half-high (centered) dot. When multiplying numbers only the multiplication sign × should be used. When dividing the values of quantities using a solidus, brackets are used to avoid ambiguity.

The use of the correct symbols for SI units, and for units in general, is mandatory. In this way ambiguities and misunderstandings in the values of quantities are avoided. Symbols for quantities are generally single letters set in an italic font, although they may be qualified by further information in subscripts or superscripts or in brackets. Recommended names and symbols for quantities are listed in many standard references (such as the IUPAC, Quantities, Units and Symbols in Physical Chemistry, fourth Edition, 2023, which is used in this book). However, symbols for quantities are recommendations (in contrast to symbols for units, for which the use of the correct form is mandatory). In certain circumstances, authors may wish to use a symbol of their own choice for a quantity, for example to avoid a conflict arising from the use of the same symbol for two different quantities. In such cases, the meaning of the symbol must be clearly stated.

Example 2.1. Use and misuse of SI units

In the table that follows, several examples are presented with erroneous use of SI units.

Example	Rule(s) violated
$T = 20\ ^{\circ}\mathrm{C}$	Rule 1: Unit symbols are printed in upright type Rule 9: A space is always used to separate the unit from the number
$T = 20\ ^{\circ}\mathrm{C}$	Rule 9: A space is always used to separate the unit from the number
$T = 20\ ^{\circ}\mathrm{C}$	Rule 1: Unit symbols are printed in upright type
$T = 20\ ^{\circ}\mathrm{C}$	Correct
$L = 0.234\ \mathrm{m}$	Rule 9: A space is always used to separate the unit from the number
$L = .234\ \mathrm{m}$	Rule 11: If the number is between +1 and −1, then the decimal marker is always preceded by a zero.
$L = 0.234\ \mathrm{m}$	Correct
$u = 2\ \mathrm{m/sec}$	Rule 6: It is not permissible to use abbreviations for unit symbols or unit names. The use of the correct symbols for SI units, and for units in general, is mandatory. The unit for time is s, not sec.
$m = 2\ \mathrm{kgr}$	Rule 1: Unit symbols are printed in upright type The use of the correct symbols for SI units, and for units in general, is mandatory. The unit for mass is kg, not kgr.
$m = 1\ \mu\mathrm{kg}$	Rule 3: Compound prefixes are never used. Two prefixes are used μ and k. The unit for mass that is "prefixed" is the g.
$P = 1\ \mathrm{kwh}$	

—continued

Example	Rule(s) violated
	Rule 2: Unit symbols are printed in lower-case letters unless they are derived from a proper name
$P = 2$ k Pa	Rule 3: A Prefix is part of the unit and precedes the unit symbol without a separator
$P = 2$ kPas	The use of the correct symbols for SI units, and for units in general, is mandatory. The unit for pressure is Pa, not Pas. Units do not have plurals.
$L = 0.000, 001$ m	Rule 12: Neither dots nor commas are inserted in the spaces between groups of three.
$L = 0.000\ 001$ m	Correct
$R = 8.314$ J/kmol/K	Rule 5: A solidus must not be used more than once in a given expression without brackets to remove ambiguities.
R = 8.314 J/(kmol K)	Correct
R = 8.314 J kmol^{-1} K^{-1}	Correct
$N = 10$ kgmol	The correct SI unit is mol (not gmol); the use of the correct symbols for SI units, and for units in general, is mandatory.
$N = 10$ kmole	The correct SI unit is mol (not mole). The use of the correct symbols for SI units, and for units in general, is mandatory.
$N = 10$ gmol	The correct SI unit is mol (not gmol); the use of the correct symbols for SI units, and for units in general, is mandatory.
$N = 1000$ mol	Correct
$N = 1$ kmol	Correct

2.2.4 U.S. Customary units

American engineers are probably more familiar with the **U.S. customary units** than with SI units. The United States uses customary units in commercial activities, as well as for personal and social use. In science, medicine, many sectors of industry, and some government and military areas, metric units are used while the U.S. National Institute of Standards and Technology (NIST) promotes the use of the SI system. The United States customary system emanated from British units used in the British Empire before U.S. independence.

The base units for this system are the

- foot (ft) for length,
- the pound-mass (lb_m) for mass,
- the second (s) for time, and
- the Fahrenheit scale (°F) for temperature,

and the following units are also in common use

- the pound force (lb_f) for force,
- the British thermal unit (Btu) for heat.

The main shortcomings with the American system are the need to define the Newton's law proportionality factor ($g_c = 32.174\ lb_m{\cdot}ft{\cdot}s^{-2}{\cdot}lb_f^{-1}$) and the fact that it is a duodecimal (base-12) system (12 inches equals a foot). In this book the SI system will be used exclusively and the only skill that you need to develop is to be able to change the units from other systems of units to the SI.

2.2.5 Conversion of units and conversion factors

Think about the following: what is the result if we divide 1 m by 100 cm:

$$\frac{1\ \text{m}}{100\ \text{cm}} = ?$$

The answer is 1, as 1 m equals 100 cm. This might look strange or even counter-intuitive, but if we write the same thing in a slightly different way, it becomes self-evident

$$1\ \text{m} = 100\ \text{cm} \overset{\div 100\ \text{cm}}{\widehat{\Rightarrow}} \frac{1\ \text{m}}{100\ \text{cm}} = 1$$

The ratio 1 m/100 cm is an example of a **conversion ratio** or **conversion factor**. A **conversion factor** is an expression for the relationship between units that is used to change the units of a measured quantity without changing the value. A conversion ratio (or unit factor) always equals one (1), where the numerator and the denominator have **the same value expressed in different units**. This allows the conversion of units by multiplying the initial measurement by one (or more) forms of the number 1. While the multiplication by one does not change the value of the measurement, it does change the measurement units.

We can systematically apply the unit conversion process to solve conversions within or between measurement systems. It may be necessary to multiply by more than one conversion ratio in more complex conversions. Use these steps to set up a unit conversion problem where one or more units cancel out, ensuring that only the desired unit remains:

Step 1. Identify the unit you have. These are the Starting Units.
Step 2. Identify the unit you want. These are the Desired Units.
Step 3. Identify appropriate unit conversion factor(s).
Step 4. Cancel units and perform the math calculations.
Step 5. Evaluate the result. Does the answer make sense?

Table 2.6 can be useful when performing unit conversion between the U.S. Customary units and the SI units. Extensive tables with unit conversions can be found in the literature.

Example 2.2. Units conversion

Convert the following to the SI units indicated

a. 1 Btu → ? kJ
b. 3600 kJ → ? kWh
c. 1000 Btu → ? kWh
d. 1000 lb_m → ? t
e. 100 gal/min → ? L/s
f. 50 mi/h → ? m/s

Solution.

a. $1\ \text{Btu} \times \frac{1055.056\ \text{J}}{1\ \text{Btu}} \times \frac{1\ \text{kJ}}{1000\ \text{J}} = 1.055\ 056\ \text{kJ}.$

b. $3600\ \text{kJ} \times \frac{1\ \text{kW}}{1\ \frac{\text{kJ}}{\text{s}}} \times \frac{1\ \text{h}}{3600\ \text{s}} = 1\ \text{kWh}.$

Table 2.6 Some useful unit conversions.

Dimension	SI Unit	Other Unit	Equivalence
Time	s	s	–
Length	m	ft	1 ft = 0.3048 m
		In	1 in = 0.0254 m
		mi	1 mi = 1609.344 m
Volume	m^3, L	gal (US)	1 gal = 3.785411784 L
Mass	kg	lb_m	1 lb_m = 0.45359237 kg
Force	N	lb_f	1 lb_f = 4.4482216152605 N
pressure	Pa	psi	1 psi = 6894.757 Pa
temperature	degree °C	degree °F	1 Δ°C = 1.8 Δ°F
Absolute temperature	K	degree °R	1 ΔK = 1.8 Δ°R
Power	W	hp	1 hp = 745.699872 W
heat	J	Btu	1 Btu = 1055.056 J
		cal_{IT}	1 cal = 4.1868 J

c. $1000\ \text{Btu} \times \frac{1.055\ 056\ \text{kJ}}{1\ \text{Btu}} \times \frac{1\ \text{kWh}}{3600\ \text{kJ}} = 0.293\ 0711\ \text{kWh}$

d. $1000\ \text{lb}_\text{m} \times \frac{0.453\ 592\ 37\ \text{kg}}{1\ \text{lb}_\text{m}} \times \frac{1\ \text{t}}{1000\ \text{kg}} = 0.453\ 592\ 37\ \text{t}$

e. $100\,\frac{\text{gal}}{\text{min}} \times \frac{3.785411784\ \text{L}}{1\ \text{gal}} \times \frac{1\ \text{min}}{60\ \text{s}} = 6.309\ 02\ \frac{\text{L}}{\text{s}}$

f. $50\ \frac{\text{mi}}{\text{h}} \times \frac{1\ \text{h}}{3600\ \text{s}} \times \frac{1609.344\ \text{m}}{1\ \text{mi}} = 22.352\,\frac{\text{m}}{\text{s}}$

2.3 Uncertainty in measurements and uncertainty propagation

All measurements, however careful and scientific, are subject to uncertainty. Error analysis is the study and evaluation of these uncertainties, and the aim is to estimate how large the uncertainties are, and to handle the propagation of uncertainty in scientific calculations. Error in a scientific measurement means the inevitable uncertainty in all measurements. It is important to understand that errors are not mistakes; we cannot eliminate them by careful planning of the experiments. The best we can ensure is that errors are as small as reasonably possible and to have a reliable estimate of their size.

To illustrate the inevitable occurrence of uncertainties, we have only to examine any everyday measurement carefully. Consider, for example, the case presented in Fig. 2.2, where we try to measure the length of a pencil using a ruler. Let us assume that the ruler measures centimeters. In Fig. 2.2a we use a ruler that has a mark for every cm. Then, as the beginning of the pen is perfectly aligned with the zero mark, we need only one measurement at the end of the tip of the pencil. We can observe that the tip of the pencil is somewhere between the 14 and 15 cm marks. However, it would be ridiculous to give our estimation as 14.523 456 782 cm. The measuring device has 1 mark for every cm and the accuracy of our reading is hampered by the accuracy of the ruler used. A proper answer would be that the length of the pencil is between 14 and 15 cm or we may write

$$l_\text{pencil} = (14.5 \pm 0.5)\ \text{cm}$$

In this way we state that our best estimate of the length of the pencil is 14.5 cm and the probable range is 14–15 cm. In general, we can state the result of our measurement as:

$$y = y_\text{best estimate} \pm \delta y \tag{2.2.6}$$

This statement means, first, that the experimenter's best estimate for the quantity concerned is the number $y_\text{best estimate}$, and

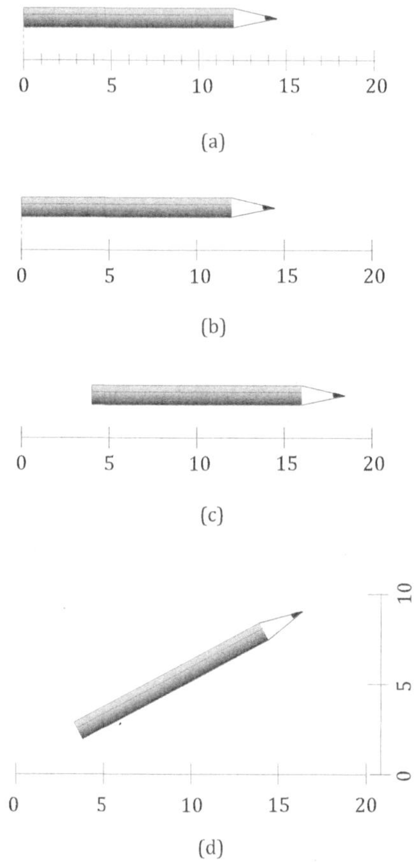

Figure 2.2 Four different settings for measuring the length of a pencil.

second, that he or she is reasonably confident the quantity lies somewhere between $y_{\text{best estimate}} + \delta y$ and $y_{\text{best estimate}} - \delta y$. The number δy is called the uncertainty, or error, or margin of error in the measurement of y. For convenience, the uncertainty is always defined to be positive, so that $y_{\text{best estimate}} + \delta y$ is always the highest probable value of the measured quantity and $y_{\text{best estimate}} - \delta y$ the lowest.

We now consider Fig. 2.2b. The ruler used in this case has a mark for every 5 cm. It is therefore apparent that the accuracy at which we can measure the length of the pencil is reduced. A reasonable estimation of the length of the pencil based on Fig. 2.2b is the following

$$l_{\text{pencil}} = (14 \pm 1) \text{ cm}$$

We move on and now examine Fig. 2.2c. The beginning of the ruler is not aligned with the beginning of the pencil and we now need to make two measurements: one for the beginning of the pencil and one for the end

$$l_{\text{pencil,begin}} = (3.5 \pm 1)\ \text{cm}$$

$$l_{\text{pencil,end}} = (18 \pm 1)\ \text{cm}$$

To estimate the length of the pencil we need to subtract the two numbers

$$l_{\text{pencil}} = l_{\text{pencil,end}} - l_{\text{pencil,begin}}$$

The best estimate is the difference between the two best estimates for the beginning and the end of the pencil. But what about the uncertainty? The answer is that when we add uncertain numbers then the uncertainty in the results equals the sum of the uncertainties (but this turns out to be conservative in most cases, as we will see shortly). We therefore write in this case

$$l_{\text{pencil}} = (14.5 \pm 2)\ \text{cm}$$

We finally consider Fig. 2.2d. We now need to make four measurements (see Fig. 2.3).

- the beginning of the pencil in the x direction $x_0 \pm \delta x_0$
- the end of the pencil in the x direction $x_f \pm \delta x_f$
- the beginning of the pencil in the y direction $y_0 \pm \delta y_0$
- the end of the pencil in the x direction $y_f \pm \delta y_f$

We then determine the sides of the right-angled triangle $x \pm \delta x$ and $y \pm \delta y$ and we finally determine the hypotenuse through the Pythagorean Theorem, as shown in Fig. 2.3. We have now run into the problem of error propagation as we perform a series of calculations, with the final one (square root of the sum of squares)

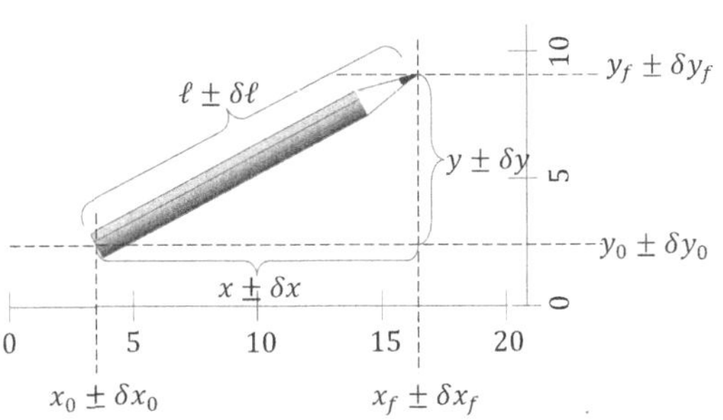

Figure 2.3 Measuring the length of a pencil in a complicated setting.

being nonlinear in nature. We need a general rule that can help us with such complicated situations.

We will present the rules without proof, and the interested reader can consult the references given at the end of the chapter.

Uncertainty in sums and differences. Suppose that $x_1, x_2, \ldots, x_n$ and $y_1, y_2, \ldots, y_m$ are measured with independent and random uncertainties $\delta x_1, \delta x_2, \ldots, \delta x_n$ and $\delta y_1, \delta y_2, \ldots, \delta y_m$ then for the uncertainty in $z = (x_1 + x_2 + \ldots x_n) \pm (y_1 + y_2 + \ldots + y_m)$ the following holds true

$$\delta z = \sqrt{(\delta x_1)^2 + (\delta x_2)^2 + \ldots + (\delta x_n)^2 + (\delta y_1)^2 + (\delta y_2)^2 + \ldots + (\delta y_m)^2} \tag{2.3.1}$$

It also holds true that a conservative estimate for the uncertainty in z is given by

$$\delta z \leq (\delta x_1 + \delta x_2 + \ldots + \delta x_n) + (\delta y_1 + \delta y_2 + \ldots + \delta y_m) \tag{2.3.2}$$

Uncertainty in products and quotients. Suppose that $x_1, x_2, \ldots, x_n$ and $y_1, y_2, \ldots, y_m$ are measured with independent and random uncertainties $\delta x_1, \delta x_2, \ldots, \delta x_n$ and $\delta y_1, \delta y_2, \ldots, \delta y_m$ then for the uncertainty in $z = (x_1 x_2 \ldots x_n)/(y_1 y_2 \ldots y_m)$ the following holds true

$$\frac{\delta z}{|z|} = \sqrt{\left(\frac{\delta x_1}{x_1}\right)^2 + \left(\frac{\delta x_2}{x_2}\right)^2 + \ldots + \left(\frac{\delta x_n}{x_n}\right)^2 + \left(\frac{\delta y_1}{y_1}\right)^2 + \left(\frac{\delta y_2}{y_2}\right)^2 + \ldots + \left(\frac{\delta y_m}{y_m}\right)^2} \tag{2.3.3}$$

It also holds true that a conservative estimate for the uncertainty in z is given by

$$\frac{\delta z}{|z|} \leq \left(\frac{x_1}{|x_1|} + \frac{x_2}{|x_2|} + \ldots \frac{x_n}{|x_n|}\right) + \left(\frac{y_1}{|y_1|} + \frac{y_2}{|y_2|} + \ldots + \frac{y_m}{|y_m|}\right) \tag{2.3.4}$$

Uncertainty in a real function of several variables. Suppose that $x_1, x_2, \ldots, x_n$ are measured with independent and random uncertainties, then for the uncertainty in $z = f(x_1, x_2, \ldots, x_n)$ the following holds true

$$\delta z = \sqrt{\left(\frac{\partial f}{\partial x_1}\delta x_1\right)^2 + \left(\frac{\partial f}{\partial x_2}\delta x_2\right)^2 + \ldots + \left(\frac{\partial f}{\partial x_n}\delta x_n\right)^2} \tag{2.3.5}$$

It also holds true that a conservative estimate for the uncertainty in z is given by

$$\delta z \leq \left|\frac{\partial f}{\partial x_1}\right|\delta x_1 + \left|\frac{\partial f}{\partial x_2}\right|\delta x_2 + \ldots + \left|\frac{\partial f}{\partial x_n}\right|\delta x_n \qquad (2.3.6)$$

Let's return now to the calculation of the length of the pencil in Fig. 2.3. We first estimate that

$$x_0 \pm \delta x_0 = (3.5 \pm 1)\ \text{cm}$$

$$x_f \pm \delta x_f = (16.5 \pm 1)\ \text{cm}$$

and

$$x \pm \delta x = \left(13.0 \pm \sqrt{1^2 + 1^2}\right)\ \text{cm} = \left(13.0 \pm \sqrt{2}\right)\ \text{cm}$$

In a similar way

$$y_0 \pm \delta y_0 = (2.5 \pm 1)\ \text{cm}$$

$$y_f \pm \delta y_f = (9 \pm 1)\ \text{cm}$$

and

$$y \pm \delta y = \left(6.5 \pm \sqrt{1^2 + 1^2}\right)\ \text{cm} = \left(6.5 \pm \sqrt{2}\right)\ \text{cm}$$

The best estimate of the length of the pencil follows from the Pythagorean Theorem

$$\ell = \sqrt{13.0^2 + 6.5^2} = 6.5\sqrt{5} \approx 14.5$$

To calculate the uncertainty, we need to apply Eq. (2.3.5). We first calculate the partial derivatives

$$\frac{\partial f}{\partial x} = \frac{\partial}{\partial x}\left(\sqrt{x^2 + y^2}\right) = \frac{x}{\sqrt{x^2 + y^2}} = \frac{13}{6.5\sqrt{5}} = \frac{2}{\sqrt{5}}$$

$$\frac{\partial f}{\partial y} = \frac{\partial}{\partial y}\left(\sqrt{x^2 + y^2}\right) = \frac{y}{\sqrt{x^2 + y^2}} = \frac{1}{\sqrt{5}}$$

and finally

$$\delta\ell = \sqrt{\left(\frac{2}{\sqrt{5}}\sqrt{2}\right)^2 + \left(\frac{1}{\sqrt{5}}\sqrt{2}\right)^2} = \sqrt{\left(4\frac{2}{5}\right) + \left(1\frac{2}{5}\right)} = \sqrt{2}$$

Note that Eq. (2.3.6) is satisfied

$$\delta\ell = \sqrt{2} \leq \frac{2}{\sqrt{5}}\sqrt{2} + \frac{1}{\sqrt{5}}\sqrt{2} = \frac{3}{\sqrt{5}}\sqrt{2}$$

as $\sqrt{5} < 3$. Our final answer is

$$l_{\text{pencil}} = \left(14.5 \pm \sqrt{2}\right) \text{ cm}$$

That was a long but scientifically correct way to determine an estimation for the length of the pencil in Fig. 2.3. We now have enough tools to handle almost any problem in the propagation of errors. Any complex calculation can be decomposed down into a sequence of simple steps, involving

a. sums and differences;
b. products and quotients; and
c. computation of a nonlinear function of one or several variables.

If we cannot justify that the individual errors are random and independent, then we can use the simpler but conservative estimates.

There are two issues that need to be resolved to complete this section. The first one is to give a definite answer on what is the result of the calculation

$$\ell = \sqrt{13.0^2 + 6.5^2} = 6.5\sqrt{5} \approx ?$$

MATLAB gives the following result.

```
>> format long g
>> sqrt(13^2+6.5^2)

ans =

          14.5344418537486
```

Given the fact that both 13.0 and 6.5 are known within the accuracy of the measuring tape used, we need to decide how many digits we can include in our final answer.

The second issue is related to the fact that one way to reduce uncertainty is to take a large number of independent measurements of the same quantity. The question raised in this case is how we organize our experimental data and how we systematically report our results using a statistical model.

Example 2.3. Multiplying two uncertain numbers

The mass and the velocity of an object are measured and the following measurements are available

$$m = (1.59 \pm 0.03) \text{ kg}$$

$$v = (3.1 \pm 0.1) \ \frac{\text{m}}{\text{s}}$$

Give your best estimate for momentum $p = mv$ with the accuracy clearly indicated.

Solution.

We first calculate the best point estimate of the momentum

$$p = (1.59\ \text{kg}) \times \left(3.1\,\frac{\text{m}}{\text{s}}\right) = 4.929\ \text{kg}\cdot\frac{\text{m}}{\text{s}} = 4.9\ \frac{\text{kg}\cdot\text{m}}{\text{s}}$$

We now calculate the fractional errors in mass and velocity

$$\frac{\delta m}{m} = \frac{0.03}{1.59} = 0.02$$

$$\frac{\delta v}{v} = \frac{0.1}{3.1} = 0.03$$

The conservative estimate of uncertainty is (see Eq. (2.3.4))

$$\frac{\delta p}{|p|} \le \frac{m}{|m|} + \frac{v}{|v|} = 0.02 + 0.03 = 0.05\ \text{ή}\ 5\%$$

and

$$\delta p = \frac{\delta p}{|p|}p = 0.05 \times 4.929\ \text{kg}\cdot\frac{\text{m}}{\text{s}} = 0.245\ \text{kg}\cdot\frac{\text{m}}{\text{s}}$$

Our conservative estimate is that

$$p = (4.9 \pm 0.2)\ \text{kg}\cdot\frac{\text{m}}{\text{s}}$$

Using Eq. (2.3.3) we obtain

$$\frac{\delta p}{|p|} = \sqrt{\left(\frac{m}{|m|}\right)^2 + \left(\frac{v}{|v|}\right)^2} = \sqrt{\left(\frac{0.03}{1.59}\right)^2 + \left(\frac{0.1}{3.1}\right)^2} = 0.037$$

and

$$\delta p = \frac{\delta p}{|p|}p = 0.037 \times 4.929\ \text{kg}\cdot\frac{\text{m}}{\text{s}} = 0.183\ \text{kg}\cdot\frac{\text{m}}{\text{s}}$$

Our estimate is that

$$p = (4.9 \pm 0.2)\ \text{kg}\cdot\frac{\text{m}}{\text{s}}$$

Example 2.4. Propagation of errors

In a lab experiment two solids are mixed and an estimation of their mass fractions in the mixture is needed. The solid samples were placed in certain containers whose masses were measured when full and empty, as follows.

Solid A: mass of container one when full: (1080 ± 20) g, when empty: (50 ± 1) g.

Solid B: mass of container two when full: (1880 ± 40) g, when empty: (100 ± 1) g.

Give your best estimate for the mass fraction of solid A in the mixture.

Solution.

The mass of solid A is (see Eq. (2.3.1))

$$m_A = \left[(1080 - 50) \pm \sqrt{20^2 + 1^2}\right] \text{ g} = (1030 \pm 20) \text{ g}$$

The total mass of the mixture is (see Eq. (2.3.1))

$$m = \left[(1080 - 50) + (1880 - 100) \pm \sqrt{20^2 + 1^2 + 40^2 + 1^2}\right] \text{ g}$$
$$= (2810 \pm 45) \text{ g}$$

The mass fraction of A is $w_A = m_A/m$ and by applying Eq. (2.3.3) we obtain

$$\frac{\delta m_A}{m_A} = \frac{20}{1030} \approx 2\%$$

$$\frac{\delta m}{m} = \frac{45}{2810} \approx 2\%$$

$$\delta w_A = \frac{\delta w_A}{|w_A|} w_A = \sqrt{\left(\frac{\delta m_A}{|m_A|}\right)^2 + \left(\frac{\delta m}{|m|}\right)^2} w_A$$
$$= \left[\sqrt{\left(\frac{20}{1030}\right)^2 + \left(\frac{45}{2810}\right)^2}\right] \times \frac{1030}{2810} = 0.02517 \times 0.36655 \text{ kg} \cdot \frac{\text{m}}{\text{s}}$$
$$= 0.009225 \approx 1\%$$

We finally have that

$$w_A = 0.37 \pm 0.01$$

If you are confused with the way we have dealt with the decimal point and the truncation applied to the results do not worry; that was deliberate to prepare you for the section that follows.

2.4 Significant figures

Significant figures, also referred to as significant digits, are specific digits within a number that carry both reliability and necessity in conveying a particular quantity. When presenting the outcome of a measurement, if the number of digits exceeds what the measurement instrument can resolve, only the number of digits within the resolution's capability are dependable and therefore considered significant. Any measurement using the ruler shown in Fig. 2.2a can have up to three significant figures (for instance, the length of the pencil equals to 14.5 cm) while for the ruler in Fig. 2.2b can only have two significant figures (for instance, the length of the pencil equals to 14 cm). Using more significant figures cannot be justified from the resolution of the measuring device used.

When we are given a measurement, we can use the following rules to decide what the number of significant figures is:

Rule 1. All nonzero digits are significant: 1.234 g has four significant figures; 1.2 g has two significant figures.

Rule 2. Zeroes between nonzero digits are significant: 10.04 kg has four significant figures; 1.03 mL has three significant figures.

Rule 3. Zeroes to the left of the first nonzero digits are not significant; such zeroes merely indicate the position of the decimal point: 0.004 N has only one significant figure; 0.034 g has two significant figures.

Rule 4. Zeroes to the right of a decimal point in a number are significant: 0.034 m has two significant figures and 0.200 bar has three significant figures.

Rule 5. When a number ends in zeroes that are not to the right of a decimal point, the zeroes are not necessarily significant: 120 kg may be two or three significant figures, 10,300 Pa may be 3, 4, or 5 significant figures. The potential ambiguity in the last rule can be avoided by the use of standard exponential, or "scientific," notation. For example, depending on whether 3, 4, or 5 significant figures is correct, we could write.

10,300 Pa:

1.03×10^4 Pa (3 significant figures), or

1.030×10^4 Pa (4 significant figures), or

1.0300×10^4 Pa (5 significant figures).

It is reminded that **scientific notation** is a way to express numbers as the product of two numbers: a coefficient m and the number 10 raised to a power n: $m \times 10^n$. It is a very useful tool for working with numbers that are either very large or very small. As an example, the Earth's circumference measured around the equator is approximately 40 080 000 m—a very large distance. In scientific notation the Earth's circumference is written as 4.008×10^7 m. The coefficient is the 4.008 and must be a number greater than or equal to 1, and less than 10 (i.e., $1 \leq m \leq 10$). The power of 10, or exponent, is seven because you would have to multiply 4.008 by 10^7 to get the correct number (or move the decimal point seven positions to the right). The radius of the hydrogen atom is approximately 0.000 000 000 040 m which is a very small number. In scientific notation the radius of the hydrogen atom is written 4×10^{-11} m.

In carrying out calculations, the general rule is that the accuracy of a calculated result is limited by the least accurate measurement involved in the calculation.

Addition and subtraction. The result is rounded off to the last common digit occurring furthest to the right in all components.

For example, 123 (assume three significant figures) + 12.345 (5 significant figures) = 135.345, which should be rounded to 135 (3 significant figures).

Multiplication and division. The result should be rounded off to have the same number of significant figures as in the component with the least number of significant figures.

For example, 1.23 (3 significant figures) × 12.34 (4 significant figures) = 15.1782 which should be rounded off to 15.2 (3 significant figures).

When we round off numbers the following rules can be used:

Rule 1. If the digit to be dropped is greater than 5, the last retained digit is increased by one. For example, 45.6 is rounded to 46.

Rule 2. If the digit to be dropped is less than 5, the last remaining digit is left as it is. For example, 45.4 is rounded to 45.

Rule 3. If the digit to be dropped is 5, and if any digit following it is not zero, the last remaining digit is increased by one. For example, 45.51 is rounded to 46.

Rule 4. If the digit to be dropped is five and is followed only by zeroes, the last remaining digit is increased by one if it is odd, but left as it is if even. For example, 45.5 is rounded to 46, 46.5 is rounded to 46. This rule means that if the digit to be dropped is five followed only by zeroes, the result is always rounded to the even digit. The rationale is to avoid bias in rounding: half of the time we round up, half the time we round down.

Example 2.5. Significant figures and scientific notation.

Express the numbers given below in scientific notation and indicate how many significant figures appear in each case with a short explanation:

a. 0.012340

b. 12345
c. 12345.0
d. 0.0123×10^{-5}
Solution
a. $0.012340 = 1.2340 \times 10^{-2}$, the number has five significant figures (see Rule 4)
b. $12345 = 1.2345 \times 10^{4}$, the number has five significant figures (see Rule 1)
c. $12345.0 = 1.23450 \times 10^{4}$, the number has six significant figures (see Rule 4)
d. $0.01230 \times 10^{-5} = 1.230 \times 10^{-7}$, the number has three significant figures (see Rules 3 & 4)

Example 2.6. Significant figures in a sum or product.

What is the result that you will report (in scientific notation) in the following problems? Give a short explanation. Use MATLAB to perform the calculation indicated.

a. $1234 \times 1.234 \times 0.012 = ?$
b. $1234 \times 1.234 \times 0.012/12.34 = ?$
c. $1.2 + 1.234 + 0.012 = ?$
Solution

a.

```
>> format long
>> 1234*1.234*0.012
ans =
  18.273072000000003
```

1234 and 1.234 have four significant figures. However, 0.012 has only two significant figures. We must report the result by rounding the number using two significant figures (i.e., 18).

b.

```
>> 1234*1.234*0.012/12.34
ans =
   1.480800000000000
```

1234, 12.34, and 1.234 have four significant figures. However, 0.012 has only two significant figures. We must report the result by rounding the number using two significant figures (i.e., 1.5).

c.

```
>> 1.2+1.234+0.012
ans =
   2.446000000000000
```

1.2 and 0.012 have two significant figures while 1.234 has four significant figures. We must report the result by rounding the number using two significant figures (i.e., 2.4).

2.5 Organizing multiple experimental measurements

Assume that you run the following experiment: you place a measuring cylinder filled with 1 L of seawater (measured with accuracy ±1 mL) on a precision balance (accuracy ±1 g) and repeat the experiment 10 times. The measurement of the precision balance equals the density of the seawater in g/L. The results of such experiment are reported in Table 2.7.

Why do we get different measurements in each, presumably identical, experiment? We borrow the answer from Heraclitus: "No man ever steps in the same river twice, for it is not the same river and he is not the same man." There are numerous sources of uncertainty that we cannot eliminate, no matter how carefully we conduct even such a simple experiment. You can perform the experiment to convince yourself. However, the important question is still the same: what is the density of seawater? We need as (future) engineers to come up with a unique and definite answer. Statistics comes to our rescue. If the errors in the experimental measurement are independent and random, then the best estimate of the density of the seawater equals the sample average. If we have n measurements of the same variable x_i then the sample average or mean value $\overline{x}$ is defined as

Table 2.7 Experimental estimation of seawater density generated in MATLAB by sampling from a normal distribution with mean of 1030 and standard deviation of 5 (using `rng(3)` for reproducibility).

Experiment	1	2	3	4	5
Seawater density (g/L)	1030.343	1034.756	1028.276	1030.180	1035.611

Experiment	6	7	8	9	10
Seawater density (g/L)	1031.618	1027.944	1026.791	1026.693	1028.874

$$\overline{x} = \frac{x_1 + x_2 + \cdots + x_n}{n} \tag{2.5.1}$$

For the experimental values of seawater, we have that

$$\overline{\rho} = \frac{\rho_1 + \rho_2 + \cdots + \rho_{10}}{10} = \frac{1030.343 + 1034.756 + \cdots + 1028.874}{10}$$
$$= 1030.108\ \frac{\text{g}}{\text{L}}$$

The mean value gives us a point estimate of the true value of density. What about the spread of its variation? Statistics offers again a simple answer under the assumptions of independent and random errors in the measurements: a measure of the spread is the standard deviation s of the sample defined as

$$\widehat{s}^2 = \frac{n}{n-1} s^2 \tag{2.5.2}$$

where s^2 is the sample variance

$$s^2 = \frac{(x_1 - \overline{x})^2 + (x_2 - \overline{x})^2 + \ldots + (x_n - \overline{x})^2}{n} = \frac{1}{n}\sum_{i=1}^{n}(x_i - \overline{x})^2 \tag{2.5.3}$$

$\widehat{s}^2$ is an unbiased estimator of the true (but unknown) variance σ^2 of the population. A large standard deviation indicates that the data points can spread very far from the sample mean, and a small standard deviation indicates that they are clustered closely to the mean. For example, each of the three populations {0, 0, 20, 20}, {2.5, 7.5, 12.5, 17.5}, and {9, 10, 10, 11} has a mean of 10. Their standard deviations are 11.5, 6.5, and 0.8, respectively. The third population has a much smaller standard deviation than the other two because its values are clustered close to 10. It is important to note that standard deviations have the same units as the data points themselves. If, for instance, the data set {9, 10, 10, 11} represents the mass of four consecutive samples in g then the standard deviation is 0.8 g. For the case of seawater density measurements, we have that

$$\widehat{s}^2 = \frac{(\rho_1 - \overline{\rho})^2 + (\rho_2 - \overline{\rho})^2 + \cdots + (\rho_{10} - \overline{\rho})^2}{n-1} = 3.1^2\ (\text{g/L})^2 \tag{2.5.4}$$

And we now come to the important result: standard deviation of the sample mean, often called the standard error of the mean, provides a measure of uncertainty in our reported value of the seawater density. To quantify this, we can rely on the exact form of the normal (or Gaussian) distribution, which is simply a statistical model for experimental measurements. One of the most important theorems in statistics, the Central Limit Theorem, states that the sum or average of many independent, identically distributed (iid) random variables with finite variance tends to follow the familiar bell-shaped normal distribution with a probability density function of

$$f(x;\mu,\sigma) = \frac{1}{\sigma\sqrt{2\pi}}\exp\left[-\frac{1}{2}\left(\frac{x-\mu}{\sigma}\right)^2\right] \tag{2.5.5}$$

where σ is the standard deviation and μ is the mean. Table 2.8 gives the % of measurements that are expected to appear in the interval

$$[\mu - k\sigma, \mu + k\sigma] \tag{2.5.6}$$

as a function of k. It is important to note that (see also Fig. 2.4).

- two thirds (68.3%) of the values will appear in the interval $\mu \pm \sigma$
- 95.5% of the values will appear in the interval $\mu \pm 2\sigma$
- 99.7% of the values will appear in the interval $\mu \pm 3\sigma$

All these remain true only when the sample size is large (say more than 100 measurements). For small samples, however, the sample mean does not follow a normal distribution but (the standardized statistics) follows the Student t-distribution with $n-1$ degrees of freedom, and the confidence interval for the sample mean is

Table 2.8 Proportion of measurements within $x_{mean} \pm ks$ as a function of k.

% Confidence	68.27%	90%	95.45%	99%	99.73%	99.999%
Normal ($n > 100$) k	1.000	1.644	2.000	2.576	3.000	4.417
$t_{\frac{a}{2},n-1}$, $n = 2$	∞	6.314	13.968	63.657	235.784	63,662
$t_{\frac{a}{2},n-1}$, $n = 3$	1.837	2.920	4.527	9.925	19.206	316.225
$t_{\frac{a}{2},n-1}$, $n = 10$	1.067	1.833	2.320	3.250	4.094	8.8275
$t_{\frac{a}{2},n-1}$, $n = 30$	1.018	1.699	2.090	2.756	3.280	5.335
$t_{\frac{a}{2},n-1}$, $n = 100$	1.005	1.660	2.026	2.626	3.078	4.657

a, confidence in fraction

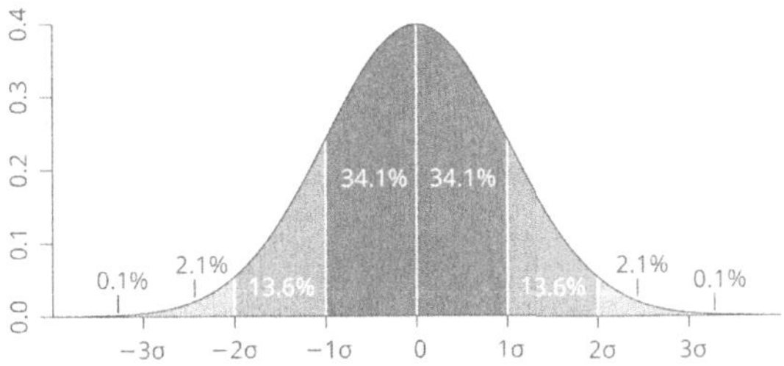

Figure 2.4 A plot of normal or Gaussian distribution.

$$\left[\bar{x} - t_{a/2,n-1}\frac{\hat{s}}{\sqrt{n}}, \bar{x} + t_{a/2,n-1}\frac{\hat{s}}{\sqrt{n}}\right] \tag{2.5.7}$$

where $t_{a/2,n-1}$ can be obtained from Table 2.8 (or calculated in software).

For the case of the seawater data, we have already calculated the sample mean and we use Eq. (2.5.7) to determine the confidence interval for the mean

$$\bar{x} \pm t_{a/2,n-1}\frac{\hat{s}}{\sqrt{n}} = 1030.108 \pm \underbrace{1.833}_{\substack{\text{see Table 2.8} \\ \text{for } n=10 \\ \text{and 90\% confidence}}} \times \frac{3.1}{\sqrt{10}} = 1030.108 \pm 1.796\ \frac{\text{g}}{\text{L}} \tag{2.5.8}$$

i.e., we are 90% confident that the population mean lies in the interval [1028.311, 1031.905] g/L. The statistical terminology used in this discussion has been intentionally simplified to emphasize the main concepts. A more rigorous treatment can be found in standard references on statistics and uncertainty analysis.

Let us summarize our discussion thus far. In engineering practice, measurements are often limited in number, sometimes consisting of only a few observations. Consequently, the ability to make precise inferences about the underlying population is limited. Rather than attempting a complete statistical characterization of the population, the focus shifts to estimating a parameter of interest, most commonly the population mean, together with a quantitative measure of uncertainty. Accordingly, we report the sample mean along with a confidence interval corresponding to a specified confidence level, most commonly 90% or 95%. This approach conveys both the central tendency of the data and the uncertainty associated with the estimate.

The generation of data for the seawater density application and the statistical analysis in MATLAB were performed as shown below.

```
clear all
% Parameters of the Gaussian distribution
mu = 1030;          % mean
sigma = 5;          % standard deviation
% Number of samples
n = 10;
% Fix the random seed so the sample is always the same
rng(3);      % You can choose any integer seed
% Generate random samples from N(mu, sigma^2)
rho_meas   = mu+sigma*randn(n, 1); % density measurements in g
per L
% calculate the sample mean and the sample standard deviation
rho_mean   = mean(rho_meas)
rho_sigma  = std(rho_meas,0)
% calculate the interval for 90% confidence
confidence = 0.9;
alpha      = 1-confidence;
nu         = n-1;
CI_low     = rho_mean+tinv(  alpha/2, nu)*rho_sigma/sqrt(n)
CI_up      = rho_mean+tinv(1-alpha/2, nu)*rho_sigma/sqrt(n)

rho_mean  =    1030.10843668129
rho_sigma =       3.09932233605398
CI_low    =    1028.31181776694
CI_up     =    1031.90505559564
```

The MATLAB commands that follow generate Fig. 2.5, which illustrates the Student's t-distribution used for inference on the population mean. The critical values of this distribution are used to quantify the uncertainty associated with the estimated mean and to construct the corresponding 90% confidence interval. The figure also displays the individual sample measurements, the sample mean, and the corresponding 90% confidence limits.

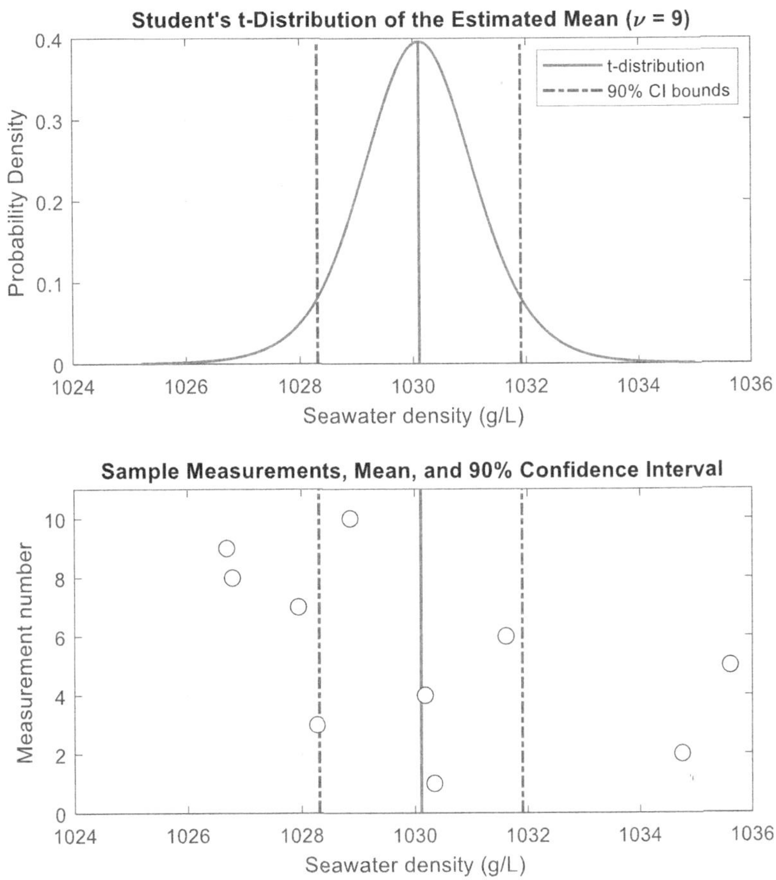

Figure 2.5 Results of the statistical analysis of the seawater density measurements.

```
% Standard error of the mean
SE = rho_sigma/sqrt(n);

% Generate x-values around the estimated mean
x = (rho_mean - 5*SE) : 0.05 : (rho_mean + 5*SE);

% Student's t-distribution for inference on the mean
z = (x - rho_mean)/SE;
y = (1/SE)*tpdf(z,nu);

figure(1); clf;

subplot(2,1,1)

plot(x,y,'b','LineWidth',1.5)
grid on
hold on

% Show confidence interval limits
plot([CI_low CI_low],[0 max(y)],'b-.','LineWidth',1.5)
plot([CI_up  CI_up ],[0 max(y)],'b-.','LineWidth',1.5)
plot([rho_mean rho_mean],[0 max(y)],'b-','LineWidth',1.5)

xlabel('Seawater density (g/L)')
ylabel('Probability Density')
title(['Student''s t-Distribution for the Estimated Mean (\nu = '
...
       num2str(nu) ')'])
axis([1024 1036 0 0.42])
legend('t-distribution','90% CI bounds','Location','best')

subplot(2,1,2)

plot([rho_mean rho_mean],[0 11],'b-','LineWidth',1.5)
grid on
hold on
plot([CI_low CI_low],[0 11],'b-.','LineWidth',1.5)
plot([CI_up CI_up],[0 11],'b-.','LineWidth',1.5)

plot(rho_meas,linspace(1,n,n),'o', ...
     'MarkerSize',8, ...
     'MarkerFaceColor','w', ...
     'MarkerEdgeColor','k')

axis([1024 1036 0 11])

xlabel('Seawater density (g/L)')
ylabel('Measurement number')
title('Sample Measurements, Mean, and 90% Confidence Interval')
```

2.6 Dimensional homogeneity

Any equation in science and engineering contains terms on both sides. The **principle of dimensional homogeneity** requires that the dimensions of all terms in an equation must be identical for the equation to be physically meaningful. If this condition is satisfied, we then say that the equation is dimensionally homogeneous. Stating the same thing in a slightly different way: only commensurable quantities (physical quantities having the same dimension) may be compared, equated, added, or subtracted. Let's consider the theoretical equation that describes the distance traveled by a free-falling body of an initial vertical velocity u_0 if the acceleration of gravity is g (m/s^2), then

$$x = x_0 + u_0 t - \frac{1}{2} g t^2 \tag{2.6.1}$$

The left-hand side has dimensions of length, i.e., $[x] = \mathrm{L}$. In order for Eq. (2.6.1) to be dimensionally homogeneous, each term on the right-hand side must also have dimensions of length. This is obvious for x_0. We note that

$$[u_0 t] = [u_0][t] = \left(\mathrm{LT}^{-1}\right)(\mathrm{T}) = \mathrm{L}$$

$$\left[\frac{1}{2} g t^2\right] = \left[\frac{1}{2}\right][\mathrm{g}]\left[t^2\right] = (1)\left(\mathrm{LT}^{-2}\right)\left(\mathrm{T}^2\right) = \mathrm{L}$$

In this example we used the fact that the dimension of the product of n variables is equal to the product of the dimensions of the n variables:

$$[x_1 x_2 \ldots x_n] = [x_1][x_2]\ldots[x_n] \tag{2.6.2}$$

The following also hold true

$$\left[\frac{x_1 x_2 \ldots x_n}{y_1 y_2 \ldots y_m}\right] = \frac{[x_1][x_2]\ldots[x_n]}{[y_1][y_2]\ldots[y_m]} \tag{2.6.3}$$

$$[x^n] = [x]^n \tag{2.6.4}$$

$$\left[\frac{dy}{dx}\right] = \frac{[y]}{[x]} \tag{2.6.5}$$

$$\left[\int f(x)dx\right] = [f(x)dx] = [f(x)][x] \tag{2.6.6}$$

It is important to note at this point that in every **analytically derived equation**, both sides of the equation must have identical dimensions and all numbers appearing in this equation must be dimensionless constants, i.e., they must have the dimension of 1. This is the reason why the constant 1/2 appearing in Eq. (2.6.1) has the dimension of 1.

Exponents (such as the n in x^n), transcendental functions (sin, log, exp) and arguments of transcendental functions (such as x in

$e^x, \sin x, \log x$) must be dimensionless. To understand why, consider the sine function. Its Taylor series expansion is

$$\sin(x) = x - \frac{x^3}{3!} + \frac{x^5}{5!} - \frac{x^7}{7!} + \cdots$$

We can see clearly that if x is not dimensionless, the terms $x, x^3, x^5, \ldots$ cannot have identical dimensions. Also, the units of x must be the same with the units of $\sin(x)$. It then follows that both x and $\sin(x)$ are dimensionless (having dimension of 1). The same holds true for $\cos(x)$, $\exp(x)$, $\log(x)$. Since the terms in a sum must have identical dimensions, every term in this series must possess the same dimensions. However, if x carries dimensions, then x^2, x^3, x^5, and so on have different dimensions and therefore cannot be added together. The only way for the series to be dimensionally consistent is for x to be dimensionless. Consequently, sin(x) is also dimensionless.

Although the numbers that appear in analytically derived equations are dimensionless constants, the same does not hold true for empirical equations. The following empirical equation, for instance, has been proposed by A.V. Hill to calculate the mass of a dolphin m (in kg) as a function of its length L (in m):

$$m = 10L^3 \tag{2.6.7}$$

In this case we have that

$$[m] = [10][L]^3 \quad or \quad [10] = \frac{[m]}{[L]^3} = \mathrm{M\,L^{-3}}$$

It follows that the numerical constant has dimension of density or mass per unit volume.

Example 2.7. Find the correct equation by inspection

For the volume of a cylindrical tank with conical bottom you have located the following potential equations

a. $V = \pi\left(H\frac{h}{3}\right)r^2$

b. $V = \pi\left(\frac{1}{H}+\frac{h}{3}\right)r^2$

c. $V = \pi\left(H^2+\frac{h}{3}\right)r^2$

d. $V = \pi\left(H+\frac{h}{3}\right)r^2$

e. $V = \pi\left(H + \frac{h}{3}H\right)r^2$

where h is the height of the conical bottom, H is the height of the cylindrical part of the tank, and r is the radius of the tank.

Can you locate the equation(s) that might be the correct one(s)?

Solution.

We note that the left-hand side of all potential equations is the volume of the tank and therefore the units are m^3. In the right-hand sides of the equations we observe that we have three factors. The first factor is the dimensionless constant π while the last factor has units of area (m^2). In the correct equation(s) the second factor (term inside the parenthesis) must have units of length (m).

Case a is rejected because the parenthesis has units of area (m^2).

Case b is rejected because the parenthesis does not have consistent units: the first term has units m^{-1} and the second m.

Case c is also rejected as the two terms that are added are inconsistent.

Case d is dimensionally consistent as the term in the parenthesis has units (m).

Case e is rejected.

We therefore have only one potentially valid equation.

Example 2.8. Units of the numerical constants in an empirical equation

An equation that has been proposed for the calculation of the heat transfer coefficient h (in kW/(m^2 K)) is the following

$$h = 51.1 + 2.044T$$

where T is the temperature in K. What are the units of the numerical constants?

Solution

We first note that

$$[h] = [51.1]$$

$$[h] = [2.044][T_s]$$

We see that the constant 51.1 has the same units as the heat transfer coefficient, i.e., kW/(m^2 K).

For the constant 2.044 it follows that

$$[2.044] = \frac{[h]}{[T_s]}$$

It then follows that the units of the numerical constant 2.044 are

$$\frac{\text{kW}}{\text{m}^2\ \text{K}}/\text{K} = \frac{\text{kW}}{\text{m}^2\ \text{K}^2}$$

Example 2.9. Units of the numerical constants in an empirical equation

One of the most well-known equations in chemical engineering is the Sieder and Tate equation for calculating the heat transfer coefficient h (in W/(m^2 K)) in circular pipes at turbulent conditions. The Sieder and Tate equation can be written as follows

$$\frac{hD}{k} = 0.027\left(\frac{uD}{\nu}\right)^n\left(\frac{\nu}{a}\right)^m$$

where h is the heat transfer coefficient in W·m^{-2}K^{-1}

D is the inside diameter of the pipe in m

k is the thermal conductivity of the fluid in W·m^{-1} K^{-1}

u is the velocity of the fluid in m·s^{-1}

ν is the kinematic viscosity in m^2 s^{-1}

a is the thermal diffusivity of the fluid in m^2 s^{-1}

n and m are positive constants. Verify that the equation is dimensionally homogeneous. What are the units of the numerical constant 0.027? What are its units in the American system of units?

Solution

We note that for the equation to be consistent, the units of the left–hand side must be the same as the units of the right–hand side. We find the units of the left–hand side by replacing the units of the variables into the left–hand side:

$$\overbrace{\frac{\text{W}}{\text{m}^2\ \text{K}}}^{\text{units of } h} \cdot \overbrace{\text{m}}^{\text{units of } D} \cdot \overbrace{\frac{\text{m}\cdot\text{K}}{\text{W}}}^{\text{units of } \frac{1}{k}} = 1$$

We conclude the left–hand side has unit dimensions (i.e., it is dimensionless). We conclude that the right–hand side must have unit dimensions as well.

We calculate the units of the term in the first parenthesis

$$\overbrace{\frac{\text{m}}{\text{s}}}^{\text{units of } u} \cdot \overbrace{\text{m}}^{\text{units of } D} \cdot \overbrace{\frac{\text{s}}{\text{m}^2}}^{\text{units of } \frac{1}{\nu}} = 1$$

For the term in the second parenthesis, it immediately follows that is also dimensionless. We conclude that for the equation to be dimensionally consistent, numerical constant 0.023 must be dimensionless as well. Since the equation has been presented in dimensionless form, it is also applicable in all systems of units with the same value of the numerical constant.

2.7 Important process variables

2.7.1 Temperature

Temperature, as a measure of how hot or cold something is, needs no explanation as it is a matter of common experience. To make scientific measurements of temperature requires, however, an appropriate scale that attaches numbers to the perception of hot and cold. This scale must also extend far beyond the range of temperatures we experience every day. Establishing such a scale and developing measuring instruments (thermometers) based on this scale has a long history. The simplest thermometer is the common alcohol thermometer, wherein the liquid expands when heated. For the working temperature range, the meniscus or interface between the liquid and gas/vapor is within the capillary and when temperature increases, the volume of liquid expands and the meniscus moves up the capillary. The position of the meniscus shows the temperature against an inscribed scale. Apparently, we need to define a scale and devise a methodology to "calibrate" the instruments.

The **Celsius scale** was established early and remains in common use throughout most of the world. Its calibration is based on the ice point of water (0 °C) and the normal boiling point of pure water (100 °C). Thus, calibrating a thermometer is easy as we immerse the thermometer in an ice bath and mark the zero point and then immerse in boiling water and mark the 100 point. We finally divide the length between these marks into 100 equal spaces, called degrees. We then extend outside this range by keeping the size of the degree constant.

Other types of temperature measuring instruments also exist. Thermocouples are based on the principle that when two dissimilar metals are joined, an electromotive force develops which is primarily a function of temperature. The most used pairs of metals are platinum/alloy of platinum and rhodium, copper/constantan (alloy of copper and nickel), and iron/constantan. Temperature sensors that are based on the fact that the electrical resistance of various materials changes with temperature are called electrical resistance thermometers. Resistance temperature detectors and thermistors belong to this class. Temperature sensors that measure temperature by sensing radiation are known by terms such as radiation thermometers and optical pyrometers. An important advantage of pyrometers is that they do not come in contact with the body whose temperature is to be determined and can be used at extremely high temperatures.

It is desirable to have a procedure for assigning temperature values that does not depend on the properties of any particular substance or class of substances. Such a scale is called a thermodynamic temperature scale. The **Kelvin scale** is an **absolute thermodynamic temperature scale** that provides a continuous definition of temperature, valid over all ranges of temperature. The Celsius temperature scale uses the unit degree Celsius (°C). As shown in Fig. 2.6, temperature differences are identical on both the Kelvin scale and the Celsius temperature scale. It is important to note that temperature scales are defined based on the triple point of water (point at which ice, liquid water, and steam coexist). This point is, for convenience, assigned to 273.16 K. Then the scales shown in Fig. 2.6 follow naturally. It is also reminded that the kelvin is the SI unit for temperature.

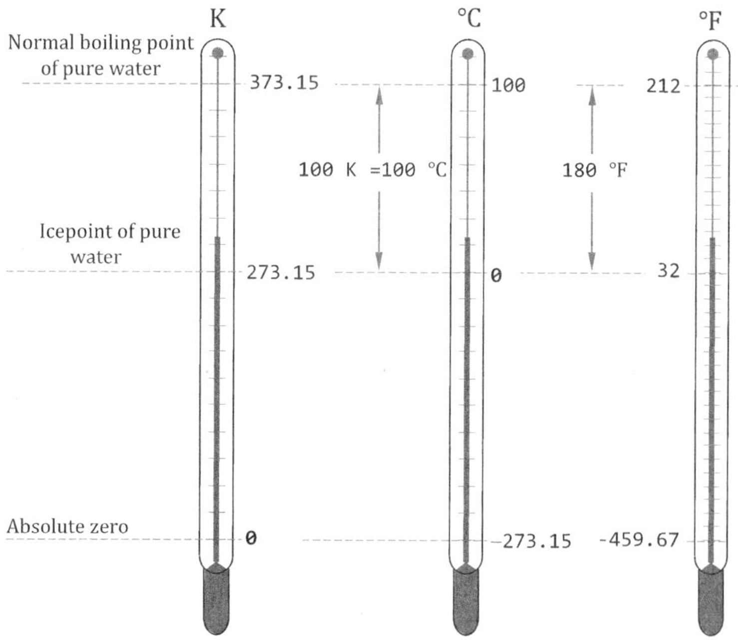

Figure 2.6 Comparison of temperature scales (*left*: Kelvin, middle: Celsius, *right*: Fahrenheit).

The temperature conversions can be performed easily using the following equations

$$T(\mathrm{K}) = T(^{\circ}\mathrm{C}) + 273.15 \tag{2.7.1}$$

$$T(^{\circ}\mathrm{C}) = \frac{5}{9}(T(^{\circ}\mathrm{F}) - 32) \tag{2.7.2}$$

For unit conversion it is also important to remember that when the temperature changes by 1 K in the Kelvin scale, it changes by 9/5 = 1.8 degrees in the Fahrenheit scale, and the corresponding conversion factor is

$$\frac{\Delta\mathrm{K}}{1.8\ \Delta^{\circ}\mathrm{F}} \tag{2.7.3}$$

The Rankine scale shares the same zero point as the Kelvin scale, as it is also an absolute temperature scale, but the size of a degree in the Rankine scale equals that in the Fahrenheit scale ($\Delta^{\circ}\mathrm{F} = \Delta^{\circ}\mathrm{R}$), or

$$T(^{\circ}\mathrm{R}) = T(^{\circ}\mathrm{F}) + 459.67 \tag{2.7.4}$$

There are several variables that need special attention while changing units from the Fahrenheit (or Rankine) scale to the Kelvin (or Celsius) scale. The two most important ones are the specific heat capacity and the convective heat transfer coefficients. Although their units involve temperature units in the denominator, the actual unit is change in the temperature (ΔT) and as a result the conversion factor given by Eq. (2.7.3) should be used (when converting from °F to °C or K). The symbol used in this book for temperature is T (upper case) irrespective of the temperature scale (note that some authors use the lower–case t to denote temperature in the Celsius scale).

Example 2.10. Units conversion for specific heat capacity

The specific heat capacity in the U.S. customary and British Engineering units is given in Btu/($\mathrm{lb_m}$·°F). Find the conversion factor for the SI system of units (J/(kg·K)).

Solution

We use the conversion factors (see Table 2.6): 1 Btu = 1055.056 J, 1 $\mathrm{lb_m}$ = 0.4536 kg and 1 ΔK = 1.8 Δ°F. We then have the following

$$1\ \frac{\mathrm{Btu}}{\mathrm{lb_m}\cdot{}^{\circ}\mathrm{F}} \times \frac{1055.056\ \mathrm{J}}{1\ \mathrm{Btu}} \times \frac{1\ \mathrm{lb_m}}{0.4536\ \mathrm{kg}} \times \frac{1.8\ {}^{\circ}\mathrm{F}}{1\ \mathrm{K}} = 4186.73\ \frac{\mathrm{J}}{\mathrm{kg}\cdot\mathrm{K}}$$

Example 2.11. Units conversion for heat transfer coefficient

The convective heat transfer coefficient in the U.S. customary and British Engineering units is given in Btu/(h·ft^2·°F). Find the conversion factor for the SI system of units (J/(s·m^2·K)).

Solution

We use the conversion factors (see Table 2.6): 1 Btu = 1055.056 J, 1 1h = 3600 s, 1 ft = 0.3048 m and 1 ΔK = 1.8 Δ°F. We then have the following

$$1\ \frac{\text{Btu}}{\text{h}\cdot\text{ft}^2\cdot{}^\circ\text{F}} \times \frac{1055.056\ \text{J}}{1\ \text{Btu}} \times \frac{1\ \text{h}}{3600\ \text{s}} \times \left(\frac{1\ \text{ft}}{0.3048\ \text{m}}\right)^2 \times \frac{1.8\ {}^\circ\text{F}}{1\ \text{K}}$$
$$= 5.6783\ \frac{\text{J}}{\text{s}\cdot\text{m}^2\cdot\text{K}}$$

Keep the results of Examples 2.10 and 2.11 in mind as they can be very useful.

2.7.2 Pressure

Pressure is defined as the force applied perpendicularly on a surface per unit area. The symbol for pressure recommended by IUPAC is "p" (lower-case p). However, upper-case P is widely used and accepted. Pressure is a scalar quantity. In the famous Torricelli experiment a simple barometer is used to measure the pressure of air by filling it with mercury. The barometer is placed inverted on the dish full of mercury and this causes the mercury in the tube to fall down until the difference between the mercury on the surface and in the tube is about 760 mm. Even when the tube is shaken or tilted, the difference between the surface and in the tube is not affected due to the influence of atmospheric pressure. This experiment, conducted in the

Figure 2.7 Practical applications of the U-tube manometer.

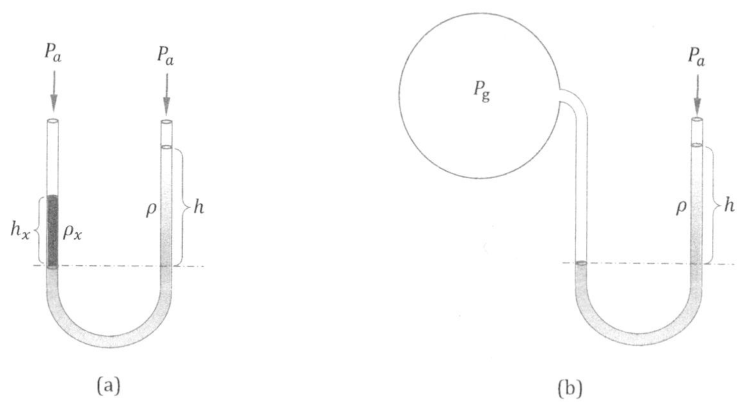

17th century, established what is today known as the standard atmosphere

$$1 \text{ atm} = 760 \text{ mmHg}$$

Such a device, called a **barometer** (from the Greek βάρος, "weight"), is still used to measure atmospheric pressure.

An important idea when dealing with pressure is that of **hydrostatic pressure**, the pressure due to fluids that are static. If a fluid has constant density ρ, and the local acceleration of gravity is g, then the pressure due to a static fluid of height h is given by

$$P = \rho g h \tag{2.7.5}$$

Using the density of mercury (13600 kg/m^3), the acceleration of gravity (9.8 m/s^2), and the fact that the height of the mercury is equal to 0.76 m, it follows that the pressure that balances the hydrostatic pressure of the mercury in the inverted tube is

$$P = 1.36 \times 10^4 \frac{\text{kg}}{\text{m}^3} \times 9.8 \frac{\text{m}}{\text{s}^2} \times 0.76 \text{ m} = 101.3 \times 10^3 \frac{\text{kg}}{\text{m s}^2} \tag{2.7.6}$$

Using Table 2.3 we see that $\text{kg}/(\text{m s}^2) = 1$ Pa, which is the SI unit for pressure. It follows that the atmospheric pressure (the pressure due to the weight of the static air from the sea level to the end of the atmosphere) is 101.3 kPa and the equivalence of the pressure units is

$$101.325 \text{ kPa} = 1.01325 \text{ bar} = 1 \text{ atm} = 760 \text{ mmHg} = 760 \text{ Torr}$$

Another unit for pressure that is in use is the psi (pound per square inch) for which the equivalence is 100 kPa = 14.50377 psi.

Hydrostatic pressure is a very useful idea in practical applications. Let's consider the U-tube manometer shown in Fig. 2.7a. A U-tube manometer with both arms open to the atmosphere is used to measure the unknown density ρ_x of a fluid, which when placed in the left arm has a height h_x above the interface (horizontal dash-dotted line). The second fluid (the manometric fluid) is of known density ρ and has heigh h above the interface. At static equilibrium the pressure on the same horizontal level must be the same and it follows that (P_a is the pressure of the atmosphere)

$$P_a + \rho_x g h_x = P_a + \rho g h \tag{2.7.7}$$

Solving for the unknown density

$$\rho_x = \rho \frac{h}{h_x} \tag{2.7.8}$$

In the second practical application shown in Fig. 2.7b, the left leg of the U-tube manometer is connected to a process vessel filled with gas in which we want to measure the pressure. As the system is in static equilibrium it follows that

$$P_g = P_a + \rho g h \tag{2.7.9}$$

The last equation gives a direct way to determine the pressure of the gas in the process vessel.

An important equation that can be derived from the equation of hydrostatic pressure is known as the **barometric equation** and gives the atmospheric pressure above sea level. The starting equation is the differential form of Eq. (2.7.5)

$$dP = -\rho g dh \tag{2.7.10}$$

The negative sign was introduced as the pressure is decreasing by increasing the altitude. We substitute the density of the air calculated from the ideal gas law where we also assume constant temperature T_0 (as we assume no further information on that matter)

$$dP = -M_{air}\frac{P}{RT_0}gdh \tag{2.7.11}$$

where M_{air} is the molar mass of air and R is the ideal gas constant. The last equation can also be written as

$$\frac{dP}{P} = -a\, dh \tag{2.7.12}$$

where $a = \frac{M_{air}g}{RT_0}$. We use the fact that $d\ln P = dP/P$ to obtain

$$\int_{P_0}^{P_H} d\ln P = -a\int_0^H dh$$

whence

$$\frac{P_H}{P_0} = \exp(-aH) \tag{2.7.13}$$

H is the elevation above the sea level and P_H is the corresponding pressure. We use the data

$$M_{air} = 0.028964\,\frac{\text{kg}}{\text{mol}}, \qquad g = 9.8\,\frac{\text{m}}{\text{s}^2}, \qquad R = 8.314\,\frac{\text{J}}{\text{mol K}},$$

$$T_0 = 288.15\ \text{K}$$

to obtain

$$a = \frac{M_{\text{air}}g}{RT_0} = \frac{0.028964\ \frac{\text{kg}}{\text{mol}} \times 9.8\ \frac{\text{m}}{\text{s}^2}}{8.314\ \frac{\text{J}}{\text{mol K}} \times 288.15\ \text{K}} = 1.18483 \times 10^{-4}\ \frac{1}{\text{m}}$$

where we have used the definition of $\text{J} = \text{kg m}^2/\text{s}^2$. Eq. (2.7.13) predicts that at elevation of 5000 m the pressure is

$$P_H = 101325\ \text{Pa}\exp\left(-1.18483 \times 10^{-4} \times 5000\right) = 56031.7\ \text{Pa}$$

The CRC Handbook of Chemistry and Physics reports the value of 54050 Pa.

Example 2.12. The total mass of the atmosphere

Given the fact that the pressure at sea level is 0.101325×10^6 Pa (1 Pa $= \frac{\text{kg}}{\text{m s}^2}$) and that the average diameter of Earth is 12.765×10^6 m, estimate the mass of the atmosphere.

Solution

Pressure is the force divided by the area and the force is the mass of the atmosphere times g

$$P = \frac{m_{\text{atm}}g}{\text{A}_{\text{earth}}}$$

whence

$$m_{\text{atm}} = P\frac{A_{\text{earth}}}{g} = \frac{P}{g}\left(\pi D^2_{\text{earth}}\right)$$

We substitute the numerical values

$$\begin{aligned} m_{\text{atm}} &= \frac{P}{g}\left(\pi D^2_{\text{earth}}\right) \\ &= \frac{0.101325 \times 10^6 \frac{\text{kg}}{\text{m s}^2}}{9.80665\ \frac{\text{m}}{\text{s}^2}}\left(3.1415926 \times 12.765^2 \times 10^{12}\right)\ \text{m}^2 \\ &= 5.2892 \times 10^{18}\ \text{kg} \end{aligned}$$

2.7.3 Density and specific volume

The **density** (ρ) of a pure substance is the mass per unit volume. Density normally refers to mass density. Molar density is the number of moles in a unit volume. The **specific volume** (v) of a substance is the volume occupied by a unit mass of the substance and the specific molar volume is the volume occupied by 1 mole of substance. Density and specific volume are the inverse of each other. Indicative values of the density of several substances are given in Table 2.9. Note that the density of gases at near–ambient conditions is close to 1 kg/m^3, the density of liquids is close to 1000 kg/m^3 while pure solids have densities ranging from 2000 kg/m^3 to 20 000 kg/m^3. Keep these numbers in mind to avoid mistakes. If you are looking for the density of oxygen at near–ambient conditions you must have an estimation of its value. If a colleague of yours finds the density of oxygen gas as 1000 then it is either wrong or the units are different from kg/m^3.

Densities of pure solids are essentially independent of pressure and temperature. The variation with temperature may be

Table 2.9 Indicative values of density of several pure components (STP:0 °C, 1 bar).

Element	Note	Density (kg/m^3)
Gas		
Hydrogen	STP	0.090
Helium	STP	0.178
Nitrogen	STP	1.251
Air	STP	1.293
Carbon dioxide	STP	1.977
Liquid		
Ethanol	ATM	810
Benzene		900
Water	20 °C	998
Water	4 °C	1000
Solids		
Coal		1200
Granite		2650
Aluminum		2700
Steel/iron		7800
Copper		8500
Lead		11,300
Gold		19,300
Platinum		21,400

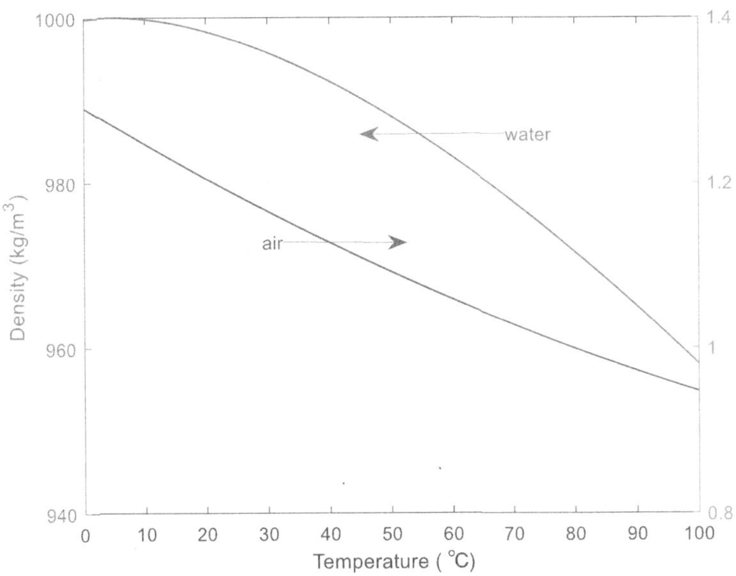

Figure 2.8 The density of water and air at 1 atm as a function of temperatures.

either positive or negative. The density of liquid water, for example, increases from 999.87 kg/m^3 at 0 °C, exhibits a maximum density of 1000 kg/m^3 at 4 °C, and then decreases to 958.4 kg/m^3 at 100 °C, as shown in Fig. 2.8. It is interesting to note that ice has a density around 917 kg/m^3 (at 0 °C). Densities of many pure compounds, including solutions, and mixtures may be found in

- Perry's Chemical Engineers' Handbook,
- CRC Handbook of Chemistry and Physics,

and in many web pages, such as

- Dortmund Data Bank (DDB) (http://www.ddbst.com/)
- Journal of Chemical & Engineering Data (https://pubs.acs.org/)
- NIST Chemistry Webbook (https://webbook.nist.gov/chemistry/)

The density of water and air at atmospheric pressure are shown in Fig. 2.8. As water and air appear in many practical applications it is important to have relatively simple equations to calculate their densities. There are numerous such equations available, including very complex or very simple ones. The following equations are accurate enough and simple enough and can be used.

$$\rho_{\text{water}}\left(\frac{\text{kg}}{\text{m}^3}\right) = 1001.33 - 9.46612\left(\frac{\text{T}}{100}\right) - 34.3129\left(\frac{\text{T}}{100}\right)^2, \text{T in °C}, 5\text{ °C} < \text{T} < 100\text{ °C} \tag{2.7.14}$$

$$\rho_{\text{air}}\left(\frac{\text{kg}}{\text{m}^3}\right) = 1.2904 - 0.45012\left(\frac{\text{T}}{100}\right) + 0.10742\left(\frac{\text{T}}{100}\right)^2, \text{T in °C}, 0\text{ °C} < \text{T} < 100\text{ °C} \tag{2.7.15}$$

The error of these equations is less than ± 0.6 kg/m^3 for water and ± 0.002 kg/m^3 for air.

An important question is the following: if m_A kg of substance A with density ρ_A is mixed with m_B kg of substance B with density ρ_B, what is the density ρ_{mix} of the resulting mixture? Assume that substances A and B are either solids or completely miscible liquids that form ideal mixtures (mixing is not accompanied by contraction or expansion phenomena—nonidealities). To answer the question, we use the definition of density and apply the definition to the mixture as follows

$$\rho_{\text{mix}} = \frac{m_A + m_B}{V_A + V_B}$$

At this point we note that from the definition of density of pure substances we have that $\rho_A = m_A/V_A$ and $\rho_B = m_B/V_B$. We use these results to eliminate the volume of the substance from the denominator to obtain

$$\rho_{\text{mix}} = \frac{m_A + m_B}{\dfrac{m_A}{\rho_A} + \dfrac{m_B}{\rho_B}} = \frac{1}{\left(\dfrac{m_A}{m_A + m_B}\right)\dfrac{1}{\rho_A} + \left(\dfrac{m_B}{m_A + m_B}\right)\dfrac{1}{\rho_B}}$$

whence

$$\frac{1}{\rho_{\text{mix}}} = \frac{w_A}{\rho_A} + \frac{w_B}{\rho_A} \tag{2.7.16}$$

w_A, w_B are the weight fractions of A and B. In the general case of a mixture formed by mixing n substances

$$\frac{1}{\rho_{\text{mix}}} = \sum_{j=1}^{n} \frac{w_j}{\rho_j} \tag{2.7.17}$$

It is reminded that the mass fraction of a substance A is the mass of substance A in the mixture divided by the overall mass of the mixture

$$w_A = \frac{m_A}{m_A + m_B + \ldots + m_n} = \frac{m_A}{\sum_{j=1}^{n} m_j} \tag{2.7.18}$$

Example 2.13. A common misconception

A common mistake in calculating the density of a mixture is the use of the following equation

$$\rho_{\text{mix}} = w_A \rho_A + w_B \rho_B + \ldots + w_n \rho_n$$

Consider the case where we mix two solids, A and B. Solid A has a density of 9000 kg/m^3 and solid B has a density of 1000 kg/m^3. Calculate the error in the estimation of the density of mixture.

Solution

We will use the case where the mass fraction of A is $w_A = 0.25$ and we perform the calculations. From Eq. (2.7.17) we have that

$$\rho_{\text{mix}} = \left(\frac{w_A}{\rho_A} + \frac{w_B}{\rho_B}\right)^{-1} = \left(\frac{0.25}{9000} + \frac{0.75}{1000}\right)^{-1} = 1285.7 \, \frac{\text{kg}}{\text{m}^3}$$

From the erroneous equation we obtain

$$\rho_{\text{mix}} = w_A \rho_A + w_B \rho_B = 0.25 \cdot 9000 + 0.75 \cdot 1000 = 3000 \, \frac{\text{kg}}{\text{m}^3}$$

The relative error is

$$\text{error in density estimation } \% = 100 \cdot \left(\frac{3000 - 1285.7}{1285.7}\right) = 133.3 \, \%$$

Can you see what is the reason behind this huge error and why many use the erroneous equation?

2.7.4 The definition of mole

The name mole is a translation of the German unit Mol, coined by the chemist Wilhelm Ostwald in 1894 from the German word Molekül (molecule). Before the 2019 revision of the SI, the mole was defined as **the amount of substance of a system that contains as many elementary entities as there are atoms in** 12 g **of carbon-12** (most common isotope of carbon). This is now only an approximation. The mole (symbol mol) is now (after the 2019 revision) the base unit in the SI for amount of substance: 1 mole is **an aggregate of exactly 6.022 140 76** $\times$ **10**^{23} **elementary entities**, which can be atoms, molecules, ions, ion pairs, or other particles. The number of particles in a mole equals the Avogadro number (note also that the Avogadro constant, N_A, has the units mol^{-1}—see Table 2.4).

In chemical engineering practice the kilomole (kmol) adopts the SI convention for standard multiples of metric units (kmol means 1000 mol). This is equivalent to the use of kg instead of g. The use of kmol is not only for "magnitude convenience" but also **makes the equations used for modeling chemical engineering systems coherent** (a coherent set of units is a system of measurement where all derived units are consistently defined in terms of a set of base units—without needing additional conversion factors). For example, the conversion of a mass flowrate of kg/s to molar flowrate of kmol/s only requires dividing by the molar mass in g/mol (or kg/kmol).

You should be aware that most professional chemical process simulation packages have adopted the notation kgmole or gmole instead of the correct notation kmol or mol in SI. You should avoid using this notation in your reports as it does not follow the rules of the SI system of units. For convenience in the US customary units, some engineers have also adopted the pound-mole (lb-mol or lbmol), which is defined as the number of entities in 12 lb of ^{12}C. One lb-mol is equal to 453.59237 g-mol, which is the same numerical value as the number of grams in an international avoirdupois pound. Again, this is a violation of the SI rules but can be accepted, provided that you are only using this notation to avoid confusion.

We now come to the important practical (nearly exact) relationship. The mole was (prior to 2019) defined in such a way that **the molar mass of a compound, in** g/mol, **is numerically equal to the average mass of one molecule or formula unit, in** daltons (Da). It was exactly equal before the redefinition of the mole in 2019, and is now only approximately equal, **but the difference is negligible for all practical purposes**. Thus, for example, **the average mass of a molecule of water is about 18.0153** Da, **and the molar mass of water is about 18.0153** g/mol. It is reminded that the dalton or unified atomic mass unit (Da or u, respectively) is a unit of mass defined as 1/12 of the mass of carbon-12. It is a non-SI unit accepted for use with SI.

The atomic mass constant, denoted m_u, is defined identically, giving $m_u = 1/12\ m(^{12}C) = 1$ Da. To conclude, the practical rule is, despite all the above, relatively simple:

Let M denote the molar mass of a substance, then the mass of 1 kmol of the substance is numerically equal to M kg.

If we have m kg of a substance or a mass flowrate of $\dot{m}$ kg/s of a substance then divide by the molar mass in g/mol (or equivalently in kg/kmol) to obtain the kmol of the substance or the molar flowrate in kmol/s.

When a mixture is formed by mixing n substances, then Eq. (2.7.17) applies for the case of the molar mass of the mixture

$$\frac{1}{M_{\text{mix}}} = \sum_{j=1}^{n} \frac{w_j}{M_j} \tag{2.7.19}$$

where M_j is the molar mass of entity j and w_j is the mass fraction of entity j. However, if the mole fraction is used then the equation is fundamentally different:

$$M_{\text{mix}} = \sum_{j=1}^{n} M_j x_j \tag{2.7.20}$$

The mole fraction is defined as

$$x_A = \frac{n_A}{n_A + n_B + \ldots + n_n} = \frac{n_A}{\sum_{j=1}^{n} n_j} \tag{2.7.21}$$

where n_j denotes the moles of entity j in the mixture. It is noted that in this book we will try to follow the rules proposed by IUPAC when we select specific symbols for quantities.

Example 2.14. The average molar mass of air and its composition (mole and mass fractions)

The main constituents of dry atmospheric air have the following volume % analysis:

N_2 (28.0134): 78.084%
O_2 (31.999): 20.946%
Ar (39.948): 0.934%
CO_2 (44.009): 0.042%
Ne (20.1797): 0.0018%
He (4.0026): 0.000,524%
CH_4 (16.04): 0.000,192%
Kr (83.798): 0.000,114%
H_2 (2.016): 0.00,005%

N_2O (44.013): 0.00,005%

The number in the parenthesis is the molar mass of the corresponding gas.

Calculate the average molar mass of dry atmospheric air.

Solution

The calculation of the average molar mass of dry atmospheric air is performed as follows. We consider 100 mol of dry air. The % by volume is identical to the molar % composition. We then multiply the moles of each gas in 100 mol of air by its molar mass to calculate the mass of each gas in g (in 100 mol of air). The average molar mass can then be calculated using Eq. (2.7.19). The calculations are extensive and are summarized in the table that follows.

GAS	mol of gas in 100 mol dry air	Molar Mass (g/mol)	Mass of gas (in g) In 100 mol dry air	Mass Fractions	Average molar mass (w_i/M_i)
N_2	78.084000	28.01340	2187.398326	0.755092	0.02695469
O_2	20.946000	31.99900	670.251054	0.231371	0.00723058
Ar	0.934000	39.94800	37.311432	0.012880	0.00032242
CO_2	0.042000	44.00900	1.848378	0.000638	0.00001450
Ne	0.001800	20.17970	0.036323	0.000013	0.00000062
He	0.000524	4.00260	0.002097	0.000001	0.00000018
CH_4	0.000192	16.04000	0.003080	0.000001	0.00000007
Kr	0.000114	83.79800	0.009553	0.000003	0.00000004
H_2	0.000055	2.01600	0.000111	0.000000	0.00000002
N_2O	0.000033	44.01300	0.001452	0.000001	0.00000001
TOTAL	**100.0**		**2896.861806**	**1.000000**	**28.96609280**

Note that the fourth column gives the mass of the mixture that corresponds to 100 mol of the mixture. The average molar mass of dry air is 28.97 g/mol.

2.7.5 Expressions of concentrations

Solutions are composed of two parts: the substance that has been dissolved is the solute and the substance in which the solute dissolves is the solvent. It is not necessary to have a liquid as a solvent, even though the term solution probably suggests a solid solute, like salt, dissolved in a liquid solvent, like water.

Air is a solution where the solvent is N_2 and gold used in jewelry is an alloy (a mixture of metals) of gold (solvent) and one or more metals (solutes), including copper and nickel. When physical units are used, the concentrations of solutions are generally expressed by:

1. the mass of solute per unit volume of solution (as, 2 g NaCl per L solution),
2. the mass units of solute per 100 mass units of solution,
3. the mass of solute per mass of solvent (as, 2 g NaCl in 100 g H_2O),
4. the parts (mass or volume) of solute per million (ppm) or billion (ppb) parts of solution.

The mass concentration is defined as the mass of a constituent m_j divided by the volume of the solution V

$$\text{mass concentration} = \frac{m_j}{V}$$

The molar concentration is the moles of an entity n_j divided by the volume of the solution

$$\text{molar concentration} = \frac{n_j}{V} = \frac{\text{mass concentration}}{M_j}$$

where M_j is the molar mass of entity j.

In science and engineering parts-per notation is a set of pseudo-units used to describe the small values of miscellaneous dimensionless quantities, e.g., mole fraction or mass fraction. Since these fractions are quantity-per-quantity measures, they are pure numbers with no associated units of measurement. Commonly used are

- parts-per-million—ppm, 10^{-6}
- parts-per-billion—ppb, 10^{-9}

The quantity "1 ppm" can be used, for instance, for a mass fraction if a water-borne pollutant is present at one-millionth of a gram per gram of sample solution. In this case, the letter "w" (standing for "weight") is sometimes added to the abbreviation (e.g., ppmw, ppbw). When working with aqueous solutions, it is common to assume that the density of water is 1.00 g/mL. Therefore, it is common to equate 1 kg of water with 1 L of water. Consequently, 1 ppm corresponds to 1 mg/L and 1 ppb corresponds to 1 μg/L. The quantity "1 ppm" is also used for a gas mixture where the gas constituent is present at one-millionth of a volume unit per unit volume of a constituent. In this case, the letter "v" (standing for "volume") is sometimes added to the abbreviation (e.g., ppmv, ppbv). We normally assume that volume fractions in gas mixtures are equal to mole fractions.

Although the International Bureau of Weights and Measures (BIPM) recognizes the use of parts-per notation, it is not formally part of the International System of Units (SI). Note that although "percent" (%) is not formally part of the SI, both the BIPM and the International Organization for Standardization (ISO) take the position that "in mathematical expressions, the internationally recognized symbol % (percent) may be used with the SI to represent the number 0.01″ for dimensionless quantities. According to IUPAP, "a continued source of annoyance to unit purists has been the continued use of percent, ppm, ppb, and ppt". Although SI-compliant expressions should be used as an alternative, the parts-per notation remains nevertheless widely used in technical disciplines. The main problems with the parts-per notation are:

- the named numbers starting with "billion" have different values in different countries,
- "ppt" usually means "parts per trillion"; however, it occasionally means "parts per 1000,"
- the parts-per notation may refer to mass fraction, mole fraction, or volume fraction and it is usually not stated which quantity is used, creating confusion.

Example 2.15. Mass of carbon dioxide in the atmosphere

In Earth's atmosphere, carbon dioxide is a trace gas that plays an integral part in the greenhouse effect, carbon cycle, photosynthesis, and oceanic carbon cycle. It is one of the main greenhouse gases in the atmosphere of Earth (the others are water vapor, methane, nitrous oxide, and ozone). The concentration of atmospheric carbon dioxide (CO_2) measured at the Mauna Loa Observatory in Hawaii reached 427 ppmv in 2024. What is the total mass of CO_2 in the atmosphere?

Solution

ppmv is equivalent to volume of a particular gas in 10^6 volumes of atmospheric air. For gases this is equivalent to 1 mol of a particular gas in 10^6 mol of atmospheric air, i.e.,

$$\text{mass fraction } CO_2 \text{ in the atmosphere} = 427\ \frac{\text{mol } CO_2}{10^6 \text{ mol air}}$$
$$\times \frac{1 \text{ mol air}}{28.966 \text{ g air}} \times \frac{44.009 \text{ g } CO_2}{1 \text{ mol } CO_2}$$
$$= 648.76 \times 10^{-6}\ \frac{\text{g } CO_2}{\text{g atmospheric air}}$$

where the average molar mass of atmospheric air calculated in Example 2.14, has been used. In Example 2.12 we calculated the mass of air in the atmosphere (5.2892×10^{18} kg). The total mass of CO_2 in the atmospheric air is

$$\text{mass of } CO_2 \text{ in the atmosphere}$$

$$= 648.76 \times 10^{-6} \frac{\text{g } CO_2}{\text{g atmospheric air}} \times 5.2892 \times 10^{18} \text{ kg} \times \frac{10^3 \text{ g}}{1 \text{ kg}}$$

$$= 3.43 \times 10^{18} \text{ g } CO_2 \times \frac{1 \text{ kg}}{10^3 \text{ g}} \times \frac{1 \text{ t}}{10^3 \text{ kg}} = 3.43 \times 10^{12} \text{ t } CO_2$$

$$= 3.43 \text{ Tt } CO_2$$

Example 2.16. Converting concentrations into ppm

The threshold limit value (TLV) is a level of occupational exposure to a hazardous substance where it is believed that nearly all healthy workers can repeatedly experience at or below this level of exposure without adverse effects. Strictly speaking, TLV is a reserved term of the American Conference of Governmental Industrial Hygienists (ACGIH), who determines and publishes TLVs annually. TLVs issued by the ACGIH are the most widely accepted occupational exposure limits both in the United States and most other countries. The TLV for ammonia is 25 ppmv (all TLVs are reported at 25 °C and 1 atm). In a lab, 0.1 kg of ammonia is stored. The volume of the lab is 1000 m^3. If the ammonia is released instantly in the closed lab, what will the concentration of ammonia in ppmv be? Develop a general expression for converting concentrations expressed in mg/m^3 to ppmv.

Solution

We initially establish an equation for converting concentrations in mg/m^3 to ppmv. According to the definition of ppmv

$$\text{concentration in ppmv} = \frac{\text{volume of gas } j}{10^6 \text{ volumes of gas mixture}} = \frac{V_j \text{ m}^3 \text{ of } j}{10^6 \text{ m}^3 \text{ mixture}}$$

where V_j is the volume of j in the gas mixture. When the pressure and the temperature are known then the volume of an ideal gas is given by

$$V_j = \frac{n_j RT}{P} = \left(\frac{m_j}{M_j}\right)\frac{RT}{P}$$

where m_j is the mass of j, M_j is the molar mass of j, R is the ideal gas constant, T is the absolute temperature, and P is the pressure. We replace into the definition of ppmv to obtain

$$\text{concentration in ppmv} = \frac{1}{10^6}\left(\frac{R}{M_j}\right)\frac{T}{P}\left(\frac{m_j}{\text{m}^3\ \text{mixture}}\right)$$

If m_j is expressed in mg instead of kg (10^6 mg = 1 kg), then

$$\text{concentration in ppmv} = \left(\frac{1}{M_j}\right)\frac{RT}{P}\left(\text{concentration in } \frac{\text{mg}}{\text{m}^3}\right)$$

If the 0.1 kg of ammonia (=10^5 mg) is suddenly released in the lab space of 1000 m^3 then the concentration of ammonia will be

$$\text{concentration of NH}_3 \text{ in } \frac{\text{mg}}{\text{m}^3} = \frac{10^5\ \text{mg}}{10^3\ \text{m}^3} = 10^2\ \frac{\text{mg}}{\text{m}^3}$$

$$\text{concentration in ppmv} = \left(\frac{1}{17.031\ \frac{\text{g}}{\text{mol}}}\right) \times \frac{0.082057\frac{\text{m}^3\ \text{atm}}{\text{mol K}} \times 298.15\ \text{K}}{1\ \text{atm}} 10^2\ \frac{\text{mg}}{\text{m}^3} = 143.615\ \text{ppmv}$$

We see that ammonia concentration will exceed the TLV limit and this is a safety issue for the laboratory, and special procedures for handling the ammonia container need to be established.

2.8 Energy, work, and heat

2.8.1 Definitions and terms used

Energy is classified in three specific forms.

a. The macroscopic **kinetic energy**, denoted by E_K, is the energy associated with the macroscopic motion of a system. An object of mass m moving at velocity v has a specific kinetic energy given by:

$$e_K = \frac{E_K}{m} = \frac{1}{2}v^2 \qquad (2.8.1)$$

b. The macroscopic **potential energy**, denoted by E_P, is the energy associated with the position of the system in a potential field. For example, an object in the Earth's gravitational field has a potential energy given by:

$$e_P = \frac{E_P}{m} = gz \tag{2.8.2}$$

c. The **internal energy**, U, (with specific internal energy $u = U/m$) is the energy associated with the motion, position, and chemical-bonding configuration of the individual molecules of the substances within the system.

Energy is not an absolute quantity, but it is defined relative to a reference state that we need to define and identify. If you are sitting at your desk reading this text think about your speed. Is it zero? Relative to what? What is your reference system?

As we mentioned, internal energy, U, is arguably the most important form of energy for chemical engineering applications. Internal energy is related to all forms of molecular energy, including the kinetic and potential energies of the molecules themselves. However, we detect a change in internal energy using macroscopic terms. More specifically, changes in internal energy can have the following macroscopic manifestations:

1. temperature changes,
2. phase changes,
3. changes in chemical structure through chemical reactions.

A change in internal energy that leads to a change in temperature is often termed sensible heat while a change that results in phase transformations as latent heat.

Work and **heat** are two terms that refer to the **transfer of energy** between the surroundings and a system. If a system does not exchange mass with the surroundings (called a **closed system**) then the transfer of energy between the system and the surroundings can be accomplished by heat or by work only. Heat is the transfer of energy caused by a temperature difference, whereas all other forms of energy transfer in a closed system occur via work. We generally associate work with something useful being done by (or to) the system.

There are many forms of work, but the forms that are relevant to chemical engineering applications are mechanical (expansion/compression, rotating shaft) and electrical. The most common case of work is when a force causes a displacement in the boundary of a system. In the case of expansion, for example, the system needs to push the surroundings. Thus, the system exchanges energy with the surroundings in the form of work. The work can be described mathematically as follows (see Fig. 2.9):

Figure 2.9 A process of gas in cylinder expansion against the surroundings. (a) Physical description of the process, and (b) representation in pressure-volume diagram. The work done on the surroundings from the system is the area below the line shown on (b).

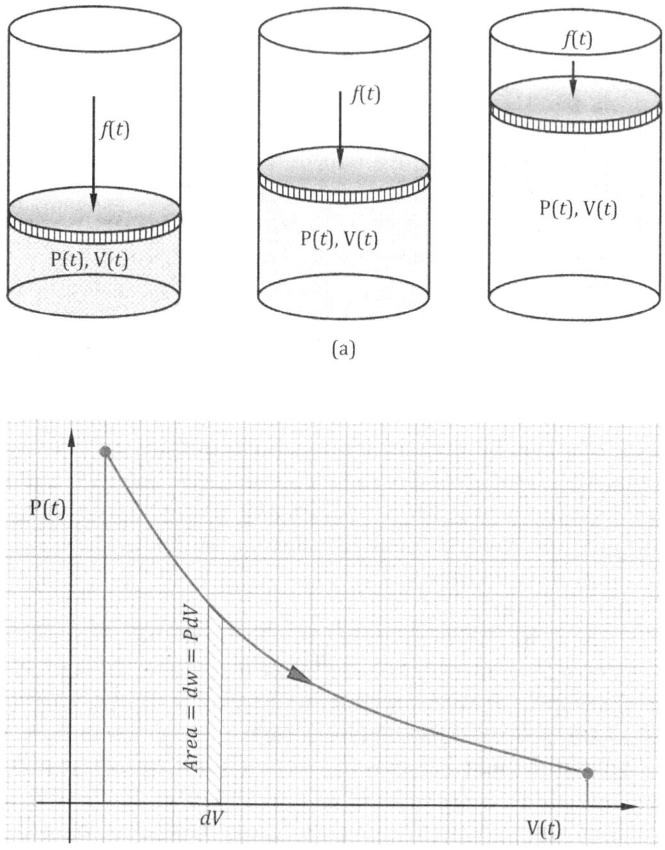

$$W = \int_{X(0)}^{X(t)} f dx = \int_{X(0)}^{X(t)} \left(\frac{f}{A}\right) d(Ax) = \int_{V(0)}^{V(t)} PdV \tag{2.8.3}$$

where A is the cross-sectional area of the cylinder. Using the fact that the total volume of the confined fluid is its mass times its specific volume it follows that

$$w = \frac{W}{m} = \int_{V(0)}^{V(t)} Pdv \tag{2.8.4}$$

Note that the symbol v is unfortunately used to denote specific volume but is also used to denote velocity.

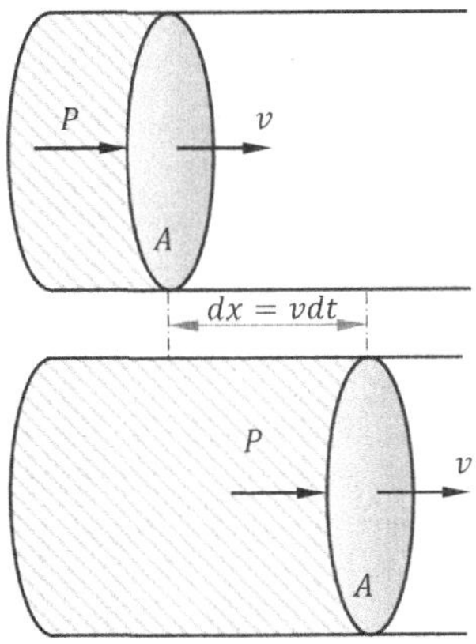

Figure 2.10 A process of a fluid flowing into a pipe and pushing the fluid ahead forward. The fluid is performing work on the environment called flow work.

An equally important case of work is the one shown in Fig. 2.10 where a fluid that flows inside a pipe pushes forward the fluid ahead and thus performs work on its surroundings. This type of work is called **flow work** and is calculated as follows (note that A is the cross-sectional area of the pipe, v is the velocity of the fluid, P is the upstream pressure, and dx is the distance traveled by the fluid in differential time dt)

$$W = \int f dx = \int (PA)dx = P(Av)\int dt = P\dot{V}\Delta t = P\frac{\dot{M}}{\rho}\Delta t \tag{2.8.5}$$

where $\dot{V}$ denotes the volumetric flowrate (in m^3/s) and $\dot{M}$ is the mass flowrate (in kg/s). Using the fact that the specific volume is the inverse of the density it follows that

$$w = \frac{W/\Delta t}{\dot{M}} = P\frac{1}{\rho} \tag{2.8.6}$$

It is important to note at this point that **enthalpy**, an important form of energy mainly for systems that exchange mass with the surroundings (**open systems**), is defined as

$$H = U + PV \tag{2.8.7}$$

and the specific enthalpy (enthalpy per mass or mole)

$$h = u + P\frac{1}{\rho} = u + Pv \tag{2.8.8}$$

i.e., enthalpy is the sum of the internal energy and the product of the pressure and specific volume of the fluid. Note that we use capital letters for extensive quantities and the corresponding lower–case letters for the specific quantities (i.e., H for the total enthalpy and h for the specific enthalpy).

Heat refers to the transfer of energy between the surroundings and the system due to a temperature difference (or temperature gradient). It is common sense that thermal energy will transfer spontaneously from higher-temperature areas to lower-temperature areas. The symbol Q is used to denote heat. If a system is well insulated and no heat interaction is observed between the system and the surroundings, then the system is called (actually, the process that the system undergoes) **adiabatic**. A closed system that has no work or heat interaction with its surroundings is called **isolated**. Although heat (as you will discover in the Transport Phenomena courses) is transferred by radiation, conduction, and convection, only the latter mechanism will be useful in this book. The idea of heat transfer by convection is easy to understand by imagining a horizontal surface over which a fluid (gas or liquid) flows (see Fig. 2.11). If the surface is hot with temperature T_s ("s" stands for surface) and the fluid has a much lower temperature T_f ("f" stands for fluid) then heat will flow from the hot solid surface to the cold fluid. The rate at which heat is transferred from the hot surface to the cold fluid is analogous to the temperature difference $(T_s - T_f)$ and the surface area A. The exact mathematical expression is as follows

$$\dot{Q} = hA(T_s - T_f) \tag{2.8.9}$$

where the coefficient h is called the heat transfer coefficient and has units of energy per unit time and per unit area (it is unfortunate that the symbol h is traditionally used for both the specific enthalpy and the heat transfer coefficient). It is estimated through carefully performed experiments, but there are also cases for

Figure 2.11 Convective heat transfer between a hot surface and a cold fluid.

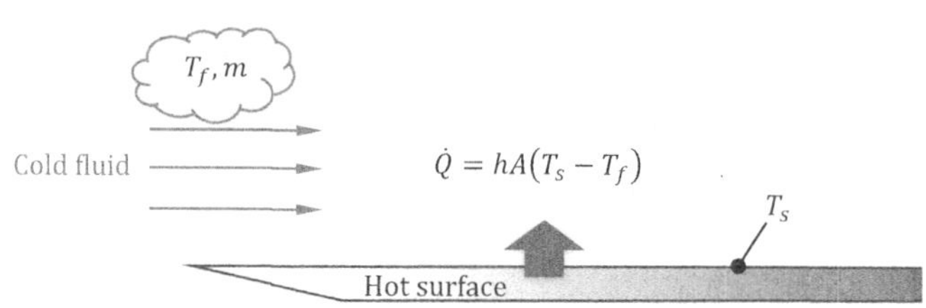

which it can be determined theoretically. The same idea can be used to understand the heating process of the contents of a process vessel, where the fluid inside the vessel is heated up by the hot surface of the vessel (heated externally by a process stream) or from a coil immersed in the fluid where process steam is fed internally.

We now come to an important decision regarding the sign convention for heat and work. In this text, both heat and work are defined as positive when they represent energy transfer from the surroundings to the system. Conversely, both heat and work are defined as negative when they represent energy transfer from the system to the surroundings. This sign convention is arbitrary; alternative conventions exist, provided they are applied consistently throughout the analysis. We choose this convention to be consistent with today's convention, i.e., the one adopted from IUPAC. However, when this convention was initially used it was done in the context of powering the steam engine, and the opposite sign convention was used for work (the engineering objective was to get work out of the system to power a train by supplying heat/coal!). In many textbooks, this obsolete convention is still used and you need to be careful. Fig. 2.12 can be used to reinforce the sign convention used in this text. It should be noted that, based on this convention, a negative sign must be added in both Eqs. (2.8.4) and (2.8.6) to account for the fact that work is, in both cases, negative.

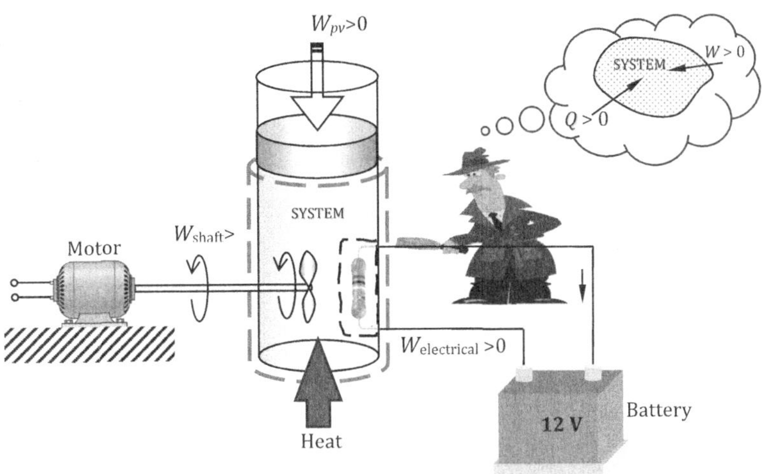

Figure 2.12 The sign convention used in this text: work and heat are positive when work is done on the system and heat "flows" from the surroundings to the system.

Example 2.17. Calculating work done by an expanding gas by numerical means

A gas in a cylinder closed by a piston is expanding from an initial pressure of 1.1×10^5 Pa and initial volume of 0.1 m^3 to a final pressute of 1.01×10^5 Pa and volume of 0.6 m^3. The exact way in which the gas is expanding is not known apart from one intermediate measurement: when the pressure is 1.03×10^5 Pa the gas volume is 0.35 m^3 and that the variation of pressure and volume is smooth and is performed very slowly. Can you give an estimate of the work done by the gas to the surroundings?

Solution

An approximate representation of the process is shown in the *P–V* diagram of Fig. 2.13. The continuous ABC line gives an estimation of the real process. The two straight dashed line segments (AB and BC) show the variation of the system based on the three points available. The work done by the expanding gas to the piston, when the pressure is decreased from the initial pressure of 1.1×10^5 Pa to the intermediate pressure of 1.03×10^5 Pa, equals the area of the trapezoid shown in Fig. 2.13, denoted by I_1. The work done by the expanding gas when the process A→B takes place, is the following

$$I_1 = W_{AB} = \text{area below AB} = \int_A^B P(V)dV \approx \text{area of trapezoid}$$
$$= \frac{1}{2}(P_A + P_B)(V_B - V_A)$$

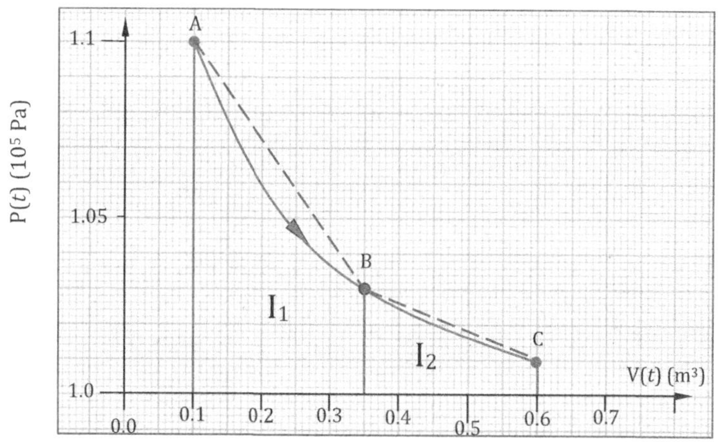

Figure 2.13 The process of gas expansion against the surroundings studied in Example 2.17.

In a similar way

$$I_2 = W_{BC} = \text{area below BC} = \int_B^C P(V)dV \approx \text{area of trapezoid}$$
$$= \frac{1}{2}(P_B + P_C)(V_C - V_B)$$

We replace the numerical values to obtain the following.

$$-W = W_{AB} + W_{BC} = \frac{1}{2}(P_A + P_B)(V_B - V_A) + \frac{1}{2}(P_B + P_C)(V_C - V_B)$$
$$= \left[\frac{1}{2}(1.1 + 1.03)(0.35 - 0.1) + \frac{1}{2}(1.03 + 1.01)(0.6 - 0.35)\right] \times 10^5 \text{ Pa} \cdot \text{m}^3$$
$$= 52125 \text{ J}$$

The negative sign denotes that the work is negative as it is performed by the system on its environment (see Fig. 2.13).

2.8.2 Nonconventional units used in energy systems

In the field of study of energy systems, several nonconventional units are used that need to be discussed. The unit cal is frequently used for which

$$1 \text{ cal} = 4.1868 \text{ J}$$

Another unit is the **watt-hour** (Wh). This is an energy unit for which

$$1 \text{ Wh} = 3600 \text{ J}$$

The equivalence might seem strange, but it can be proven as follows

$$1 \text{ J} = 1 \text{ J} \cdot \frac{1 \text{ h}}{3600 \text{ s}} = \frac{1}{3600} \frac{\text{J}}{\text{s}} \cdot 1 \text{ h} = \frac{1}{3600} \text{ Wh}$$

The **tonne of oil equivalent** (abbreviated toe) is a unit of energy defined as the amount of energy released by burning one tonne of crude oil. The International Energy Agency (IEA) (a Paris-based autonomous intergovernmental organization, established in 1974, that provides policy recommendations, analysis, and data on the global energy sector with 31 member countries) defines one ton of oil equivalent (toe) to be equal to

- 1 toe = 11.63 MWh
- 1 toe = 41.868 GJ

- 1 toe = 10 gigacalories (Gcal)—using the international steam table calorie (cal_{IT})
- 1 toe = 39,683 207.2 Btu
- 1 toe = 1/0.7 t of coal equivalent (tce)
- 1 toe = 7.33 barrel of oil equivalent

The **barrel of oil equivalent** (boe) is a unit of energy based on the approximate energy released by burning one barrel (42 US gal or about 159 L) of crude oil. The boe is used by oil and gas companies in their financial statements as a way of combining oil and natural gas reserves and production into a single measure, although this energy equivalence does not take into account the lower financial value of energy in the form of gas. The U.S. Energy Information Administration defines the barrel of oil equivalent as about 6 GJ (1.7 MWh; 5.7 Btu). It is recommended to use the IEA conversion factors given above, from which it follows that.

1 boe = 11.63/7.33 = 1.58,663 MWh

A **tonne of coal equivalent** (tce), is a conventional value, based on the amount of energy released by burning one tonne of coal. According to the IEA one tce corresponds to 0.700 t of oil equivalent (toe).

The **therm** (symbol, thm) is a non-SI unit of heat energy equal to

- 1 therm = 100,000 Btu
- 1 therm = 105.506 MJ
- 1 therm = 29.3072 kWh,
- 1 therm = 25,200 kcal

One therm is the energy content of approximately 100 ft^3 (2.83 m^3) of natural gas at standard temperature and pressure. However, the Btu is not standardized worldwide, with slightly different values in the EU, UK, and United States, meaning that the energy content of the therm also varies by territory.

A quad is a unit of energy equal to 10^{15} Btu or 1.055×10^{18} J (1.055 EJ or EJ) in SI units. The unit is used by the U.S. Department of Energy in discussing world and national energy budgets. The global primary energy production in 2022 was 637.8 quad, i.e., 672.9 EJ.

2.9 System definition, processes and the ideal gas law

2.9.1 Open and closed systems

Two basic types of systems are considered in this text. These are the **closed systems** and the **open systems**. In a closed system a fixed amount of matter is studied while in an open system a fixed control volume is defined as a region of space through which mass may flow. There is no mass transfer across the system boundary in a closed system while mass may cross the boundary of an open system. The distinction between closed and open systems is based on the transfer of mass through the system boundary. Both closed and open systems may exchange energy with the surroundings. A special type of closed system that does not exchange energy with its surroundings is called an **isolated system**.

In subsequent sections of this text, system analyses are made of devices such as process vessels, heat exchangers, and reactors through which mass flows. These analyses are often conducted by studying a selected region of space through which mass flows. With this approach, a region of space, called a **control volume**, within a prescribed boundary is studied. A dashed line that will define a control volume that surrounds the system under study will be shown. This line indicates the control volume surface. In general, the choice of the control volume is not unique but the aim of the analysis task indicates what is an appropriate control volume for the system under study. An example is shown in Fig. 2.14. Three potential control volumes are shown in the figure, although many others could be defined. If the aim of the study is to perform analysis of the cooling water used then the control volume II around the condenser might be the most appropriate. If the aim of the study is to analyze the composition of the residue

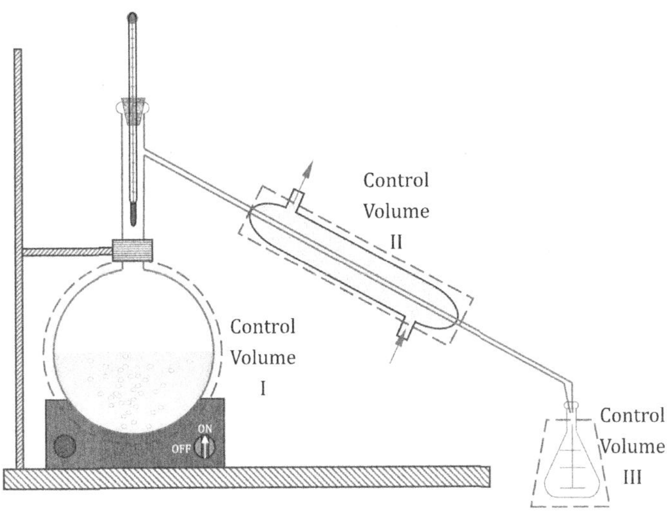

Figure 2.14 A system of an experimental fractional distiller where several control volumes can be defined based on what is the aim of the analysis.

in the pot, then the control volume I might be the most appropriate.

2.9.2 Properties and processes

To describe a system and predict its behavior through mathematical models requires knowledge of its properties. A **property** is a macroscopic characteristic of a system, such as temperature, pressure, volume, mass, or energy, that has a well-defined value when the system is in a given state and whose value depends only on the current state of the system, not on the path by which that state was reached. Systems engineering also deals with quantities that are not properties, such as mass flow rates and energy interactions by work and heat. The term "**state of a system**" refers to the condition of a system as described by its properties. Since there are relations among the properties of a system, the state often can be specified by providing the values of a minimal subset of the properties. All other properties can then be determined in terms of these few, using **constitutive equations** or **equations of state**. These are either phenomenological (empirical) equations or equations derived from first principles. Thermodynamics is the main source of constitutive equations in chemical engineering. Eqs. (2.7.14) and (2.7.15) for calculating the density of water and air is an example of empirical constitutive equations.

A system is said to be at **steady state** if none of its properties change with time. This is a theoretical state that cannot exist in the real world as all physical systems change with time. It is a very powerful idealization that helps to simplify the study of physical systems. When an external stimulus causes any of the properties of a system to change, the state changes and the system is said to have undergone a **process**. A process is a transformation from one state to another. A **thermodynamic cycle** is a sequence of processes that begins and ends at the same state. Thermodynamic cycles that are repeated periodically play prominent roles in many areas of application such as air conditioning systems and electricity production in power plants.

At a given state each property has, according to their definition, a definite value that does not depend on how the system arrived at that state. Water at 10 °C, 1 bar has a specific density and enthalpy irrespective of how water ended up in these conditions (by cooling down water from 20 °C to 1 bar or by heating ice from −100 °C and 100 bar). Therefore, the change in value of any property in a process from one state to another is determined exclusively by the two end states and is independent of the process. If the change in a quantity between two states is independent of the path taken, then the quantity is a property. This provides a definite test for determining whether a quantity is a property: **a quantity is a property if its change in value between two states is independent of the process**. It follows that if the value of a particular quantity depends on the details of the process, and not solely on the end states, that quantity cannot be a property. **Work and heat are the two important variables that are not properties** as they strongly depend on the process. The work you perform differs significantly if you go to your university every morning by using the bus or on foot. Density, specific volume, energy, enthalpy, etc. are properties.

Thermodynamic properties can be either **extensive** or **intensive**. Extensive properties depend on the size of the system while intensive properties do not. In other words, **extensive properties are additive**. An easy thought experiment that can help you to decide whether a property is intensive or extensive is to ask: Would the value of this property remain unchanged for each part if the system is divided into identical subsystems? If the answer is "yes," the property is intensive. If the answer is "no," the property is extensive. Intensive variables are the temperature, pressure, density, specific enthalpy, etc. Total mass, total enthalpy, and total volume are extensive variables.

The adjective "specific" before the name of an extensive quantity is used to mean divided by mass. When the symbol for the extensive quantity is a capital letter, the symbol used for the specific quantity is often the corresponding lower-case letter. An example is the total volume, V, and specific volume, $v = V/m$. Another example is the heat capacity at constant pressure, C_p, and specific heat capacity at constant pressure, $c_p = C_p/m$. ISO and the Chemistry and Human Health Division of IUPAC recommend systematic naming of physical quantities derived by division with mass and volume by using the attributes specific (or massic) and volumic, respectively. Thus, for example, the specific volume could be called massic volume. The adjective molar before the name of an extensive quantity generally means divided by amount of substance. The subscript "m" on the symbol for the extensive quantity denotes the corresponding molar quantity. An example is the volume, V and the molar volume, $V_m = V/n$, another example is the enthalpy, H and the molar enthalpy, $H_m = H/n$. It is sometimes convenient to divide all extensive quantities by amount of substance, so that all quantities become intensive; the subscript m may then be omitted if this convention is stated and there is no risk of ambiguity. In this text, we will use capital letters for all extensive quantities and lower-case letters for either the specific (massic or molar) quantity with the understanding that the general setting indicates which one is used. We will not be using the subscript "m" for the molar properties as this violates one of the most fundamental restrictions of the SI system: "neither the name of a quantity, nor the symbol used to denote it, should imply a particular name of a unit."

2.9.3 The ideal gas model

Thermodynamic properties provide a systematic tool for analyzing systems and making engineering calculations. Since they are independent of the process, we can use data or models that are available to characterize processes or equilibrium states of interest. The important observation that makes this possible is that once we know the value of a certain number of properties of the system, all the other properties can be calculated. This principle is known as **the state postulate**, and for a pure substance can be stated as follows: If we have a pure substance, its thermodynamic state and all its intensive thermodynamic properties can be determined from two independent intensive properties.

We call the two intensive properties we select to describe the state of a system the independent variables. All other intensive properties are then dependent variables. We are free to decide which independent intensive properties we choose. For example, the molar volume (dependent variable) of a pure substance is completely specified once we know its temperature and pressure (the two independent variables we have selected). In mathematical terms the state postulate implies the following when temperature and pressure have been chosen as independent (intensive) variables

$$v = f(T, P) \tag{2.9.1}$$

The ideal gas law is the simplest known equation of state that satisfies Eq. (2.9.1)

$$Pv = \mathrm{RT} \tag{2.9.2}$$

where R is the ideal gas constant, which is equal to the product of the Boltzmann constant k and the Avogadro constant N_A

$$R = kN_A = 1.380\,649 \times 10^{-23}\,\frac{\mathrm{J}}{\mathrm{K}} \times 6.022\,140\,76 \times 10^{23}\,\mathrm{mol}^{-1}$$
$$= 8.314\,463\,\frac{\mathrm{J}}{\mathrm{mol}\cdot\mathrm{K}}$$

Note that

$$1\,\text{J} = 1\frac{\text{kg}\cdot\text{m}^2}{\text{s}^2} = 1\frac{\text{kg}}{\text{m s}^2}\text{m}^3 = 1\,\text{Pa}\cdot\text{m}^3$$

and

$$R = 8.314\,463\,\frac{\text{Pa}\cdot\text{m}^3}{\text{mol}\cdot\text{K}}$$

The development of the ideal gas law was made possible by the collective work of Boyle, Gay-Lussac, Avogadro, and Charles (as it was proposed by Clapeyron in 1839). It can also be derived from simple arguments through the kinetic theory of gases (as it was shown by Krönig in 1856 and Clausius in 1857). The molecules of the ideal gas are point particles with negligible volume that exhibit no intermolecular attractions. The ideal gas law is most accurate for monatomic gases at high temperatures and low pressures. The neglect of molecular size becomes less important for lower densities, i.e., for larger volumes at lower pressures, because the average distance between adjacent molecules becomes much larger than the molecular size. The relative importance of intermolecular attractions diminishes with increasing thermal kinetic energy, i.e., with increasing temperatures.

The model of the ideal gas is complemented by a method for calculating the internal energy of the ideal gas. Internal energy includes molecular kinetic contributions and intermolecular potential energy contributions. Temperature is directly related to the average translational kinetic energy of the molecules. Because an ideal gas has no intermolecular forces, its molecular potential energy is constant and may be chosen as a reference zero. Therefore, the internal energy of the ideal gas, u^{ig}, depends only on the temperature. Hence,

$$u^{ig} = u(T) \tag{2.9.3}$$

It then follows that the same holds true for the specific enthalpy of the ideal gas (see Eq. (2.8.8))

$$h^{ig} = u^{ig} + Pv = u(T) + RT \tag{2.9.4}$$

2.9.4 The specific heat capacities

Two important properties related to internal energy and enthalpy are known as specific heats. The specific heats are particularly useful for thermodynamic calculations involving the ideal gas model introduced earlier. The specific heats are intensive properties and are the **specific heat capacity under constant volume**, denoted by c_v, and the specific **heat capacity under constant pressure**, denoted by c_p. These properties are defined for pure, simple compressible substances as partial derivatives of the functions $u(T, v)$ and $h(T, p)$, respectively

$$c_v = \frac{\partial u(T, v)}{\partial T}\bigg|_v \tag{2.9.5}$$

$$c_p = \frac{\partial h(T, p)}{\partial T}\bigg|_p \tag{2.9.6}$$

The specific internal energy of the ideal gas is only a function of temperature and therefore

$$c_v^{ig} = \frac{du^{ig}(T)}{dT} \tag{2.9.7}$$

$$c_p^{ig} = \frac{dh^{ig}(T)}{dT} \tag{2.9.8}$$

Combining Eqs. (2.9.4), (2.9.7) and (2.9.8) we obtain the following result

$$c_p^{ig} = c_v^{ig} + R \tag{2.9.9}$$

The ratio of the two heat capacities is denoted by γ and

$$\gamma = \frac{c_p}{c_v} \tag{2.9.10}$$

Values for c_v^{ig} and c_p^{ig} can be obtained via statistical mechanics using spectroscopic measurements or can be determined macroscopically through property measurements. They are available in many sources, but the web page by NIST (NIST Chemistry WebBook, webbook.nist.gov) will be used in this text. Some indicative data for the ideal gas heat capacity at constant pressure are presented in Table 2.10. The data correspond to the following polynomial form of the heat capacity equation

$$c_p^{ig}(T) = A + BT + CT^2 + DT^3 + ET^{-2} \tag{2.9.11}$$

The variation of the specific heat capacity at constant pressure of several representative gases as a function of absolute

Table 2.10 Data for the specific heat capacity equation (from NIST Chemistry WebBook), c_p in kJ/(kmol·K), T in K (see Eq. (2.9.11)).

Gas	A	$B \times 10^3$	$C \times 10^6$	$D \times 10^9$	$E \times 10^{-3}$	T_{min}(K)	T_{max}(K)
CO_2	24.99735	55.186960	−33.691370	7.948387	−136.638	298	1200
H_2	33.06618	−11.363417	11.432816	−2.772874	−158.558	298	1000
H_2O	30.09200	6.832514	6.793435	−2.534480	82.139	500	1700
O_2	31.32234	−20.23531	57.866440	−36.506240	−7.374	100	700
N_2	28.98641	1.85398	−9.647459	16.635370	0.117	100	500
CH_4	−0.70303	108.47730	−42.521570	5.862788	678.565	298	1300
C_2H_2	40.68697	40.73279	−16.178400	3.669741	−658.411	298	1100
C_2H_4	−6.38788	184.40190	−112.97180	28.495930	315.540	298	1200

For monatomic gases such as He, Ne, Ar $c_p^{ig} = 5R/2$.

temperature is shown in Fig. 2.15. The data from Table 2.10 has been used. From Fig. 2.15 it follows that when temperature variations of 100 K are considered, using a constant value for the specific heat capacity will result in significant errors, for reasons that will be explained in what follows.

The practical usefulness of the definitions given by Eqs. (2.9.7) and (2.9.8) is significant as these equations, in their integrated form, are used to calculate the specific internal energy or the specific enthalpy of a pure gaseous component. For the specific enthalpy we have that

$$\Delta h^{ig} = h^{ig}(T_f) - h^{ig}(T_{\text{ref}}) = \int_{T_{\text{ref}}}^{T_f} c_p(T)dT \qquad (2.9.12)$$

where T_f is the temperature at which we need to determine the specific enthalpy and T_{ref} is the reference temperature at which we assign a reference value for the specific enthalpy $(h^{ig}\left(T_{\text{ref}}\right))$. Using Eq. (2.9.11) we obtain

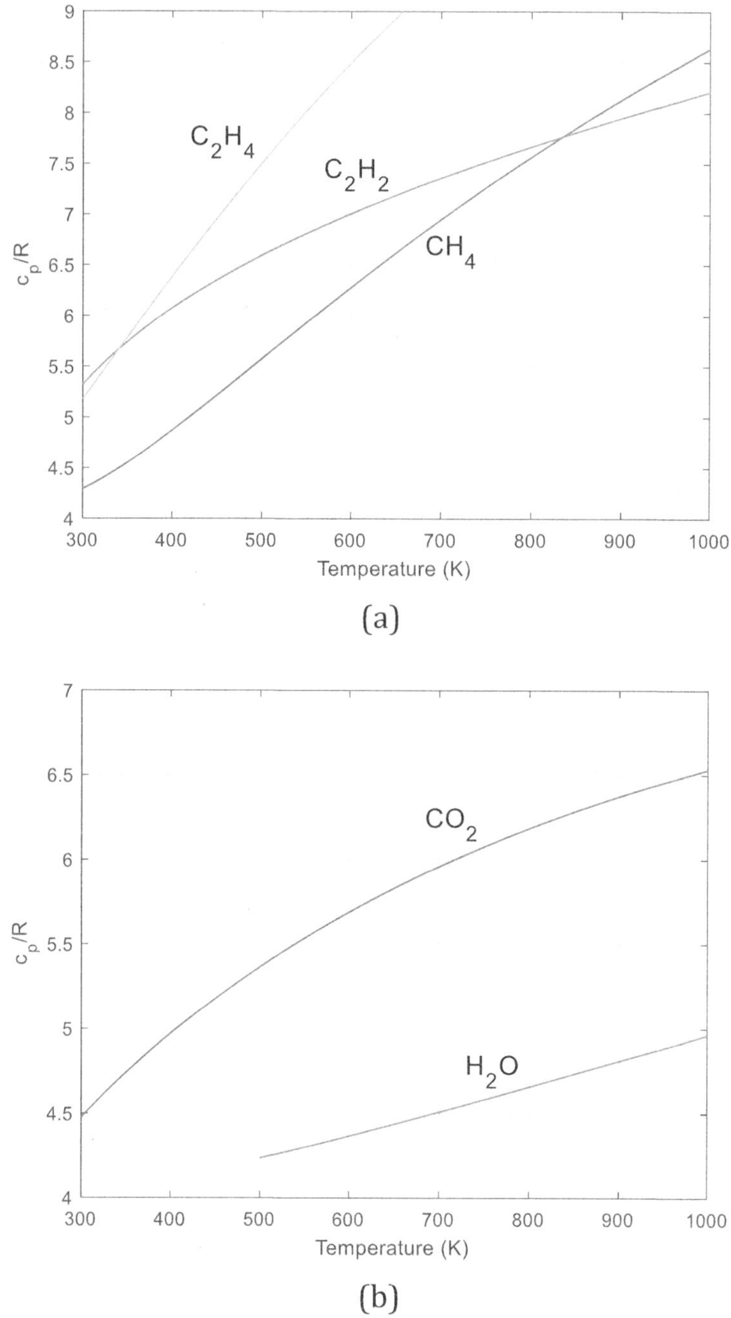

Figure 2.15 Variation of specific heat capacity at constant pressure as a function of absolute temperature.

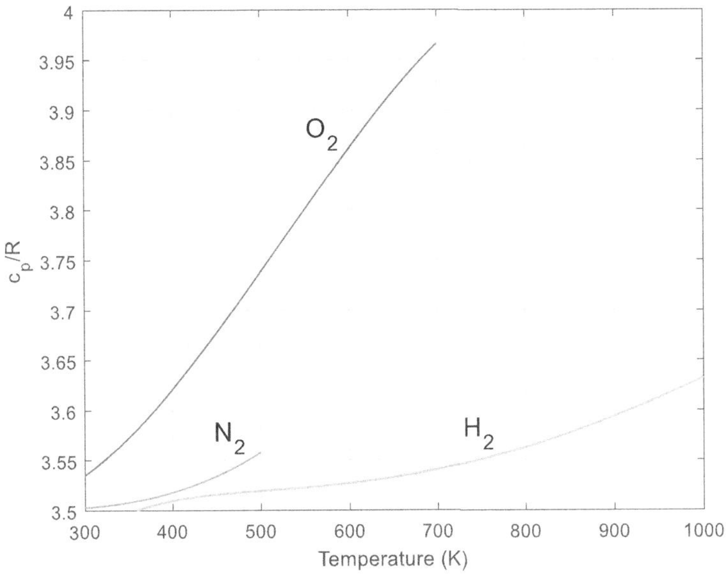

(c)

Figure 2.15 continued

$$
\begin{aligned}
\Delta h^{ig} &= h^{ig}(T_f) - h^{ig}(T_{\text{ref}}) = \int_{T_{\text{ref}}}^{T_f} (A + BT + CT^2 + DT^3 + ET^{-2})\,dT \\
&= A(T_f - T_{\text{ref}}) + \frac{B}{2}\left(T_f^2 - T_{\text{ref}}^2\right) + \frac{C}{3}\left(T_f^3 - T_{\text{ref}}^3\right) + \frac{D}{4}\left(T_f^4 - T_{\text{ref}}^4\right) \\
&\quad - E\left(\frac{1}{T_f} - \frac{1}{T_{\text{ref}}}\right)
\end{aligned} \tag{2.9.13}
$$

Example 2.18. Calculating heat transfer rate necessary to superheat steam

Steam with mass flowrate of $\dot{m}_{\text{steam}} = 1$ kg/s and temperature 500 K, must be heated to 600 K in a heat exchanger. If the steam is assumed to be an ideal gas, estimate the rate at which heat is transferred in the heat exchanger $\dot{Q}$. Use the fact that the heat transfer rate is equal to the change in the specific enthalpy of the gas stream times its mass flowrate, i.e.,

$$\dot{Q} = \dot{m}_{\text{steam}} \Delta h^{ig}_{\text{steam}}$$

Solution

As the reference temperature in Eqs. (2.9.12) and (2.9.13) is arbitrary, we assign the initial temperature of the steam (500 K) as the reference temperature. The change in the specific enthalpy of the steam can be calculated by the direct application of Eq. (2.9.13) where the data from Table 2.10 for water can be used.

$$\begin{aligned}\dot{Q} &= \dot{m}_{\text{steam}}\left(h^{ig}(T_f) - h^{ig}(T_0)\right)\\ &= \dot{m}_{\text{steam}}\left\{A(T_f - T_0) + \frac{B}{2}\left(T_f^2 - T_0^2\right) + \frac{C}{3}\left(T_f^3 - T_0^3\right)\right.\\ &\quad \left.+\frac{D}{4}\left(T_f^4 - T_0^4\right) - E\left(\frac{1}{T_f} - \frac{1}{T_0}\right)\right\}\\ &= 1\frac{\text{kg}}{\text{s}}\left[30.092(600-500) + \frac{6.832514\times10^{-3}}{2}\left(600^2 - 500^2\right)\right.\\ &\quad +\frac{6.793435\times10^{-6}}{3}\left(600^3 - 500^3\right)\\ &\quad -\frac{2.534480\times10^{-9}}{4}\left(600^4 - 500^4\right) - 0.082139\times10^3\\ &\quad \left.\left(\frac{1}{600} - \frac{1}{500}\right)\right]\frac{\text{kJ}}{\text{kmol}}\cdot\frac{1\ \text{kmol}}{18.015\ \text{kg}}\\ &= 196.98\ \frac{\text{kJ}}{\text{s}} = 198\ \text{kW}\end{aligned}$$

The sign is positive as steam is heated by its environment.

Example 2.19. Calculating the final temperature of a gas stream with known input temperature and known rate of heat transfer

If the steam in Example 2.18 is heated by a stream of helium with mass flowrate of $\dot{m}_{\text{helium}} = 1$ kg/s and initial temperature of 1000 K, find the final temperature of helium.

Solution

Helium is a monatomic gas for which (see Table 2.10) the heat capacity at constant pressure is constant and equal to $5R/2$. For monatomic gases, Eq. (2.9.13) simplifies to

$$h^{ig}(T_f) - h^{ig}(T_{\text{ref}}) = \int_{T_{\text{ref}}}^{T_f} c_p dT = c_p \int_{T_{\text{ref}}}^{T_f} dT = \frac{5R}{2}\left(T_f - T_{ref}\right)$$

We use again the fact that the reference temperature is arbitrary and we now set the reference temperature at $T_0 = 1000$ K. We have that

$$-\dot{Q} = \dot{m}_{\text{helium}} \Delta h^{ig}_{\text{helium}} = \dot{m}_{\text{helium}} \frac{5R}{2} (T_f - T_0)$$

The negative sign is due to the fact that the heat is transferred from the helium to its environment. We finally have that

$$T_f = T_0 - \frac{\dot{Q}}{\dot{m}_{\text{helium}}} \frac{2}{5R}$$
$$= 1000 \text{ K} - \frac{198}{1} \frac{\text{kJ/s}}{\text{kg/s}} \times \frac{4.0026 \text{ kg}}{\text{kmol}} \frac{2}{5 \times 8.314 \dfrac{\text{kJ}}{\text{kmol·K}}} = 962 \text{ K}$$

Example 2.20. Calculating the volume of a tank for storing high–pressure hydrogen

A tank of volume 50 L is to be used to store compressed hydrogen at 100 atm. If hydrogen is assumed to behave as an ideal gas, even at this high pressure, what is the mass of hydrogen that can be stored in the tank? If the tank is to be used in a hydrogen car which consumes 0.5 kg H_2/100 km, what is the distance that can be traveled by the hydrogen car (compare with an internal combustion engine car that can travel distances up to 700 km with a liquid fuel tank of 50 L). Assume that the temperature is 15 °C.

Solution

The ideal gas law can be written as (see Eq. (2.9.2))

$$PV = \left(\frac{m}{M}\right) RT$$

or

$$m = \frac{MPV}{RT} = \frac{2.0157 \dfrac{\text{g}}{\text{mol}} \times 100 \text{ atm} \times \dfrac{101\,325 \text{ Pa}}{1 \text{ atm}} \times 0.050 \text{ m}^3}{8.314\,463 \dfrac{\text{Pa·m}^3}{\text{mol·K}} \times 288.15 \text{ K}}$$
$$= 426.25 \text{ g} = 0.43 \text{ kg}$$

The distance that can be traveled by the hydrogen car is 0.43/0.5 = 86 km! This is an order of magnitude less than the distances covered by an internal combustion engine car with the same volume of fuel tank. To achieve practical driving ranges, hydrogen must be stored at very high pressures, typically around 700 bar in modern fuel cell vehicles.

2.10 MATLAB and Python

2.10.1 Simple calculations, data manipulations and 2D plots in MATLAB

Ten students are asked to measure the length of a table using a 30 cm long ruler. The estimations of the students are as follows.

Student	Measurement (cm)
1	139
2	137
3	136
4	139
5	138
6	138
7	140
8	141
9	137
10	139

To calculate the mean and standard deviation of the measurements in MATLAB we can use the commands.

```
mean(x)    : is the mean value of the elements in x, x is a
vector
std(x)     : returns the standard deviation of x when x is a
vector
```

The following script calculates the mean and the standard deviation of the measurements and also plots the measurements for inspection (Fig. 2.16).

```
clear all
student    = [1 2 3 4 5 6 7 8 9 10];                    % stu-
dents
length     = [139 137 136 139 138 138 140 141 137 139 ];% meas-
urements
mean_length = mean(length) % sample mean
std_length  = std(length)  % standard deviation of the sample

plot(student,length,'ro','MarkerFaceColor','w')
grid on
xlabel('Student')
ylabel('Length (cm)')
axis([0 11 135 142])
```

We will now see how we can perform the following calculation in MATLAB (calculating the integral of a real value function)

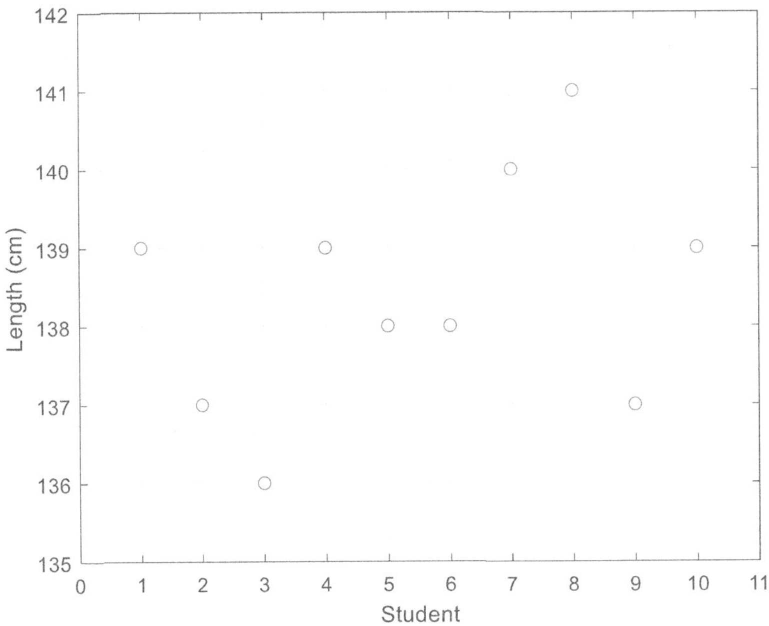

Figure 2.16 Example of a 2D plot in MATLAB.

$$I = \int_{x_0}^{x_f} f(x)dx$$

A classical application of this calculation is shown in Fig. 2.9, where the work performed by the gas in a gas-in-a-piston system is calculated. The area between the line denoting the $f(x)$ and the x-axis ($f(x) = 0$) in the interval $[x_0, x_f]$ equals the (definite) integral. If $f(x)$ is a relatively simple equation, then the integral can be calculated analytically. Numerical calculation has the advantage that it only needs understanding of the physical meaning of the mathematical operation of integration and can be performed easily in MATLAB with acceptable accuracy even for extremely complicated functions.

We will assume that a gas undergoes the process shown in Fig. 2.9 and that

$$PV^d = c$$

If we substitute into Eq. (2.8.3) to eliminate P, we obtain

$$W = \int_{V(0)}^{V(t)} PdV = \int_{V(0)}^{V(t)} \frac{c}{V^d} dV \tag{2.10.1}$$

We consider a particular case study where $P(0) = 10^6$ Pa, $V(0) = 0.01\ m^3$, $V\left(t_f\right) = 0.0681\ m^3$, $d = 1.2$, and $c = 3981.3$ Pa$\cdot\left(m^3\right)^{1.2}$.

To calculate the integral numerically in MATLAB we first define anonymous functions in MATLAB. An anonymous function is a function that is not stored in a program file, but is associated with a variable whose data type is `function_handle`. Anonymous functions can accept multiple inputs and return one output. They can contain only a single executable statement. In the case of the function appearing in Eq. (2.10.1), the function we seek to calculate the integral is

$$f(V) = \frac{c}{V^d} = \frac{3981.3}{V^{1.2}}$$

We can use the following anonymous function.

```
>> f = @(V) 3981.3./V.^1.2
```

It is now easy to calculate the integral using the `integral` command in MATLAB

```
>> W = integral(f,0.01,0.0681)

W =
    1.5933e+04
```

The work performed by the gas is, therefore, W = 15933 J.

The calculation of the integral in MATLAB is a very handy and efficient method for determining the heat that needs to be transferred to a gas to change its temperature. This calculation is performed as shown in Eq. (2.9.13)

$$Q = h^{ig}\left(T_f\right) - h^{ig}\left(T_{ref}\right) = \int_{T_{ref}}^{T_f} \left(A + BT + CT^2 + DT^3 + ET^{-2}\right) dT$$

The function that is integrated in this case is the following

$$f(T) = A + BT + CT^2 + DT^3 + ET^{-2}$$

The calculation performed in Example 2.18 can be done in MATLAB as follows.

```
clear all
A = 30.092;
B = 6.832514E-3;
C = 6.793435E-6;
D = -2.53448E-9;
E = 82.139E3;
f = @(T) A+B*T+C*T.^2+D*T.^3+E./T.^2;
Dh = integral(f,500,600,'AbsTol',1e-6,'RelTol',1e-5)
Q = Dh/18.015

Dh =
   3.5759e+03

Q =
  198.4968
```

As a final example, we consider the example of the barometric equation given by Eq. (2.7.11)

$$dP = -M_{\text{air}}\frac{P}{RT_0}gdh$$

where the temperature is considered independent of the elevation, resulting in an estimation of the temperature which deviates from the actual temperature. We now use the fact that the temperature is decreasing at a known rate (lapse rate) of $L = -0.0065$ K/m (from sea level up to 11 km). We then write Eq. (2.7.11) as follows

$$\int_{P_0}^{P_H} d\ln P = \ln\frac{P_H}{P_0} = -\frac{M_{\text{air}}g}{R}\int_0^H \left(\frac{1}{T_0 - Lh}\right)dh$$

or

$$P_H = P_0 \exp\left[-\left(\frac{M_{\text{air}}g}{R}\right)\int_0^H \left(\frac{1}{T_0 - Lh}\right)dh\right]$$

The calculation of the pressure at 5000 m can be performed in MATLAB as follows.

```
clear all
Mair = 0.028964; % kg/mol
g    = 9.8;      % m/s^2
R    = 8.314;    % J/(mol K)
T0   = 288.15;   % K
L    = 0.0065;   % K/m
P0   = 101325;   % Pa

f = @(H) 1./(T0-L*H);
I = integral(f,0,5000,'AbsTol',1e-6,'RelTol',1e-5)
PH = P0*exp(-(Mair*g/R)*I)

PH =

   5.4042e+04
```

The calculation result in an estimation of the pressure at 5000 m above the sea level of 54042 Pa. The CRC Handbook of Chemistry and Physics reports the value of 54050 Pa.

2.10.2 Unit manipulation in Python

`sympy.units` is a module in the SymPy library (a Python library for symbolic mathematics) that provides tools for working with physical quantities and units—such as meter, second, kilogram, etc.—in a symbolic way. Another similar module is `Pint`. `sympy.units` allows you to express quantities with units (e.g., 10 * meter/second), supports unit conversion and dimensional analysis, and is useful for solving engineering problems symbolically while keeping track of units. In what follows a few examples are shown of using unit conversion and computations using `sympy.units`. However, no use will be made of this module in this text.

```
import sympy.physics.units as u

# define mass in kg
m_kg=5*u.kg
print("     mass in kg:",m_kg,"")
     mass in kg: 5*kilogram

# convert to g
m_g=u.convert_to(m_kg,u.g).n()
print("same mass in g :",m_g)
same mass in g : 5000.0*gram
# convert to tonne
m_t=u.convert_to(m_g,u.t).n()
print("same mass in t :",m_t)
same mass in t : 0.005*tonne
# pressure conversion
p_psi=1*u.psi
p_Pa=u.convert_to(p_psi,u.kPa).n()
print("Pressure conversion:",p_psi," = ",p_Pa)
Pressure conversion: psi  =  6.89475729316836*kilopascal
# speed conversion
distance = 100 * u.m
time = 10 * u.s
v = distance / time
v_km_per_h=u.convert_to(v,u.km/u.h).n()
print("Speed of ",v," is equal to",v_km_per_h)
Speed of  10*meter/second  is equal to 36.0*kilometer/hour
# import ideal gas constant and select specific units
from sympy.physics.units import R, m, L, Pa, bar, atm, mol, K,
J
Rg1=u.convert_to(R, [L, Pa, mol, K])
print("Ideal gas constant =",Rg1)
Ideal gas constant = 8314.46261815324*liter*pascal/(kelvin*mole)
Rg2=u.convert_to(R, [m, atm, mol, K])
print("Ideal gas constant =",Rg2)
Ideal gas constant = 8.20573660809597e-5*atmosphere*meter**3/
(kelvin*mole
Rg3=u.convert_to(R, [J, mol, K])
print("Ideal gas constant =",Rg3)
Ideal gas constant = 8.31446261815324*joule/(kelvin*mole)
Rg4=u.convert_to(R, [m, bar, mol, K])
print("Ideal gas constant =",Rg4)
Ideal gas constant = 8.31446261815324e-5*bar*meter**3/
(kelvin*mole)
''' use ideal gas constant to calculate the mass
    in g of steam in a vessel of 1000 L volume
    and specific temperature and pressure
'''
P=100000*u.Pa
T=300*u.K
V=1000*u.L
N=P*V/(Rg1*T)
print("Stored steam in mol :",N)
```

```
Stored steam in mol : 40.090785014242*mole

MW=18.02*u.g/u.mol
M=N*MW
print("Stored steam in g   :",M)
Stored steam in g   : 722.435945956641*gram
```

2.10.3 Simple calculations, data manipulation, and 2D plots in Python

The calculation of the mean and standard deviation of the measurements by 10 students can be performed as follows (Fig. 2.17).

```
import numpy as np
import matplotlib.pyplot as plt

student=np.array([1, 2, 3, 4, 5, 6, 7, 8, 9, 10])
length =np.array([139, 137, 136, 139, 138, 138, 140, 141, 137,
139])
mean_length =np.mean(length)
std_length  =np.std(length)

plt.plot(student, length,'ro', markersize=12)
plt.xlabel("Student")
plt.ylabel("Length (cm)")
plt.grid()
plt.xlim([0, 11])
plt.ylim([135, 142])
```

The calculation of definite integrals can be performed using the `scipy.integrate` module. The function `quad` is provided to

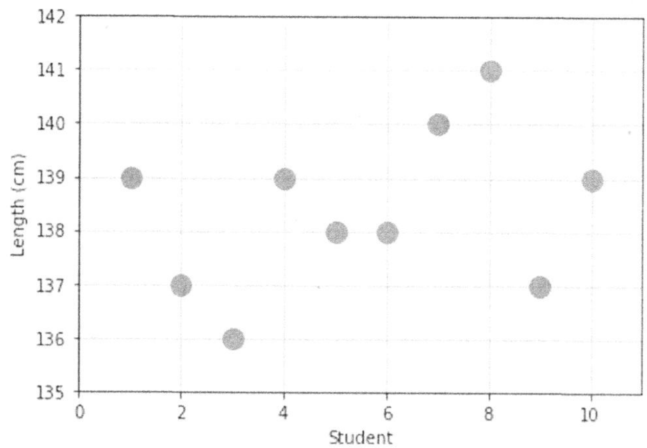

Figure 2.17 Example of a 2D plot in Python.

integrate a function of one variable between two points. The points can be (±inf) to indicate infinite limits. To calculate the pressure at elevation of 5000 m from sea level we can use the following.

```
from scipy.integrate import quad
import numpy as np

Mair = 0.028964 # molecular mass in kg/mol
g    = 9.8      # acceleration of gravity in m/s^2
R    = 8.314    # ideal gas constant in J/(mol K)
T0   = 288.15   # temperature at sea level in K
L    = 0.0065   # lapse rate in K/m
P0   = 101325   # pressure at sea level in Pa

def f(H): return 1/(T0-L*H)

I = quad(f,0,5000)
PH=P0*np.exp(-(Mair*g/R)*I[0])

print('Pressure at altitude of 5000 m is',PH,'Pa')

Pressure at altitude of 5000 m is 54042.11922053971 Pa
```

An important element of both MATLAB and Python is the symbolic manipulation and symbolic calculations, including the symbolic calculation of definite and indefinite integrals and the solution of ordinary differential equations. The calculation of the integral given by Eq. (2.9.13) can be performed as follows.

```
from sympy import *
from sympy.abc import A, B, C, D, E, T, t
T_ref=Symbol('T_ref')
print(integrate(A+B*t+C*t**2+D*t**3+E/t**2, (t,T_ref,T)))

A*T - A*T_ref + B*T**2/2 - B*T_ref**2/2 + C*T**3/3 -
C*T_ref**3/3 + D*T**4/4 - D*T_ref**4/4 + E/T_ref - E/T
```

Another important capability of sympy is the solution of ordinary differential equations, as shown below for the differential Eq. (2.7.12)

```
from sympy import *
# define symbols to be used
h, a, P0 = var('h, a, P0')
P = Function('P')
# solve differential equation
eq = Derivative(P(h),h) + a * P(h)
solution = dsolve( eq, ics = {P(0): P0})
# print the result
print(solution)
```

2.10.4 Creating your unit conversion table using scipy in Python

The `scipy.constants` module is part of the SciPy library and provides a comprehensive collection of physical constants, unit definitions, and unit conversion factors commonly used in scientific and engineering calculations. This module allows you to perform precise, standards-based computations without manually typing or memorizing fundamental constants or unit conversions. All values are based on the latest CODATA (Committee on Data for Science and Technology) recommendations, ensuring accuracy and consistency across scientific disciplines. A demonstration of the capabilities offered by `scipy.constants` is given in the Python program that follows.

```
from scipy.constants import R,g,c,h,k,N_A, foot,inch, li-
ter,gallon_US, gram,lb,metric_ton, bar,atm,psi,mmHg,
Btu,calorie, hp, minute,hour, zero_Celsius, yot-
ta,giga,mega,kilo,milli,micro,nano
print('USEFULL CONSTANTS .............................')
print(f"Gas constant               : {R:.8e} J.mol^(-1) K^(-1))")
print(f"Standard accel.of grav.: {g:.8e} m.s^(-2)")
print(f"Speed of light             : {c:.8e} m.s^(-1)")
print(f"Planck constant            : {h:.8e} J·s")
print(f"Boltzmann constant         : {k:.8e} J.K^(-1)")
print(f"Avogadro's number          : {N_A:.8e} mol^(-1)")
print('LENGTH ........................................')
print(f"1 ft                       : {foot:.8e} m")
print(f"1 in                       : {inch:.8e} m")
print('VOLUME ........................................')
print(f"1 liter                    : {liter:.8e} m^{3}")
print(f"1 US gal                   : {gallon_US:.8e} m^{3}")
print('MASS  ........................................')
print(f"1 gram                     : {gram:.8e} kg")
print(f"1 lb                       : {lb:.8e} kg")
print(f"1 t                        : {metric_ton:.8e} kg")
print('PRESSURE ......................................')
print(f"1 bar                      : {bar:.8e} Pa")
print(f"1 atm                      : {atm:.8e} Pa")
print(f"1 psi                      : {psi:.8e} Pa")
print(f"1 mmHg                     : {mmHg:.8e} Pa")
print(f"1 Btu                      : {Btu:.8e} J")
print('ENERGY  ......................................')
print(f"1 cal                      : {calorie:.8e} J")
print(f"1 hp                       : {hp:.8e} J")
print('TIME  ........................................')
print(f"1 min                      : {minute:.8e} s")
print(f"1 h                        : {hour:.8e} s")
print('ZERO IN CELSIUS SCALE AT K......................')
print(f"0 °C                       : {zero_Celsius:.8e} K")
print('PREfixes.......................................')
print(f"Y                          : {yotta:.1e} ")
print(f"G                          : {giga:.1e} ")
print(f"M                          : {mega:.1e} ")
print(f"k                          : {kilo:.1e} ")
print(f"m                          : {milli:.1e} ")
print(f"µ                          : {micro:.1e} ")
print(f"n                          : {nano:.1e} ")
```

The result is the following.

```
In [1] : ….
USEFULL CONSTANTS ............................
Gas constant             : 8.31446262e+00 J.mol^(-1) K^(-1))
Standard accel.of grav.: 9.80665000e+00 m.s^(-2)
Speed of light           : 2.99792458e+08 m.s^(-1)
Planck constant          : 6.62607015e-34 J·s
Boltzmann constant       : 1.38064900e-23 J.K^(-1)
Avogadro's number        : 6.02214076e+23 mol^(-1)
LENGTH ..........................................
1 ft                     : 3.04800000e-01 m
1 in                     : 2.54000000e-02 m
VOLUME ..........................................
1 liter                  : 1.00000000e-03 m^3
1 US gal                 : 3.78541178e-03 m^3
MASS   ..........................................
1 gram                   : 1.00000000e-03 kg
1 lb                     : 4.53592370e-01 kg
1 t                      : 1.00000000e+03 kg
PRESSURE ........................................
1 bar                    : 1.00000000e+05 Pa
1 atm                    : 1.01325000e+05 Pa
1 psi                    : 6.89475729e+03 Pa
1 mmHg                   : 1.33322368e+02 Pa
1 Btu                    : 1.05505585e+03 J
ENERGY  .........................................
1 cal                    : 4.18400000e+00 J
1 hp                     : 7.45699872e+02 J
TIME   ..........................................
1 min                    : 6.00000000e+01 s
1 h                      : 3.60000000e+03 s
ZERO IN CELSIUS SCALE AT K......................
0 °C                     : 2.73150000e+02 K
PREfixes........................................
Y                        : 1.0e+24
G                        : 1.0e+09
M                        : 1.0e+06
k                        : 1.0e+03
m                        : 1.0e-03
µ                        : 1.0e-06
n                        : 1.0e-09
```

Let's take advantage of Python's capabilities to answer an intriguing question: how many water molecules are there in the smallest possible raindrop? Raindrop volumes can vary widely, from as little as 0.00004 mL for a light drizzle to as much as 2 mL for large drops in a heavy downpour. For this calculation, we will assume a temperature of 15 °C. Remarkably, we can perform this computation entirely in Python using only basic physical properties—the molecular mass and density of water—without needing any external data sources for unit conversion.

```
from scipy.constants import N_A,liter,kilo,milli
# how many molecules in a glass of water
V    =   0.00004 # volume of the smallest raindrop (drizzle)
in mL
mm   =  18.0153  # molecular mass of water in g.mol^(-1)
dens = 999.103   # density of water at 15 °C in kg.m^(-3)

V    = V*milli   # volume of the smallest raindrop in L
V    = V*liter   # volume of the smallest raindrop in m^3
m    = V*dens    # mass   of the smallest raindrop in kg
m    = m*kilo    # mass   of the smallest raindrop in g
n    = m/mm      # mol of water the smallest raindrop
N    = n*N_A     # # of molecules in the smallest raindrop

print(f"the smallest raindrop contains {N:.6e} molecules of
water")

the smallest raindrop contains 1.335918e+18 molecules of wa-
ter
```

Chapter summary

In this chapter the consistent use of the SI units is presented. Special attention needs to be paid to Tables 2.1–2.5. You should consult these tables every time you have doubts about using SI units, together with Rules 1–13 of Section 2.2.3. Use Table 2.6 to perform unit conversion.

Use Eqs. (2.3.1)(2.3.6) when performing calculations using uncertain numbers and never forget that the least certain number restricts the accuracy of any result and try to use as many significant figures as possible to denote clearly the uncertainty in your results. In Section 2.7 the main variables, such as temperature, pressure, and density, that appear in engineering calculations are reviewed. Section 2.8 reviews the basic definitions related to energy, work, and heat, and the sign convention used. The notion of system and of process are reviewed in Section 2.9 together with the idea of the ideal gas law.

Basic calculations in MATLAB and Python, relative to data manipulation and numerical and symbolic calculation of integrals, are presented.

Problems

2.1. Explain what is incorrect in the statements that follow and correct them.

	Statement	Corrected
1	My height is 174 cm	
2	My height is 1.74 m	
3	My height is 1.74 meter	
4	My height is 1.74m	
5	My height is 1.74 m	
6	My weight is 87 kgs	
7	My weight is 87 kg	
8	My weight is 87 kilos	
9	My weight is 87,000 g	
10	The speed of the car is 50 km/hour	
11	The speed of the car is 50 Km/h	
12	The speed of the car is 50 Mmm/h	
13	The speed of the car is 50 kmh^{-1}	
14	The pressure is 5 bara	
15	The pressure is 5 bars	
16	The pressure is 5 bar	
17	The temperature is 30 °C	
18	The temperature is 303°C	
19	The temperature is 30 °C	
20	The temperature is 30 °C	
21	The temperature is 400 K	
22	The temperature is 400 K	
23	The temperature is 400 K	
24	The temperature is 400 K	
25	The dimension of force is kg·m/s/s	
26	The dimension of force is mkg/ s^2	
27	The dimension of force is kg·m/h	
28	The engine's power is 3600 J/second	
29	The engine's power is 3600 W	
30	The engine's power is 3600 watts	
31	The engine's power is 36 hw	

2.2. Convert the following to SI units

a. 50 ft^3
b. 25 psi
c. 100 Btu
d. 5 in
e. 20 mi
f. 100 ft
g. 10 gal
h. 20 lbm

i. 32 °F
j. 100 hp

2.3. Convert the following to SI units
k. 50 ft^3/h
l. 25 mi/h
m. 100 Btu/h
n. 5 ft/min
o. 20 Btu/(lb_m·°C)
p. 100 Btu/(lbmol·°F)
q. 10 Btu/(h·ft^2·°F)

2.4. The dynamic viscosity of water at 55 °C is found in a handbook to be 10^{-5} lb_f·s·ft^{-2}. What is the value of the viscosity in g·m^{-1} s^{-1}?

2.5. The ideal gas constant is the product of the Avogadro constant and the Boltzmann constant. Calculate the ideal gas constant R and express the result in all combinations of m^3, L, bar, and atm.

2.6. Give the number of significant figures
a. 1.23
b. 0.012
c. 12.34
d. 0.1004
e. 123
f. 6 students

2.7. Complete the calculations below and give your answer with the appropriate number of significant digits.
a. 12.34 + 1.03,056,789
b. 12.34,567 + 12.34
c. 5×12.345
d. 2×1.234 + 12.34
e. 123.4 + 2×12.34
f. $\frac{123.4}{2} - 12.3$
g. $\frac{12.345}{1.234} - 1.2$
h. $123.4 - \frac{123.45}{1.23}$

2.8. What are the units of the constants a and b appearing in the van der Waals equation?

$$\left(P+\frac{a}{v^2}\right)(v-b) = RT$$

where P is the pressure, T is the absolute temperature, v is the molar volume, and R is the ideal gas constant.

2.9. What are the units of the constants A, B, and C appearing in the Antoine equation?

$$\ln\left(\frac{P^{sat}}{\bar{P^{o}}}\right) = A - \frac{B}{C+T}$$

where P^{sat} is the saturation pressure, $\bar{P^{o}}$ is a reference pressure, and T is the absolute temperature.

2.10. Choked flow is a compressible flow effect. The parameter that becomes "choked" or "limited" is the fluid velocity. When a flowing fluid at a given pressure and temperature passes through a constriction into a lower pressure the fluid velocity increases. At initially subsonic upstream conditions, the conservation of energy principle requires the fluid velocity to increase as it flows through the smaller cross-sectional area of the constriction. Choked flow is a limiting condition where the mass flow cannot increase with a further decrease in the downstream pressure. When the gas velocity is choked, one can obtain the mass flowrate (in g/s) as a function of the upstream pressure from the following equation

$$\dot{M} = C_d S \sqrt{\gamma \rho_0 P_0 \left(\frac{2}{\gamma+1}\right)^{\frac{\gamma+1}{\gamma-1}}} = C_d S \frac{P_0}{\sqrt{RT_0}} \sqrt{\gamma \left(\frac{2}{\gamma+1}\right)^{\frac{\gamma+1}{\gamma-1}}}$$

where

Symbol	Description	Units
C_d	Discharge coefficient	?
S	Discharge hole cross-sectional area	m^2
γ	Heat capacity ratio (c_p/c_v)	Dimensionless
ρ_0		$\frac{kg}{m^3}$

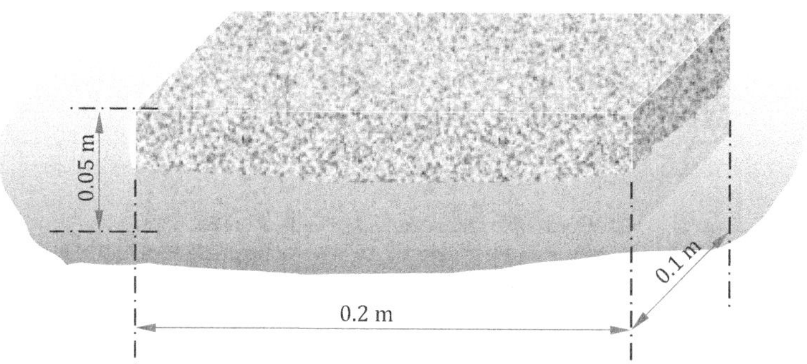

Figure P2.15 Immersed material into mercury for estimating density.

	Real gas density at pressure P_0 and temperature T_0	
P_0	Absolute upstream pressure of the gas	Pa
T_0	Absolute upstream temperature of the gas	K
R	Ideal gas constant for the gas	$\frac{m^3\ Pa}{g \cdot K}$

Find what are the units of the discharge coefficient C_d.

2.11. The following equation is frequently used to determine the convective heat transfer coefficient h for the laminar condensation of a vapor on the outer surface of a horizontal tube

$$h = C\left[\frac{g\rho_L(\rho_L - \rho_G)k_L^3\Delta h_{vap}}{\mu_L(T_{sat} - T_s)D_t}\right]^{1/4}$$

where

Symbol	Description	Units
h	Convective heat transfer coefficient	$\frac{J}{s \cdot m^2 \cdot K}$
g	Acceleration of gravity	$\frac{m}{s^2}$
ρ_L	Density of condensate	$\frac{kg}{m^3}$
ρ_G	Density of vapor	$\frac{kg}{m^3}$
k_L	Thermal conductivity of condensate	$\frac{J}{s \cdot m \cdot K}$
Δh_{vap}	Enthalpy of evaporation	$\frac{J}{kg}$
μ_L	Viscosity of condensate	$\frac{kg}{m \cdot s}$
T_{sat}	Absolute temperature of saturation	K
T_s	Absolute temperature of the surface	K
D_t	External diameter of the horizontal tube	m

What are the units of constant C?

2.12. The Antoine equation for water is

$$\ln P^{sat} = 18.3036 - \frac{3816.44}{T - 46.13}$$

where T is the absolute temperature in K and P^{sat} the saturation pressure in mmHg. Can we express the equation in °C and bar or kPa?

2.13. The following heat capacity equation is given for a substance

$$c_p = A + BT + CT^2$$

Where c_p is the heat capacity in Btu/(lbmol·°F) and T is the absolute temperature in °F.

a. What would be the units of the constants A, B, and C in SI?

b. Make all changes necessary to express the polynomial in SI units (T in K and c_p in J/(mol·K))?

2.14. The density of a liquid is approximated by the empirical equation

$$\rho = 1.6\,\exp(6.9 \cdot P)$$

where ρ is the density in kg/L and P is the pressure in mPa (millipascal).

a. What are the units of the coefficients 1.6 and 6.9?

b. Express the equation in units of $\mathrm{lb_m/ft^3}$ for density and $\mathrm{lb_f/in^2}$.

It is given that $14.696\,\frac{\mathrm{lb_f}}{\mathrm{in^2}} = 101325$ Pa.

2.15. You have, most probably, learned in high school that according to Archimedes' principle the upward buoyant force that is exerted on a body immersed in a fluid, whether fully or partially, is equal to the weight of the fluid that the body displaces.

You have immersed a block of material with unknown density into mercury at 20 °C (density of mercury 13600 kg/m^3). The block has dimensions of L×W×H = 0.2 ×0.1 ×0.05 m (as shown in Fig. P2.15).

a. If half of the block is submerged what is the density of the material in kg/m^3?

b. The purpose of this experiment was to find what is the composition of the block, which consists of unknown proportions of aluminum (density of 2700 kg/m^3) and steel (density of 7800 kg/m^3). Can you determine the composition?

c. What is the mass fraction and volume fraction of aluminum in the block?

2.16. A particular flue gas has composition (by volume %): 15% CO_2, 75% N_2, 8% O_2, and 2% H_2O. What are the mole fractions, mass fractions, and what is the average molecular mass?

2.17. A gas in a cylinder undergoes a compression process. During a "polytropic" compression the following pressure–volume relationship is satisfied

$$PV^n = \text{constant}$$

where $n > 0$ is called the polytropic index. The process $(P_0, V_0) \rightarrow (P_f, V_f)$ $(P_f > P_0)$ is performed.

a. Calculate the work done from the environment to the gas and derive a general equation that involves initial pressure and temperature and final pressure only.

b. If 1 kg of air at initial pressure of 1 bar and initial temperature of 300 K is compressed to final pressure of 5 bar with $n = 1.3$ calculate the work done to the gas.

2.18. An ideal gas in a cylinder undergoes a compression process. During an adiabatic, reversible compression the following pressure–volume relationship is satisfied

$$PV^\gamma = \text{constant}, \gamma = \frac{c_p}{c_v}$$

The process $(P_0, V_0, T_0) \rightarrow (P, V, T)$ $(P > P_0)$ is performed. In such a process

$$W = Nc_p(T - T_0)$$

a. Calculate the work done from the environment to the gas and derive a general equation that involves initial pressure and temperature and final pressure only.

b. If 1 kg of air at initial pressure of 1 bar and initial temperature of 300 K is compressed to final pressure of 5 bar with $\gamma = 1.4$ calculate the work done to the gas.

2.19. Steam with molar flowrate of 1 kmol/s at temperature 450 °C is cooled down to 220 °C at constant pressure. If steam is assumed to be an ideal gas, calculate the enthalpy change using the data for the specific heat capacity equation given in Table 2.10.

2.20. Most areas of Earth receive a mean annual solar radiation at ground level between 100 and 300 W/m^2. Greece receives an average yearly sum of 1400 kWh/m^2 of solar irradiation.

a. Assume an average efficiency of photovoltaic cells of 10%, estimate what is the necessary land area for covering the full energy consumption in Greece. The energy consumption in Greece is approximately 1.5 quad/y. (Area of Greece 131957 km^2.)

b. If the cost of installed photovoltaic cells is 1000 €/kW of rated capacity, what is the necessary investment? (GDP of Greece for 2024 was 250 billion US$).

c. Repeat the calculation for your country, obtain the necessary data from the internet.

2.21. (Based on Problems 2.17 and 2.18) Calculate the answer to Problem 2.17.b numerically in both MATLAB and Python. It is reminded that you need to calculate the integral:

$$W = -N\int_{V_0}^{V_f} P(V)dV, PV^n = P_0V_0^n$$

2.22. From Wikipedia we obtain the following: "*At the end of the 19th century, physicists were unable to explain why the observed spectrum of black-body radiation, which by then had been accurately measured, diverged significantly at higher frequencies from that predicted by existing theories. In 1900, German physicist Max Planck heuristically derived a formula for the observed spectrum by assuming that a hypothetical electrically charged oscillator in a cavity that contained black-body radiation could only change its energy in a minimal increment, E, that was proportional to the frequency of its associated electromagnetic wave. While Planck originally regarded the hypothesis of dividing energy into increments as a mathematical artifice, introduced merely to get the correct answer, other physicists including Albert Einstein built on his work, and Planck's insight is now recognized to be of fundamental importance to quantum theory.*" Planck heuristically derived (!) the following equation for blackbody radiation

$$E(\lambda, T) = \frac{2\pi hc^2}{\lambda^5\left(\exp\left(\frac{hc/k_B}{\lambda T}\right) - 1\right)} = \frac{C_1}{\lambda^5\left(\exp\left(\frac{C_2}{\lambda T}\right) - 1\right)}$$

where

Symbol	Explanation	Units
$E(\lambda, T)$	Electromagnetic radiation emitted by a black body	$W \cdot m^{-2} \cdot nm^{-1}$
h	The Planck constant (see Table 2.4)	J·s
c	Speed of light in vacuum (see Table 2.4)	m/s

k_B	The Boltzmann constant (see Table 2.4)	J/K
λ	Wavelength	nm
T	Absolute temperature	K

a. Find the numerical values of the constants C_1 and C_2 and their units (consistent with the units of $E(\lambda, T)$).

b. Plot $E(\lambda, T)$ as a function of λ, for $0 \leq \lambda \leq 2000$ nm.

c. Find the numerical values and units of constant σ in Stefan's law by numerical integration (units of $W/(m^2\ K^4)$)

$$\int_0^{\infty} E(\lambda, T) d\lambda = \sigma T^4$$

Compare with the analytically derived value

$$\sigma = \frac{2\pi^5 k_B^4}{15 c^2 h^3}$$

2.23. A closed tank with a total volume of $V = 1000\ m^3$ is used to store a liquid petroleum product. At the initial conditions of temperature $T_0 = 293.15$ K and pressure $P_0 = 100$ kPa, the liquid occupies 90% of the tank volume.
The temperature of the tank may vary seasonally: it can drop to $T_{min} = 273.15$ K in winter and rise to $T_{max} = 333.15$ K on a hot summer day. Because the tank is closed, the vapors trapped above the liquid are compressed or expanded as the liquid volume changes with temperature. This results in variations of internal pressure that could affect the structural integrity of the tank. The liquid density is assumed to vary linearly with temperature according to

$$\rho(T) = \rho(T_0)\,[1 - \beta(T - T_0)]$$

where $\beta = 0.001\ K^{-1}$ is the volumetric thermal expansion coefficient. Determine the maximum and minimum pressures that can develop inside the tank as the temperature varies between T_{min} and T_{max}.

2.24. Mount Olympus, with an elevation of 2917 m, is the highest peak in Greece. Beyond its mythological significance as the home of Zeus and the Olympian gods, it provides an interesting example of the effect of altitude on phase equilibrium. Determine the temperature at which water boils at the summit of Mount Olympus. Assume standard atmospheric pressure at sea level is 1.013×10^5 Pa, and take into account the decrease in pressure with altitude. Neglect humidity effects. Repeat for Mount Everest, which has a

peak elevation of 8849 m. Can you think of an implication of this reduction in the water boiling point temperature?
The saturation pressure of water is given by the following Antoine Equation:

$$\log_{10} P^{sat}[\text{bar}] = 6.20963 - \frac{2354.731}{T[^{\circ}\text{C}] + 280.71}$$

Further reading

Many definitions have been retrieved from the Wikipedia, the free encyclopedia.

Use of the SI system of units

Brett, M.A., Frey, J.G., Hinde, R., Kuroda, Y., Marquardt, R., Pavese, F., Quack, M., Stohner, J., Thor, A.J., 2023. Quantities, Units and Symbols in Physical Chemistry, 4th Abridged Version, Royal Society of Chemistry. International Union of Pure and Applied Chemistry.

The International System of Units (SI), ninth ed., 2019. Bureau International des Poids et Mesures (BIPM).

Dimensional homogeneity

Conejo, A.N., 2021. Fundamentals of Dimensional Analysis Theory and Applications in Metallurgy. Springer Nature Singapore Pte Ltd.

Hill, A.V., 1950. The dimensions of animals and their muscular dynamics. Sci. Prog. 150, 209–230.

Ipsen, D.C., 1960. Units, Dimesnsions and Dimensionless Numbers. McGraw Hill, NY.

Langhaar, H.L., 1951. Dimensional Analysis and Theory of Models. John Wiley & Sons, London.

Books on material and energy balances

Andersen, L.B., Wenzel, L.A., 1961. Introduction to Chemical Engineering. McGraw Hill, NY.

Duncan, T.M., Reimer, J.A., 1998. Chemical Engineering Design and Analysis. Cambridge University Press, NY.

Felder, R.M., Rousseau, R.W., 2000. Elementary Principles of Chemical Processes, third ed. John Wiley and Sons, NJ.

Ghasem, N., Henda, R., 2015. Principles of Chemical Engineering Processes, second ed. CRC Press, Boca Raton, FL.

Henley, E.J., Bieber, H., 1959. Chemical Engineering Calculations. McGraw Hill, NY.

Himmelblau, D., 1989. Basic Principles and Calculations in Chemical Engineering, fifth ed. Prentice Hall International, Editions, NY.

Hougen, O.A., Watson, K.M., Ragatz, R.A., 1943. Chemical Process Principles, Part I, Material and Energy Balances. John Wiley and Sons, NY.

Kirkbride, C.G., 1947. Chemical Engineering Fundamentals. McGraw Hill, NY.

Littlejohn, C.E., Meenaghan, G.F., 1959. An Introduction to Chemical Engineering. Reinhold Publishing Corporation, NY.

Myers, A.L., Seider, W.D., 1976. Introduction to Chemical Engineering and Computer Calculations. Prentice Hall, Inc., Englewood Cliffs, NJ.
Murphy, R.M., 2007. Introduction to Chemical Processes: Principles, Analysis, Synthesis. McGraw Hill, Singapore.
Reklaitis, G.V., 1983. Introduction to Material and Energy Balances. John Wiley and Sons, NY.
Russell, T.W.F., Denn, M.M., 1972. Introduction to Chemical Engineering Analysis. John Wiley and Sons, NY.
Simpson, R., Sastry, S.K., 2013. Chemical and Bioprocess Engineering. Springer, NY.

Thermodynamics Books

Abbott, M.M., Van Ness, H.C., 1972. Theory and Problems of Thermodynamics, Schaums Outline Series. McGraw-Hill Book Company, NY.
Daubert, T.E., 1985. Chemical Engineering Thermodynamics. McGraw-Hill Book Company, NY.
Elliot, J.R., Lira, C.T., 2012. Introductory Chemical Engineering Thermodynamics, second ed. Prentice Hall, Upper Saddle River, NJ.
Gmehling, J., Kolbe, B., Kleiber, M., Rarey, J., 2012. Chemical Thermodynamics for Process Simulation. Wiley-VCH, Weinheim.
Matsoukas, T., 2012. Fundamentals of Chemical Engineering Thermodynamics. Prentice Hall, Upper Saddle River, NJ.
Moran, M.J., Shapiro, H.N., 1988. Fundamentals of Engineering Thermodynamics. John Wiley and Sons, NY.
Sandler, S.I., 1977. Chemical and Engineering Thermodynamics. John Wiley and Sons, NY.

3

General balance equations

3.1 The general balance equation

Energy and mass balances are specific applications of a broader concept known as the **material balance equation**, which is widely applicable beyond the field of chemical engineering. To illustrate the general principle of a material balance, consider a simple example: tracking the balance of money in a personal bank account.

Suppose that a week ago (Day 0), the account held \$999.50. To simplify the scenario, we assume that a daily interest of 0.05% is applied to the account at 24:00 each day, calculated on the balance at the end of the day. Additionally, we have records of daily expenses and income from Day 1 to Day 7, as shown in Table 3.1.

Our goal is to determine the account balance at the end of each day throughout the week. How can we approach this?

We begin our analysis on Day 0, where the known account balance at the end of the day is \$999.50. To compute the interest earned, we apply the daily rate of 0.05%:

$$\frac{0.05}{100} \times 999.50\ \$ = 0.50\ \$$$

Adding this interest to the balance gives:

$$999.50\ \$ + 0.50\ \$ = 1000.00\ \$$$

This is the available balance at the start of Day 1.

On Day 1, we spend \$98.00 and receive no income. The balance at the end of the day (which becomes the starting point for Day 2) is calculated using a form of the material balance

$$\begin{aligned}\text{end} - \text{of} - \text{day balance} &= \text{Starting balance} + \text{Income} \\ &\quad - \text{Expenses} + \text{Interest}\end{aligned}$$

or

$$\begin{aligned}\text{end} - \text{of} - \text{day balance} &= 1000\ \$ + 0.00\ \$ - 98.00\ \$ + \text{Interest} \\ &= 902.00 + \text{Interest}\end{aligned}$$

Introduction to Process Systems Engineering. https://doi.org/10.1016/B978-0-443-45444-8.00009-9

Table 3.1 Data available for our money balance.

Day	Balance Beginning of day ($)	Income ($)	Expenses ($)	Balance End of day ($)	Interest (0.05%) ($)
0				999.50	
1			98.00		
2			82.30		
3			87.10		
4			107.40		
5		750.00	56.70		
6			120.20		
7			130.30		

Now we calculate the interest earned on Day 1:

$$\text{Interest} = \frac{0.05}{100} \times 902.00\ \$ = 0.45\ \$$$

So, the final balance at the end of Day 1 is:

$$902.00 + 0.45 = 902.45\ \$$$

This procedure is repeated for each subsequent day (Day 2 through Day 7), using the same balance equation. The computed daily balances are summarized in Table 3.2. You are encouraged to reproduce these results yourself to ensure a solid understanding of the calculation process.

Now, consider the following question: What is the general equation that describes the balance of money in a system over time?

Before formulating the equation, we first define the necessary notation. We use the symbol t to represent time, where $t = \{\text{Day } 0, \text{Day } 1, \ldots, \text{Day } 7\}$. We also denote the amount of money by the symbol $M(t)$ ("M" stands for money). We can now express the balance of money with the following general rule:

$$\text{money available at the beginning of next day} = \text{money available at the beginning of the current day} + \text{Income} - \text{Expenses} + \text{Interest}$$

Table 3.2 Detailed calculation of our money balance for a week.

Day	Balance Beginning of day ($)	Income ($)	Expenses ($)	Balance End of day ($)	Interest (0.05%) ($)
0				999.50	0.50
1	1000.00	0.00	98.00	902.00	0.45
2	902.45	0.00	82.30	820.15	0.41
3	820.56	0.00	87.10	733.46	0.37
4	733.83	0.00	107.40	626.43	0.31
5	626.74	750.00	56.70	1320.04	0.66
6	1320.70	0.00	120.20	1200.50	0.60
7	1291.10	0.00	130.30	1160.80	0.58

In mathematical terms

$$M(t+1) = M(t) + M_{in}(t) - M_{out}(t) + M_{gen}(t) \qquad (3.1.1)$$

where $M(t)$: money available at time t

$M_{in}(t)$: income (inflow) at time t

$M_{out}(t)$: expenses (outflow) at time t

$M_{gen}(t)$: money generated at time t (e.g., interest)

Eq. (3.1.1) is a generally valid equation which can now be stated in a more general setting

$$\begin{bmatrix} \text{Quantity} \\ \text{at time} \\ \text{t}+1 \end{bmatrix} = \begin{bmatrix} \text{Quantity} \\ \text{at time} \\ \text{t} \end{bmatrix} + \begin{bmatrix} \text{inflow of} \\ \text{quantity} \\ \text{between t and t}+1 \end{bmatrix} - \begin{bmatrix} \text{outflow of} \\ \text{quantity} \\ \text{between t and t}+1 \end{bmatrix} + \begin{bmatrix} \text{generation of} \\ \text{quantity} \\ \text{between t and t}+1 \end{bmatrix} - \begin{bmatrix} \text{consumption of} \\ \text{quantity} \\ \text{between t and t}+1 \end{bmatrix} \qquad (3.1.2)$$

Stated more compactly

$$\text{New value} = \text{Old value} + \text{input} - \text{output} + \text{generation} - \text{consumption} \tag{3.1.3}$$

To show the general application of Eqs. (3.1.2) and (3.1.3) we consider the standard Susceptible-Infected-Recovered (SIR) model for the recent infectious disease caused by the severe acute respiratory syndrome novel coronavirus (SARS-CoV-2). In this model the host population is divided into susceptible, infected and recovered individuals, denoted by $S(t)$, $I(t)$, and $R(t)$, respectively (see Fig. 3.1).

The quantities $S(t)$, $I(t)$, and $R(t)$ track the numbers of individuals in each compartment over different time periods. The SIR model describes the time evolution of three populations:
Susceptible individuals, $S(t)$: individuals who are not infected, however, could become infected. A susceptible individual may become infected or remain susceptible. As the virus spreads more individuals will become infected.
Infected individuals, $I(t)$: individuals who have already been infected by the virus and can transmit it to those individuals who are susceptible. An infected individual may remain infected and can be removed from the infected population to recover.
Removed individuals, $R(t)$: individuals who have recovered from the virus and are assumed to be immune.

In the standard SIR (Susceptible–Infectious–Recovered) model, it is typically assumed that the time scale under consideration is short enough that natural births and deaths (i.e., those unrelated to the disease) can be neglected. Additionally, it is assumed that the number of disease-related deaths is small relative to the total living population and can therefore be ignored in the basic formulation.

The population is also assumed to be isolated from the external environment, meaning there is no inflow or outflow of individuals—a condition known in chemical engineering as a **batch system**. This assumption is often justified by quarantine or lockdown measures designed to prevent the spread of the virus.

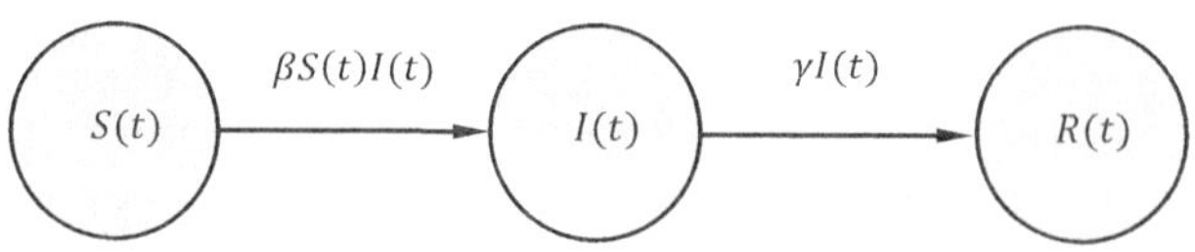

Figure 3.1 Compartments used in the SIR model for infectious disease.

Under these assumptions, the dynamics of the SIR model are governed by two key processes:

1. Infection: The rate at which susceptible individuals become infected is proportional to both the number of susceptible individuals, $S(t)$, and the number of infected individuals, $I(t)$. This interaction is captured by the term

$$\beta I(t)S(t) \tag{3.1.4}$$

   Here, β is the transmission rate, defined as the average number of susceptible individuals infected by one infectious individual per unit time. If this expression seems abstract, consider the analogy to a chemical reaction of the form $I+S\rightarrow 2I$, where the reaction rate depends on the product of the concentrations of I and S.
2. Recovery: The rate at which infected individuals recover is assumed to be proportional to the number of currently infected individuals. This is expressed as:

$$\gamma I(t) \tag{3.1.5}$$

where γ is the average number of infected individuals recovered per unit of time (recovery rate).

We first write the balance equation for the susceptible individuals. We need to express all terms appearing in Eq. (3.1.3) in terms of the variables used ($S(t)$, $I(t)$, and $R(t)$). We have that

$$S(t+\Delta t) = \text{new value}$$

$$S(t) = \text{old value}$$

$$\text{input} = \text{output} = \text{generation} = 0$$

$$\text{consumption} = (\beta\Delta t)I(t)R(t)$$

We replace into Eq. (3.1.3) to obtain:

$$S(t+\Delta t) = S(t) - (\beta\Delta t)I(t)S(t) \tag{3.1.6}$$

In a similar way we obtain

$$I(t+\Delta t) = I(t) + (\beta\Delta t)I(t)S(t) - (\gamma\Delta t)I(t) \tag{3.1.7}$$

and

$$R(t+1) = R(t) + (\gamma\Delta t)I(t) \tag{3.1.8}$$

Table 3.3 Data and parameters for the SIR model.

Parameter	Symbol	Value	Units
Initial population of susceptible individuals	$S(0)$	10,000	#
Initial population of infected individuals	$I(0)$	1	#
Initial population of recovered individuals	$R(0)$	0	#
Transmission rate	β	0.0001	$(\#)^{-1}\ \mathrm{d}^{-1}$
Recovery rate	γ	0.2	d^{-1}
#= number of individuals			

where $\Delta t = 1$ d (1 day). The data for a particular case are presented in Table 3.3.

We can determine the individual population at $t = 1$ as follows

$$S(1) = S(0) - (\beta\Delta t)I(0)S(0) = 10000 - \left(10^{-4} \times 1\right) \times 1 \times 10^4 = 9999\ \#$$

$$I(1) = I(0) + (\beta\Delta t)I(0)S(0) - (\gamma\Delta t)I(0) = 1 + \left(10^{-4} \times 1\right) \times 1 \times 10^4 - (0.2 \times 1) \times 1 = 1.8\ \#$$

$$R(1) = R(0) + (\gamma\Delta t)I(0) = 0 + (0.2 \times 1) \times 1 = 0.2\ \#$$

where # denotes number of individuals. Note that the total population remains constant.

While it is possible to perform the calculations manually, this task is far better suited to a computer. To streamline and automate the process, we use MATLAB to carry out the computations. The complete script file in MATLAB is as follows:

```
clear all
% note that vectors in MATLAB start from 1 not 0
beta = 0.0001;  % rate constant in 1/(# d)
gama = 0.200;   % rate constant in 1/d
time(1)=   0;   % time in d
S(1)   = 10000; % susceptibles in #
I(1)   =    1;  % infectives in #
R(1)   =    0;  % removed in #
% simulate dynamics
for t=1:30     % run for 30 days
    time(t+1)=t;
    S(t+1) = S(t) - beta*S(t)*I(t);
    I(t+1) = I(t) + beta*S(t)*I(t) - gama*I(t);
    R(t+1) = R(t)                  + gama*I(t);
end

figure(1)
plot(time,S,'b-o','Linewidth',1.2)
hold on
plot(time,I,'r-s','Linewidth',1.2)
plot(time,R,'g-^','Linewidth',1.2)
xlabel('Time (d)')
ylabel('S(t), I(t), R(t) (#)')
grid on
```

Let us now explain the script file line by line. We first clear the memory

```
clear all
```

Then we declare all variables and initial conditions appearing in the model

```
beta = 0.0001;  % rate constant in 1/(# d)
gama = 0.200;   % rate constant in 1/d
time(1)=   0;   % time in d
S(1)   = 10000; % susceptibles in #
I(1)   =    1;  % infectives in #
R(1)   =    0;  % removed in #
```

Finally we run a **for loop** over 30 days and apply Eqs. (3.1.6) to (3.1.8)

```
for t=1:30     % run for 30 days
    time(t+1)=t;
    S(t+1) = S(t) - beta*S(t)*I(t);
    I(t+1) = I(t) + beta*S(t)*I(t) - gama*I(t);
    R(t+1) = R(t)                  + gama*I(t);
end
```

The results of the calculations are shown in Fig. 3.2.

Note that the number of infected individuals (squares) initially increases, reaches a peak, and then gradually declines as individuals recover. This characteristic rise and fall in

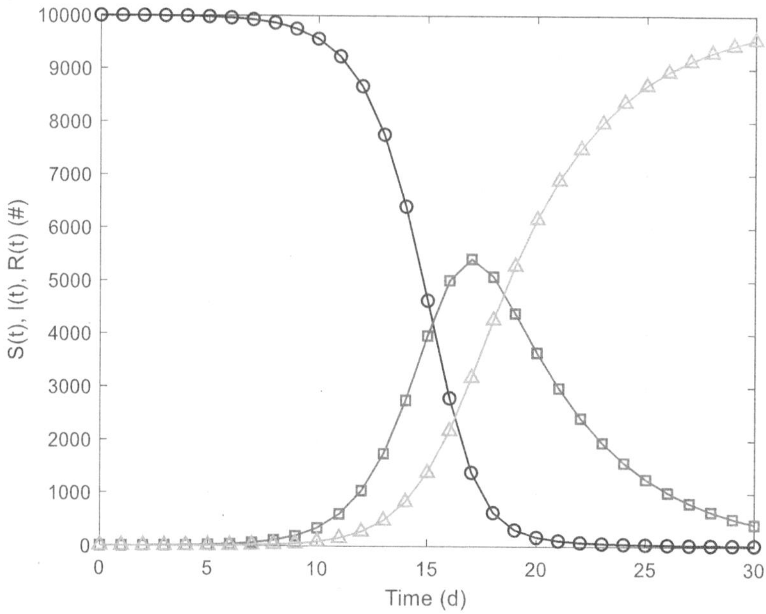

Figure 3.2 Population dynamics predicted by the SIR model when 1 infected individual arrives at an isolated community with 10000 susceptible individuals. circles: S(t), squares: I(t), triangles: R(t).

infections is typical of epidemic dynamics described by the SIR model.

A particularly interesting observation from the simulation is that, even after running the program over a long period, approximately 20 individuals remain uninfected. This highlights an important feature of the model: not everyone is necessarily infected during the course of an outbreak.

The dynamics of the infectious population are strongly influenced by the basic reproduction number, denoted as R_0, which is defined as:

$$R_0 = \frac{\beta}{\gamma} R(0) \tag{3.1.9}$$

This quantity represents the average number of new (secondary) infections generated by a single infected individual in a population where everyone is initially susceptible. In the current example, the basic reproduction number is calculated as $R_0 = 5$, indicating that each infected person, on average, transmits the infection to five others in the early stages of the outbreak.

In the infectious disease case study, we have chosen a time step of 1 day ($\Delta t = 1$day), which is a common choice in epidemiological studies of human populations. Similarly, the introductory example involving the money account also uses the same relatively long time interval.

The choice of time step is closely related to the rate at which the variables of interest change over time. In chemical engineering, system dynamics can vary widely—ranging from fractions of a second to days, months, or even years. However, when models are expressed as differential equations, the precise choice of time step becomes less critical. Differential equations describe continuous-time behavior, allowing us to avoid concerns about discretization and time scale selection. To apply this to the SIR model, we start from Eq. (3.1.6) and rewrite it in a slightly different form to facilitate this continuous-time representation

$$\frac{S(t+\Delta t)-S(t)}{\Delta t} = -\beta I(t)S(t) \tag{3.1.10}$$

We then use the definition of the derivative of a real–valued function of a real variable

$$\frac{df(t)}{dt} = \lim_{\Delta t \to dt \to 0} \frac{f(t+\Delta t)-f(t)}{\Delta t}$$

and we obtain the following differential equation

$$\frac{dS(t)}{dt} = -\beta I(t)S(t) \tag{3.1.11}$$

In a similar way we also obtain the following

$$\frac{dI(t)}{dt} = \beta I(t)S(t) - \gamma I(t) \tag{3.1.12}$$

$$\frac{dR(t)}{dt} = \gamma I(t) \tag{3.1.13}$$

Example 3.1. Who wants to be a millionaire?

A financial advisor presents you with the following investment plan: you deposit a fixed amount of money, denoted as A dollars, at the end of each year for 50 years, earning an annual interest rate of i%. At the end of this period, you are promised a substantial lump sum payment.

To make the proposal more concrete and more attractive, the advisor provides a specific example: by depositing $859.17 at the end of each year for 50 years, with an annual interest rate of 10%, you will receive a lump sum of \$1,000,000 at the end of the 50–year period!

The question is: Can you verify the advisor's claim? Is the calculation accurate, and does the math check out?

Solution

If we once again let $M(t)$ represent the amount of money accumulated in the investment after t years, we can use Eq. (3.1.1)—the general balance equation—to evaluate and validate the financial advisor's claim

$$M(t+1) = M(t) + M_{in}(t) - M_{out}(t) + M_{gen}(t)$$

where $M(0) = 0$,

$$M_{in}(t) = A, t = 1, 2, 3, \ldots, 50$$

$$M_{out}(t) = 0, t = 1, 2, 3, \ldots, 50$$

$$M_{gen}(t) = \frac{i\,\%}{100} \times M(t), t = 1, 2, 3, \ldots, 50$$

Putting everything together

$$M(t) = M(t-1) + A + \frac{i\,\%}{100} \times M(t-1), \quad M(0) = 0, t = 1, 2, 3, \ldots, 50$$

For $t = 1$: $M(1) = 0 + 859.17 + 0.1 \times 0 = 859.17$ \$.

For $t = 2$: $M(2) = 859.17 + 859.17 + 0.1 \times 859.17 = 1804.26$ \$.

For $t = 3$: $M(3) = 1804.26 + 859.17 + 0.1 \times 1804.26 = 2843.86$ \$.

Manually calculating the accumulated amount year by year up to $t = 50$ is tedious and time-consuming. Instead, we can use MATLAB to efficiently perform all the calculations and simulate the investment over the entire 50-year period:

```
clear all
A=859.17; % annual payment
i=10/100; % interest rate
M(1) = A;
for t=2:50
    M(t)=M(t-1)+M(t-1)*i+A;
end
figure(1)
plot(linspace(1,50,50),M,'o')
xlabel('Time (y)')
ylabel('$')
```

The results of the MATLAB calculations are presented in Fig. 3.3, and they confirm the financial advisor's claim: the investment indeed grows to approximately \$1 million over 50 years under the conditions given.

However, before accepting this offer as a compelling investment opportunity, it's important to reflect on two critical questions:

a. What is the present value of \$1 million received 50 years from now—in today's money?

b. Is a 10% annual interest rate realistic in the context of today's financial environment?

Before closing this section, we return to Eq. (3.1.2). This is a general equation that accounts for possible generation or consumption of an extensive quantity. However, most of the quantities of interest in chemical engineering, such as total mass, energy, and momentum, are "conserved" quantities and cannot be "produced" or "consumed." The only non-conserved quantity of interest in this text is the mass of chemical species, which can be consumed or produced through chemical reactions. Eq. (3.1.2) can be simplified for the **conserved quantities**

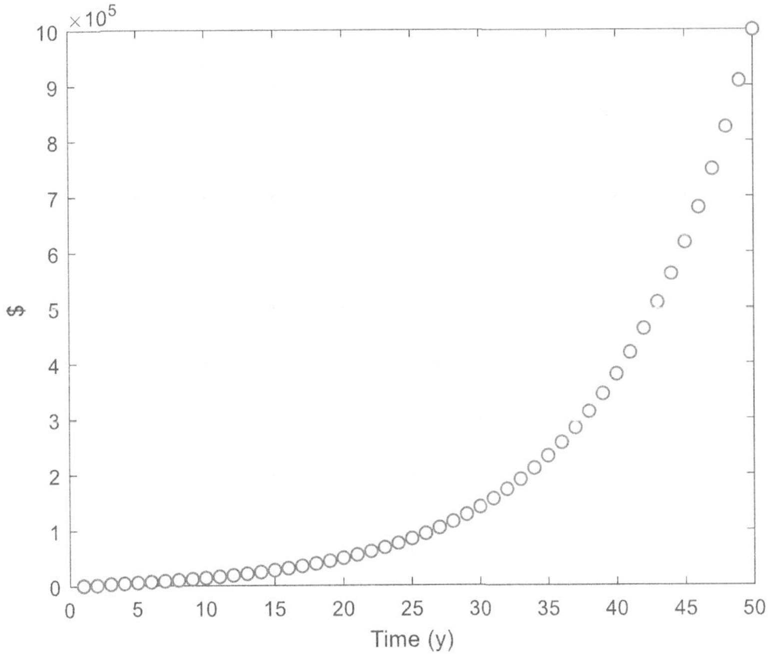

Figure 3.3 Deposit 859.17 \$ each year, for 50 years with an interest rate of 10% in a savings account and you will become a millionaire.

$$\begin{bmatrix}\text{(extensive) quantity}\\ \text{at time}\\ \text{t+1}\end{bmatrix} - \begin{bmatrix}\text{(extensive) quantity}\\ \text{at time}\\ \text{t}\end{bmatrix} = \begin{bmatrix}\text{inflow of}\\ \text{(extensive) quantity}\\ \text{between t and t+1}\end{bmatrix} - \begin{bmatrix}\text{outflow of}\\ \text{(extensive) quantity}\\ \text{between t and t+1}\end{bmatrix} \tag{3.1.14}$$

A further simplification can be obtained if we restrict attention to **steady state** operation, i.e., for systems that do not change with time, in which case Eq. (3.1.14) is simplified further to (the system does not change with time or it is time invariant, and time plays no role)

$$\begin{bmatrix}\text{inflow}\\ \text{of (extensive) quantity}\end{bmatrix} = \begin{bmatrix}\text{outflow}\\ \text{of (extensive) quantity}\end{bmatrix} \tag{3.1.15}$$

3.2 Total mass balance

When we perform **total mass balances**, we investigate the overall mass (an extensive quantity) within the system of interest. We will be using the symbol $M(t)$ to denote the total mass within a system. We state the overall mass balance using Eq. (3.1.14), as total mass is a conserved quantity

$$\begin{bmatrix}\text{total mass}\\ \text{at time}\\ \text{t+1}\end{bmatrix} = \begin{bmatrix}\text{total mass}\\ \text{at time}\\ \text{t}\end{bmatrix} + \begin{bmatrix}\text{inflow of}\\ \text{mass}\\ \text{between}\\ \text{t and t+1}\end{bmatrix} - \begin{bmatrix}\text{outflow of}\\ \text{mass}\\ \text{between}\\ \text{t and t+1}\end{bmatrix} \tag{3.2.1}$$

or

$$M(t+dt) = M(t) + \dot{M}_{in}(t)dt - \dot{M}_{out}(t)dt \tag{3.2.2}$$

where the overdot denotes rate, i.e., mass flowrate (mass per unit time). To understand why (3.2.1) is written as in (3.2.2) using instantaneous mass flowrate terms multiplied by a differential time interval, you need first to understand that the product of instantaneous mass flowrate and a differential time interval gives the differential mass added to the system during this infinitesimal interval. Let's discuss this first.

If you are filling a tank with a variable mass flowrate then the mass added in the tank ΔM_{in} between t and $t + \Delta t$, where Δt is arbitrary and can be very large, is given by the exact equation

$$\Delta M_{in} = \int_{t}^{t+\Delta t} \dot{M}_{in}(\tau)d\tau \tag{3.2.3}$$

This is shown in Fig. 3.4a. When, however, Δt becomes arbitrarily small then, as shown in Fig. 3.4b, the mass fed is a differential quantity equal to

$$dM_{in} = \lim_{\Delta t \to dt \to 0} \int_{t}^{t+dt} \dot{M}_{in}(\tau)d\tau \underbrace{=}_{\substack{\text{see Figure} \\ \text{3.4.b}}} \dot{M}_{in}(t) \int_{t}^{t+dt} d\tau = \dot{M}_{in}(t)dt \tag{3.2.4}$$

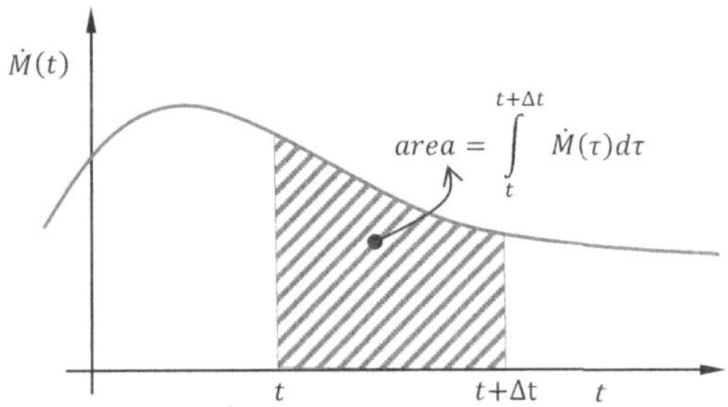

(a)

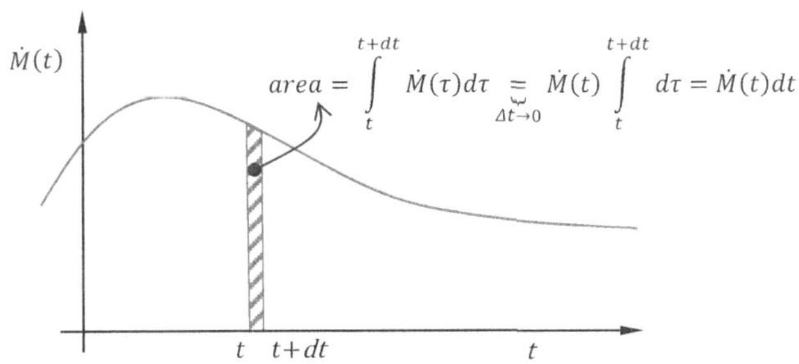

(b)

Figure 3.4 Integrating the mass flowrate between two arbitrary time points gives the total mass flowing through a stream between these two time points (a). When the time points come close together and their distance is arbitrarily small, then the total mass is equal to the instant mass flowrate times the differential time (b).

We return to Eq. (3.2.2) which can be written as

$$\frac{M(t+\Delta t)-M(t)}{dt}=\dot{M}_{in}(t)-\dot{M}_{out}(t) \tag{3.2.5}$$

We note that as dt tends to zero the left-hand side becomes the derivative of the total mass inside the system (control volume)

$$\frac{dM(t)}{dt}=\dot{M}_{in}(t)-\dot{M}_{out}(t) \tag{3.2.6}$$

This is the differential form of the macroscopic, overall mass balance equation. It can be applied to any well-defined system where the terms $\dot{M}_{in}(t)$ and $\dot{M}_{out}(t)$ need to be repeated for each incoming or outgoing stream that crosses the boundary of the control volume (that defines the system). It can be stated in words as follows

$$\begin{bmatrix}\text{rate of}\\ \text{accumulation}\\ \text{of total mass}\end{bmatrix}=\begin{bmatrix}\text{rate of}\\ \text{inflow}\\ \text{of total mass}\end{bmatrix}-\begin{bmatrix}\text{rate of}\\ \text{outflow}\\ \text{of total mass}\end{bmatrix} \tag{3.2.7}$$

Each term has units of mass per unit time (kg/s).

We may express the mass flowrate terms in the right-hand side of Eq. (3.2.6) in the following equivalent way

$$\dot{M}(t)=\rho\dot{F}(t) \tag{3.2.8}$$

where ρ is the ("massic") density (kg/m^3) and $\dot{F}$ is the volumetric flowrate (m^3/s). The same holds true for the total mass within the system boundaries

$$M=\rho V \tag{3.2.9}$$

where V is the volume of the system (in m^3). In this case the overall material balance can be written as

$$\frac{d(\rho V)}{dt}=\rho_{in}\dot{F}_{in}-\rho_{out}\dot{F}_{out} \tag{3.2.10}$$

If the density is constant, then $\rho=\rho_{in}=\rho_{out}$ and Eq. (3.2.10) simplifies further as it becomes a "volume balance"

$$\frac{dV}{dt}=\dot{F}_{in}-\dot{F}_{out} \tag{3.2.11}$$

When there is no chemical reaction, then the total moles are also conserved and we may express the total mass balance in an equivalent form known as total mole balance

$$\frac{dN(t)}{dt} = \dot{N}_{in}(t) - \dot{N}_{out}(t) \quad (3.2.12)$$

where $N(t)$ are the total mole within the system boundaries (in mol)

$\dot{N}_{in}(t)$ is the molar flowrate of the incoming stream (in mol/s)

$\dot{N}_{out}(t)$ is the molar flowrate of the outgoing stream (in mol/s)

The molar flowrate terms need to be repeated as many times as there are streams crossing the system boundaries. Eq. (3.2.12) can be transformed into Eq. (3.2.10) with the only difference that the density is now the molar density of the corresponding stream (in kmol/m^3) and not the "massic" density (in kg/m^3).

If we restrict attention to steady state conditions, all derivatives with respect to time become zero and we receive the following equivalent total mass and mole balances at steady state

$$\dot{M}_{in}(t) = \dot{M}_{out}(t) \quad (3.2.13)$$

$$\dot{N}_{in}(t) = \dot{N}_{out}(t) \quad (3.2.14)$$

$$\rho_{in}\dot{F}_{in} = \rho_{out}\dot{F}_{out} \quad (3.2.15)$$

Please note that

a. Eqs. (3.2.13)–(3.2.15) are valid at steady state,

b. Eq. (3.2.14) is valid in the absence of chemical reactions,

c. Eq. (3.2.15) has units dictated by the units of density (either mass or molar density).

The form of the total mass or mole balance that is more appropriate for a particular case study is dictated by the data of the problem, so do not worry about that (having options is not always a nightmare!).

Example 3.2. Emptying a cylindrical tank

A cylindrical tank with a cross-sectional area of $S = 0.1$ m^2 and a height of $H = 1$ m is filled with water to a height of $h = 0.8$ m. The tank is open to the atmosphere as shown in Fig. 3.5. At the bottom of the tank there is a hole of area $s = 0.001$ m^2. The velocity of the liquid through the hole is given by the following equation

$$u(t) = kh(t) \quad (3.2.16)$$

where $k = 0.1$ 1/s. Assume that water has a constant density of 1000 kg/m^3. Determine the time necessary for emptying the tank, which is practically achieved when $h(t) = 0.01h(0)$.

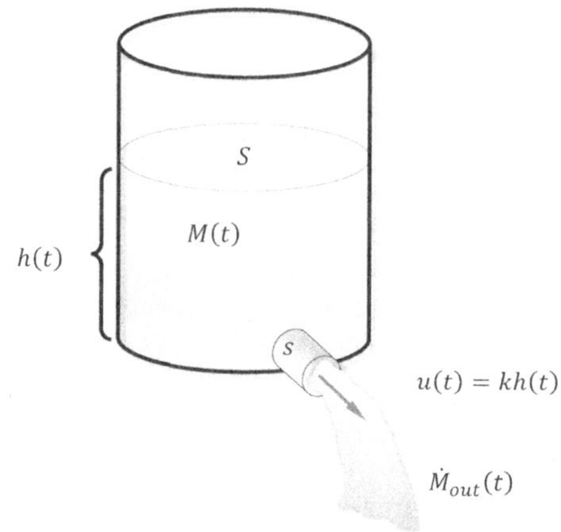

Figure 3.5 Emptying a cylindrical liquid storage tank.

Solution
We start from Eq. (3.2.6)

$$\frac{dM(t)}{dt} = \dot{M}_{in}(t) - \dot{M}_{out}(t)$$

where

$$M(t) = \rho V(t) = \rho S h(t)$$

$$\dot{M}_{in}(t) = 0$$

$$\dot{M}_{out}(t) = \rho s u(t) = \rho s k h(t)$$

We combine the equations to obtain

$$\frac{d(\rho S h(t))}{dt} = 0 - \rho s k h(t)$$

or

$$\frac{dh(t)}{dt} = -\left(\frac{s}{S}k\right)h(t)$$

To solve this differential equation, we note that

$$\frac{1}{h}dh = d(\ln h)$$

and it follows that

$$d \ln h(t) = -\left(\frac{s}{S}k\right)dt$$

We integrate to obtain the final solution

$$\int_{h(0)}^{h(t)} d \ln h = -\left(\frac{s}{S}k\right)\int_{0}^{t} d\tau$$

or

$$\ln h(t) - \ln h(0) = -\left(\frac{s}{S}k\right)t$$

which can also be written as

$$\frac{h(t)}{h(0)} = \exp\left[-\left(\frac{s}{S}k\right)t\right]$$

Note that $h(0) = 0.8$ m and $\frac{s}{S}k = 10^{-3}\frac{1}{s}$. The evolution of the liquid level in the tank is shown in Fig. 3.6. The tank is practically empty when $\frac{h(t)}{h(0)} = 0.01$, and the necessary time is

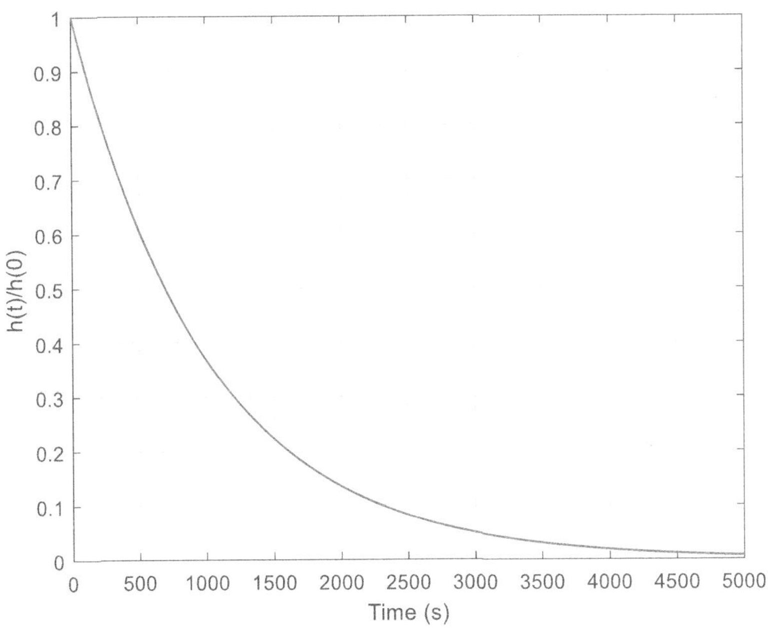

Figure 3.6 Heigh of water in the cylindrical tank as a function of time.

$$t = -\left(\frac{S}{sk}\right)\ln\left(\frac{h(t)}{h(0)}\right) = -10^3\ln(0.01)\ \text{s} = 4605.17\ \text{s}$$

Example 3.3. Emptying a conical tank

A conical water tank (vertex down – see Fig. 3.7) is filled with water up to a height $h(0)$. The velocity of the liquid through the hole at the bottom of the tank is given by the following equation

$$u(t) = k\sqrt{h(t)} \tag{3.2.17}$$

where k is a constant. Assume that water has a constant density and determine the time necessary for emptying the tank. All symbols appearing in Fig. 3.7 have the same meaning as in Fig. 3.5 and Example 3.2. Assume, to simplify calculations, that the diameter of the missing vertex of the cone is so small that can be neglected.

Solution

We first note that for a cone, based on the similar triangles formed, the following holds true

$$\frac{r}{h} = \frac{R}{H}$$

The volume of a cone with height h and radius r of the base is given by

$$V = \pi r^2 \frac{h}{3}$$

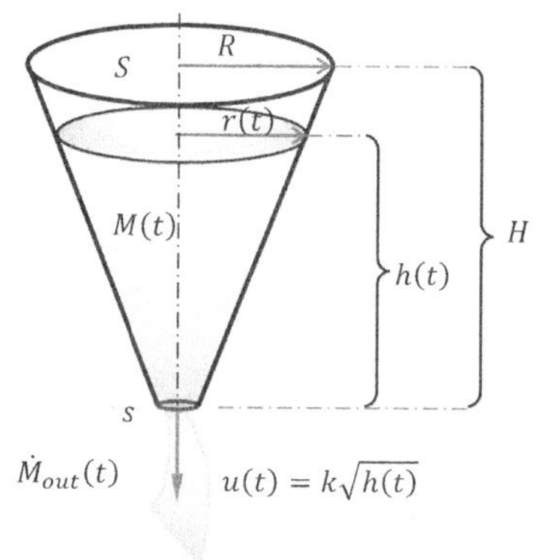

Figure 3.7 Emptying a cylindrical liquid storage tank.

We can eliminate $r(t)$ to obtain volume as a function of height $h(t)$ only

$$V(h) = \frac{\pi}{3}\left(\frac{R}{H}\right)^2 h^3$$

Note that

$$dV = \pi\left(\frac{R}{H}\right)^2 h^2 dh$$

As density is constant the material balance simplifies to "volume balance":

$$\frac{dV}{dt} = -F_{out}$$

or

$$\frac{\pi R^2}{H^2} h^2 \frac{dh}{dt} = -sk\sqrt{h}$$

Finally

$$h^{3/2} dh = -H^2\left(\frac{s}{S}k\right)dt$$

We note that

$$dh^{5/2} = \frac{5}{2} h^{3/2} dh$$

whence

$$dh^{5/2} = -\frac{5}{2} H^2\left(\frac{s}{S}k\right)dt$$

We integrate to obtain

$$(h(t))^{5/2} - (h(0))^{5/2} = -\frac{5}{2} H^2\left(\frac{s}{S}k\right)t$$

The tank is empty when $h(t) = 0$ and the time needed to empty the tank is

$$t = \left(\frac{S}{sk}\right)\frac{2}{5H^2}(h(0))^{5/2}$$

In the special case where $h(0) = H$

$$t = \frac{2}{5}\left(\frac{S}{sk}\right)\sqrt{H}$$

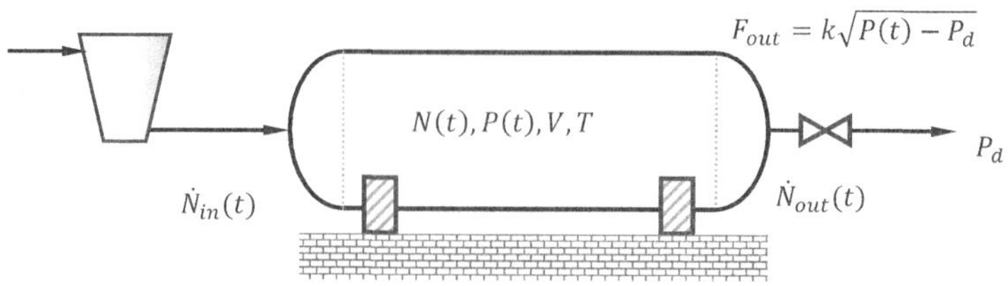

Figure 3.8 Compressed air storage tank.

Example 3.4. Mole balance in a compressed air tank

A tank is used to store compressed air as shown in Fig. 3.8. The volume of the tank is V and the incoming molar flowrate of the compressed air is $\dot{N}_{in}(t)$. The volumetric flowrate of the air leaving the tank depends on the pressure drop across the valve at the exit pipeline

$$F_{out}(t) = k\sqrt{P(t) - P_d}$$

where $P(t)$ is the pressure of air in the tank and P_d is the downstream pressure. Assume that the ideal gas law is valid for this process, despite the increased pressure, and derive the differential equation that describes the pressure in the tank as a function of time.

Solution

We note that the feed flowrate is expressed as molar flowrate while the ideal gas law is usually expressed in molar units. The appropriate form of balance equation is the mole balance equation (Eq. (3.2.12))

$$\frac{dN(t)}{dt} = \dot{N}_{in}(t) - \dot{N}_{out}(t)$$

From the ideal gas law, we have that

$$N(t) = \frac{V}{RT}P(t)$$

For the molar flowrate of the product stream

$$\dot{N}_{out}(t) = \rho F_{out} = \frac{P(t)}{RT}k\sqrt{P(t) - P_d}$$

We replace into the differential equation

$$\left(\frac{V}{RT}\right)\frac{dP(t)}{dt} = \dot{N}_{in}(t) - \left(\frac{P(t)}{RT}\right)k\sqrt{P(t) - P_d}$$

3.3 Analytic solution of first order differential equations

We will present the general solution of linear **first order differential equations** that appear very frequently in chemical engineering applications. Our choice in this text is to solve all differential equations using numerical methods in either MATLAB or Python. The ability to solve the simplest possible form of differential equations is necessary to be able to analyze and validate the results of the numerical calculations.

The first form of first order differential equations that we have already encountered (see Example 3.2) is the **separable equations**. These are first order differential equations that can be written in the form

$$f(y)dy + g(t)dt = 0 \tag{3.3.1}$$

Note that $f(y)$ depends only on the dependent variable y while $g(t)$ depends only on the independent variable t. The general solution is obtained by integration to obtain

$$\int f(y)dy + \int g(t)dt = \mathrm{C} \tag{3.3.2}$$

where C is a constant that can be determined from the initial condition.

Consider the following example:

$$\frac{dy}{dt} = \frac{t^2}{y}$$

which can be written in the form of Eq. (3.3.1)

$$ydy = t^2 dt$$

or

$$\int ydy = \int t^2 dt + C$$

After performing the integration, we obtain

$$\frac{y^2}{2} = \frac{t^3}{3} + C$$

Here C is a constant. The explicit form of the solution is

$$y(t) = \pm\sqrt{2\left(\frac{t^3}{3} + \mathrm{C}\right)}$$

If we are also given the initial condition $y(0) = 1$, then it follows that $C = \frac{1}{2}$ and finally

$$y(t) = \sqrt{2\left(\frac{t^3}{3} + \frac{1}{2}\right)} = \sqrt{\frac{2}{3}}\sqrt{t^3 + \frac{3}{2}}$$

The same solution can be obtained in symbolic form in MATLAB

```
>> clear all
>> syms y(t)
>> ode = diff(y,t)==t^2/y

ode(t) =
diff(y(t), t) == t^2/y(t)

>> cond = y(0) == 1;
>> ySol(t) = dsolve(ode,cond)

ySol(t) =
(2^(1/2)*3^(1/2)*(t^3 + 3/2)^(1/2))/3
```

The second form of first order differential equations that appear very frequently is the **linear and homogeneous** first order differential equation

$$\frac{dy}{dt} + Py = 0 \tag{3.3.3}$$

Note that Eq. (3.3.3) is a special case of (3.3.1) with $f(y) = \frac{1}{y}$ and $g(x) = P$. To solve this problem, we may use (3.3.2), but we will follow a different approach. We multiply both sides of Eq. (3.3.3) by e^{Pt} to obtain

$$e^{Pt}\frac{dy}{dt} + Pe^{Pt}y = 0 \tag{3.3.4}$$

We then note that Eq. (3.3.4) can be written as

$$e^{Pt}\frac{dy}{dt} + \underbrace{\frac{d(e^{Pt})}{dt}}_{Pe^{Pt}}y = 0$$

or

$$\frac{d(e^{Pt}y)}{dt} = 0 \tag{3.3.5}$$

Since the derivative of $e^{Pt}y$ is zero it follows that $e^{Pt}y$ is constant, i.e.,

$$e^{Pt}y(t) = C \tag{3.3.6}$$

We apply the initial conditions ($t = 0$) to obtain $C = y(0)$ and the solution is

$$\frac{y(t)}{y(0)} = e^{-Pt} \tag{3.3.7}$$

As an example, we consider again the differential mass balance of the cylindrical, water tank for which we derived the following equation in Example (3.2):

$$\frac{dh(t)}{dt} + \left(\frac{s}{S}k\right)h(t) = 0$$

It follows that $P = \left(\frac{s}{S}k\right)$ and from Eq. (3.3.7), we obtain the solution

$$h(t) = h(0)\exp\left[-\left(\frac{sk}{S}\right)t\right]$$

which agrees with the solution obtained in Example 3.2.

We now consider the slightly more general form where the right-hand side of Eq. (3.3.3) is nonzero

$$\frac{dy}{dt} + Py = Q \tag{3.3.8}$$

We again multiply both sides of Eq. (3.3.8) by the so–called **integrating factor** e^{Pt}

$$e^{Pt}\frac{dy}{dt} + Pe^{Pt}y = Qe^{Pt} \tag{3.3.9}$$

or

$$\frac{d(e^{Pt}y)}{dt} = Qe^{Pt} \tag{3.3.10}$$

We integrate both sides to obtain the general solution

$$e^{Pt}y(t) - y(0) = \int_0^t Qe^{P\tau}d\tau$$

or

$$y(t) = y(0)e^{-Pt} + e^{-Pt}\int_0^t Qe^{P\tau}d\tau \tag{3.3.11}$$

As P and Q are constants the last equation simplifies further

$$y(t) = y(0)e^{-Pt} + \frac{Q}{P}e^{-Pt}\int_0^t de^{P\tau}$$

or

$$y(t) = y(0)e^{-Pt} + \frac{Q}{P}\left(1 - e^{-Pt}\right) \tag{3.3.12}$$

Consider the differential equation

$$\frac{dy(t)}{dt} + 2y(t) = 1$$

where $P = 2$ and $Q = 1$. The integrating factor is $e^{\int 2dt} = e^{2t}$ and we multiply the differential equation by the integrating factor to obtain

$$e^{2t}\frac{dy(t)}{dt} + \left[2e^{2t}\right]y(t) = e^{2t}$$

or

$$d\left[e^{2t}y(t)\right] = e^{2t}dt = \frac{1}{2}d\left[e^{2t}\right]$$

Integrating both sides gives

$$e^{2t}y(t) = \frac{1}{2}e^{2t} + C$$

whence

$$y(t) = \frac{1}{2} + Ce^{-2t}$$

If we are also given the initial condition $y(0) = 1$, then it follows that $C = \frac{1}{2}$ and finally

$$y(t) = \frac{1}{2}\left(1 + e^{-2t}\right)$$

The same solution can also be obtained symbolically using MATLAB

```
>> clear all
>> syms y(t)
>> ode = diff(y,t)==-2*y+1

ode(t) =
diff(y(t), t) == 1 - 2*y(t)

>> ySol(t) = dsolve(ode)

ySol(t) =
(C1*exp(-2*t))/2 + 1/2

>> cond = y(0) == 1;
>> ySol(t) = dsolve(ode,cond)

ySol(t) =
exp(-2*t)/2 + 1/2
```

We finally consider the more general form in which the right-hand side of Eq. (3.3.3) is different from zero and both P and Q are functions of the independent variable t

$$\frac{dy}{dt} + P(t)y = Q(t) \tag{3.3.13}$$

The steps to obtain the solution are the same as before with the only difference that the integrating factor is now defined as

$$e^{\int P(t)dt} \tag{3.3.14}$$

The final solution is again given by Eq. (3.3.11) with the appropriate integrating factor

$$y(t) = e^{-\int P(t)dt}\left(\int Q(\tau)e^{\int P(\eta)d\eta}d\tau + C\right) \tag{3.3.15}$$

Consider the differential equation

$$\frac{dy(t)}{dt} + \underbrace{\left(\frac{2}{t}\right)}_{P(t)} y(t) = \underbrace{t}_{Q(t)}$$

We first calculate the integrating factor

$$\exp\int\frac{2dt}{t} = e^{2\ \ln|t|} = t^2$$

We replace in Eq. (3.3.15)

$$y(t)t^2 = \int \tau^3 d\tau + C$$

or

$$y(t) = \frac{t^2}{4} + \frac{C}{t^2}$$

The same solution is obtained using MATLAB

```
>> clear all
>> syms y(t)
>> ode = diff(y,t)==t-(2/t)*y;
>> ySol(t) = dsolve(ode)

ySol(t) =
C1/t^2 + t^2/4
```

or Python

```
from sympy import symbols, Function, dsolve
t     = symbols('t')
y     = Function('y')(t)
dydt  = y.diff(t)
ode   = dydt + (2/t)*y - t
# Solve the ODE
ySol = dsolve(ode,y)
# Print the solution
print(ySol.rhs)

(C1 + t**4/4)/t**2
```

3.4 Numerical solution of first order differential equations

In the section that follows we will see how MATLAB and Python can be used to solve differential equations of arbitrary complexity, alleviating the burden to rely on the analytic solutions. The use of numerical methods is the preferred method for solving differential equations for the following reasons

- we do not care about deriving the analytic solution,
- we become more productive and better problem solvers as we decouple problem formulation from its solution,
- we develop models that are more understandable and more amenable to generalizations and improvements as we are

not restricted by the constraints inherent to equations that have analytic solutions,
- we can solve problems with many differential equations in the same way as we solve a single differential equation.

But how is the solution of differential equations achieved numerically? The simplest method is to use the definition of the first order derivative. Assume that we wish to solve the differential equation governing the cylindrical water storage tank

$$\frac{dh(t)}{dt} + \left(\frac{s}{S}k\right)h(t) = 0 \tag{3.4.1}$$

where $h(0)$ is known (initial condition). We define the constant $a = sk/S$ to simplify notation

$$\frac{dh(t)}{dt} = -ah(t) \tag{3.4.2}$$

We now remind ourselves of the definition of the derivative

$$\frac{dh(t)}{dt} \approx \frac{h(t+\Delta t) - h(t)}{\Delta t} \tag{3.4.3}$$

Eq. (3.4.3) is only an approximation when Δt is small but not infinitesimally small. We now replace Eq. (3.4.3) into Eq. (3.4.2) to obtain

$$\frac{h(t+\Delta t) - h(t)}{\Delta t} = -ah(t), \quad h(0) \text{ known} \tag{3.4.4}$$

We rearrange the terms to obtain

$$h(t+\Delta t) = (1 - a\Delta t)h(t), \quad h(0) \text{ known} \tag{3.4.5}$$

We use the same numerical values for the parameters as in Example (3.2)

$$a = \frac{sk}{S} = \frac{1}{1000}\,\text{s} \times \frac{3600\ \text{s}}{1\ \text{h}} = 3.6\,\frac{1}{\text{h}}$$

and the same initial condition $h(0) = 0.8$ m. We select $\Delta t = 0.1$ h and we calculate the solution for $t = 0.1$ h

$$h(0.1) = (1 - 3.6 \times 0.1)h(0) = 0.64 \times 0.8 = 0.512\,000\ \text{m}$$

The exact solution gives

$$h(\Delta t) = e^{-a\times\Delta t}h(0) = e^{-0.36} \times 0.8 = 0.558\,141\ \text{m}$$

The fact that the exact solution is significantly different from the approximate solution is a matter of concern.

How can we improve our numerical solution? We remind ourselves that in the definition of the first order derivative, the step Δt needs to be differentiable (very small) for the definition to be valid. We therefore decrease the step to $\Delta t = 0.01$ h, and note that

$$h(k\Delta t) = (1-a\Delta t)h((k-1)\Delta t) = (1-a\Delta t)(1-a\Delta t)h(k-2\Delta t)$$
$$= \ldots = (1-a\Delta t)^k h(0)$$

As $0.1 = 10\times 0.01$ it follows that $k = 10$ and

$$h(0.1) = h(10\times 0.01) = (1-3.6\times 0.01)^{10}h(0) = 0.964^{10}\times 0.8$$
$$= 0.554\ 447\ \text{m}$$

We have improved our estimation of the solution, which is now exact in two decimal places, but we can do better by selecting $\Delta t = 0.001$ h

$$h(0.1) = h(100\times 0.001) = (1-3.6\times 0.001)^{100}h(0)$$
$$= 0.9964^{10}\times 0.8 = 0.557\ 778\ \text{m}$$

We can keep on decreasing the size of the step but the calculations that need to be performed also increase. However, the conclusion is simple: by decreasing the size of the integration step Δt the accuracy of the numerical solution is improved and, at least for the case study we have selected, the accuracy in the solution is of the same order as the integration step.

The method we have just seen for solving differential equations numerically is known as Euler's method. The basic idea in Euler's method is to solve the differential equation

$$\frac{dy(t)}{dt} = f(y(t)) \tag{3.4.6}$$

using the approximation

$$\frac{y(t+\Delta t)-y(t)}{\Delta t} = f(y(t)) \tag{3.4.7}$$

A better idea is to use the following approximation (Implicit Euler's method)

$$\frac{y(t+\Delta t)-y(t)}{\Delta t} = f(y(t+\Delta t)) \tag{3.4.8}$$

and an even better idea is to use the approximation (trapezoidal method)

$$\frac{y(t+\Delta t)-y(t)}{\Delta t}=\frac{f(y(t))+f(y(t+\Delta t))}{2} \tag{3.4.9}$$

The problem with the approximations in Eqs. (3.4.8) and (3.4.9) is that the unknown $y(t+\Delta t)$ appears in both sides of the equations and the solution involves solving a nonlinear equation. To see what are the benefits, let us use Eq. (3.4.9) to solve the cylindrical water tank problem with a time step of $\Delta t = 0.1$ h. At $t = 0$ we have that $h(0) = 0.8$ and $f(h(0)) = -ah(0) = -3.6 \times 0.8 = -2.88$. We replace into Eq. (3.4.9)

$$\frac{h(0.1)-h(0)}{\Delta t}=\frac{-ah(0)-ah(0.1)}{2}$$

or

$$h(0.1)=\frac{\left(1-\frac{\Delta t}{2}a\right)}{\left(1+\frac{\Delta t}{2}a\right)}h(0)=\frac{\left(1-\frac{0.1}{2}\times 3.6\right)}{\left(1+\frac{0.1}{2}\times 3.6\right)}0.8=0.555\,932\text{ m}$$

Compare with the exact solution $h(0.1) = 0.558\,141$ m. The improvement is impressive. By decreasing the step size to $\Delta t = 0.001$ the solution obtained by the approximate method is

$$h(0.1)=\left[\frac{1-\frac{\Delta t}{2}a}{1+\frac{\Delta t}{2}a}\right]^{100}h(0)=\left[\frac{1-\frac{0.001}{2}3.6}{1+\frac{0.001}{2}3.6}\right]^{100}0.8=0.558\,141\text{ m}\,!$$

The numerical solution is exact to at least 6 decimal digits!

MATLAB and Python provide highly efficient numerical solvers for ordinary differential equations which, apart from using advanced methods for the numerical solution of differential equations, also use adaptive (variable) step-size control that enhances computational efficiency. It is fair to say that the available solvers in MATLAB and Python are so efficient and easy to use that they make analytical solutions obsolete, and the effort to derive them using common approximations is often difficult to justify.

3.5 MATLAB and Python for solving ordinary differential equations

3.5.1 Solving ordinary differential equations in MATLAB

In this section a detailed description of how ordinary differential equations (ODEs) are solved in MATLAB is presented. It is important that you understand all the details of this section. You are going to invest a lot of time and effort, but your investment will pay off in the chapters that follow. You may not realize this, but after completing this section you will have made a gigantic step toward acquiring the problem-solving capabilities that will allow you to face realistic process engineering problems. The skills that you will develop are also transferable skills that will help you in other courses where ODEs are of paramount importance (chemical kinetics and reactor design, process control and dynamics and unit operations).

The solvers that are available in MATLAB for solving ODEs have the following structure

```
[T,Y] = ode_solver_name(ODEFUN(t,y),[t0 tf],y0)
```

which integrates the system of differential equations $dy/dt = f(t, y)$ from time `t0` to `tf` with initial conditions `y0`. `ODEFUN` is a function handle. For a scalar t and a vector y, `ODEFUN(t,y)` must return a column vector corresponding to $f(t, y)$. Each row in the solution array **Y** corresponds to a time returned in the column vector **T**. The two most efficient ODE solvers in MATLAB are the **ode113** and **ode15s**, the latter being our preference in this text. The best way to present the use of the **ode15s** is through specific examples.

The model for the cylindrical water storage tank has been presented in Example 3.2. The dynamic model of the tank consists of one ordinary differential equation and three algebraic equations (see Example 3.2)

$$\frac{dM(t)}{dt} = \dot{M}_{in}(t) - \dot{M}_{out}(t)$$

where

$$M(t) = \rho S h(t)$$

$$\dot{M}_{in}(t) = 0$$

$$\dot{M}_{out}(t) = \rho skh(t)$$

To simulate the system in MATLAB we first define the model in an m-file assuming that the m-file receives the time t and the $M(t)$ as input arguments and returns the $f(t, M(t))$, i.e., the right–hand side of the ODE $dy/dt = f(t, y)$. For the case of the cylindrical water storage tank this is achieved as follows

```
function dMdt = CylWaterTank(t,M)
% m-file to simulate the dynamics of a water storage tank
% dMdt = CylWaterTank(t,M)
% t is the time in h
% M is the mass of water in the tank in kg

%% constant parameters
% water density in kg per m^3
rho_w = 1000;
% cross-sectional area of the tank in m^2
S = 0.1;
% cross-sectional area of the hole in m^2
s = 0.001;
% outflow model constant in 1/h
k = 0.1*3600;

%% incoming mass flowrate in kg/h (input)
M_dot_in = 0;

%% calculate the height of the liquid in the tank in m
h = M/(rho_w*S);

%% calculate the mass flowrate of the product stream in kg/h
M_dot_out = rho_w*s*(k*h);

%% ODE
dMdt = M_dot_in - M_dot_out;
end
```

Let us discuss this m-file line by line. The first line

```
function dMdt = CylWaterTank(t,M)
```

simply states that this is a new MATLAB function. The commands and functions that comprise the new function must be put in a file whose name defines the name of the new function, with a filename extension of '**.m**' (**CylWaterTank.m** in our case).

The first line is followed by comments

```
% m-file to simulate the dynamics of a water storage tank
% dMdt = CylWaterTank(t,M)
% t is the time in h
% M is the mass of water in the tank in kg
```

When the user types ">> `help CylWaterTank`" in the command line, these four lines are reported to the user to inform him/her for the syntax definition of the new function and its functionality (what it does).

What follows next is the definitions of all parameters appearing in the model with a clear indication of the units used

```
%% constant parameters
% water density in kg per m^3
rho_w = 1000;
% cross-sectional area of the tank in m^2
S = 0.1;
% cross-sectional area of the hole in m^2
s = 0.001;
% outflow model constant in 1/h
k = 0.1*3600;

%% incoming mass flowrate in kg/h (input)
M_dot_in = 0;
```

The lines that follow solve the algebraic equations for the algebraic variables $(h, \dot{M}_{out})$ of the model

```
%% calculate the height of the liquid in the tank in m
h = M/(rho_w*S);

%% calculate the mass flowrate of the product stream in kg/h
M_dot_out = rho_w*s*(k*h);
```

The only thing remaining is to calculate the right-hand side of the ODE

```
%% ODE
dMdt = M_dot_in - M_dot_out;
```

To use this function to simulate the system dynamics we create the following script file

```
clear all
t0     = 0;          % initial simulation time in h
tf     = 2;          % final simulation time in h
h0     = 0.8;        % initial liquid level in the tank in m
S      = 0.1;        % cross-sectional area of the tank in m^2
s      = 0.001;      % cross-sectional area of the hole in m^2
k = 0.1*3600;        % outflow model constant in 1/h
rho_w = 1000;        % density of liquid water in kg/m^3
M0     = rho_w*S*h0; % initial condition M(t=t0)
[T,M]=ode15s(@CylWaterTank,[t0 tf],M0)

figure(1)
% plot numerical solution
plot(T,M,'bo')
hold on
% plot analytic solution
plot(T,rho_w*S*h0*exp(-(k*s/S)*T),'r')
axis([0 1.5 0 100])
grid on
xlabel('Time (h)')
ylabel('Mass of water in the tank (kg)')
```

An explanation follows. Initially we clear the workspace (`clear all`) and then assign values to all constants appearing in the model including the initial conditions

```
t0     = 0;     % initial simulation time in h
tf     = 2;     % final simulation time in h
h0     = 0.8;   % initial liquid level in the tank in m
S      = 0.1;   % cross-sectional area of the tank in m^2
s      = 0.001;% cross-sectional area of the hole in m^2
k = 0.1*3600; % outflow model constant in 1/h
rho_w = 1000; % density of liquid water in kg/m^3
M0     = rho_w*S*h0;
```

We then use the ODE solver in MATLAB to obtain the solution

```
[T,M]=ode15s(@CylWaterTank,[t0 tf],M0)
```

and finally plot the results and the analytic solution for comparison

```
figure(1)
% plot numerical solution
plot(T,M,'bo')
hold on
% plot analytic solution
plot(T,rho_w*S*h0*exp(-(k*s/S)*T),'R')
axis([0 1.5 0 100])
grid on
xlabel('Time (h)')
ylabel('Mass of water in the tank (kg)')
```

When you run the simulation, you obtain the results shown in Fig. 3.9

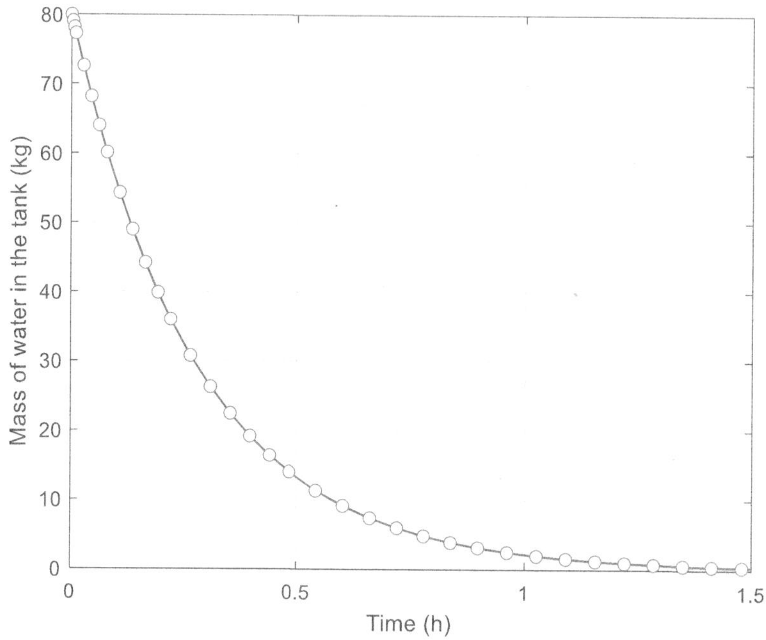

Figure 3.9 Mass of water in the cylindrical tank as a function of time. Comparing numerical (circles) and analytic solution (continuous line) in MATLAB.

It is important to note that the model of the water storage tank, for zero mass flowrate of the feed, can be written in the simplified form (see Eq. (3.4.2))

$$\frac{dh(t)}{dt} = -ah(t)$$

where

$$a = \frac{sk}{S} = 3.6\,\frac{1}{\mathrm{h}}$$

Based on this observation we may obtain the solution of the water storage tank by simply typing in the command window the following.

```
>> clear all
>> f=@(t,h) -3.6*h

f =

  function_handle with value:
    @(t,h)-3.6*h

>> ode15s(f,[0 2],0.8)
```

While the shortcut method offers a compact and efficient way to reach the same solution, we strongly recommend using the detailed and systematic approach presented earlier in this section. This is due to several key advantages:

- It avoids unnecessary substitutions, keeping the model clean and transparent.
- Any changes to the incoming or outgoing stream expressions can be easily implemented, typically by modifying just a single line.
- Errors or inconsistencies in units are easier to identify and correct.
- The model files created with this approach are modular and can be readily extended to include additional equations.
- These files are also adaptable, allowing the model to be reused or modified for similar case studies (e.g., a conical liquid storage tank).

Before closing this section, and to demonstrate the progress that we have made in terms of our problem-solving skills, we will consider the compressed air storage tank of Fig. 3.8. The corresponding model is the following (see Example 3.4)

$$\frac{dN(t)}{dt} = \dot{N}_{in}(t) - \dot{N}_{out}(t)$$

$$N(t)RT = VP(t)$$

$$\rho = \frac{N(t)}{V}$$

$$\dot{N}_{out}(t) = \rho F_{out} = \rho k\sqrt{P(t) - P_d}$$

We consider the usual case of the periodic operation of the air storage tank, in which the compressor in the feed is operating for 600 s (10 min) and then it stops for another 600 s. The following parameters are considered

$$\dot{N}_{in} = 1\ \frac{\text{mol}}{\text{s}}$$

$$V = 1\ \text{m}^3$$

$$T = 288.15\ \text{K}\ (15\ ^\circ C)$$

$$k = 8.74778 \times 10^{-6}\ \frac{\text{m}^3}{\text{Pa}^{0.5}\ \text{s}}$$

$$P_d = 2 \times 10^5\ \text{Pa}$$

The model is implemented in MATLAB through the following script file

```
function dNdt=ComprAirTank(t,N)
% m-file to simulate the dynamics of a compresed air tank
% dMdt = ComprAirTank(t,N)
% t is the time in h
% N are the mole of air in the tank

%% constant parameters
% tank volume in m^3
V = 1;
% temperature in the tank in K
T = 288.15;
% ideal gas constant in Pa m^3 /( mol K)
R = 8.314;
% pressure downstream in Pa
P_d = 200000;
% constant for the valve model in kPa^(-0.5)  s^(-1)
k = 8.74778e-6;

%% incoming molar flowrate in mol/s (input)
N_dot_in = 1*heaviside(600-t); % type >>help Heaviside for more info

%% algebraic equations
rho_v = N/V;          % gas density in mol/m^3

%% calculate the pressure in the tank in kPa
P     = rho_v * R*T

%% calculate the mass flowrate of the product stream in kg/h
N_dot_out = rho_v*k*sqrt(max(sqrt(eps),P-P_d));

%% ODE
dNdt= N_dot_in - N_dot_out;
end
```

The problem is solved as follows

```
clear all
t0  = 0;      % initial simulation time in h
tf  = 1200;   % final simulation time in h
V   = 1;      % volume of compressed air tank in m^3
T   = 288.15;% temperature in the tank in K
R   = 8.314; % ideal gas constant in Pa m^3 /(mol K)
P_d = 200000;% pressure downstream in Pa
N0  = P_d*V/(R*T);
options = odeset('RelTol',1e-8,'AbsTol',1e-8)
[T,N]=ode15s(@ComprAirTank,linspace(t0,tf,101),N0,options);

figure(1)
plot(T,N,'bo')
grid on
xlabel('Time (s)')
ylabel('mole of air in the tank (mol)')
```

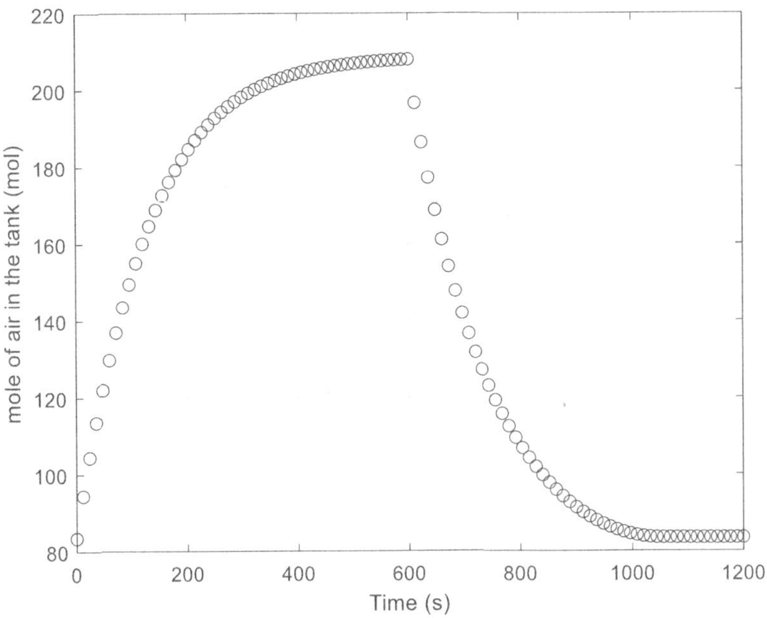

Figure 3.10 Dynamic variation of the molar holdup of compressed air in the air storage tank.

The solution is shown in Fig. 3.10. For the 10 min that the compressor is operating the molar holdup of the air is increasing. When the compressor is shut down the molar holdup is decreasing and is stabilized to a value dictated by the downstream pressure. Apparently, the accumulation of compressed air is not enough to sustain continuous feeding of the downstream units with air.

3.5.2 Solving ordinary differential equations in Python

The integration of ODEs (initial value problems) can be performed in Python using the function **solve_init** available in the module **scipy.integrate** (scipy: stands for scientific python). The process of emptying a tank of cylindrical shape can be simulated in Python as follows

```
import matplotlib.pyplot as plt
import numpy as np
from scipy.integrate import solve_ivp

t0    = 0.0         # initial simulation time in h
tf    = 1.5         # final simulation time in h
h0    = 0.8         # initial liquid level in the tank in m
S     = 0.1         # cross-sectional area of the tank in m^2
s     = 0.001       # cross-sectional area of the hole in m^2
k     = 0.1*3600    # outflow model constant in 1/h
rho_w = 1000        # density of liquid water in kg/m^3
M0    = rho_w*S*h0# initial conditions

def CylWaterTank(t, M):
    # incoming mass flowrate in kg/h (input)
    M_dot_in = 0
    # calculate the height of the liquid in the tank in m
    h = M/(rho_w*S)
    # calculate the mass flowrate of the product stream in kg/h
    M_dot_out = rho_w*s*(k*h);
    # ODE
    dMdt = M_dot_in - M_dot_out;
    return dMdt

sol = solve_ivp(CylWaterTank,[t0, tf],[M0],method='BDF', rtol=1e-6)

plt.plot(sol.t,sol.y[0],'r-o', mec = 'b',mfc = 'w')
plt.grid()
plt.xlabel('Time (h)')
plt.ylabel('Mass of water in the tank (kg)')
plt.axis([0, 1.5, 0, 80])
```

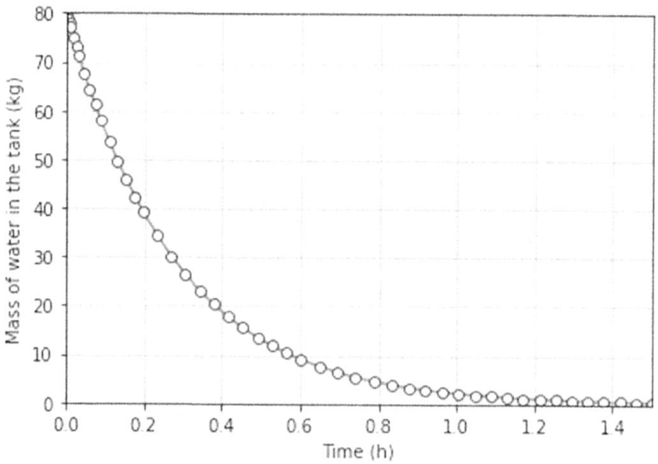

Figure 3.11 Mass of water in the cylindrical tank as a function of time. Comparing numerical (circles) and analytic solution (continuous line) in Python.

The results are presented in Fig. 3.11.

Integration method to use: it is proposed to use BDF (Implicit multi-step variable-order 1–5) method based on a backward differentiation formula for the derivative approximation, as described in L. F. Shampine, M. W. Reichelt, The MATLAB ode suite, SIAM J. Sci. Compute., 18(1), pp. 1-22, 1997. This method works fine in almost all cases. Among Runge-Kutta methods, 'DOP853' is recommended for solving with high precision (low values of rtol and atol). Other choices are Runge-Kutta methods such as 'RK45' (default), 'RK23', 'Radau' (implicit Runge-Kutta method of the Radau IIA family), and 'LSODA'. BDF will be used almost exclusively in this text.

Chapter summary

The general balance equation is the following

$$\begin{bmatrix}\text{extensive quantity}\\ \text{at time}\\ \text{t}+1\end{bmatrix} - \begin{bmatrix}\text{extensive quantity}\\ \text{at time}\\ \text{t}\end{bmatrix}$$

$$= \begin{bmatrix}\text{inflow of}\\ \text{extensive quantity}\\ \text{between t and t}+1\end{bmatrix} - \begin{bmatrix}\text{outflow of}\\ \text{extensive quantity}\\ \text{between t and t}+1\end{bmatrix}$$

$$+ \begin{bmatrix}\text{generation of}\\ \text{extensive quantity}\\ \text{between t and t}+1\end{bmatrix} - \begin{bmatrix}\text{consumption of}\\ \text{extensive quantity}\\ \text{between t and t}+1\end{bmatrix}$$

Stated more compactly

$$\text{new value} - \text{old value} = \text{input} - \text{output} + \text{generation} - \text{consumption}$$

The total material balance can be written as

$$\frac{dM(t)}{dt} = \dot{M}_{in}(t) - \dot{M}_{out}(t)$$

or

$$\begin{bmatrix}\text{rate of}\\ \text{accumulation}\\ \text{of total mass}\end{bmatrix} = \begin{bmatrix}\text{rate of}\\ \text{inflow}\\ \text{of total mass}\end{bmatrix} - \begin{bmatrix}\text{rate of}\\ \text{outflow}\\ \text{of total mass}\end{bmatrix}$$

Each term has units of mass per unit time (kg/s).

The analytic solution of the most basic types of first order ODEs:

ODE General Solution

$$f(y)dy = g(t)dt \rightarrow \int f(y)dy = \text{constant} + \int g(t)dt$$

$$\frac{dy}{dt} + Py = 0 \quad \rightarrow y(t) = y(0)e^{-Pt}$$

$$\frac{dy}{dt} + Py = Q \rightarrow y(t) = y(0)e^{-Pt} + \frac{Q}{P}(1 - e^{-Pt})$$

Solution of arbitrary nonlinear ODEs in MATLAB can be performed with ode15s function and in Python with the **solve_ivp** function (from **scipy.integrate** module). There are some alternatives, but these are going to be our choice in this text.

Problems

3.1 A spherical tank of diameter D is partially full of a liquid (see Fig. P3.1). At $t = 0$, the liquid escapes from a hole at the bottom of the tank with volumetric flowrate

$$\dot{V}(t) = k\sqrt{h(t)}$$

Find the time necessary to drain the spherical tank.

3.2 A cylindrical, horizontal tank of diameter D and length H, is partially full of a liquid (see Fig. P3.2). At $t = 0$, the liquid escapes from hole at the bottom of the tank and with volumetric flowrate

$$\dot{V}(t) = k\sqrt{h(t)}$$

Find the time necessary to drain the tank.

3.3 The tank pictured in Fig. P3.3, is partially full with a liquid. At $t = 0$, the liquid escapes from a hole at the bottom of the tank with volumetric flowrate

$$\dot{V}(t) = k\sqrt{h(t)}$$

Find the time necessary to drain the tank.

3.4 A tank of square cross section is filled to a height $h(0)$ with water (see Fig. P3.4). At $t = 0$ the liquid escapes from the tank through a hole at the side close to the bottom of the tank. The volumetric flowrate of the escaping liquid water is believed to be given by

$$\dot{V}(t) = k\sqrt{h(t)}, [k] = \frac{\text{cm}^3}{\text{min}\cdot\text{cm}^{1/2}}$$

where k is a constant. The liquid level above the hole is measured at several times and the measurements are presented in Table P3.4. Determine, using MATLAB, the value of the constant k.

3.5 In 1978 in the British medical journal, *The Lancet*, there was a report with detailed statistics of a flu epidemic in a boys' boarding school with a total of 763 boys. 512 boys were confined to bed during the epidemic caused by one infected boy who initiated the epidemic. Use the SIR model and MATLAB or Python to predict the evolution of the epidemic with $a = 0.0022$, $\gamma = 0.47$, $S(0) = 762$ and $I(0) = 1$. The statistical data are summarized in Table P3.5 (Figs. P3.1, P3.2, and P3.4).

Table P3.4 Data for emptying a tank with constant, square cross section.

t (min)	$h(t)$ cm
0	40.0
5	32.9
10	24.4
15	18.5
20	12.9
25	9.5
30	6.3
35	3.2
40	2.2

Table P3.5 Flu epidemic in a boys' boarding school studied in Problem 3.5.

t (day)	$I(t)$	t (day)	$I(t)$
1	3	8	235
2	8	9	190
3	28	10	125
4	75	11	70
5	221	12	28
6	291	13	12
7	255	14	5

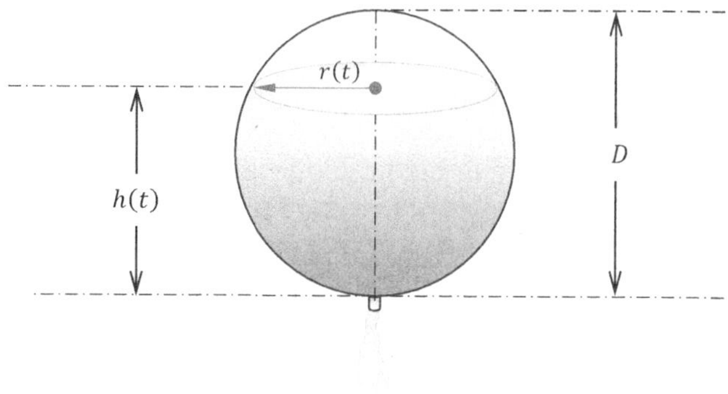

Figure P3.1 Emptying a spherical tank.

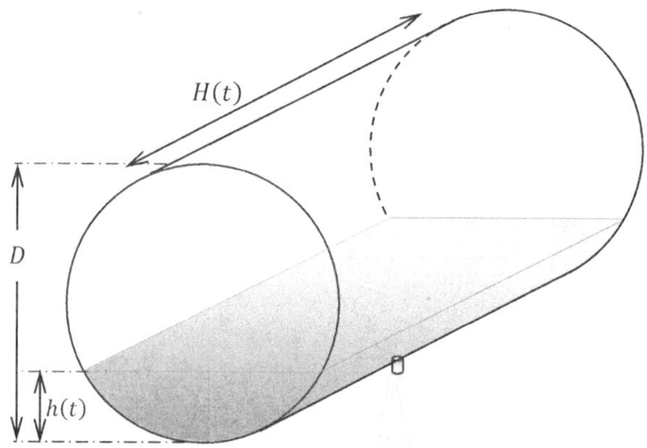

Figure P3.2 Emptying a cylindrical tank.

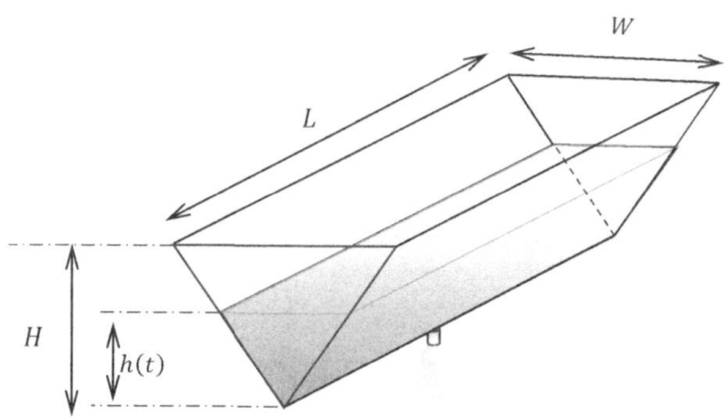

Figure P3.3 Emptying a conical tank.

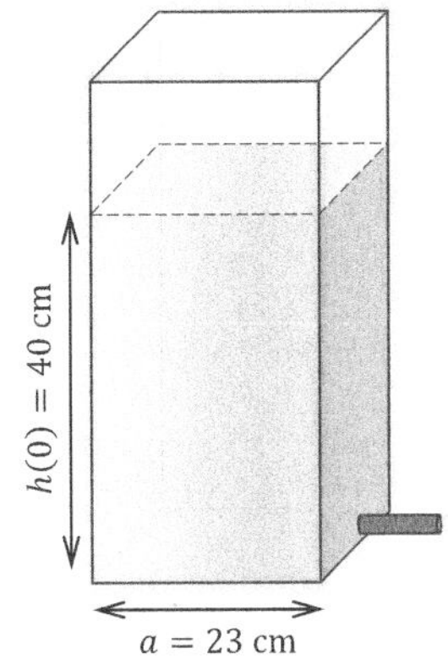

Figure P3.4 Emptying a constant, square cross–sectional tank.

3.6 In Fig. P3.6, a two-tank process commonly used in chemical engineering laboratories for process control experiments is shown. The process consists of two interacting tanks of square cross section, where a stream of flowrate $\dot{F}_{in}$ is fed to the first tank (this can be changed using a computer-controlled system to a desired value). There is a connecting pipe between the two tanks and the flowrate between these two tanks is

$$\dot{F}_{12}(t) = k_{12}(h_1(t) - h_2(t)), k_{12} = 4\ \frac{\text{cm}^2}{\text{s}}$$

where k is a constant, h_1 is the level of the liquid in the first tank and h_2 is the level of the liquid in the second tank. The liquid outflow from the second tank is given by

$$\dot{F}_2(t) = k_2 h_2(t), [k_2] = 8\ \frac{\text{cm}^2}{\text{s}}$$

The initial incoming flowrate is $\dot{F}_{in} = 80\ \text{cm}^3/\text{s}$. Answer the following questions

a. What are the total material balances for the two tanks?
b. What are h_1 and h_2 at steady state?
c. Simulate the starting up of the system with $h_1(0) = h_2(0) = 0$ in MATLAB or Python (Fig. P3.6).

3.7 In Fig. P3.7, a two-tank process commonly used in chemical engineering laboratories for process control experiments is shown. The process consists of two noninteracting tanks of square cross section in series, where a stream of flowrate $\dot{F}_{in}$ is fed to the first tank (this can be changed using a

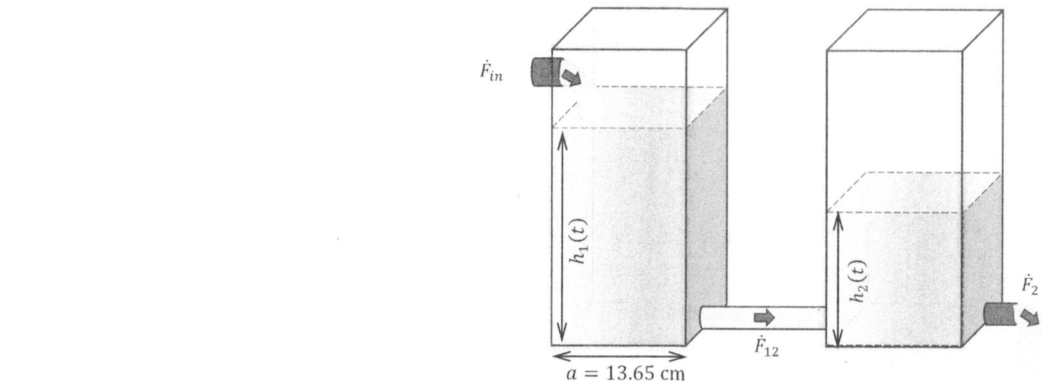

Figure P3.6 Two interconnected tanks with constant, square cross section.

computer-controlled system to a desired value). The flow-rate between these two tanks is

$$\dot{F}_{12}(t) = k_{12}h_1(t), k_{12} = 8\ \frac{\text{cm}^2}{\text{s}}$$

where k is a constant, h_1 is the level of the liquid in the first tank and h_2 is the level of the liquid in the second tank. The liquid outflow from the second tank is given by

$$\dot{F}_2(t) = k_2h_2(t), k_2 = 8\ \frac{\text{cm}^2}{\text{s}}$$

The initial incoming flowrate is $\dot{F}_{in} = 80\ \text{cm}^3/\text{s}$. Answer the following questions

a. What are the total material balances for the two tanks?
b. What are h_1 and h_2 at steady state?
c. Simulate the starting up of the system with $h_1(0) = h_2(0) = 0$ in MATLAB or Python.

3.8 Consider a five noninteracting tanks in series process where the outflow from each tank is

$$\dot{F}_i(t) = kh_i(t), k = 8\ \frac{\text{cm}^2}{\text{s}}$$

The cross section of all tanks is square with side $a = 23$ cm. The incoming flowrate is $\dot{F}_{in} = 80\ \text{cm}^3/\text{s}$. Answer the following questions

a. What are the total material balances for the tanks?
b. Calculate h_i at steady state?

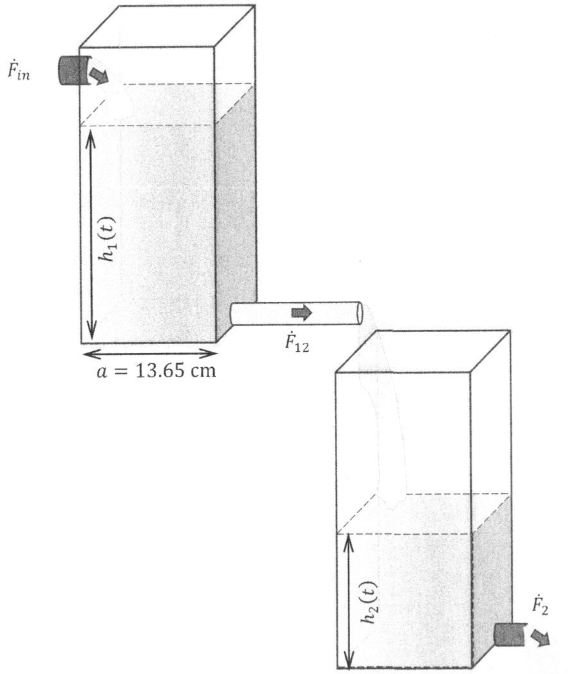

Figure P3.7 Two noninteracting tanks with constant, square cross section.

c. Simulate the starting up of the system with $h_i(0) = 0$ in MATLAB or Python (Fig. P3.7).

3.9 A tank of square cross section is filled to a height $h(0) = 40$ cm with water (see Fig. P3.9). At $t = 0$ the liquid escapes from the tank from two holes: one close to the bottom of the tank and one at height $H = 20$ cm. The volumetric flowrate of the escaping liquid water is believed to be given by

$$\dot{F}_1(t) = k\sqrt{\max\{0, h - H\}}$$

$$\dot{F}_2(t) = k\sqrt{h(t)}$$

with $k = 100 \ \frac{\text{cm}^3}{\text{s}\cdot\text{cm}^{1/2}}$.

What is the time necessary to drain the tank? Use MATLAB or Python to answer the question (Fig. P3.9).

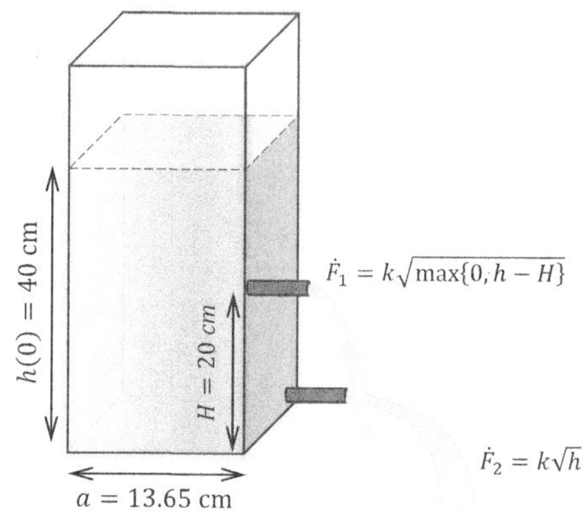

Figure P3.9 Emptying a constant, square cross–sectional tank.

3.10 A four–tank process is shown in Fig. P3.10. The four tanks are identical and have square cross sections. The volumetric flowrate of the escaping liquid water from each tank is believed to be given by

$$\dot{F}_i(t) = k\sqrt{h_i(t)}$$

Develop the dynamic material balances of the four tanks. The parameters of the dynamic total material balance model are given in Table P3.10. Use these numbers and MATLAB or Python to simulate the starting up of the four-tank process system.

3.11 The simple, batch distillation apparatus in Fig. P3.11 was first quantified by Lord Rayleigh and is referred to as a differential distillation. Consider that a binary mixture is initially fed to the still. Then heat is supplied and the mixture starts to boil. At any instant, vapor leaving the still pot with composition $y(t)$ (mole fraction of volatile component in the vapor) is assumed to be in equilibrium with liquid (residue) in the still with composition $x(t)$ (mole fraction of volatile component in the liquid). The more volatile component always has a higher composition in the vapor compared to that in the liquid. Assume also that a constant boilup rate V is achieved. The molar holdup of the residue in the still is $N(t)$. For ideal binary mixtures with constant relative volatility α the mole fraction of the more volatile component in the vapor phase and in the liquid phase are related through the following equation

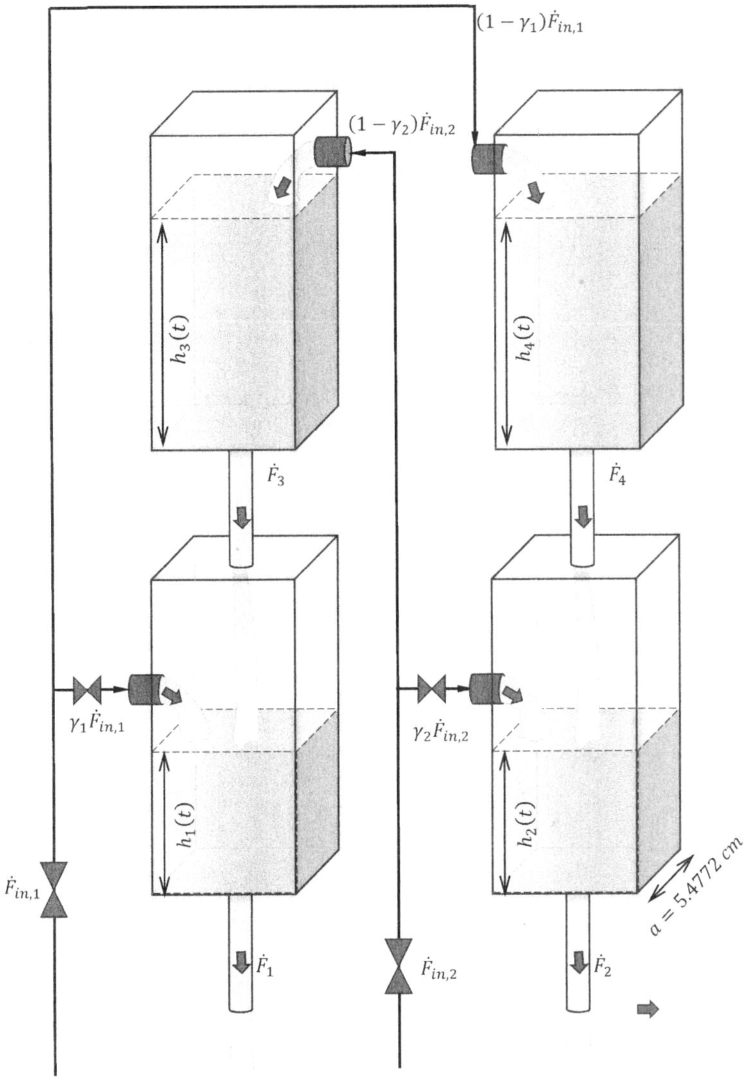

Figure P3.10 A four-tank process.

Table P3.10 Parameters of the four-tank process.

Parameter		Numerical value/units
S	Cross–sectional area	30 cm^2
k	Constant	6 $cm^{2.5}/s$
γ_1		0.35
γ_2		0.45
$\dot{F}_{in,1}$	Feed volumetric flowrate	30 cm^3/s
$\dot{F}_{in,2}$	Feed volumetric flowrate	40 cm^3/s

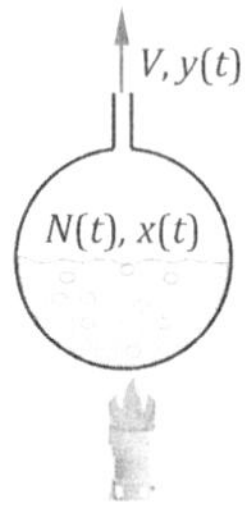

Figure P3.11 A simple distillation apparatus.

$$y = \frac{ax}{1 + (a-1)x}$$

a. Show that the mathematical model consists of two equations:
Total mole balance

$$\frac{dN(t)}{dt} = -V$$

Volatile component mole balance

$$\frac{d[N(t)x(t)]}{dt} = -Vy(t)$$

b. Show that the model can be rearranged in the form

$$N(t)\frac{dx(t)}{dt} = V(a-1)x(t)\frac{(x(t)-1)}{[1+(a-1)x(t)]}$$

c. Show that the differential equation given in (b) can be integrated to obtain:

$$\frac{1}{(a-1)}\left[\ln\left(\frac{x(t)}{x_0}\right) - a\ln\left(\frac{1-x(t)}{1-x_0}\right)\right] = \ln\left(\frac{N(t)}{N_0}\right)$$

Where $N(t) = N_0 - V_t$.

d. Propose a way to present the composition of the volatile component in the residue as a function of the time t.

e. Solve the differential equations numerically in MATLAB or Python to validate your results.

Use the following numerical values: $a = 3$, $N_0 = 100$ mol, $x_0 = 0.5$, $V = 1$ mol/min.

References

Many definitions have been retrieved from the Wikipedia, the free encyclopedia.

Books on material and energy balances

Andersen, L.B., Wenzel, L.A., 1961. Introduction to Chemical Engineering. McGraw Hill, NY.

Duncan, T.M., Reimer, J.A., 1998. Chemical Engineering Design and Analysis. Cambridge University Press, NY.

Henley, E.J., Bieber, H., 1959. Chemical Engineering Calculations. McGraw Hill, NY.

Hougen, O.A., Watson, K.M., Ragatz, R.A., 1943. Chemical Process Principles, Part I, Material and Energy Balances. John Wiley and Sons, NY.

Luyben, W.L., Wenzel, L.A., 1988. Chemical Process Analysis: Mass and Energy Balances. Prentice Hall, Englewood Cliffs, NJ.

Myers, A.L., Seider, W.D., 1976. Introduction to Chemical Engineering and Computer Calculations. Prentice Hall, Inc., Englewood Cliffs, NJ.

Reklaitis, G.V., 1983. Introduction to Material and Energy Balances. John Wiley and Sons, NY.

Russell, T.W.F., Denn, M.M., 1972. Introduction to Chemical Engineering Analysis. John Wiley and Sons, NY.

Thermodynamics Books

Abbott, M.M., Van Ness, H.C., 1972. Theory and Problems of Thermodynamics, Schaums Outline Series. McGraw-Hill Book Company, NY.

Elliot, J.R., Lira, C.T., 2012. Introductory Chemical Engineering Thermodynamics, second ed. Prentice Hall, Upper Saddle River, NJ.

Matsoukas, T., 2012. Fundamentals of Chemical Engineering Thermodynamics. Prentice Hall, Upper Saddle River, NJ.

Sandler, S.I., 1977. Chemical and Engineering Thermodynamics. John Wiley and Sons, NY.

Books on numerical methods

Clough, D.E., Chapra, S.C., 2023. Introduction to Engineering and Scientific Computing with Python. CRC Press, Boca Raton, FL.

Dorfman, K.D., Daoutidis, P., 2017. Numerical Methods with Chemical Engineering Applications. Cambridge University Press.

Mitsotakis, D., 2023. Computational Mathematics: An Introduction to Numerical Analysis and Scientific Computing with Python. CRC Press, Boca Raton, FL.

Mostoufi, N., Constantinides, A., 2023. Applied Numerical Methos for Chemical Engineers. Academic Press, London.

4

Component mass and mole balances

4.1 Review of the general balance equation

In Chapter 3 we developed the general "material balance" equation which can be stated in words as follows

$$\begin{bmatrix}\text{extensive quantity}\\ \text{at time}\\ t+1\end{bmatrix} - \begin{bmatrix}\text{extensive quantity}\\ \text{at time}\\ t\end{bmatrix}$$
$$= \begin{bmatrix}\text{inflow of}\\ \text{extensive quantity}\\ \text{between t and t}+1\end{bmatrix} - \begin{bmatrix}\text{outflow of}\\ \text{extensive quantity}\\ \text{between t and t}+1\end{bmatrix} \quad (4.1.1)$$
$$+ \begin{bmatrix}\text{generation of}\\ \text{extensive quantity}\\ \text{between t and t}+1\end{bmatrix} - \begin{bmatrix}\text{consumption of}\\ \text{extensive quantity}\\ \text{between t and t}+1\end{bmatrix}$$

or, stated more compactly as

$$\text{new value} - \text{old value} = \text{input} - \text{output} + \text{generation} - \text{consumption} \quad (4.1.2)$$

It was mentioned in Chapter 3 that this form of a balance equation is appropriate for nonconserved quantities for which there is a possibility of consumption or production. This is true for the case of chemical species as they can be produced or consumed due to chemical reactions. As chemical reactions are expressed in terms of moles of reactants and products and reaction rate terms are also expressed in molar units, in this chapter we will mainly use mole balances.

We have also expressed the general mass balance equation for the extensive quantity $M(t)$ as follows

Introduction to Process Systems Engineering. https://doi.org/10.1016/B978-0-443-45444-8.00001-4

$$M(t+dt) = M(t) + \dot{M}_{\text{in}}(t)dt - \dot{M}_{\text{out}}(t)dt \tag{4.1.3}$$

where the overdot denotes rate, i.e., mass flowrate (mass per unit time). For the case of a nonconserved quantity Eq. (4.1.3) needs to be modified to account for production/consumption

$$\begin{aligned} M(t+dt) = M(t) + \dot{M}_{\text{in}}(t)dt - \dot{M}_{\text{out}}(t)dt + \dot{M}_{\text{gen}}(t)dt \\ - \dot{M}_{\text{cons}}(t)dt \end{aligned} \tag{4.1.4}$$

Eq. (4.1.4) can be written in differential form by dividing the equation with dt and taking the limit as $dt \rightarrow 0$, to obtain

$$\frac{dM(t)}{dt} = \dot{M}_{\text{in}}(t) - \dot{M}_{\text{out}}(t) + \dot{M}_{\text{gen}}(t) - \dot{M}_{\text{cons}}(t) \tag{4.1.5}$$

Eq. (4.1.5) is the general form of a differential balance equation that applies to conserved and nonconserved quantities. All terms of the equation are rate terms with units of mass or mole per unit time.

4.2 The component mass and mole balance equations

When we restrict attention to the mass of a specific chemical component, then the extensive quantity is the mass of this particular chemical component or the moles of the same component. We will be using molar units so we select the symbol N for the extensive quantity and use the upper case letter for the total moles (involving all components) and the lower case letter n_i when we refer to a particular component (denoted by the index i). We may therefore write the general mole balance equation for component i in the following form

$$\frac{dn_i(t)}{dt} = \dot{n}_{\text{in},i}(t) - \dot{n}_{\text{out},i}(t) + \dot{n}_{\text{gen},i}(t) - \dot{n}_{\text{cons},i}(t) \tag{4.2.1}$$

where $n_i(t)$ denotes the moles of component i in the control volume

$\dot{n}_{\text{in},i}(t)$ denotes the rate at which moles of component i enter the control volume

$\dot{n}_{\text{out},i}(t)$ denotes the rate at which moles of component i exit the control volume

$\dot{n}_{\text{gen},i}(t)$ denotes the rate at which moles of component i are generated within the control volume

$\dot{n}_{\text{cons},i}(t)$ denotes the rate at which moles of component i are consumed within the control volume

The left-hand side of Eq. (4.2.1) is called the (rate of) accumulation term. Eq. (4.2.1) can be stated in words as follows

$$\begin{bmatrix} \text{rate of} \\ \text{accumulation} \\ \text{of component i} \end{bmatrix} = \begin{bmatrix} \text{rate of} \\ \text{inflow} \\ \text{of component i} \end{bmatrix} - \begin{bmatrix} \text{rate of} \\ \text{outflow} \\ \text{of component i} \end{bmatrix} + \begin{bmatrix} \text{rate of} \\ \text{generation} \\ \text{of component i} \end{bmatrix} - \begin{bmatrix} \text{rate of} \\ \text{consumption} \\ \text{of component i} \end{bmatrix} \tag{4.2.2}$$

Each term has units of mass per unit time (kg/s) or moles per unit time (mol/s).

We start by first restricting ourselves to cases where there is no chemical reaction, in which case the production/consumption terms are zero. The general mass or mole balance equation of a chemical species is, in the absence of a chemical reaction(s), like that of the total mass balance. We will then present some fundamental equations that are usually used to model the general production/consumption terms. We will finally face the general problem that incorporates production/consumption terms.

4.3 The component balance equations without chemical reactions

In the absence of chemical reaction(s) the component mole balance can be expressed in words as

$$\begin{bmatrix} \text{rate of} \\ \text{accumulation} \\ \text{of component i} \end{bmatrix} = \begin{bmatrix} \text{rate of} \\ \text{inflow} \\ \text{of component i} \end{bmatrix} - \begin{bmatrix} \text{rate of} \\ \text{outflow} \\ \text{of component i} \end{bmatrix} \tag{4.3.1}$$

or in mathematical terms

$$\frac{dn_i(t)}{dt} = \dot{n}_{\text{in},i}(t) - \dot{n}_{\text{out},i}(t) \tag{4.3.2}$$

This is pictured in Fig. 4.1. All terms are rate terms with units of moles per unit time. If we multiply both sides of Eq. (4.3.2) with

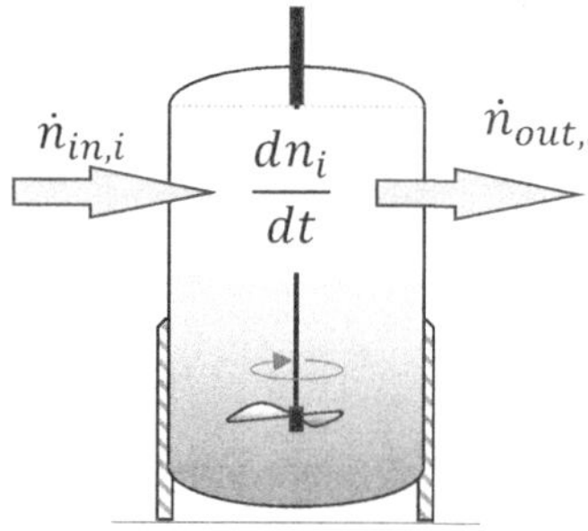

Figure 4.1 Mole balance of a chemical component: The rate at which the component enters the control volume minus the rate at which it exits equals the rate of accumulation.

the molar mass of component i ($m_{m,i}$), we obtain the mass balance equation of component i.

$$\frac{dm_i(t)}{dt} = \dot{m}_{\text{in},i}(t) - \dot{m}_{\text{out},i}(t) \tag{4.3.3}$$

with all rate terms expressed in mass per unit time (i.e., kg/s).

The rate terms can be expressed in several alternative and equivalent ways. If the total molar flowrate $\dot{N}$ of a stream is used, then the molar flowrate of component i is simply the product of the total molar flowrate and the mole fraction x_i of component i

$$\dot{n}_i = \dot{N}x_i \tag{4.3.4}$$

The corresponding form of the mole balance of component becomes

$$\frac{d(Nx_i)}{dt} = \dot{N}_{\text{in}}x_{\text{in},i} - \dot{N}_{\text{out}}x_{\text{out},i} \tag{4.3.5}$$

where N are the total molar holdup in the control volume, $\dot{N}_{\text{in}}$ is the total molar flowrate of the feed stream, $x_{\text{in},i}$ is the mole fraction of component i in the feed, $\dot{N}_{\text{out}}$ is the total molar flowrate of the product stream, and $x_{\text{out},i}$ is the mole fraction of component i in the product stream.

It is important to make a key observation at this point. In macroscopic material balances, control volumes are typically assumed to have **uniform properties** throughout. This assumption is often represented symbolically by including an **agitator** in the diagram of the control volume, as illustrated in Fig. 4.1. The agitator signifies that the **contents are continuously stirred** and well mixed, implying uniform composition, temperature, and other relevant properties within the volume. Mathematically, this means that all variables are either constant in time (steady state) or vary only with time and not with spatial coordinates inside the control volume. Therefore, whenever you assume uniform properties in a control volume, it is good practice to include an agitator in your schematic to clearly indicate this assumption. However, in the case of gas mixtures, the use of an agitator symbol is generally unnecessary. Gases tend to mix spontaneously and rapidly due to their high molecular mobility, so uniformity can typically be assumed without mechanical mixing.

The practical implication of the uniformity of the properties within the control volume is that a product stream has the same properties as the contents of the control volume. In our case this is equivalent to stating that $x_{\text{out},i} = x_i$ and Eq. (4.3.5) simplifies to

$$\frac{d(Nx_i)}{dt} = \dot{N}_{\text{in}} x_{\text{in},i} - \dot{N}_{\text{out}} x_i \tag{4.3.6}$$

If the total volumetric flowrate $\dot{V}$ of a stream is used, then the molar flowrate of component i is simply the product of the total volumetric flowrate (m^3/s) times the molar concentration C_i of component i (mol/m^3)

$$\dot{n}_i = \dot{V} C_i \tag{4.3.7}$$

and the component mole balance becomes

$$\frac{d(VC_i)}{dt} = \dot{V}_{\text{in}} C_{\text{in},i} - \dot{V}_{\text{out}} C_i \tag{4.3.8}$$

The steady state is defined as an ideal state where all variables obtain constant values that do not depend on time, that is, all derivatives with respect to time become zero. The component mole balance takes, at steady state, the following alternative forms.

$$\dot{n}_{\text{in},i} = \dot{n}_{\text{out},i} \tag{4.3.9}$$

$$\dot{N}_{\text{in}} x_{\text{in},i} = \dot{N}_{\text{out}} x_i \tag{4.3.10}$$

$$\dot{V}_{\text{in}} C_{\text{in},i} = \dot{V}_{\text{out}} C_i \tag{4.3.11}$$

It is reminded that in these equations the terms on the left-hand side and right-hand side **must be repeated for each feed stream** (left-hand side) **or product stream** (right-hand side). This is shown in Fig. 4.2. We avoid using the corresponding

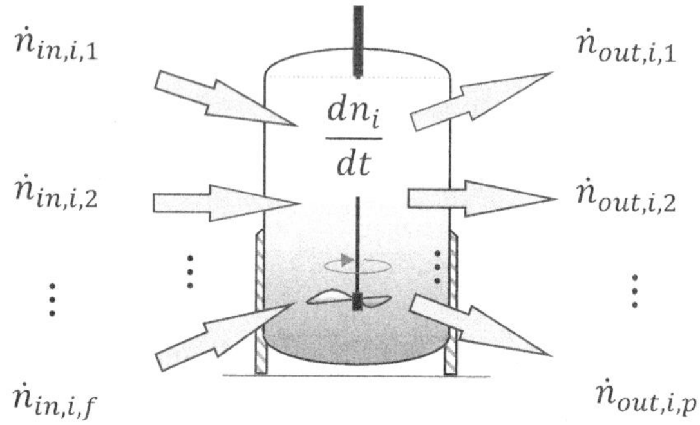

Figure 4.2 Mole balance of a chemical component: there are usually multiple feed and/or product streams.

mathematical equation which becomes messy as we must include all feed and/or product streams.

Another important note is that **we can write as many independent component mole balances as there are components.** Let's add all component mole balances. Starting from Eq. (4.3.6) we have for the case of NC components

$$\frac{d(Nx_1)}{dt} = \dot{N}_{\text{in}}x_{\text{in},1} - \dot{N}_{\text{out}}x_1$$

$$\frac{d(Nx_2)}{dt} = \dot{N}_{\text{in}}x_{\text{in},2} - \dot{N}_{\text{out}}x_2$$

$$\frac{d(Nx_{NC})}{dt} = \dot{N}_{\text{in}}x_{\text{in},NC} - \dot{N}_{\text{out}}x_{NC}$$

We add them

$$\frac{d[N(x_1 + x_2 + \ldots + x_{NC})]}{dt} = \dot{N}_{\text{in}}(x_{\text{in},1} + x_{\text{in},2} + \ldots + x_{\text{in},NC}) - \dot{N}_{\text{out}}(x_1 + x_2 + \ldots + x_{NC})$$

Note that from the definition of mole fraction

$$\begin{aligned} x_1 + x_2 + \ldots + x_{NC} &= \frac{n_1}{n_1 + n_2 + \ldots + n_{NC}} + \frac{n_2}{n_1 + n_2 + \ldots + n_{NC}} \\ &\quad + \ldots + \frac{n_{NC}}{n_1 + n_2 + \ldots + n_{NC}} \\ &= \frac{n_1 + n_2 + \ldots + n_{NC}}{n_1 + n_2 + \ldots + n_{NC}} = 1 \end{aligned}$$

The sum of all component mole balances becomes

$$\frac{dN}{dt} = \dot{N}_{\text{in}} - \dot{N}_{\text{out}} \tag{4.3.12}$$

which is the overall mole balance. Mathematically speaking, we can say that **the component material balances and the overall material balance are $NC + 1$ linearly dependent equations.** Any subset of NC equations are linearly independent.

Example 4.1. Mixing liquids in a tank

The cylindrical tank shown in Fig. 4.3 has a cross-sectional area of $S = 1\ \text{m}^2$ and a height of $H = 2\ \text{m}$, and is initially filled with pure water to a height of $h = 1\ \text{m}$. Methanol is then fed to the tank for 10 min with a molar flowrate of $n_{\text{in},M} = 0.01\ \text{kmol/s}$. Answer the following questions:

a. What is the initial volume and molar holdup of water in the tank?

b. Write the independent mole balances of the system for the period that methanol is fed to the tank. Solve the model to obtain the evolution of the molar holdup and the methanol mole fraction in the tank. Calculate their values at $t = 10$ min.

c. What is the variation of the volume of the liquid in the tank? What is the volume at $t = 10$ min?

It is given that the density of water (W - molar mass 18.02) is $\rho_W = 998\ \text{kg/m}^3$ and the density of methanol (M – molar mass 32.04) is $\rho_M = 795\ \text{kg/m}^3$.

Solution

a. The initial volume of the tank is

$$V(0) = Sh(0) = 1\ \text{m}^2 \times 1\ \text{m} = 1\ \text{m}^3$$

and the initial molar holdup

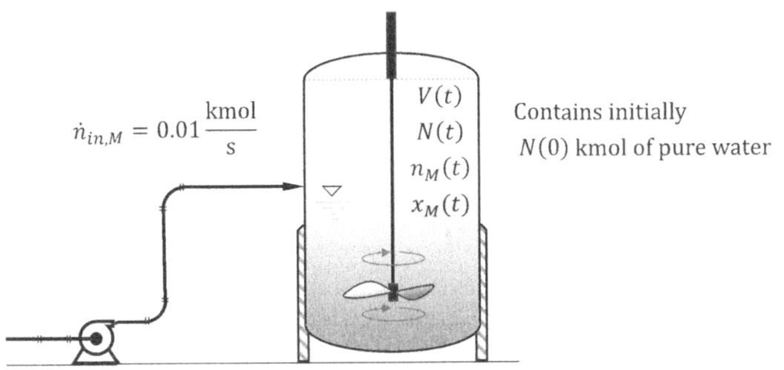

Figure 4.3 The mixing tank studied in Example 4.1.

$$N(0) = V(0)\rho_W = 1\ \text{m}^3 \times 998\ \frac{\text{kg}}{\text{m}^3} \times \frac{1\ \text{kmol}}{18.02\ \text{kg}} = 55.383\ \text{kmol}$$

b. The overall mole balance is the following

$$\frac{dN(t)}{dt} = \dot{n}_{\text{in},M}, \quad 0 < t \le 600\ \text{s} \tag{a}$$

The mole balance of water (W) can be written as

$$\frac{dn_W(t)}{dt} = 0, \quad \forall t > 0 \tag{b}$$

with $n_W(0) = N(0)$. The mole balance of methanol (M) can be written as

$$\frac{dn_M(t)}{dt} = \dot{n}_{\text{in},M}, \quad 0 < t \le 600\ \text{s} \tag{c}$$

with $n_M(0) = 0$.

Note that if we add (b) and (c) we obtain

$$\frac{dn_W(t)}{dt} + \frac{dn_M(t)}{dt} = \dot{n}_{\text{in},M}$$

or

$$\frac{d(n_W(t) + n_M(t))}{dt} = \dot{n}_{\text{in},M}$$

but $n_W(t) + n_M(t) = N(t)$, and the overall balance is obtained (Equation (a)). We select to use the overall mole balance and the mole balance of water. We then have

$$\frac{dn_W(t)}{dt} = 0 \quad \Rightarrow n_W(t) = n_W(0) = N(0)$$

$$\frac{dN(t)}{dt} = \dot{n}_{\text{in},M} \quad \Rightarrow N(t) = N(0) + \dot{n}_{\text{in},M}\, t$$

The mole fraction of water at $t = 10$ min is

$$x_W = \frac{n_W(t)}{n_W(t) + n_M(t)} = \frac{n_W(t)}{N(t)} = \frac{N(0)}{N(0) + \dot{n}_{\text{in},M} t}$$

$$= \frac{55.383}{55.383 + 0.01 \times 600} = 0.902$$

The mole fraction of methanol is

$$x_M = 1 - x_W = 1 - \frac{N(0)}{N(0) + \dot{n}_{\text{in},M}t} = \frac{\dot{n}_{\text{in},M}t}{N(0) + \dot{n}_{\text{in},M}t} = 0.098$$

c. To determine the evolution of the volume of the liquid in the tank we use the following equation

$$\frac{1}{\rho} = \frac{x_W}{\rho_W} + \frac{x_M}{\rho_M} = \left(\frac{N(0)}{N(0) + \dot{n}_{\text{in},M}t}\right)\frac{1}{\rho_W} + \left(\frac{\dot{n}_{\text{in},M}t}{N(0) + \dot{n}_{\text{in},M}t}\right)\frac{1}{\rho_M}$$
$$= \frac{1}{(N(0) + \dot{n}_{\text{in},M}t)}\left(\frac{N(0)}{\rho_W} + \frac{\dot{n}_{\text{in},M}t}{\rho_M}\right)$$

To calculate the volume of the liquid mixture we have

$$V(t) = \frac{N(t)}{\rho(t)} = \frac{N(0)}{\rho_W} + \frac{\dot{n}_{\text{in},M}t}{\rho_M} = \frac{55.383}{998/18.02} + \frac{6}{795/32.04}$$
$$= 1.2418 \text{ m}^3$$

Example 4.2. Sweep-through purging

Inerting refers to the introduction of an inert (non-combustible) gas such as nitrogen into a closed system (e.g., a container or a process vessel) to make a flammable atmosphere oxygen deficient and nonignitable. The cylindrical storage vessel of Fig. 4.4 has a volume of $V = 100$ m^3 and contains pure air (21% oxygen) at $T = 288.15$ K and 10^5 Pa (1 bar). The vessel is to be used to store a flammable liquid. Before filling the vessel with the flammable liquid, we need to flush the air using nitrogen, so that the vessel is inerted. At the particular case it is desirable to reduce the % oxygen in the vessel to 0.01% by volume. If the volumetric flowrate of nitrogen that can be achieved is $\dot{V}_{\text{in}} = 0.1$ m^3/s at the conditions of the vessel, how much time is needed and how much nitrogen will be used in kg ?

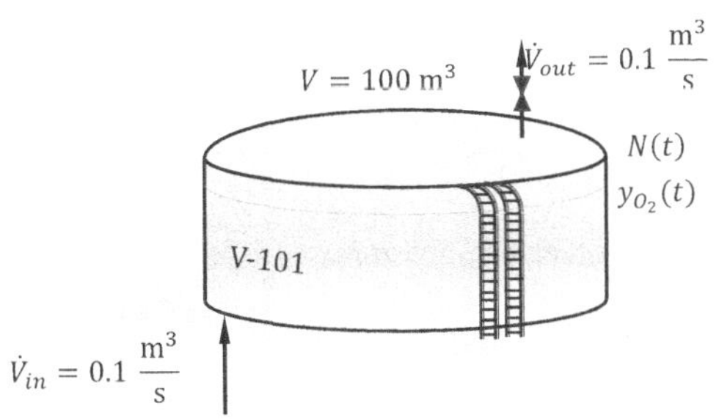

Figure 4.4 Sweep-through purging for inerting a storage vessel of flammable liquid.

Solution.
We will use Eq. (4.3.8) to solve the problem

$$\frac{d(VC_i)}{dt} = \dot{V}_{in}C_{in,i} - \dot{V}_{out}C_i$$

It is known that the volume of the storage vessel is $V = 100\ m^3$.
The incoming volumetric flowrate equals the outgoing volumetric flowrate and $\dot{V}_{in} = \dot{V}_{out} = \dot{V} = 0.1\ m^3/s$.
We write the mole balance for nitrogen to obtain

$$\frac{d(VC_{N_2})}{dt} = \dot{V}_{in}C_{in,N_2} - \dot{V}_{out}C_{N_2}$$

The volume of the vessel is constant and the volumetric flowrates of the incoming and outgoing gas are equal. We therefore have that

$$\frac{dC_{N_2}}{(C_{in,N_2} - C_{N_2})} = \left(\frac{\dot{V}}{V}\right)dt$$

or

$$d\ln(C_{in,N_2} - C_{N_2}) = -\left(\frac{\dot{V}}{V}\right)dt$$

We integrate from the initial time (t_0) to the final time (t_f) to obtain

$$\int_{C_{N_2}(t_0)}^{C_{N_2}(t_f)} d\ln(C_{in,N_2} - C_{N_2}) = -\left(\frac{\dot{V}}{V}\right)\int_{t_0}^{t_f} dt$$

and finally

$$\frac{C_{in,N_2} - C_{N_2}(t_f)}{C_{in,N_2} - C_{N_2}(t_0)} = \exp\left[-\left(\frac{\dot{V}}{V}\right)(t_f - t_0)\right]$$

The concentration of nitrogen in the feed stream is

$$C_{in,N_2} = \frac{\dot{n}_{N_2}}{\dot{V}} = \frac{P}{RT} = \frac{10^5\ \text{Pa}}{8.314\ \frac{\text{m}^3\ \text{Pa}}{\text{mol K}} \times 288.15\ \text{K}} = 41.742\ \frac{\text{mol}}{\text{m}^3}$$

The initial concentration of nitrogen in the vessel is (79% N_2)

$$C_{N_2}(0) = \frac{n_{N_2}(0)}{V} = \frac{n_{N_2}(0)}{n(0)}\frac{P}{RT} = y_{N_2}(0)C_{\text{in},N_2}$$

Our aim is to increase the nitrogen in the vessel to 99.99% by volume which, for ideal gases, is equivalent to mole percent. The desired final concentration of nitrogen is

$$C_{N_2}(t_f) = \frac{n_{N_2}(t_f)}{V} = y_{N_2}(t_f)C_{\text{in},N_2}$$

We replace into the solution to obtain

$$\frac{1 - y_{N_2}(t_f)}{1 - y_{N_2}(0)} = \exp\left[-\left(\frac{\dot{V}}{V}\right)(t_f - t_0)\right]$$

We solve for the time (note that $t_0 = 0$)

$$t_f = -\left(\frac{V}{\dot{V}}\right)\ln\left(\frac{1 - y_{N_2}(t_f)}{1 - y_{N_2}(0)}\right) = -\left(\frac{100}{0.1}\right)\ln\left(\frac{1 - 0.9999}{1 - 0.7900}\right)\text{ s}$$
$$= 7649.7\text{ s} = 127.5\text{ min} = 2.125\text{ h}$$

The amount of nitrogen that has been fed to the vessel is

$$m_{N_2} = m_{m,N_2}\dot{V}_{\text{in}}C_{\text{in},N_2}t_f = 28.02\frac{\text{g}}{\text{mol}} \times 0.1\frac{\text{m}^3}{\text{s}} \times 41.742\frac{\text{mol}}{\text{m}^3}$$
$$\times 7649.7\text{ s}$$
$$= 894\,717.2\text{ g N}_2 = 0.895\text{ t N}_2$$

Example 4.3. Adding a drop of ink in a well-mixed tank

A drop of ink (tracer) of mass equal to $m_{\text{ink}} = 0.1$ kg is added instantaneously in a 1 m^3, constant-volume, well-mixed tank. The tank has a continuous feed of pure water with a volumetric flowrate of $\dot{V}_{\text{in}} = 0.1\ m^3/\text{min}$. Assume that the density of water is constant and equal to $\rho_w = 1000\ \text{kg}/m^3$. Determine the evolution of the concentration of ink in the tank. Find also the time necessary for 99% of the ink to be flushed out of the system. A schematic representation of the setup is provided in Fig. 4.5.

It is emphasized that the ink is added instantaneously and the water in the tank is well mixed. This is equivalent to saying that the concentration of the ink obtains instantaneously a nonzero initial value which is uniform inside the tank.

Solution.

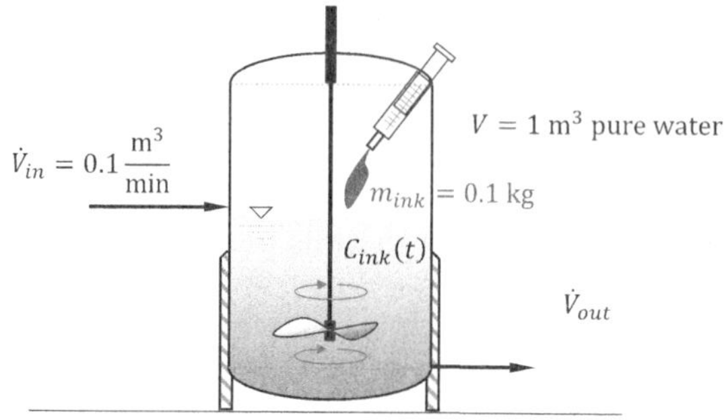

Figure 4.5 A droplet of ink is instantaneously added to a continuous, well-stirred tank of constant volume.

We can write two independent material balances, one for water and one for the ink.

Water material balance

$$\frac{dM(t)}{dt} = \dot{M}_{\text{in}} - \dot{M}_{\text{out}}$$

or

$$\rho_w \frac{dV}{dt} = \rho_w \dot{V}_{\text{in}} - \rho_w \dot{V}_{\text{out}}$$

The volume is constant and the time derivative of the volume is, therefore, zero. We obtain that

$$0 = \rho_w \dot{V}_{\text{in}} - \rho_w \dot{V}_{\text{out}} \Rightarrow \dot{V}_{\text{in}} = \dot{V}_{\text{out}}$$

Material balance of the ink

$$\frac{d[VC_{\text{ink}}(t)]}{dt} = \dot{V}_{\text{in}} C_{\text{in,ink}} - \dot{V}_{\text{out}} C_{\text{ink}}(t)$$

We note that the volume is constant

$$V \frac{dC_{\text{ink}}(t)}{dt} = \dot{V}_{\text{in}} C_{\text{in,ink}} - \dot{V}_{\text{out}} C_{\text{ink}}(t)$$

The volumetric flowrate is constant

$$V \frac{dC_{\text{ink}}(t)}{dt} = \dot{V}_{\text{in}} (C_{\text{in,ink}} - C_{\text{ink}}(t))$$

The feed is pure water ($C_{\text{in,ink}} = 0$) and the water in the tank is well mixed ($C_{\text{ink}}(t) = C_{\text{out,ink}}(t)$)

$$\frac{dC_{\text{ink}}(t)}{dt} = -\left(\frac{\dot{V}_{\text{in}}}{V}\right) C_{\text{ink}}(t)$$

We integrate to obtain

$$C_{\text{ink}}(t) = C_{\text{ink}}(0)\exp\left[-\left(\frac{\dot{V}_{\text{in}}}{V}\right)t\right] = \left(\frac{m_{\text{ink}}}{V}\right)\exp\left[-\left(\frac{\dot{V}_{\text{in}}}{V}\right)t\right]$$

Which can be solved also for time to determine the time that 99% of ink has been flushed out of the tank (or 1% is still in the tank)

$$t = -\frac{V}{\dot{V}_{\text{in}}}\ln\left(\frac{VC_{\text{ink}}}{m_{\text{ink}}}\right) = -\frac{1}{0.1}\ln 0.01 = 46.05 \text{ min}$$

Example 4.4. Carbon dioxide emissions retained in the atmosphere (carbon sink)

In Fig. 4.6 the total CO_2 emissions to the atmosphere for the period 1959–2023 are shown. The evolution of the total annual CO_2 emissions can be approximated by the straight line (data obtained from the World Bank Organization and other sources)

$$\dot{n}_{CO_2}\left[\frac{\text{Tmol}}{\text{y}}\right] = 190.51 + 10.336(t - 1958)$$

In Fig. 4.7 the concentration of the CO_2 in atmosphere in ppmv is also shown (data from the NOAA/GLM NASA). The

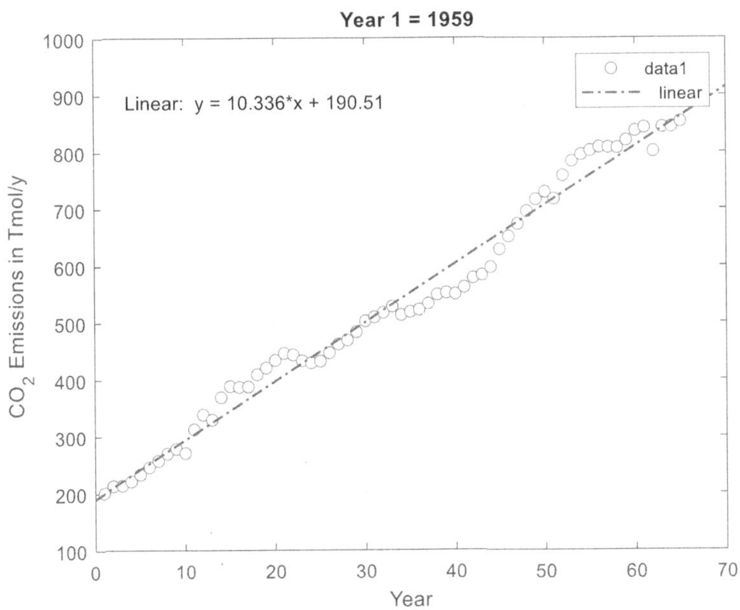

Figure 4.6 CO_2 emissions to atmospheric air for the years 1959–2023.

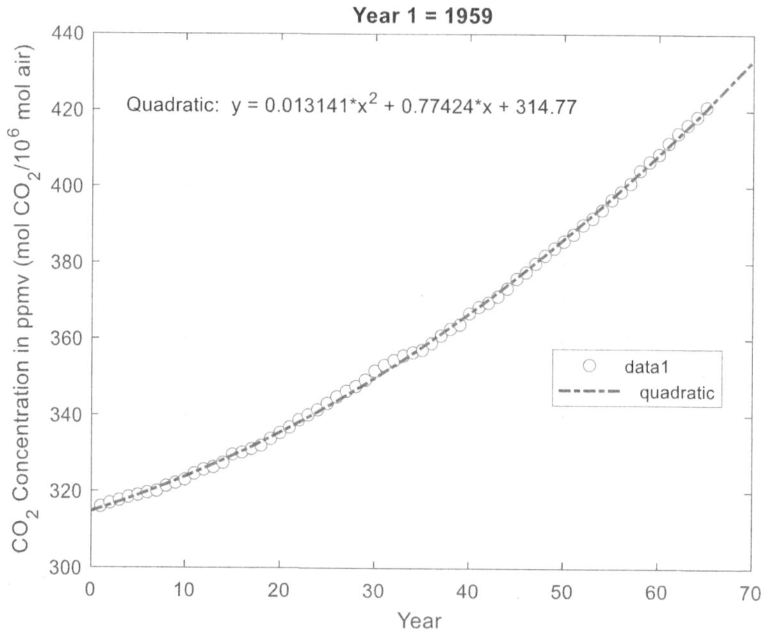

Figure 4.7 CO_2 concentration in atmospheric air for the years 1959–2023.

concentration of atmospheric CO_2 can be described by the following quadratic polynomial.

$$C_{CO_2}[\text{ppmv}] = 314.77 + 0.77424(t - 1958) + 0.013141(t - 1958)^2$$

Current CO_2 concentrations are higher than at any time over at least the past two million years.

A carbon sink is a natural or artificial carbon sequestration process that removes a greenhouse gas (an aerosol or a precursor of a greenhouse gas) from the atmosphere. A carbon sink is a type of carbon pool that has the capability to take up more carbon from the atmosphere than it releases. Globally, the two most important carbon sinks are vegetation and the ocean. Land and ocean sinks have taken up a near-constant proportion (approximately 50% per year) of CO_2 emissions from human activities over the past 6 decades.

Can you estimate the percentage of CO_2 emissions from human activities over the past 6 decades that have been taken up by carbon sinks using the data in Figs. 4.6 and 4.7?

Solution.

First, note that from the definition of ppmv (μmol per mol) it follows that the total molar amount of CO_2 in the atmosphere is given by

$$n_{CO_2} = \left(\frac{N_{air}}{10^6}\right) \times \text{ppm}, N_{air} \text{ in mol}$$

In Example 2.12 we have estimated the mass of atmospheric air (5.29×10^{21} g) and in Example 2.14 we have estimated the average molar mass of air (28.966 g/mol). The total moles of atmospheric air can be estimated

$$N_{air} = \frac{5.29 \times 10^{21}\ \text{g}}{28.966\dfrac{\text{g}}{\text{mol}}} = 1.8263 \times 10^{20} \text{ mol Air}$$
$$= 182.63 \times 10^6 \text{ Tmol Air}$$

The moles of CO_2 in the atmosphere can be calculated by multiplying the measurements of CO_2 concentrations in atmosphere in ppmv (quadratic polynomial given) by $N_{air}/10^6$.

$$n_{CO_2}(t) = \underbrace{182.63 \text{ Tmol Air}}_{\left(\frac{N_{air}}{10^6}\right)} \cdot \underbrace{\left[314.77 + 0.77424(t-1959) + 0.013141(t-1959)^2\right]}_{\text{ppmv } CO_2} \text{Tmol } CO_2$$

$$= 57486.45 + 141.4(t-1959) + 2.4(t-1959)^2 \text{ Tmol } CO_2$$

From Fig. 4.6 we note that the rate at which CO_2 from human-related activities crosses the boundary of the atmosphere is

$$\dot{n}_{in,CO_2} = 190.51 + 10.336(t-1959)\ \frac{\text{Tmol } CO_2}{\text{y}}$$

The mole balance of the CO_2 in the atmosphere can be written as

$$\frac{dn_{CO_2}}{dt} = \lambda \dot{n}_{in,CO_2}$$

where λ is the fraction of the CO_2 emissions that are retained in the atmosphere (and 1-λ is the fraction taken up by the carbon sinks). We integrate the differential equation between an initial time t_0 and a final time t_f to obtain

$$\int_{n_{co_2}(t_0)}^{n_{co_2}(t_f)} dn_{CO_2} = \lambda \int_{t_0}^{t_f} \dot{n}_{in,CO_2} dt$$

or

$$n_{CO_2}(t_f) - n_{CO_2}(t_0) = \lambda \int_{t_0}^{t_f} (190.51 + 10.336(t - 1959))dt$$

$$= \lambda\left[190.51(t_f - t_0) + 5.168(t_f - t_0)^2\right]$$

We set $t_0 = 1959$ and we obtain

$$n_{CO_2}(t_f) = 57486.45 + \lambda\left[190.51(t_f - t_0) + 5.168(t_f - t_0)^2\right], \text{Tmol CO}_2$$

which has to be compared with the fitting to the actual data

$$n_{CO_2}(t) = 57486.45 + 141.4(t - 1959) + 2.4(t - 1959)^2, \text{Tmol CO}_2$$

We could equate the two equations and solve for λ, but this will not give constant value for λ but a value that varies over the years. We can follow a different approach. We can select a constant value for λ, plot the two series of data and compare them. In this way we can find a value of λ that brings the two serial data close together. After some trial and error, we find that $\lambda = 0.56$ gives a reasonably good agreement.

```
clear all
t=linspace(1,65,65); % 1959 to 2023
%% plot the fitting to actual data
plot(t+1958,57486.45+141.4*t+2.4*t.^2,'bo')
hold on
%% select fraction of CO_2 retained in the atmosphere
lamda=0.56;
plot(t+1958,57486.45+lamda*(190.51*t+5.168*t.^2),'r--')
grid on
xlabel('Year')
ylabel('CO_{2} in the atmosphere (Tmol)')
axis([1955 2025 5e4 8e4])
```

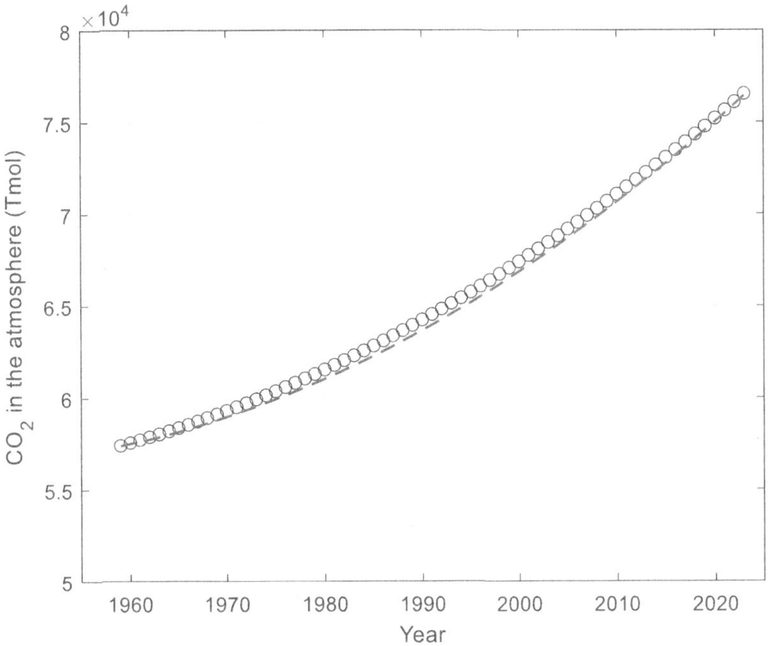

Figure 4.8 Total CO_2 in the atmosphere can be estimated with reasonable accuracy for the years 1959–2023 if a value of $\lambda = 0.56$ (56%) is selected to account for the CO_2 emitted that is taken up by the atmosphere. 44% ($1-\lambda = 0.44$) of the CO_2 emitted is taken up by the major carbon sinks.

The results are presented in Fig. 4.8. It is thus estimated that the major carbon sinks absorb 44% ($1-\lambda = 0.44$) of the total CO_2 emissions.

4.4 The component balance equations with chemical reactions

When a chemical reaction is taking place within the control volume, then Eq. (4.2.2) is applied to the system

$$\begin{bmatrix} \text{rate of} \\ \text{accumulation} \\ \text{of component i} \end{bmatrix} = \begin{bmatrix} \text{rate of} \\ \text{inflow} \\ \text{of component i} \end{bmatrix} - \begin{bmatrix} \text{rate of} \\ \text{outflow} \\ \text{of component i} \end{bmatrix} + \begin{bmatrix} \text{rate of} \\ \text{generation} \\ \text{of component i} \end{bmatrix} - \begin{bmatrix} \text{rate of} \\ \text{consumption} \\ \text{of component i} \end{bmatrix}$$

Each term has units of mole per unit time (mol/s). Before moving forward, we need to review the equations commonly used to express the generation/consumption terms.

4.4.1 Basic concepts, stoichiometric numbers, and reaction progress

Consider the following general reaction

$$\rho_1 R_1 + \rho_2 R_2 + \ldots + \rho_r R_r \leftrightarrows \pi_1 P_1 + \pi_2 P_2 + \ldots + \pi_p P_p$$

where R_i is the i-th reactant, P_j is the j-th product, ρ_i is the stoichiometric coefficient of the i-th reactant and π_j is the stoichiometric coefficient of the j-th product. All stoichiometric coefficients are positive, and this is unfortunate. We define the **stoichiometric numbers** σ_k that are equal to the stoichiometric coefficients in absolute values but are negative for the reactants and positive for the products. If we consider the reaction

$$CO_2 + 4H_2 \leftrightarrows CH_4 + 2H_2O$$

then $\sigma_{CO_2} = -1$, $\sigma_{H_2} = -4$, $\sigma_{CH_4} = +1$ and $\sigma_{H_2O} = +2$. Consider a closed system (there are no streams crossing the boundary) where a single chemical reaction may occur and, initially, there are $n_i(0)$ moles of species i present in the system. At some later time, there are $n_i(t)$ moles of the same species present. We define the **molar extent of reaction** ξ as

$$\xi(t) = \frac{n_i(t) - n_i(0)}{\sigma_i} \tag{4.4.1}$$

Due to the law of definite proportions, this equation is valid for all species. Note that the molar extent of reaction varies with time and that it is an extensive variable that is measured in moles. It is a measure of the progress of the reaction that is not tied to any specific species. We may also write Eq. (4.4.1) in a very useful way

$$n_i(t) = n_i(0) + \sigma_i \xi(t) \tag{4.4.2}$$

(do not forget that we refer to a closed system). If component i participates in many reactions, then

$$n_i(t) = n_i(0) + (\sigma_{i,1}\xi_1 + \sigma_{i,2}\xi_2 + \ldots + \sigma_{i,NR}\xi_{NR}) \tag{4.4.3}$$

where NR denotes the **n**umber of **r**eactions taking place in the control volume.

The main drawback of the extent of reaction is that it is an extensive variable. We therefore define an intensive variable, called the **fraction conversion** X, as follows

$$X(t) = \frac{\text{mole reacted}}{\text{mole fed}} = \frac{n_i(0) - n_i(t)}{n_i(0)} \tag{4.4.4}$$

Although the fraction conversion may be defined for any species, it is wise to define the fraction conversion for a reference reactant that is the **limiting reactant** (the reactant that will be consumed first as we increase the extent of reaction). If we define the fraction conversion based on the limiting reactant, then it is always limited to lie between zero and unity.

4.4.2 Reaction rate expressions and phenomenological descriptions

To define the **reaction rate**, we consider a closed, constant-temperature- and pressure-homogeneous system in which a single chemical reaction is taking place. In such a system the rate of the chemical reaction ($\dot{R}'''$) is defined as

$$\dot{R}'''(t) = \left(\frac{1}{V(t)}\right)\frac{d\xi(t)}{dt} \tag{4.4.5}$$

where $V(t)$ is the (possibly time-varying) volume of the reacting mixture. Note that the reaction rate is an intensive variable that has units of moles per unit time per unit volume (denoted by the superscript $'''$). If the extent of reaction is expressed using Eq. (4.4.1) with reference to the i-th component, then Eq. (4.4.5) can be written as

$$\dot{R}'''(t) = \frac{1}{\sigma_i}\left(\frac{1}{V(t)}\right)\frac{dn_i(t)}{dt} \tag{4.4.6}$$

It then follows that

$$\dot{R}_i'''(t) = \left(\frac{1}{V(t)}\right)\frac{dn_i(t)}{dt} = \sigma_i\dot{R}'''(t) \tag{4.4.7}$$

If component i participates in many reactions, then Eq. (4.4.7) must be extended to incorporate one term for each potential reaction.

We return now to Eq. (4.4.6) and note that if the concentration of component i is used then, as $C_i(t) = n_i(t)/V(t)$, it follows that

$$\dot{R}_i'''(t) = \left(\frac{1}{V(t)}\right)\frac{d[V(t)C_i(t)]}{dt} = \frac{dC_i(t)}{dt} + C_i(t)\left(\frac{1}{V(t)}\right)\frac{dV(t)}{dt} \tag{4.4.8}$$

Note that only when $V(t) =$ constant it follows that

$$\dot{R}_i'''(t) = \frac{dC_i(t)}{dt} \tag{4.4.9}$$

Eq. (4.4.9) must be used with extreme caution.

The reaction rate may depend on the system

- temperature,
- pressure,
- composition,
- properties of a catalyst that may be used, and
- the various physical transport processes (type of flow, mixing, heat, and mass transfer).

These variables change from location to location within a reactor, irrespective of our efforts to establish uniform properties, and the reaction rate needs to be evaluated by integrating over the reactor volume. Although this is, in general, possible, in this text we will limit ourselves to situations where physical rate processes are unimportant and all properties are uniform in the control volume and focus our concern on the main variables such as temperature, pressure, composition, and the presence or absence of catalysts in the system.

In reaction rate studies the aim is to develop a phenomenological description of a reacting system in terms of a reasonable number of empirical constants by trying to isolate the effects of the different variables and to investigate each one independently. Several observations can be made about the variables that influence reaction rates. It is generally assumed that the rate of a chemical reaction depends on the temperature, pressure, and composition of the system under investigation and the general phenomenological mathematical expression can be written as

$$\dot{R}'''(t) = k(T)f(C_i) \tag{4.4.10}$$

The coefficient k, called the **reaction rate constant**, depends on absolute temperature only, and it does not depend on the composition of the system. The dependence on the absolute temperature T is approximated through the Arrhenius equation

$$k(T) = k_0 \exp\left(-\frac{E}{RT}\right) \tag{4.4.11}$$

where E is the **activation energy** of the reaction, R is the ideal gas constant, and k_0 is the preexponential factor, called the **frequency factor**.

The second term of the phenomenological reaction rate equation (Eq. (4.4.10)) can generally be written in the form

$$f(C_i) = C_{R_1}^{\nu_1} C_{R_2}^{\nu_2} \ldots C_{R_r}^{\nu_r} C_{P_1}^{\mu_1} C_{P_2}^{\mu_2} \ldots C_{P_p}^{\mu_p} \tag{4.4.12}$$

where ν_i and μ_j are the order of the reaction with respect to the i-th reactant or j-th product, correspondingly. These can be small positive or negative integers or fractions, or zero. It must be emphasized that, in general, the individual orders of the reaction are not related to the corresponding stoichiometric coefficients, but they must be determined experimentally.

For reversible reactions the most used phenomenological expression is the following

$$\begin{aligned}\dot{R}'''(t) &= k_f(T) C_{R_1}^{\nu_1} C_{R_2}^{\nu_2} \ldots C_{R_r}^{\nu_r} - k_b(T) C_{P_1}^{\mu_1} C_{P_2}^{\mu_2} \ldots C_{P_p}^{\mu_p} \\ &= k_f(T)\left(C_{R_1}^{\nu_1} C_{R_2}^{\nu_2} \ldots C_{R_r}^{\nu_r} - \frac{C_{P_1}^{\mu_1} C_{P_2}^{\mu_2} \ldots C_{P_p}^{\mu_p}}{K_e(T)} \right)\end{aligned} \tag{4.4.13}$$

where $k_f(T)$ is the reaction rate constant for the forward reaction, $k_b(T)$ is the reaction rate constant for the reverse reaction, and $K_e(T)$ is the equilibrium constant. When a reversible reaction achieves equilibrium then the reaction becomes zero. This is equivalent to the term in the parenthesis on the right-hand side of Eq. (4.4.13) being equal to zero, or

$$K_e(T) = \frac{C_{P_1}^{\mu_1} C_{P_2}^{\mu_2} \ldots C_{P_p}^{\mu_p}}{C_{R_1}^{\nu_1} C_{R_2}^{\nu_2} \ldots C_{R_r}^{\nu_r}} \tag{4.4.14}$$

4.4.3 The component mole balance equations for reacting systems

Using the phenomenological reaction rate equations, Eq. (4.3.2) can be extended to account for the generation/consumption of component i due to chemical reaction(s)

$$\frac{dn_i(t)}{dt} = \dot{n}_{\text{in},i}(t) - \dot{n}_{\text{out},i}(t) + V(t)\sigma_i \dot{R}'''(t) \tag{4.4.15}$$

It is emphasized one more time that the terms on the right-hand side of Eq. (4.4.15) need to be repeated for all incoming or outgoing streams and for all reactions. It is also reminded that alternative expressions can be developed based on the following

$$\dot{n}_i = \dot{N}x_i = \dot{V}C_i \tag{4.4.16}$$

$$n_i = Nx_i = VC_i \tag{4.4.17}$$

Adding all component mole balances together we obtain

$$\frac{d(n_1 + n_2 + \ldots + n_{NC})}{dt} = (\dot{n}_{\text{in},1} + \dot{n}_{\text{in},2} + \ldots + \dot{n}_{\text{in},NC}) - (\dot{n}_{\text{out},1} + \dot{n}_{\text{out},2} + \ldots + \dot{n}_{\text{out},NC}) + V(t)(\sigma_1 + \sigma_2 + \ldots + \sigma_{NC})\dot{R}'''(t)$$

or

$$\frac{dN(t)}{dt} = \dot{N}_{\text{in}}(t) - \dot{N}_{\text{out}}(t) + V(t)\Sigma\dot{R}'''(t) \tag{4.4.18}$$

where

$$\Sigma = \sigma_1 + \sigma_2 + \ldots + \sigma_{NC} \tag{4.4.19}$$

Again, there are NC component mole balances and one total mole balance, but only NC are independent. You are free to select any combination of NC balances that facilitates solution.

Example 4.6. Zero-order reaction in a well-mixed batch reactor

In the constant volume V and constant temperature T, well-mixed, batch reactor of Fig. 4.9, the reaction R→P ("R" for reactant and "P" for product) takes place. The initial concentration of the reactant is $C_R(0)$. If the reaction is irreversible and the reaction rate is zeroth order with respect to the reactant, find what is the time evolution of the concentration of the reactant $C(t)$.

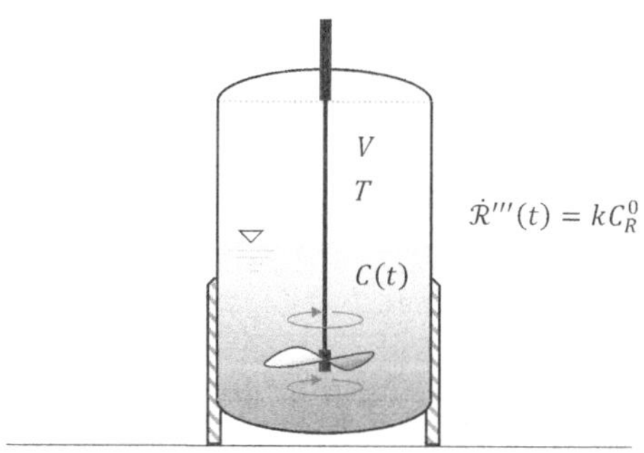

Figure 4.9 A zero-order reaction in a batch reactor.

Solution.
The component mole balance for the reactant is

$$\frac{d(VC_R(t))}{dt} = -V\dot{R}'''(t)$$

The volume is constant and the reaction rate is $\dot{R}'''(t) = kC_R^0 = k$

$$\frac{dC_R(t)}{dt} = -k$$

We integrate the last equation to obtain

$$C_R(t) = C_R(0) - kt$$

We note that the concentration of the reactant decreases linearly with time, and if we have experimental data of a zeroth-order reaction then $C_R(t)$ is a linear function of time t.

Example 4.7. First-order reaction in a well-mixed batch reactor
Repeat Example 4.6 if the reaction is irreversible and the reaction rate is first order with respect to the reactant ($\dot{R}'''(t) = kC_R(t)$), find what is the time evolution of the concentration of the reactant $C(t)$.
Solution.
The component mole balance for the reactant is

$$\frac{d(VC_R(t))}{dt} = -V\dot{R}'''(t)$$

The volume is constant and the reaction rate is $\dot{R}'''(t) = kC_R(t)$

$$\frac{dC_R(t)}{dt} = -kC_R(t)$$

We integrate the last equation to obtain

$$\ln C_R(t) = \ln C_R(t) - kt$$

or

$$\frac{C_R(t)}{C_R(0)} = \exp(-kt)$$

We note that the logarithm of the concentration of the reactant decreases linearly with time, and if we have experimental data of a first-order reaction then $\ln C_R(t)$ is a linear function of time t.

Example 4.8. Second-order reaction in a well-mixed batch reactor

Repeat Example 4.6 if the reaction is irreversible and the reaction rate is second order with respect to the reactant ($\dot{R}'''(t) = kC_R(t)^2$), find what is the time evolution of the concentration of the reactant $C(t)$.

Solution.

The component mole balance for the reactant is

$$\frac{d(VC_R(t))}{dt} = -V\dot{\mathscr{R}}'''(t)$$

The volume is constant and the reaction rate is $\dot{R}'''(t) = kC_R(t)^2$

$$\frac{dC_R(t)}{dt} = -kC_R(t)^2$$

We integrate the last equation to obtain

$$\frac{dC_R(t)}{C_R(t)^2} = -kdt$$

or

$$d\left(\frac{1}{C_R(t)}\right) = kdt$$

We integrate to obtain

$$\frac{1}{C_R(t)} = \frac{1}{C_R(0)} + kt$$

We note that the inverse of the concentration of the reactant increases linearly with time, and if we have experimental data of a second-order reaction then $1/C_R(t)$ is a linear function of time t.

Example 4.9. Estimating the reaction rate constant

The reaction

$$CH_3OCH_3 \rightarrow CH_4 + CO + H_2$$

or

$$R \rightarrow P_1 + P_2 + P_3$$

is taking place in the gas phase in a batch (closed) reactor with constant volume and temperature (Hinshelwood & Askey, Proc. Roy. Soc. London, A115, p. 215, 1927). The pressure is monitored and representative experimental data are summarized in Table

Table 4.1 Experimental data for the dimethyl ether decomposition to methane, carbon dioxide, and hydrogen.

Time (s)	Pressure (kPa)
0	56.0
57	77.9
85	88.3
114	99.1
145	108.7
182	118.8
219	127.2
261	135.1
299	140.5

4.1. Initially, only dimethyl ether (CH_3OCH_3) is fed to the reactor. According to prior information the reaction rate is first order with respect to the concentration of the dimethyl ether ($\dot{R}''' = kC_R$). Prove that this is the case and estimate the reaction rate constant k.

Solution.

The component mole balance for the reactant is

$$V\frac{dC_R}{dt} = -VkC_R$$

We use the ideal gas law ($C_R = P_R/RT$) and we express the mole balance in terms of the partial pressure of the reactant

$$\frac{dP_R}{dt} = -kP_R$$

The solution is

$$P_R(t) = P(0)\exp(-kt)$$

The mole balances of the three products are

$$V\frac{dC_{P_1}}{dt} = +VkC_R$$

$$V\frac{dC_{P_2}}{dt} = +VkC_R$$

$$V\frac{dC_{P_3}}{dt} = +VkC_R$$

or

$$\frac{dP_{P_1}}{dt} = +kP_R$$

$$\frac{dP_{P_2}}{dt} = +kP_R$$

$$\frac{dP_{P_3}}{dt} = +kP_R$$

We add the 4 mole balances to obtain

$$\frac{d(P_R + P_{P_1} + P_{P_2} + P_{P_3})}{dt} = +2kP_R$$

The sum of the partial pressures of all components present in the reactor equals the total pressure $P(t)$

$$\frac{dP(t)}{dt} = +2kP(0)e^{-kt}$$

Where we have replaced the time evolution of the partial pressure of the reactant. We note that

$$dP(t) = -2P(0)de^{-kt}$$

and we integrate to obtain

$$P(t) - P(0) = -2P(0)(e^{-kt} - 1)$$

or

$$\frac{P(t) - P(0)}{3P(0) - P(0)} = 1 - e^{-kt}$$

This result can also be written as

$$\frac{3P(0) - P(t)}{3P(0) - P(0)} = e^{-kt}$$

We take the logarithm of both sides

$$\log\left(\frac{3P(0) - P(t)}{3P(0) - P(0)}\right) = -kt$$

```
>>clear all
>>t=[57 85 114 145 182 219 261 299]; % time in s
>>P=[77.9 88.3 99.1 108.7 118.8 127.2 135.1 140.5]; % pressure in kPa
>>P0=56;
>>Y=log((3*P0-P)./(3*P0-P0));
>>plot(t,Y,'o');
```

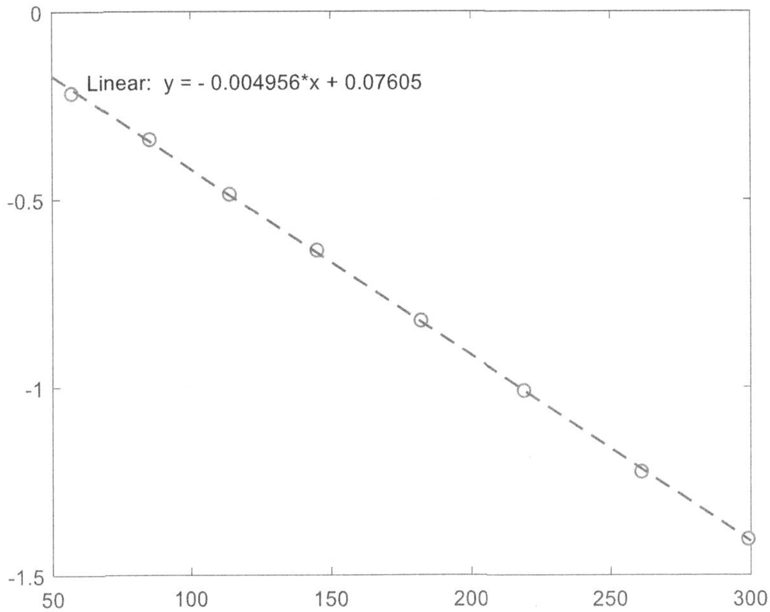

Figure 4.10 The plot of the transformed experimental data exhibits a linear behavior with respect to time that proves that the reaction is first order with respect to reactant concentration.

We use MATLAB to plot the experimental data in these new coordinates.

The results are presented in Fig. 4.10. The hypothesis is validated as the transformed data follow a linear model and the reaction rate constant is estimated to be $k = 4.96 \times 10^{-3}$ 1/s.

4.5 MATLAB and Python for solving systems of ordinary differential equations

4.5.1 Solving dynamic material balance equations in MATLAB

You'll soon recognize that we've made significant strides in developing our ability to solve complex dynamic material balance problems. MATLAB provides powerful solvers for systems of ordinary differential equations (ODEs) that are well-suited for tackling intricate and fascinating component material balances. To illustrate this, let's begin by exploring a reaction scheme involving reactions in series

$$\mathrm{R} \xrightarrow{k_1} \mathrm{P} \xrightarrow{k_2} \mathrm{B}$$

where R is the reactant, P is the targeted product, and B is a byproduct which has significantly lower value than the desirable product P. We assume the following reaction rate scheme

$$\dot{R}_1''' = k_1 C_R$$

$$\dot{R}_2''' = k_2 C_P$$

We can now write down the mole balances of the three components in a well-mixed, closed (batch) reactor at constant temperature and constant volume:

$$\frac{dC_R}{dt} = -\dot{R}_1''' = -k_1 C_R$$

$$\frac{dC_P}{dt} = +\dot{R}_1''' - \dot{R}_2''' = k_1 C_R - k_2 C_P$$

$$\frac{dC_B}{dt} = +\dot{R}_2''' = k_2 C_P$$

This is a system of ordinary differential equations (slightly more complex than a single ordinary differential equation). But this complication makes no difference when it comes to solving the systems of differential equations in MATLAB. We only need to build a proper function that returns the right-hand side of the system of differential equations. We select $C_R(0) = 1$ kmol/m^3, $k_1 = 1$ h^{− 1}, $k_2 = \frac{1}{2}$ h^{-1}. We now build the solution in MATLAB

The dynamics of the system is shown in Fig. 4.11. Note that the concentration of the desired product increases initially, then reaches a maximum and is finally consumed to produce the undesired byproduct B. The point in time at which the concentration of the desired product is maximum can be located from the simulation results or can be calculated analytically.

```
clear all
C0=[1;0;0]; %initial concentrations in kmol/m^3
t0=0;       % initial time in h
tf=7;       % final time in h
tspan=linspace(t0,tf,21);
k =[1; 1/2];% reaction rate constants in 1/h
[t,C]=ode15s(@(t,C) RxnInSeries(t,C,k),tspan,C0)
figure(1)
plot(t,C(:,1),'b-o','LineWidth',1)
hold on
plot(t,C(:,2),'b-s','LineWidth',1)
plot(t,C(:,3),'b-^','LineWidth',1)
grid on
xlabel('Time (h)')
ylabel('C (kmol/m^{3})')
text(0.4,0.72,'C_{R}(t)')
text(1.3,0.54,'C_{P}(t)')
text(3.0,0.698,'C_{B}(t)')

function dCdt=RxnInSeries(t,C,k)
% R -> P -> B, both reactions 1st order
RRate(1) = k(1)*C(1);
RRate(2) = k(2)*C(2);
dCdt(1) = -RRate(1);
dCdt(2) = +RRate(1)-RRate(2);
dCdt(3) = +RRate(2);
dCdt=dCdt(:);
end
```

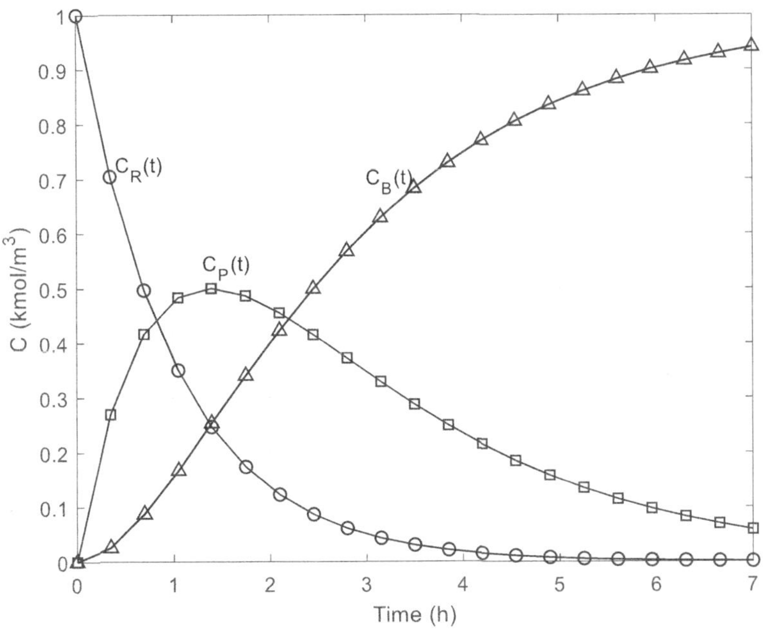

Figure 4.11 Evolution of the species concentration for the reaction scheme $R \rightarrow P \rightarrow B$ in a constant temperature, constant-volume batch reactor.

Let's now consider a fascinating example with interesting dynamics. The Lotka–Volterra equations, also known as the Lotka–Volterra predator–prey model, are a pair of first-order nonlinear differential equations, frequently used to describe the dynamics of biological systems in which two species interact, one as a predator and the other as prey.

The symbol $x(t)$ is used to denote the prey population, such as the number of rabbits per unit area. The rabbit population increases with time at a rate that is first order with respect to the number of rabbits, i.e.,

$$\frac{dx(t)}{dt} = ax(t) \tag{4.5.1}$$

If no predator is present in the area of study, then the prey population will increase in an exponential way. a is the growth rate (number of rabbits per unit area per unit time) constant (has units of inverse time). Note that the model assumes that no new populations enter or leave the area of study (a closed system). However, when predators, such as foxes, are present in the area of study with population $y(t)$ (number of foxes per unit area) then a death rate is activated (foxes eat rabbits). The death rate is first order with respect to both populations and a death rate constant β.

$$\frac{dx(t)}{dt} = ax(t) - \beta x(t)y(t) \tag{4.5.2}$$

The population dynamics of the predator is slightly different

$$\frac{dy(t)}{dt} = -\gamma y(t) + \delta x(t)y(t) \tag{4.5.3}$$

The first term on the right-hand side predicts a death rate that is analogous to the population, and if no prey is present then the population of the predator decreases exponentially. When a prey is present then a "generation" term is activated which is again first order with respect to both populations (with a growth rate constant equal to δ). Let's see what the dynamics of the two coexisting populations look like for the following parameters and initial conditions

$$x(0) = 1, a = 1, \beta = 1/2$$

$$y(0) = 2, \gamma = 3/4, \delta = 1/4$$

in appropriate units. We use MATLAB to obtain the solution.

```
clear all
x0 = 1; % initial rabbits per unit area
y0 = 2; % initial foxes per unit area
t0 = 0; % initial time
tf = 23.15;% final time
options=odeset('AbsTol',1E-6,'RelTol',1E-6)
[t,Z]=ode15s(@PredatorPrey,[t0 tf],[x0;y0],options)

plot(t,Z,'LineWidth',1.2)
grid on
xlabel('Time')
ylabel('Pupulation per unit area')
text(2,6.5,'Prey')
text(6.8,3.5,'Predator')
axis([t0 tf 0 7])

function dZdt=PredatorPrey(t,Z)
%  Lotka-Volterra or predator-prey model
alfa  = 4/4; beta  = 2/4;
gamma = 3/4; delta = 1/4;
x = Z(1); y = Z(2);
dZdt(1) =   alfa*x - beta*x*y;
dZdt(2) = -gamma*y + delta*x*y;
dZdt=dZdt(:);
end
```

The solution obtained is shown in Fig. 4.12. The dynamics is very interesting as it appears to be periodic. The prey population initially increases until there is enough food for the predator to grow. The predators dominate, causing the prey population to decrease. Decreased prey population means that there is limited food for the predators to survive and they start to decrease. This cycle is repeated indefinitely! It is also interesting that the predator–prey model just presented has been based on a work from A.J. Lotka published in 1910 on the potential of periodic solutions of autocatalytic reactions (a chemical reaction is said to be autocatalytic if one of the reaction products is also a catalyst for the same reaction). It can be shown that if reactant R is acting as a prey and is slowly solubilized (from the solid state) in a liquid solution while the autocatalytic reaction $R + P \rightarrow P + P$ is taking place (together with $P \rightarrow B$) with reaction rate kC_RC_P, then the system is described by Eqs. (4.5.2) and (4.5.3) and the potential of periodic solutions arises.

Autocatalytic reactions are well known for producing interesting dynamics. One example is shown in Fig. 4.13. The reactant R is consumed through a first-order reaction to produce an intermediate product P_1. P_1 is transformed into a second product P_2

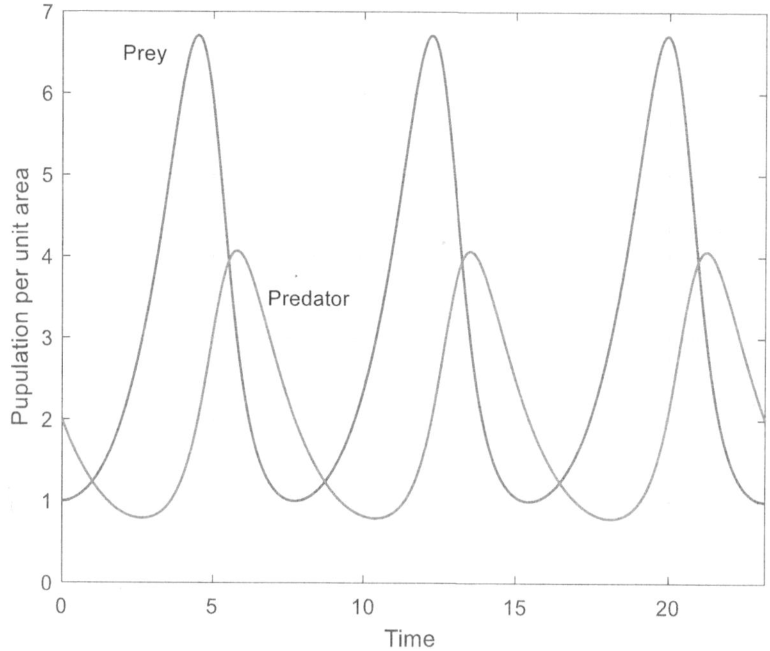

Figure 4.12 Population dynamics for rabbit and fox co-existence.

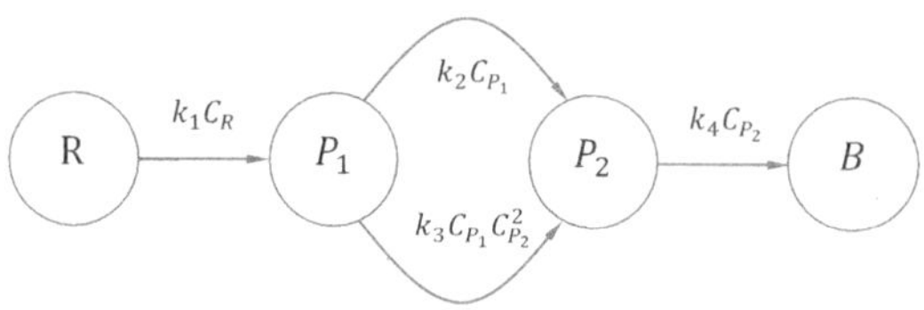

Figure 4.13 An interesting kinetics scheme involving an autocatalytic reaction.

through two parallel routes from which the first one is a usual first-order reaction while the second one is autocatalytic. P_2 is finally decomposed into a byproduct B.

You should be able to derive the following mathematical model

$$\frac{dC_R}{dt} = -k_1 C_R$$

$$\frac{dC_{P_1}}{dt} = +k_1 C_R - k_2 C_{P_1} - k_3 C_{P_1} C_{P_2}^2$$

$$\frac{dC_{P_2}}{dt} = +k_2 C_{P_1} + k_3 C_{P_1} C_{P_2}^2 - k_4 C_{P_2}$$

$$\frac{dC_B}{dt} = +k_4 C_{P_2}$$

We select the following parameters

$$C_R(0) = 500 \frac{\text{kmol}}{\text{m}^3}, C_{P_1}(0) = C_{P_2}(0) = C_B(0) = 0$$

$$k_1 = 0.002 \frac{1}{\text{s}}, k_2 = 0.08 \frac{1}{\text{s}}, k_3 = 1 \frac{\text{m}^6}{\text{kmol}^2 \cdot \text{s}}, k_4 = 1 \frac{1}{\text{s}}$$

and we then simulate the system in MATLAB.

```
clear all
C0=[500;0;0;0]; % initial concentrations in kmol/m^3
t0=0;           % initial time
tf=800;         % final time in s
options=odeset('AbsTol',1e-12,'RelTol',1E-12)
[T,Y]=ode15s(@AutoCatalytic,[t0 tf],C0,options)

figure(1)
plot(T,Y(:,2:3),'LineWidth',1)
xlabel('Time (s)')
ylabel('C_{P_{1}}, C_{P_{2}}')
grid on
axis([0 800 0 2.5])

figure(2)
plot3(Y(:,2),Y(:,3),T,'LineWidth',1.)
grid on
axis([0.3 2.4 0.15 1.6 7 700])
xlabel('C_{P_{1}}(t) (kmol/m^{3})')
ylabel('C_{P_{2}}(t) (kmol/m^{3})')
zlabel('Time (s)')

function dCdt=AutoCatalytic(t,C)
k=[0.002;0.08;1;1];
dCdt(1) = -k(1)*C(1);
dCdt(2) = +k(1)*C(1)-k(2)*C(2)-k(3)*C(2)*C(3)^2;
dCdt(3) = +k(2)*C(2)+k(3)*C(2)*C(3)^2-k(4)*C(3);
dCdt(4) = +k(4)*C(3);
dCdt = dCdt(:);
end
```

The results are shown in Fig. 4.14. The response exhibits unpredictable, almost chaotic, behavior at intermediate times. The concentration of the two intermediate products finally heads back to zero as they are consumed to produce the final product B.

By now, you should be convinced that the initial claim made in this section holds true: mastering MATLAB or Python for solving dynamic material balances significantly simplifies even the most complex problems. The real objective is not just to find solutions, but to develop the ability to formulate problems effectively and interpret the results with the goal of designing

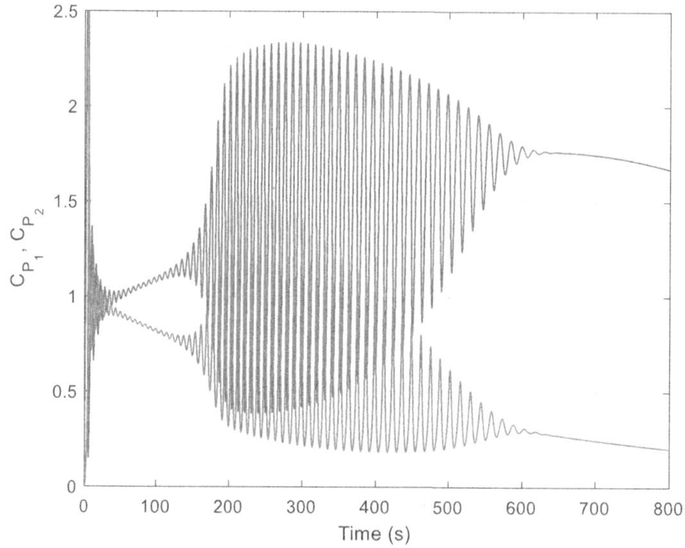

(a)

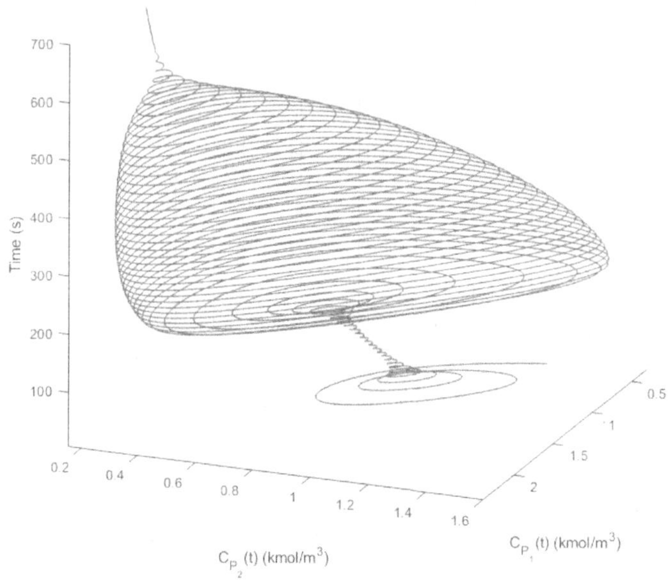

(b)

Figure 4.14 Evolution of the concentration of the two intermediate products P_1 and P_2 for the autocatalytic reaction case study.

robust, functional process systems. Let the computer handle the mathematical complexity—your role is to define the problem accurately. While many people can solve a well-defined problem, only a few can define one in the first place. These individuals are called engineers. (Take a moment to explore the origin of the word "engineer"—you might find its history quite interesting.)

4.5.2 Solving ordinary differential equations in Python

We have in Chapter 3 introduced the ODE solver available in scipy. We will consider the following reactions that take place in the continuous, stirred tank reactor pictured in Fig. 4.15

$$R + P \xrightarrow{k_1} 2P$$

$$P \xrightarrow{k_2} B$$

where the reaction rate for the first reaction is

$$\dot{R}_1''' = k_1 C_R C_P^2, \; k_1 = 10^3 \, \frac{1}{\mathrm{s}} \left(\frac{\mathrm{L}}{\mathrm{mol}}\right)^2$$

and for the second reaction

$$\dot{R}_2''' = k_2 C_P, \; k_2 = 0.2 \, \frac{1}{\mathrm{s}}$$

The volumetric flowrate of the feed is 2 L/s and the reactor volume is constant and equal to 300 L.

We first write the material balance for the reactant R:

$$V \frac{dC_R(t)}{dt} = \dot{V}_{\mathrm{in}} C_{\mathrm{in},R} - \dot{V}_{\mathrm{out}} C_R(t) - V k_1 C_R(t) C_P(t)^2$$

or

$$\frac{dC_R(t)}{dt} = \frac{\dot{V}_{\mathrm{in}}}{V} (C_{\mathrm{in},R} - C_R(t)) - k_1 C_R(t) C_P(t)^2 \tag{4.5.4}$$

In a similar way for component P we obtain

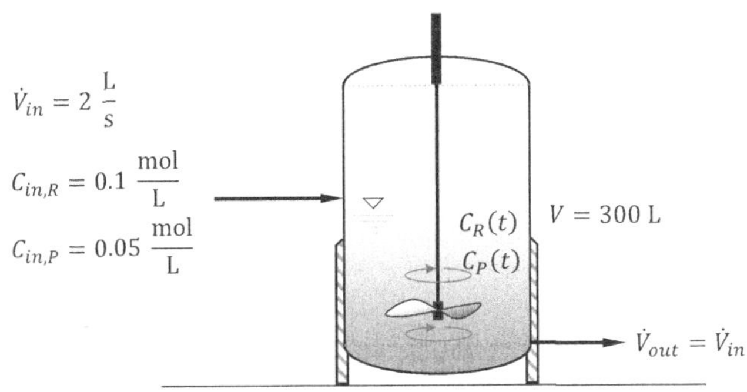

Figure 4.15 A continuous stirred tank reactor where an autocatalytic reaction is taking place.

$$\frac{dC_P(t)}{dt} = \frac{\dot{V}_{\text{in}}}{V}(C_{\text{in},P} - C_P(t)) + k_1 C_R(t) C_P(t)^2 - k_2 C_P(t) \qquad (4.5.5)$$

The two ordinary differential equations are solved in Python as follows.

```
import numpy as np
from scipy.integrate import solve_ivp

def OscCSTR(t,x):
    C_R   = x[0]     # mol/L
    C_P   = x[1]     # mol/L
    k1    = 1000     # (1/s)*(L/mol)^(2)
    k2    = 0.2      # 1/s
    F_dot_in = 2     # L/s
    V     = 300      # L
    C_in_R = 0.1     # mol/L
    C_in_P = 0.05    # mol/L
    dCdt = np.zeros(2)
    dCdt[0] = (F_dot_in/V)*(C_in_R-C_R)-k1*C_R*C_P**2
    dCdt[1] = (F_dot_in/V)*(C_in_P-C_P)+k1*C_R*C_P**2-k2*C_P
    return dCdt

t0    = 0.0          # initial simulation time in s
tf    = 1000.0       # final simulation time in s
C0    = [0.0, 0.0]   # initial conditions

sol = solve_ivp(OscCSTR, [t0, tf], C0, method='BDF', rtol = 1e-8)

plt.figure(1)
plt.subplot(2, 1, 1)
plt.plot(sol.t,sol.y[0],'b-')
plt.grid()
plt.xlabel('Time (s)')
plt.ylabel('$C_{R}$ (mol/L)')
plt.axis([0, 1000, 0, 0.040])

plt.subplot(2, 1, 2)
plt.plot(sol.t,sol.y[1],'g-')
plt.grid()
plt.xlabel('Time (s)')
plt.ylabel('$C_{P}$ (mol/L)')
plt.axis([0, 1000, 0, 0.01])

plt.figure(2)
plt.plot(sol.y[0],sol.y[1],'-')
plt.grid()
plt.xlabel('$C_{R}$ (mol/L)')
plt.ylabel('$C_{P}$ (mol/L)')
plt.axis([0, 0.04, 0, 0.01])


plt.figure(3)
ax = plt.axes(projection ='3d')
ax.plot3D(sol.y[0], sol.y[1], sol.t, 'green')
ax.set_xlabel('$C_{R}$ (mol/L)')
ax.set_ylabel('$C_{P}$ (mol/L)')
ax.set_zlabel('t (s)')
```

The solution obtained is shown in Fig. 4.16. A periodic solution is obtained after an initial, short period.

Chapter summary

The general material balance equation for component i is the following

$$\frac{dn_i(t)}{dt} = \dot{n}_{\text{in},i}(t) - \dot{n}_{\text{out},i}(t) + \dot{n}_{\text{gen},i}(t) - \dot{n}_{\text{cons},i}(t)$$

where $n_i(t)$ denotes the moles of component i in the control volume

$\dot{n}_{\text{in},i}(t)$ denotes the rate at which moles of component i enter the control volume

$\dot{n}_{\text{out},i}(t)$ denotes the rate at which moles of component i exit the control volume

$\dot{n}_{\text{gen},i}(t)$ denotes the rate at which moles of component i are generated within the control volume

$\dot{n}_{\text{cons},i}(t)$ denotes the rate at which moles of component i are consumed within the control volume

The general material balance equation for component i can be stated in words as follows

$$\begin{bmatrix}\text{rate of}\\\text{accumulation}\\\text{of component i}\end{bmatrix} = \begin{bmatrix}\text{rate of}\\\text{inflow}\\\text{of component i}\end{bmatrix} - \begin{bmatrix}\text{rate of}\\\text{outflow}\\\text{of component i}\end{bmatrix} + \begin{bmatrix}\text{rate of}\\\text{generation}\\\text{of component i}\end{bmatrix} - \begin{bmatrix}\text{rate of}\\\text{consumption}\\\text{of component i}\end{bmatrix}$$

Each term has units of mass per unit time (kg/s) or mole per unit time (mol/s).

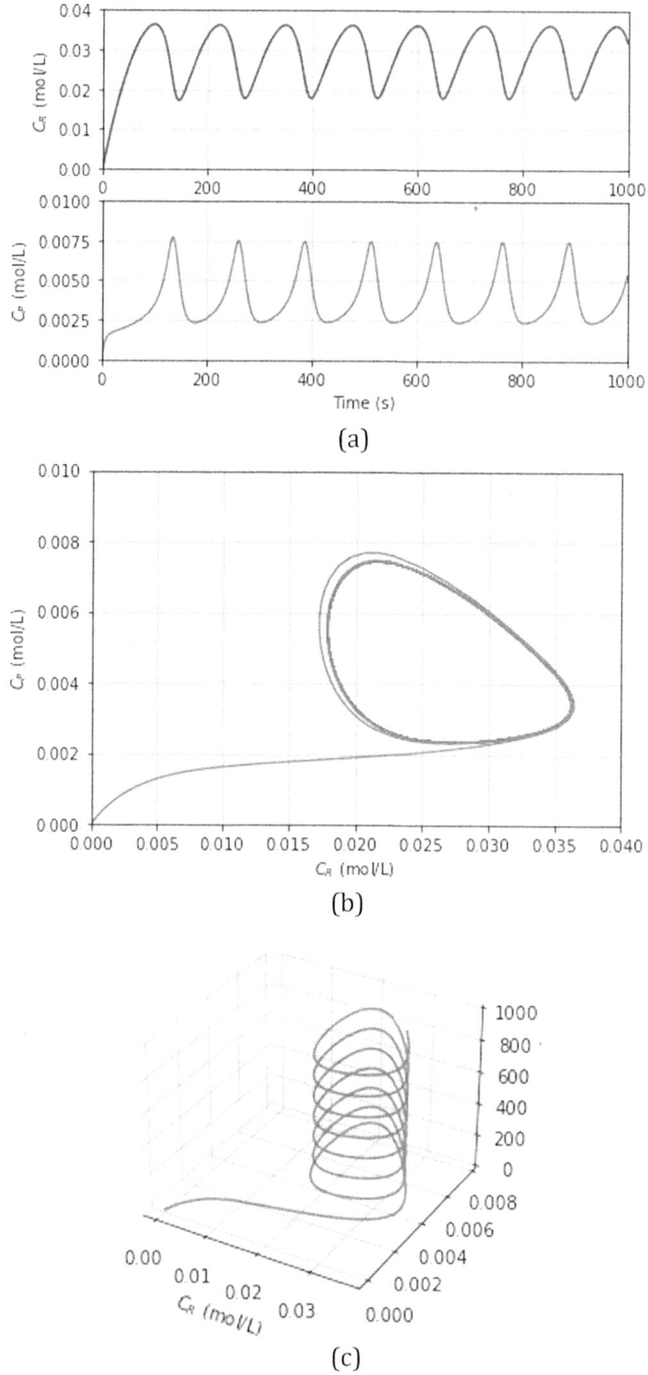

Figure 4.16 Time concentrations of reactant and product in a continuous stirred tank reactor are periodic when an autocatalytic reaction is taking place.

Problems

4.1. Consider Example 4.3. If the ink is not added instantaneously but the 0.1 kg of ink is added in 5 s, what is the response of the ink concentration in this case? Assume that the incoming volumetric flowrate is $\dot{V}_{in} = 0.1\ m^3/s$ and that the volume of the liquid in the tank is $V = 1\ m^3$.

4.2. Consider again Example 4.3. If the ink is not added instantaneously but the feed is changed from pure water to water that contains ink at a concentration of $C_{in} = 1\ kg/m^3$ of ink, what is the response of the ink concentration in this case? Assume that the incoming volumetric flowrate is $\dot{V}_{in} = 0.1\ m^3/s$ and that the volume of the liquid in the tank is $V = 1\ m^3$.

4.3. Consider Example 4.4. If the rate of increase of CO_2 emissions per year remains constant, predict the CO_2 concentration until 2100 (in ppmv). If an increase in the CO_2 concentration in the atmosphere of 100 ppmv results in an increase in the average global temperature of 1 °C, estimate the increase in the average global temperature.

4.4. Consider again Example 4.4. If the rate of increase of CO_2 emissions per year decreases at a constant rate so that the emissions in 2050 are equal to the emissions in 1958 (and become zero at a future year), predict the CO_2 concentration until 2100 (in ppmv). If an increase in the CO_2 concentration in the atmosphere of 100 ppmv results in an increase in the average global temperature of 1 °C, estimate the increase in the average global temperature.

4.5. The bromination of m-xylene ($(CH_3)_2C_6H_4$) at constant temperature is carried out by adding a small quantity of bromine into pure xylene. Data on the reduction of the bromine concentration in such an experiment are given in Table P4.5 (Hill, C.G., Jr., 1977, An Introduction to Chemical Engineering Kinetics and Reactor Design, John Wiley and Sons, NY).

The reaction rate of bromine consumption is believed to be of the form

$$\dot{R}''' = kC_{Br}^n$$

with $n = 1.545$.

Table P4.5 Experimental data for the bromination of m-xylene.

Time (min)	Bromine Concentration (mol/L)	Time (min)	Bromine Concentration (mol/L)
0	0.3335	19.60	0.1429
2.25	0.2965	27.00	0.1160
4.50	0.2660	30.00	0.1053
6.33	0.2450	38.00	0.0830
8.00	0.2255	41.00	0.0767
10.25	0.2050	45.00	0.0705
12.00	0.1910	47.00	0.0678
13.50	0.1794	57.00	0.0553
15.60	0.1632	63.00	0.0482
17.85	0.1500		

Table P4.6 Experimental data for Problem 4.6.

Time (min)	P(atm)
0.0	41.1
6.5	53.7
12.95	64.2
19.92	73.9
52.58	102.5

a. Can you determine the reaction rate constant k from the experimental data?

b. Compare the experimental data with the simulation results.

4.6. The following reaction is taking place under constant temperature and constant volume in the gas phase

$$R \rightarrow 3P$$

The reaction is believed to be first order with respect to the reactant concentration. Determine the reaction rate constant from the experimental data for the total pressure of the reacting mixture given in Table P4.6.

4.7. The reaction

$$C_2H_4O \rightarrow CH_4 + CO$$

is taking place in the gas phase at a constant temperature of 601.4 K and initial pressure of 200 kPa in a vessel with $V = 25$ L volume. The reaction is first order with respect to the concentration of ethylene oxide with a reaction rate constant $k = 3.5 \times 10^{-4}\ 1/\mathrm{s}$.

a. Calculate the moles of the chemical species in the vessel as a function of time.

b. Calculate the pressure in the vessel as a function of time.

c. Calculate the conversion as a function of time.

4.8. Most drugs are eliminated from the body by liver metabolism and/or renal excretion. The processes of drug elimination can be described by an elimination rate constant, k. The elimination rate constant connects the rate of drug elimination $\dot{R}'''_{\text{elimination}}$ and the amount of drug in the body C_B:

$$\dot{R}'''_{\text{elimination}} = kC_B$$

When a biologic compartment is sampled and analyzed for drug content, the results are reported in units of concentration instead of total amounts. The amount of drug in a given location can be related to its concentration by a proportionality constant that reflects the apparent volume of fluid in which the drug is dissolved, denoted by V_D (apparent because it does not have a true physiologic meaning in terms of an anatomic space). V_D relates the concentration of drug in plasma (C_p) and the amount of drug in the body (D_B):

$$D_B = V_D C_P$$

A single IV dose of an antibiotic was given to a woman. Blood samples were removed periodically and assayed for the parent drug, and the data in Table P4.8 were collected.

Can you determine the elimination rate constant, k?

4.9. Consider the following reactions

$$R \xrightarrow{k_1} P \xrightarrow{k_2} B_1$$

$$2R \xrightarrow{k_3} B_2$$

These reactions are taking place in a continuous stirred tank reactor at constant temperature and volume $V = 1\ \mathrm{m}^3$. The feed is pure R and the volumetric flowrate is $\dot{V}_{\text{in}} = 0.1\ \mathrm{m}^3/\mathrm{min}$

Table P4.8 Concentration of antibiotics in blood samples as a function of time after an IV dose.

Time (min)	Concentration of antibiotic In blood sample (mg/L)
0	5.0
15	4.2
30	3.5
60	2.5
120	1.25
240	0.31
360	0.08

and concentration of reactant $C_{in,R} = 10\ \text{kmol/m}^3$. The reaction rate constants are $k_1 = 1\ \frac{1}{\text{min}}$, $k_2 = 0.5\ \frac{1}{\text{min}}$ and $k_3 = 0.1\ \text{m}^3/(\text{kmol}\cdot\text{min})$. Simulate the startup of the reactor in MATLAB or Python. The reactor is initially full of reactant R ($C_R(0) = 10\ \text{kmol/m}^3$).

4.10. A bioreactor is a tank in which organisms or biochemically active substances derived from such organisms perform chemical reactions. These bioreactors are commonly cylindrical, ranging in size from a few L to thousands of m^3, and are often made of stainless steel. Based on mode of operation, a bioreactor may be operated as batch or fed-batch systems and less frequently as continuous. Organisms or biochemically active substances growing in bioreactors may be submerged in liquid medium or may be anchored to the surface of a solid medium. Suspension bioreactors can support a wide variety of organisms and produce biomass and several intracellular or extracellular products such as bioplastics, ethanol, or lactic acid. The complex biochemical reactions are usually simplified in a general description

$$\text{substrate (food and nutrients)} \xrightarrow{\text{microorganisms}} \text{Products} + \text{more microorganisms}$$

The symbol X is commonly used to denote the concentration of biomass in g/L and the symbol S is used to denote the concentration of substrate (such as glucose) in g/L. The specific growth of biomass is usually described by Monod kinetics

$$\frac{\dot{R}_X'''}{X(t)} = \frac{1}{X(t)}\frac{dX(t)}{dt} = \mu(S) = \mu_{max}\frac{S}{K_s + S}$$

where μ is the growth rate of a considered microorganism, μ_{max} is the maximum growth rate of this microorganism, S is the concentration of the limiting substrate S for growth, K_s is the saturation constant (is equal to the value of S when $\mu/\mu_{max} = 0.5$).

For the substrate consumption a yield coefficient Y is defined which relates the rate of biomass formation $\dot{R}_X'''$ to the rate of substrate consumption $\dot{R}_S'''$

$$Y = \frac{\dot{R}_X'''}{\dot{R}_S'''} = \frac{\mu(S)X(t)}{\dot{R}_S'''}$$

For a particular case study of biomass production on glucose the Monod parameters are $\mu_{max} = 0.4\ \frac{1}{h}$ and $K_s = 5\ \frac{g}{L}$ and the yield coefficient is $Y = 0.4\frac{g}{g}$.

a. For the given parameters of the Monod kinetics, plot the specific growth rate of the biomass as a function of the substrate concentration. What is an appropriate approximation for $S < K_s$ and what for $S \gg K_s$?

b. Write the material balances for the substrate and for the biomass for a batch reactor.

c. Solve the model in MATLAB or Python for a reacting volume of $V = 100$ L, an initial glucose concentration of $S(0) = 100$ g/L and a biomass concentration of $X(0) = 1$ g/L.

4.11. In many cases the growth rate of biomass is inhibited at large substrate concentrations. In a specific case the substrate inhibition can be expressed by an extended Monod model:

$$\frac{\dot{R}_X'''}{X(t)} = \frac{1}{X(t)}\frac{dX(t)}{dt} = \mu(S) = \mu_{max}\frac{S}{K_s + S}\exp\left(-\frac{S}{S_{max}}\right)$$

In such a case, the use of a large initial substrate concentration is unwise. A continuous feed of substrate is recommended, resulting in what is known as a fed-batch system. Consider the following model parameters:

$\mu_{max} = 0.4\ \frac{1}{h}$, $K_s = 5\ \frac{g}{L}$, $Y = 0.4\frac{g}{g}$, $S_{max} = 120\ \frac{g}{L}$.

Simulate the following operation strategy: the initial volume is $V(0) = 50$ L in the bioreactor and the initial substrate and biomass concentrations are $S(0) = 50\ \frac{g}{L}$ and $X(0) = 1\ \frac{g}{L}$. There is a continuous feed with volumetric flowrate of $V_{in} = 2.5\ \frac{L}{h}$ with $S_{in} = 150\ \frac{g}{L}$ for 20 h.

4.12. The reaction

$$CH_3COOH + C_2H_5OH \underset{k_b}{\overset{k_f}{\rightleftharpoons}} CH_3COOC_2H_5 + H_2O$$

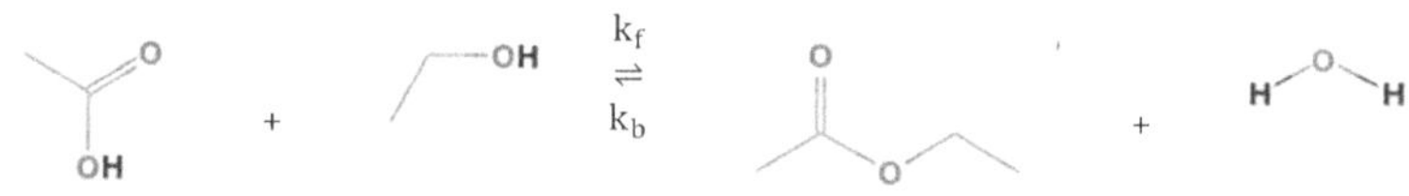

Acetic acid (AA) + Ethanol (E) Ethyl Acetate (EA) Water (W)

is taking place at constant temperature and constant volume in a batch reactor. The reaction rate is given by

$$\dot{R}''' = k_f C_{AA} C_E - k_b C_{EA} C_W$$

where

$$k_f = 8 \times 10^{-6} \frac{m^3}{kmol{\cdot}s}, k_b = 2.7 \times 10^{-6} \frac{m^3}{kmol{\cdot}s}$$

The batch reactor has volume $V = 6.85\ m^3$, the density of the reacting mixture is $\rho = 1145\ kg/m^3$ and the initial concentration of the reactants is $C_{AA}(0) = 5\ \frac{kmol}{m^3}$ and $C_E(0) = 11.96\ \frac{kmol}{m^3}$ and that of water $C_W(0) = 16.39\ \frac{kmol}{m^3}$.

Simulate the reactor operation for $t_f = 6000$ s and determine the final conversion achieved.

4.13. The Belousov-Zhabotinsky reaction involves the oxidation of an organic species, such as malonic acid (MA), by an acidified aqueous bromate solution in the presence of a metal ion catalyst such as the $Ce^{(III)}/Ce^{(IV)}$ couple. The reaction has been studied in a publication (R. Dyson et al., Simulation of Complex Chemical Kinetics, *Inorganic Reaction Mechanisms*, Vol 5, pp.39-46, 2003). The reaction scheme involves the following reactions.

R1:	$B + C \xrightarrow{k_1} D + F$	$\dot{R}_1''' = k_1[B][C], k_1 = 1.28\ \mathrm{s^{-1}M^{-1}}$
R2:	$B + D \xrightarrow{k_2} 2D + 2E$	$\dot{R}_2 = k_2[B][D], k_2 = 33.6\ \mathrm{s^{-1}M^{-1}}$
R3:	$D + C \xrightarrow{k_3} 2F$	$\dot{R}_3 = k_3[C][D], k_3 = 2.4 \times 10^6\ \mathrm{s^{-1}M^{-1}}$
R4:	$D + D \xrightarrow{k_4} B + F$	$\dot{R}_4 = k_4[D][D], k_4 = 3 \times 10^3\ \mathrm{s^{-1}M^{-1}}$
R5:	$MA + E \xrightarrow{k_5} \frac{1}{2}\ C$	$\dot{R}_5 = k_5[A][E], k_5 = 1\ \mathrm{s^{-1}M^{-1}}$

where B = BrO_3^-, C = Br^-, D = $HBrO_2$, E = $Ce^{(IV)}$ and F = HOBr. The symbol [ ] denotes concentration (in M = moles of solute/liters of solution, molarity).

In a particular experiment, the following initial concentrations are used in a constant-volume, well-mixed batch reactor:

[MA](0) = 0.275 M, [B](0) = 0.063 M and [E](0) = 0.002 M (all other concentrations are zero).

a. Calculate the concentration of E and MA for a period of 400 and 4000 s and plot the results. Compare your results presented in the above-mentioned publication. Use the ODE solver ode113 with absolute and relative tolerances of 10^{-8}.

b. Compare at least three alternative ODE solvers in either MATLAB or Python and report their performance in terms of time to integrate the system of ODEs for 36,000 s (10 h). Comment on your findings.

c. Increase the absolute and relative tolerances to 10^{-3} and repeat the simulation. What is the result?

4.14. Consider the case of an addition polymerization in a batch reactor. M_1 represents the monomer, M_n the dead (stabilized) polymer that consists of n monomers, and P_n the active (growing) polymer with n monomeric units. The kinetic scheme is described in Fig. P4.14. The polymerization starts when a monomeric unit is activated and produces an active monomer Fig. P4.14.

$$M_1 \xrightarrow{k_i} P_1$$

The rate of the initiation reaction is first order with respect to the monomer's concentration

$$\dot{R}_1''' = k_i[M_1]$$

The active polymer grows in size (propagation)

$$P_n + M_1 \xrightarrow{k_p} P_{n+1}, \quad n = 1, 2, \ldots, N - 1$$

Figure P4.14 Representation of the addition polymerization kinetic scheme.

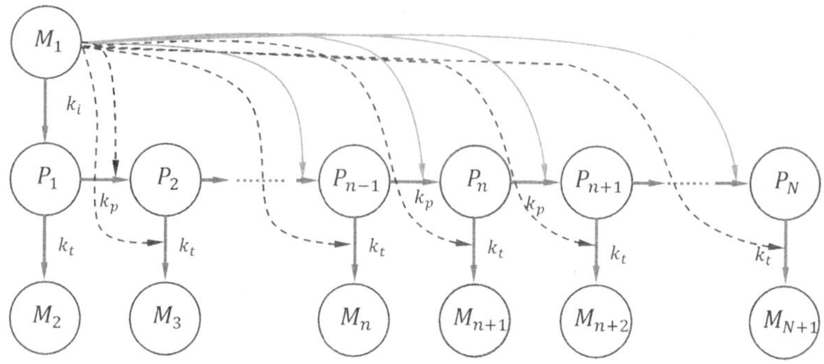

with reaction rate

$$\dot{R}_p''' = k_p[P_n][M_1]$$

or it is deactivated to produce a dead polymer chain (termination)

$$P_n + M_1 \xrightarrow{k_p} M_{n+1}, \quad n = 1, 2, \ldots, N$$

Although the chain length can grow indefinitely, we can limit the size of the polymer chain by assuming the propagation reaction stops when the degree of polymerization is N (a polymer chain with N monomeric units).

In a particular case, an initial monomer concentration of $M_1(0) = 1$ mol/L is used and the polymerization is performed in a batch, well-mixed reactor for 7 days (168 h). The reaction rate constants are

$$k_i = 0.0025\ \frac{1}{\text{h}}$$

$$k_p = 6\ \frac{1}{\text{h}} \cdot \frac{\text{L}}{\text{mol}}$$

$$k_t = 0.5\ \frac{1}{\text{h}} \cdot \frac{\text{L}}{\text{mol}}$$

Find the weight fraction % as a function of the number of monomers in the polymer chain at the end of the batch polymerization.

4.15. In a traditional chemical reactor, the reaction occurs in one vessel, and then the products are separated afterward (using distillation, absorption, etc.). In a membrane reactor, a membrane is built into the reactor itself, allowing one or more components to selectively pass through during the reaction. Special attention has attracted the case of reactions limited by equilibrium. A classical example is the following reaction (dehydrogenation of ethylbenzene to styrene):

$$C_6H_5\text{–}CH_2CH_3 \rightarrow C_6H_5\text{–}CHCH_2 + H2$$

the reaction rate of which is

$$\dot{R}'''\left[\frac{\text{kmol}}{\text{m}^3\ \text{s}}\right] = k\left(C_R - \frac{C_P C_{H_2}}{K_e}\right)$$

where R = C_6H_5–CH_2CH_3, and P = C_6H_5–$CHCH_2$.

Consider the general reaction

$$R \rightarrow P + H_2$$

that is taking place in a plug-flow reactor at steady state at constant temperature T and pressure P. Hydrogen can permeate through the membrane surface at a flux $\dot{J}''$ given by

$$\dot{J}''\left[\frac{\text{kmol}}{\text{m}^2\ \text{s}}\right] = k_M C_{H_2}$$

Based on Fig. P4.15, the following material balances can be derived

$$\frac{d\dot{n}_R}{dV} = -\dot{R}'''$$

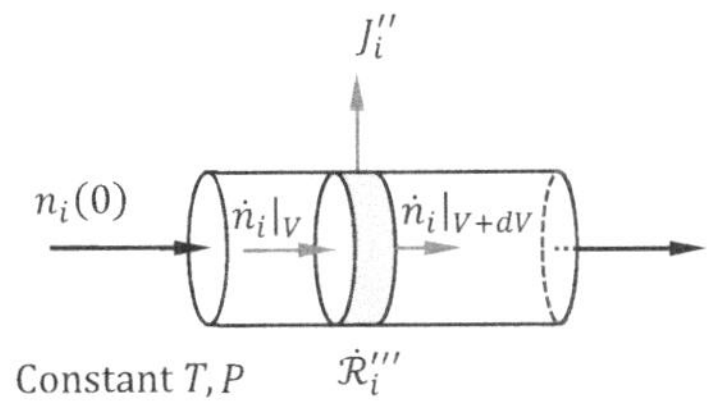

Figure P4.15 A membrane reactor.

$$\frac{d\dot{n}_P}{dV} = +\dot{R}'''$$

$$\frac{d\dot{n}_{H_2}}{dV} = +\dot{R}''' - a_V\dot{J}''$$

where $a_V = \frac{dA}{dV}$ is the available surface area of the membrane per unit volume of the membrane reactor. For a cylindrical reactor with diameter D:

$$a_V = \frac{A}{V} = \frac{\pi D \cdot \ell}{\frac{\pi D^2}{4} \cdot \ell} = \frac{4}{D}$$

Note also that

$$C_i = y_i C$$

where

$$y_i = \frac{\dot{n}_i}{\sum_{i=\{R,P,H_2\}} \dot{n}_i}$$

and

$$C = \frac{P}{RT}$$

At a particular case study, a fresh feed with molar flowrates $\dot{n}_R(0) = 1$ kmol/s, $\dot{n}_P(0) = \dot{n}_{H_2} = 0$ and the temperature is kept constant at $T = 500$ K. The following constants are also given:

$$k = 0.01\ \frac{1}{s}, \quad K_e = 0.05\ \frac{\text{kmol}}{\text{m}^3}, \quad a_V k_M = 0.1\ \frac{1}{s}$$

a. What is the equilibrium conversion of R?

b. Find what conversion is achieved for pressures up to 1000 kPa and reactor volumes up to 5000 m^3.

c. For $V = 1000$ m^3 and $P = 1000$ kPa, plot the variation of the molar flowrate of the three components across the reactor.

Further reading

Many definitions have been retrieved from the Wikipedia, the free encyclopedia

Books on material and energy balances

Andersen, L.B., Wenzel, L.A., 1961. Introduction to Chemical Engineering. McGraw Hill, NY.

Duncan, T.M., Reimer, J.A., 1998. Chemical Engineering Design and Analysis. Cambridge University Press, NY.

Foust, A.S., Wenzel, L.A., Clump, C.W., Maus, L., Andersen, L.B., 1980. Principles of Unit Operations, second ed. John Wiley & Sons, NY.

Henley, E.J., Bieber, H., 1959. Chemical Engineering Calculations. McGraw Hill, NY.

Himmelblau, D.M., Bischoff, K.B., 1968. Process analysis and simulation. In: Deterministic Systems. John Wiley & Sons, NY.

Hougen, O.A., Watson, k.m., Ragatz, R.A., 1943. Chemical Process Principles, Part I, Material and Energy Balances. John Wiley and Sons, NY.

Myers, A.L., Seider, W.D., 1976. Introduction to Chemical Engineering and Computer Calculations. Prentice Hall, Inc., Englewood Cliffs, NJ.

Reklaitis, G.V., 1983. Introduction to Material and Energy Balances. John Wiley and Sons, NY.

Russell, T.W.F., Denn, M.M., 1972. Introduction to Chemical Engineering Analysis. John Wiley and Sons, NY.

Books on chemical reaction engineering

Fogler, H.S., 2022. Elements of Chemical Reaction Engineering. Pearson, NY.

Hill Jr., C.G., Root, T.W., 2014. Introduction to Chemical Engineering Kinetics and Reactor Design, second ed. John Wiley and Sons, NY.

Thermodynamics books

Abbott, M.M., Van Ness, H.C., 1972. Theory and Problems of Thermodynamics, Schaums Outline Series. McGraw-Hill Book Company, NY.

Elliot, J.R., Lira, C.T., 2012. Introductory Chemical Engineering Thermodynamics, second ed. Prentice Hall, Upper Saddle River, NJ.

Matsoukas, T., 2012. Fundamentals of Chemical Engineering Thermodynamics. Prentice Hall, Upper Saddle River, NJ.

Moran, M.J., Shapiro, H.N., 1988. Fundamentals of Engineering Thermodynamics. John Wiley and Sons, NY.

Sandler, S.I., 1977. Chemical and Engineering Thermodynamics. John Wiley and Sons, NY.

Books on numerical methods and ordinary differential equations

Borrelli, R.L., Coleman, C.S., 1998. Differential Equations, A Modelling Perspective. John Wiley & Sons, NY.

Clough, D.E., Chapra, S.C., 2023. Introduction to Engineering and Scientific Computing with Python. CRC Press, Boca Raton, FL.

Dorfman, K.D., Daoutidis, P., 2017. Numerical Methods with Chemical Engineering Applications. Cambridge University Press.

Mostoufi, N., Constantinides, A., 2023. Applied Numerical Methos for Chemical Engineers. Academic Press, London.

Mitsotakis, D., 2023. Computational Mathematics: An Introduction to Numerical Analysis and Scientific Computing with Python. CRC Press, Boca Raton, FL.

5

Energy balance equation

5.1 The general energy balance equation

In Chapter 3 we developed the general "balance" equation for conserved quantities, which can be stated in words as follows

$$\begin{bmatrix}\text{extensive quantity}\\ \text{at time}\\ \text{t}+1\end{bmatrix}-\begin{bmatrix}\text{extensive quantity}\\ \text{at time}\\ \text{t}\end{bmatrix}=\begin{bmatrix}\text{inflow of}\\ \text{extensive quantity}\\ \text{between t and t}+1\end{bmatrix}-\begin{bmatrix}\text{outflow of}\\ \text{extensive quantity}\\ \text{between t and t}+1\end{bmatrix} \tag{5.1.1}$$

or more compactly as

$$\text{new value} - \text{old value} = \text{input} - \text{output} \tag{5.1.2}$$

This is the form of the general balance equation that applies to all conserved quantities, including the overall energy, denoted by $E(t)$. We have also expressed the general balance equation for the extensive quantity $E(t)$ in differential form as follows

$$\frac{dE(t)}{dt} = \dot{E}_{\text{in}}(t) - \dot{E}_{\text{out}}(t) \tag{5.1.3}$$

Eq. (5.1.3) is the general form of a differential energy balance equation. All terms of the equation are rate terms with units of energy per unit time.

5.2 The energy balance equation for process systems

The total energy consists of three forms of energy, as has been discussed in Chapter 2.

Introduction to Process Systems Engineering. https://doi.org/10.1016/B978-0-443-45444-8.00013-0

$$E(t) = \begin{array}{c}\text{Internal}\\ \text{Energy}\end{array} + \begin{array}{c}\text{Kinetic}\\ \text{Energy}\end{array} + \begin{array}{c}\text{Potential}\\ \text{Energy}\end{array} = U(t) + E_K(t) + E_P(t) \tag{5.2.1}$$

For the macroscopic energy balance of process systems, the kinetic and potential energy terms are usually neglected and Eq. (5.2.1) is simplified to

$$E(t) \approx \begin{array}{c}\text{Internal}\\ \text{Energy}\end{array} = U(t) \tag{5.2.2}$$

The rate at which energy enters an open system involves the internal energy, the kinetic energy, and the potential energy of the incoming streams but also involves the work done by the environment to the system and the heat transferred from the environment to the system

$$\begin{aligned}\dot{E}_{\text{in}}(t) &= \begin{array}{c}\text{Internal}\\ \text{Energy}\end{array} + \underbrace{\begin{array}{c}\text{Kinetic}\\ \text{Energy}\end{array} + \begin{array}{c}\text{Potential}\\ \text{Energy}\end{array}}_{\approx 0} + \text{Work} + \text{Heat}\\ &\approx \left[\dot{U}(t) + \dot{W} + \dot{Q}\right]_{\text{in}}\end{aligned} \tag{5.2.3}$$

The work term is decomposed in flow work ($\dot{W}_{\text{flow}}$) and all other types of work (that involve shaft work, expansion/contraction work of the open system boundaries, electrical, ...)

$$\dot{E}_{\text{in}}(t) = \left[\dot{U}(t) + \dot{W}_{\text{flow}} + \dot{W} + \dot{Q}\right]_{\text{in}} \tag{5.2.4}$$

we have kept the symbol W for all types of work other than flow work to simplify the notation used. We then use the definition of enthalpy $H = U + W_{\text{flow}}$ to obtain

$$\dot{E}_{\text{in}}(t) = \left[\dot{H}(t) + \dot{W} + \dot{Q}\right]_{\text{in}} \tag{5.2.5}$$

The same terms are involved in the rate at which energy leaves the open system

$$\dot{E}_{\text{out}}(t) = \left[\dot{H}(t) + \dot{W} + \dot{Q}\right]_{\text{out}} \tag{5.2.6}$$

We replace Eqs. (5.2.2), (5.2.5), and (5.2.6) into Eq. (5.1.3) to obtain

$$\frac{dU(t)}{dt} = \left[\dot{H}(t) + \dot{W} + \dot{Q}\right]_{\text{in}} - \left[\dot{H}(t) + \dot{W} + \dot{Q}\right]_{\text{out}} \tag{5.2.7}$$

or

$$\frac{dU(t)}{dt} = \dot{H}_{in}(t) - \dot{H}_{out}(t) + \left[\dot{W} + \dot{Q}\right]_{net} \quad (5.2.8)$$

where the subscript "net" denotes the algebraic sum of the terms within the parenthesis (net = in-out, i.e., $\dot{W}_{net} = \dot{W}_{in} - \dot{W}_{out}$, etc.).

When we refer to the energy balance equation, we will refer to the form given by Eq. (5.2.8). Eq. (5.2.8) simply states that the rate of change of the internal energy of an open system equals

- the rate at which enthalpy is added to the system through the incoming streams,
- minus the rate at which enthalpy is removed from the system through the outgoing streams,
- plus the net (input minus the output) rate at which work is performed on the system through the system boundaries,
- plus the net rate at which heat is transferred to the system through the system boundaries.

It is important to recognize that each term on the right-hand side of the equation must be included for every mass or energy stream (whether work or heat) crossing the system boundary. Also, note that the left-hand side of Eq. (5.2.8) represents the properties of the entire control volume (the system as a whole), while the right-hand side accounts for interactions occurring at the system boundary (the surface of the control volume). Lastly, regarding the sign convention for work and heat—as introduced in Chapter 2—we refer to the standard shown in Fig. 2.12. This convention treats both work and heat as positive when they are transferred into the system from the environment, and negative when transferred out of the system.

It is apparent from Eq. (5.2.8) that, to apply the macroscopic energy balance equation, we need a model for calculating the internal energy and the enthalpy of pure components and mixtures. We will start by developing a model for calculating the enthalpy of pure components, and we will then expand this model to be able to calculate the enthalpy of ideal mixtures.

5.3 Enthalpy of pure compounds

We begin with the case of pure substances, developing a model for calculating their enthalpy. Our starting point is the ideal gas, for which enthalpy depends only on temperature. Furthermore, based on the definition of specific heat capacity at constant pressure (see Eq. (2.9.6)), we can express this relationship as follows:

$$c_p = \left.\frac{\partial h(T,P)}{\partial T}\right|_P \tag{5.3.1}$$

For an ideal gas, enthalpy is a function of temperature only

$$c_p(T) = \frac{dh^{ig}(T)}{dT} \tag{5.3.2}$$

We integrate from a reference temperature (T_{ref}) to any arbitrary temperature

$$h^{ig}(T) = h^{ig}(T_{\text{ref}}) + \int_{T_{\text{ref}}}^{T} c_p(T)dT \tag{5.3.3}$$

The specific heat capacity of many pure components at ideal gas state (low pressure) can be found in many sources, with the NIST chemistry WebBook to be the most reliable and handy. The form of the specific heat capacity equation usually found in the NIST web page is the following (see also Chapter 2, Eq. (2.9.11) and Table 2.10)

$$c_p^{ig}(T) = A + BT + CT^2 + DT^3 + ET^{-2} \tag{5.3.4}$$

Warning: in NIST the polynomial is defined with respect to a variable $t = T/1000$. The integral has been calculated in Chapter 2 (see Eq. (2.9.13))

$$\begin{aligned}\int_{T_{\text{ref}}}^{T} c_p^{ig}(T)dT &= A(T - T_{\text{ref}}) + \frac{B}{2}\left(T^2 - T_{\text{ref}}^2\right) + \frac{C}{3}\left(T^3 - T_{\text{ref}}^3\right)\\ &+ \frac{D}{4}\left(T^4 - T_{\text{ref}}^4\right) - E\left(\frac{1}{T} - \frac{1}{T_{\text{ref}}}\right), \qquad A = A_{\text{NIST}}, B\\ &= \frac{B_{\text{NIST}}}{10^3}, C = \frac{C_{\text{NIST}}}{10^6}, D = \frac{D_{\text{NIST}}}{10^9}, \qquad E = E_{\text{NIST}}10^6\end{aligned} \tag{5.3.5}$$

where the subscript NIST denotes value obtained from NIST Chemistry WebBook. There are two important details that need to be clarified: what is the reference temperature (T_{ref}) and what is the value of the enthalpy at the reference temperature ($h^{ig}\left(T_{\text{ref}}\right)$)?

Most textbooks in thermodynamics advise defining the reference temperature by studying each case study and selecting the temperature that facilitates the calculations involved. Usually, the enthalpy is set equal to zero at the arbitrary reference temperature. This strategy works well in most cases but it has several disadvantages and is not recommended for process systems engineering calculations. The systematic way to define the reference temperature and the corresponding enthalpy is based on the definition of the standard enthalpy of formation: **the standard enthalpy (or heat) of formation of a compound is the change of enthalpy during the formation of** 1 mol **of the substance from its constituent elements in their reference state, with all substances in their standard states. The standard pressure value P_{ref}** $= 10^5$ Pa ($=$1 bar) **is recommended by IUPAC, although prior to 1982 the value** 1.00 atm (101 325 Pa) **was used. There is no standard temperature but, as most experimental values are reported at** 25 °C **or** 298.15 K, **it is strongly recommended to select T_{ref}** $=$ 298.15 K (25 °C). Based on this convention Eq. (5.3.3) can be written as

$$h^{ig}(T) = \Delta h_f^{ig}(298.15\ \text{K}) + \int_{298.15\ \text{K}}^{T} c_p^{ig}(T)dT \tag{5.3.6}$$

where the reference temperature is $T_{\text{ref}} = 298.15$ K and the **ideal gas state**. The enthalpy of any component is set equal to the specific enthalpy of formation at the reference conditions. This reference state has been adopted by most of the commercial process simulators.

In a final step, to calculate the enthalpy of a real (and not ideal) pure fluid we need to define the **residual property** (also referred to as departure function) as **the difference between a real fluid property and an ideal gas property**, both considered at the same pressure and temperature. For the case of enthalpy we can write the following definition of residual enthalpy:

$$h^{\text{real fluid}}(T,P) = h^{\text{ideal gas}}(T) + h^{\text{residual}}(T,P) \tag{5.3.7}$$

The detailed calculation of residual enthalpy—and other residual thermodynamic properties—will be covered later in this text and further explored in the thermodynamics course. In this chapter, however, we will use a simplified model to represent the behavior of a real fluid

- *if the pure fluid is in the gas state*

$$h(T, P, \text{state} = \text{gas}) \approx h^{\text{ideal gas}}(T) = \Delta h_f^{ig}(298.15\ \text{K}) + \int_{298.15\ \text{K}}^{T} c_p^{ig}(T)dT \quad (5.3.8)$$

- *if the pure fluid is in the liquid state*

$$h(T, P, \text{state} = \text{liquid}) \approx h^{\text{ideal gas}}(T) - \Delta h_{\text{vap}}(T)$$
$$= \Delta h_f^{ig}(298.15\ \text{K}) + \int_{298.15\ \text{K}}^{T} c_p^{ig}(T)dT - \Delta h_{\text{vap}}(T) \quad (5.3.9)$$

where $\Delta h_{\text{vap}}(T)$ represents the specific enthalpy (or heat) of vaporization—the amount of energy required to convert a unit mass of a liquid into vapor at a given temperature. The enthalpy of vaporization depends on both the pressure and temperature at which the phase change occurs. This method **provides a reasonably accurate approximation at low pressures**, typically not exceeding 10 bar. We will refer to this as the first method for calculating the specific enthalpy of pure components. Later in this chapter, we will introduce an alternative approach, referred to as the second method. Note that we have suggested that the residual enthalpy of a liquid is approximately equal to the negative of the enthalpy of vaporization.

In simple words, we calculate the specific enthalpy of a gas using the ideal gas model and we subtract the specific enthalpy of evaporation from the specific enthalpy of the ideal gas to determine the specific enthalpy of the liquid phase (at the same temperature). In other words, **the residual enthalpy of a gas is assumed negligible while the residual enthalpy of the liquid phase is assumed equal to the negative of the heat of evaporation** (calculated at the temperature of the system). The data necessary to apply this model are the following:

i. The specific, standard enthalpy of formation of ideal gas at 298.15 K $\Delta h_f^{ig}(298.15\text{ K})$.

ii. The specific heat capacity at constant (low) pressure of the ideal gas $c_p^{ig}(T)$ in the form of a polynomial in absolute temperature (such as Eq. (5.3.4)),

iii. The specific enthalpy of evaporation at the temperature of the system $\Delta h_{vap}(T)$.

All these data can be found in several handbooks such as Perry's handbook but can also be found in the NIST Chemistry WebBook (https://WebBook.nist.gov). For the specific enthalpy of evaporation NIST Chemistry WebBook proposes the use of the following approximate model

$$\Delta h_{vap}(T) = A\exp\left[-\beta\left(\frac{T}{T_c}\right)\right]\left(1-\frac{T}{T_c}\right)^{\beta} \tag{5.3.10}$$

where A, β are constants specific to each pure component and T_c is the critical temperature of the component. If only one value of the specific enthalpy of evaporation is available at a single temperature T_1, then the specific enthalpy of evaporation can be approximated at any other temperature T by the following empirical equation known as the Watson's equation

$$\frac{\Delta h_{vap}(T)}{\Delta h_{vap}(T_1)} = \left(\frac{T-T_c}{T_1-T_c}\right)^{0.38} \tag{5.3.11}$$

Example 5.1. Calculating the specific enthalpy of toluene at different temperatures and different states

Using a commercial simulator and an advanced thermodynamic model we calculate the following properties of pure toluene (C_7H_8, molar mass: 92.14) at 1 bar:

Data point	Temperature (°C/K)	State	Specific enthalpy (J/mol)
1	25 /298.15	Subcooled liquid	13028.5
2	109.8 /382.95	Saturated vapor	59857.6
3	226.9 /500	Superheated vapor	77765.4

Determine the specific enthalpy of toluene in these three states using the model proposed and compare the results with the data obtained from the commercial simulator.

Solution

We first collect all necessary data from NIST Chemistry WebBook (validate our findings):

Heat of formation of ideal gas at 298.15 K (three values are available we select the most recent one)	$\Delta h_f^{ig}(298.15\text{ K}) = 50100\,\frac{\text{J}}{\text{mol}}$
Heat of vaporization at normal boiling point ($T_{nb} = 383.8$ K)	$\Delta h_{vap}(T_{nb}) = 37300\,\frac{\text{J}}{\text{mol}}$
Constants of Eq. (5.3.10)	$A = 53090$ J/mol $\beta = 0.2774$ $T_c = 591.7$ K
Data on heat capacity T	c_p
298.15 K	103.7 J/(mol K)
300 K	104.4 J/(mol K)
400 K	139.9 J/(mol K)
500 K	170.8 J/(mol K)
600 K	196.2 J/(mol K)
Polynomial for c_p determined in MATLAB	$c_p = -32.81 + 0.53284T - 2.518 \times 10^4 T^2$

Note that the polynomial for the specific heat capacity is obtained in MATLAB using raw data from NIST Chemistry WebBook in the temperature range 250 K–600 K (using Tools/Basic Fitting and then select Quadratic):

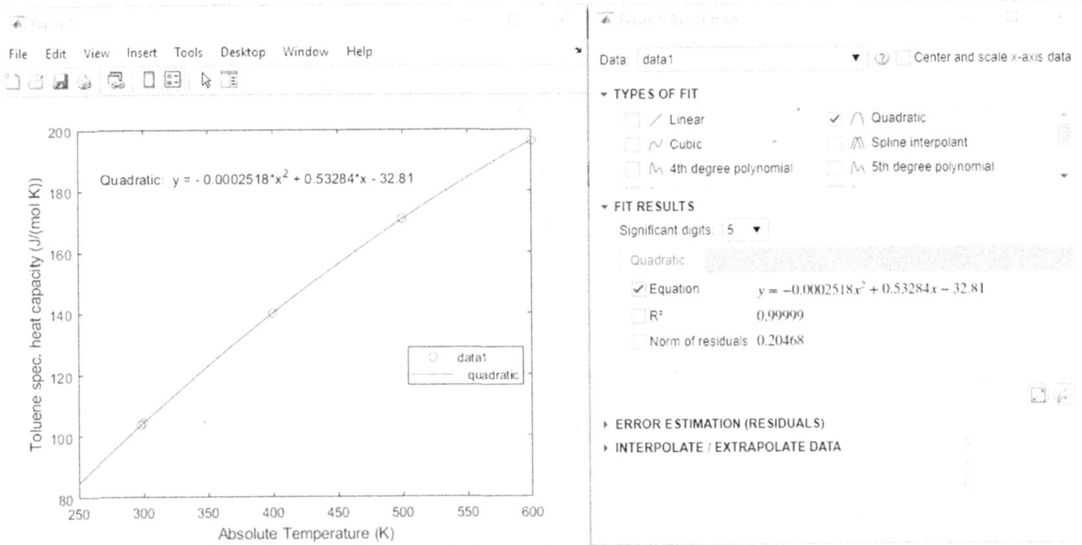

We start the calculations from the 3rd data point at 500 K. Toluene is in the gaseous state and we use Eq. (5.3.8)

$$h(T = 500\text{ K}, P = 1\text{ bar}, \text{state} = \text{gas}) = \Delta h_f^{ig}(T_{\text{ref}}) + \int_{T_{\text{ref}}}^{T} c_p^{ig}(T)dT$$

$$= \Delta h_f^{ig}(T_{\text{ref}}) - 32.81(T - T_{\text{ref}}) + \frac{0.53284}{2}\left(T^2 - T_{\text{ref}}^2\right) - \frac{2.518 \times 10^{-4}}{3}\left(T^3 - T_{\text{ref}}^3\right)$$

$$= 50100 - 32.81(500 - 298.15) + \frac{0.53284}{2}(500^2 - 298.15^2) - \frac{2.518 \times 10^{-4}}{3}\left(500^3 - 298.15^3\right) = 78\,132\ \frac{\text{J}}{\text{mol}}$$

At the second data point, the toluene is a saturated gas at its normal boiling point

$$h(T = 382.95\text{ K}, P = 1\text{ bar}, \text{state} = \text{gas})$$

$$= 50100 - 32.81(382.9 - 298.15) + \frac{0.53284}{2}(382.9^2 - 298.15^2) - \frac{2.518 \times 10^{-4}}{3}\left(382.9^3 - 298.15^3\right) = 60\,216\ \frac{\text{J}}{\text{mol}}$$

At the first data point, the toluene is a subcooled liquid and we now apply Eq. (5.3.9)

$$h(T = 298.15\ \text{K}, P = 1\ \text{bar}, \text{state} = \text{liquid}) = \Delta h_f^{ig}(T_{\text{ref}}) + \int_{T_{\text{ref}}}^{T} c_p^{ig}(T)dT - \Delta h_{\text{vap}}(T)$$

The temperature is equal to the reference temperature and the integral is zero. For the specific enthalpy of vaporization, we apply Eq. (5.3.10)

$$\Delta h_{\text{vap}}(T) = 53090 \exp\left[-0.2774\left(\frac{298.15}{591.7}\right)\right]\left(1 - \frac{298.15}{591.7}\right)^{0.2774} = 38\,007 \frac{\text{J}}{\text{mol}}$$

$$h(T = 298.15\ \text{K}, P = 1\ \text{bar}, \text{state} = \text{liquid}) = \Delta h_f^{ig}(T_{\text{ref}}) - \Delta h_{\text{vap}}(T) = 50100 - 38\,007 = 12\,093 \frac{\text{J}}{\text{mol}}$$

The calculated values and the values taken from the commercial simulator are compared in Table 5.1. The deviation is significant in the estimation of the specific enthalpy of the liquid but negligible for the gas.

An alternative approach to calculate the specific enthalpy of a pure component (second method) is based on starting the calculation from the specific enthalpy of formation at the liquid state at the reference temperature $\Delta h_{f,L}\left(T_{\text{ref}}\right)$ (we will be using the subscript L for the liquid state and subscript G for the gaseous state). To calculate the specific enthalpy for temperatures lower than the boiling temperature (T_b), such as for point C in Fig. 5.1, we may integrate the specific heat capacity of the liquid from the reference temperature (T_{ref}) (point A in Fig. 5.1) to the temperature of the system (T) (point B in Fig. 5.1)

$$h(T, \text{state} = \text{liquid}) = \Delta h_{f,L}(T_{\text{ref}}) + \int_{T_{\text{ref}}}^{T} c_{p,L}(T)dT \qquad (5.3.12)$$

However, if the component is at a gaseous state (such as point E in Fig. 5.1) then, as it is shown in Fig. 5.1, we must first add the heat of evaporation at the boiling point temperature ($\Delta h_{\text{vap}}(T_b)$) (moving from point C to point D in Fig. 5.1) and then integrate the specific heat capacity of the gas from the boiling temperature T_b

Table 5.1 Comparison of calculated values and values obtained from a commercial simulator.

Data point	Temperature (°C/K)	State	Specific enthalpy (J/mol)	Calculated enthalpy (J/mol)	% Deviation
1	25 /298.15	Liquid	13 028.5	12 093	−7.18
2	109.8 /382.95	Vapor	59 857.6	60 216	+0.60
3	226.9 /500	Vapor	77 765.4	78 132	+0.47

up to the temperature of the system T (moving from point D to point E in Fig. 5.1):

$$h(T, \text{state} = \text{gas}) = \Delta h_{f,L}(T_{\text{ref}}) + \int_{T_{\text{ref}}}^{T_b} c_{p,L}(T)dT + \Delta h_{\text{vap}}(T_b) + \int_{T_b}^{T} c_{p,G}(T)dT \tag{5.3.13}$$

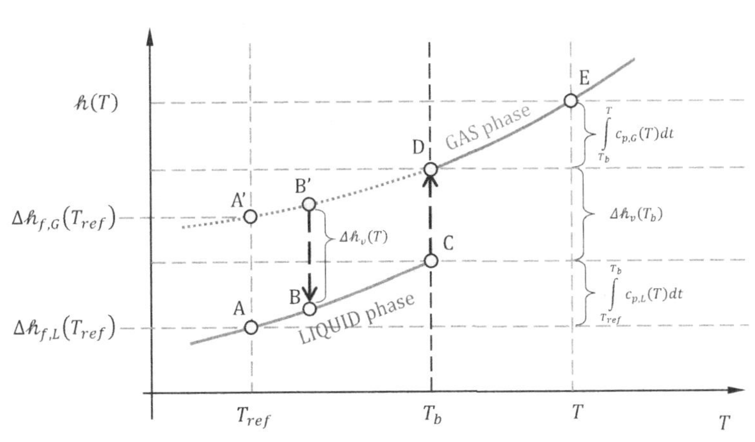

Figure 5.1 The alternative model for calculating the enthalpy of a pure component is based on both the specific heat capacity of the liquid and the gas state.

The two approaches that have been presented for calculating the enthalpy of a pure component can be explained using Fig. 5.1. If our aim is to calculate enthalpy at point B (liquid) then, in the second approach, we start from point A where we use Eq. (5.3.12):

$$h(B, \text{state} = \text{liquid}) = \underbrace{\Delta h_{f,L}(T_{\text{ref}})}_{\text{point A}} + \underbrace{\int_{T_{\text{ref}}}^{T} c_{p,L}(T)dT}_{\text{moving from A to B}}$$

Had we used the first approach then we would have used Eq. (5.3.9)

$$h(B, \text{state} = \text{liquid}) = \underbrace{\Delta h_f^{ig}(298.15\ \text{K})}_{\text{point A}'} + \underbrace{\int_{298.15\ \text{K}}^{T} c_p^{ig}(T)dT}_{\text{moving from A}' \text{ to B}'} + \underbrace{(-\Delta h_{\text{vap}}(T))}_{\text{moving from B}' \text{ to B}}$$

If our aim was to calculate enthalpy at point E (gas) then, in the second approach, we start from point A where we use Eq. (5.3.13):

$$h(E, \text{state} = \text{gas}) = \underbrace{\Delta h_{f,L}(T_{\text{ref}})}_{\text{point A}} + \underbrace{\int_{T_{\text{ref}}}^{T_b} c_{p,L}(T)dT}_{\text{moving from A to C}} + \underbrace{\Delta h_{\text{vap}}(T_b)}_{\text{moving from C to D}} + \underbrace{\int_{T_b}^{T} c_{p,G}(T)dT}_{\text{moving from D to E}}$$

Had we used the first approach then we would have used Eq. (5.3.10)

$$h(E, \text{state} = \text{gas}) = \underbrace{\Delta h_f^{ig}(298.15\ \text{K})}_{\text{point A}'} + \underbrace{\int_{298.15\ \text{K}}^{T} c_p^{ig}(T)dT}_{\text{moving from A}'\text{ to E}}$$

The first approach summarized by Eq. (5.3.8) and (5.3.9) is more efficient than the second method summarized by Eq. (5.3.12) and (5.3.13).

It is interesting to note that the specific enthalpy of formation of a component in the gas state $\Delta h_{f,G}(T, P)$ and in the liquid state $\Delta h_{f,L}(T, P)$ at the same temperature T and pressure P are related by the

$$\Delta h_{f,G}(T, P) = \Delta h_{f,L}(T, P) + \Delta h_{\text{vap}}(T, P) \tag{5.3.14}$$

Finally, in many cases the computations are simplified significantly if an average value of the specific heat capacity is used. The mean specific heat capacity between two temperatures T_1 and T_2 can be defined as

$$\bar{c}_p = \frac{1}{(T_2 - T_1)} \int_{T_1}^{T_2} c_p(T)dT \tag{5.3.15}$$

Example 5.2. Heating rate of water stored in a tank using electrical energy

Water is stored in a vessel at $T(0) = 25\,^\circ C$ as shown in Fig. 5.2. The mass of water is $M = 1000$ kg. The water starts to be heated by an electrical resistor with electrical power of $\dot{W}_{el} =$ 100 kW. Write the energy balance of the tank and calculate the heating rate of the liquid water in °C/min. Use an average specific heat capacity of liquid water and determine the analytic solution. Determine the time necessary to increase the temperature to 75 °C. For liquid water the specific heat capacity can be approximated by

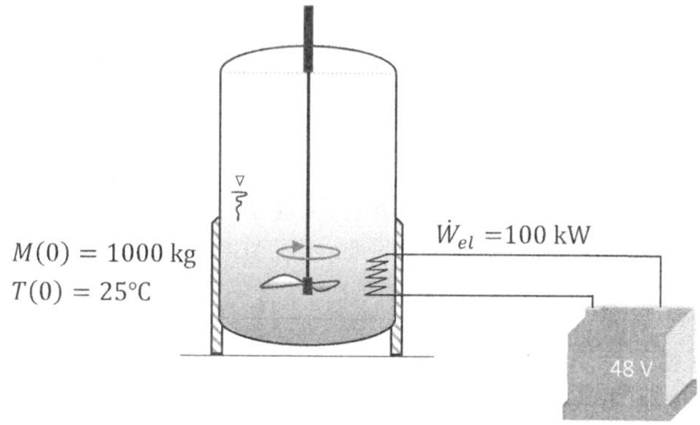

Figure 5.2 Water stored in a tank, heated by an electrical resistor.

$$c_{p,L} = -11.3018 + 0.0846T - 1.7743 \times 10^{-4}T^2 + 1.3735 \times 10^{-7}T^3 + 2.14 \times 10^5 T^{-2} \text{ in kJ/(kg·K)}.$$

Solution
We start with Eq. (5.2.8)

$$\frac{dU(t)}{dt} = \dot{H}_{\text{in}}(t) - \dot{H}_{\text{out}}(t) + \left[\dot{W} + \dot{Q}\right]_{\text{net}}$$

This is a closed system with no material stream crossing the system boundaries ($\dot{H}_{\text{in}} = \dot{H}_{\text{out}} = 0$) and there is no heat added or removed from the system ($\dot{Q} = 0$). The energy balance simplifies to

$$\frac{dU(t)}{dt} = \dot{W}_{el}$$

For the internal energy we have that $H = U + PV$, but as pressure and volume are constant it follows that

$$\frac{dU(t)}{dt} = \frac{d(H(t) - PV)}{dt} = \frac{dH(t)}{dt} - \frac{d(PV)}{dt} = \frac{dH(t)}{dt}$$

The enthalpy is now the product of the mass of water times the specific enthalpy of water

$$H(t) = Mh(t)$$

The differential energy balance is written as

$$\frac{dh(t)}{dt} = \frac{\dot{W}_{el}}{M}$$

Which can be integrated to obtain

$$h(t) - h(0) = \left(\frac{\dot{W}_{el}}{M}\right)t$$

According to the second model for calculating the specific enthalpy

$$h(T(t)) = \Delta h_{f,L}(T_{\text{ref}}) + \int_{T_{\text{ref}}}^{T(t)} c_{p,L}(\theta)d\theta$$

from which it follows that

$$h(T(t)) - h(T(0)) = \left[\Delta h_{f,L}(T_{\text{ref}}) + \int_{T_{\text{ref}}}^{T(t)} c_{p,L}(\theta)d\theta\right] - \left[\Delta h_{f,L}(T_{\text{ref}}) + \int_{T_{\text{ref}}}^{T(0)} c_{p,L}(\theta)d\theta\right] = \int_{T_{\text{ref}}}^{T(t)} c_{p,L}(\theta)d\theta - \int_{T_{\text{ref}}}^{T(0)} c_{p,L}(\theta)d\theta = \int_{T(0)}^{T(t)} c_{p,L}(\theta)d\theta$$

Using the integral of the specific heat capacity equation (see Eq. (2.9.13)) we obtain

$$A(T(t) - T(0)) + \frac{B}{2}\left(T(t)^2 - T(0)^2\right) + \frac{C}{3}\left(T(t)^3 - T(0)^3\right) + \frac{D}{4}\left(T(t)^4 - T(0)^4\right) - E\left(\frac{1}{T(t)} - \frac{1}{T(0)}\right) = \left(\frac{\dot{W}_{el}}{M}\right)t$$

However, this nonlinear equation generally requires computational tools for its solution. If we assume a constant average specific heat capacity for liquid water then

$$h(T(t)) - h(T(0)) = \int_{T(0)}^{T(t)} c_{p,L}(\theta)d\theta \approx \bar{c}_{p,L}(T(t) - T(0))$$

and the energy balance becomes

$$\bar{c}_{p,L}(T(t) - T(0)) = \left(\frac{\dot{W}_{el}}{M}\right)t$$

or

$$T(t) = T(0) + \frac{1}{\bar{c}_{p,L}}\left(\frac{\dot{W}_{el}}{M}\right)t$$

which is a linear function of time. To calculate the constant term appearing in the right-hand side we need to calculate the average specific heat capacity

$$\bar{c}_p = \frac{1}{(T_2 - T_1)}\int_{T_1}^{T_2} c_p(T)dT = \frac{1}{(T_2 - T_1)}\Bigg[A(T_2 - T_1) + \frac{B}{2}(T_2^2 - T_1^2) + \frac{C}{3}(T_2^3 - T_1^3) + \frac{D}{4}(T_2^4 - T_1^4) - E\left(\frac{1}{T_2} - \frac{1}{T_1}\right)\Bigg]$$

We replace the numerical values (note that $T_1 = 298.15$ K; $T_2 = 348.15$ K)

$$\bar{c}_p = \frac{1}{(348.15 - 298.15)}\Bigg[-11.3018 \times (348.15 - 298.15) + \frac{0.0846}{2} \times (348.15^2 - 298.15^2) - \frac{1.7743 \times 10^{-4}}{3} \times (348.15^3 - 298.15^3) + \frac{1.3735 \times 10^{-7}}{4} \times \left((348.15^4 - 298.15^4)\right) - 2.14 \times 10^5 \times \left(\frac{1}{348.15} - \frac{1}{298.15}\right)\Bigg] = 4.1957\frac{\text{kJ}}{\text{kg}\cdot\text{K}}$$

We therefore have

$$\frac{1}{\bar{c}_{p,L}}\left(\frac{\dot{W}_{el}}{M}\right) = \frac{1}{4.1957\dfrac{\text{kJ}}{\text{kg}\cdot\text{K}}}\left(\frac{100\dfrac{\text{kJ}}{\text{s}}}{1000\ \text{kg}}\right) \times \frac{60\ \text{s}}{1\ \text{min}} = 1.43\frac{\text{K}}{\text{min}}$$

The temperature of the liquid water as a function of time is given by the following equation

$$T(t) = 25 + \underbrace{1.43}_{\text{heating rate}} \; t, t \text{ in min}, T \text{ in } ^\circ\text{C}$$

The time at which the temperature becomes equal to 75 °C can be found easily

$$t = \frac{75 - 25}{1.43} \approx 35 \text{ min}$$

Example 5.3. Heating rate of water stored in a tank using mixing with hot water stream

Water is stored in a vessel at $T(0) = 25\,^\circ C$ as shown in Fig. 5.3. The mass of water is $M = 1000$ kg. A stream of saturated liquid water with mass flowrate of $\dot{M}_{\text{in}} = 1$ kg/s and temperature $T_{\text{in}} = 100$ °C is fed to the tank. Write the energy balance of the tank and calculate the variation of the temperature as a function of time. Use an average specific heat capacity of liquid water $\bar{c}_{p,L} = 4.2$ kJ/(kg K) to determine the analytic solution. Determine the time necessary to increase the temperature to 75 °C.

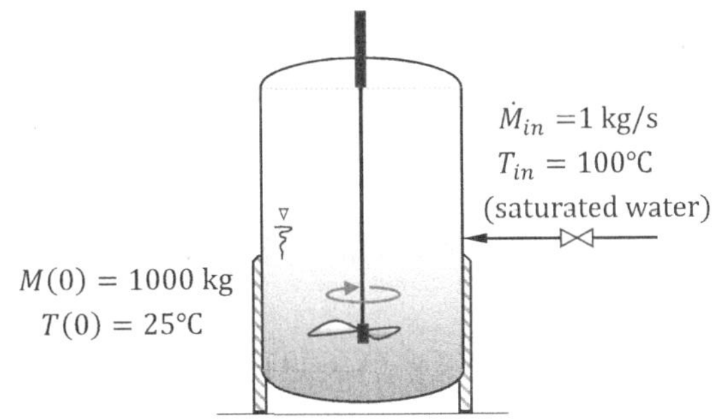

Figure 5.3 Water stored in a tank, heated by saturated water stream.

Solution

We start with Eq. (5.2.8)

$$\frac{dU(t)}{dt} = \dot{H}_{\text{in}}(t) - \dot{H}_{\text{out}}(t) + \left[\dot{W} + \dot{Q}\right]_{\text{net}}$$

There is no heat or work interaction with the environment ($\dot{W} = \dot{Q} = 0$) and no material stream is leaving the vessel. Therefore

$$\frac{dU(t)}{dt} = \dot{H}_{\text{in}}(t) = \dot{M}_{\text{in}}h(T_{\text{in}})$$

Using the steps used in Example 5.2 we obtain

$$\frac{dH(t)}{dt} = \dot{M}_{\text{in}}h(T_{\text{in}})$$

We note that

$$H(t) = M(t)h(T(t))$$

From the last two equations

$$\frac{dH(t)}{dt} = M(t)\frac{dh(t)}{dt} + h(t)\frac{dM(t)}{dt} = \dot{M}_{\text{in}}h(T_{\text{in}})$$

From the overall material balance equation we have that

$$\frac{dM(t)}{dt} = \dot{M}_{\text{in}}$$

and we finally obtain

$$M(t)\frac{dh(t)}{dt} = \dot{M}_{\text{in}}(h(T_{\text{in}}) - h(t))$$

Note that for the liquid water

$$h(t) = \Delta h_{f,L}(T_{\text{ref}}) + \int_{T_{\text{ref}}}^{T(t)} c_{p,L}(\theta)d\theta \approx \Delta h_{f,L}(T_{\text{ref}}) + \bar{c}_{p,L}(T(t) - T_{\text{ref}})$$

and

$$\frac{dh(t)}{dt} = \frac{d}{dt}(\Delta h_{f,L}(T_{\text{ref}}) + \bar{c}_{p,L}(T(t) - T_{\text{ref}})) = \bar{c}_{p,L}\frac{dT(t)}{dt}$$

while for the saturated liquid water (T_b denotes the boiling point temperature)

$$h_{\text{in}} = \Delta h_{f,L}(T_{\text{ref}}) + \int_{T_{\text{ref}}}^{T_b} c_{p,L}(\theta)d\theta \approx \Delta h_{f,L}(T_{\text{ref}}) + \bar{c}_{p,L}(T_b - T_{\text{ref}})$$

Note that

$$h(T_{\text{in}}) - h(t) = \left(\Delta h_{f,L}(T_{\text{ref}}) + \bar{c}_{p,L}(T_b - T_{\text{ref}})\right) - \left(\Delta h_{f,L}(T_{\text{ref}}) + \bar{c}_{p,L}(T(t) - T_{\text{ref}})\right) = \bar{c}_{p,L}\left(T_b - T(t)\right)$$

We replace in the differential energy balance

$$M(t)\frac{dT(t)}{dt} = \dot{M}_{\text{in}}T_b - \dot{M}_{\text{in}}T(t)$$

or

$$\frac{dT(t)}{dt} + p(t)T(t) = q(t),\, T(0) = T_0$$

where

$$p(t) = \left(\frac{\dot{M}_{\text{in}}}{M(0) + \dot{M}_{\text{in}}t}\right), \qquad q(t) = \left(\frac{\dot{M}_{\text{in}}}{M(0) + \dot{M}_{\text{in}}t}\right)T_b$$

We calculate the integrating factor

$$\int \frac{\dot{M}_{\text{in}}}{M(0) + \dot{M}_{\text{in}}t}dt = \int \frac{d\left(M(0) + \dot{M}_{\text{in}}t\right)}{M(0) + \dot{M}_{\text{in}}t}$$

$$= \int d\ln\left(M(0) + \dot{M}_{\text{in}}t\right) = \ln\left(M(0) + \dot{M}_{\text{in}}t\right)$$

It follows that

$$\exp\left(\int p(t)dt\right) = \exp\left(\ln\left(M(0) + \dot{M}_{\text{in}}t\right)\right) = M(0) + \dot{M}_{\text{in}}t = M(t)$$

We then have

$$\frac{d}{dt}(M(t)T(t)) = M(t)q(t) = T_b\dot{M}_{\text{in}}$$

and we integrate

$$\int_0^t d(M(\tau)T(\tau)) = T_b\dot{M}_{\text{in}}\int_0^t d\tau$$

to obtain

$$M(t)T(t) - M(0)T(0) = T_b\dot{M}_{\text{in}}t$$

or

$$T(t) = \frac{M(0)T(0) + \left(\dot{M}_{\text{in}}t\right)T_b}{M(0) + \dot{M}_{\text{in}}t}$$

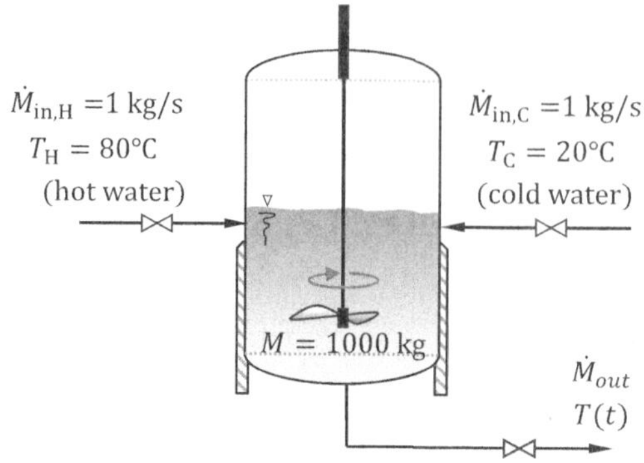

Figure 5.4 Mixing two streams of pure water to produce a stream of desired temperature.

We replace the numerical values

$$T(t) = \frac{1000 \times 298.15 + 373.15t}{1000 + t}$$

To determine the time that the temperature obtains a specific value we need to solve for the time

$$t = 1000\left(\frac{T(t) - 298.15}{373.15 - T(t)}\right)$$

The temperature becomes equal to 75 °C when the time is equal to

$$t = 1000\left(\frac{348.15 - 298.15}{373.15 - 348.15}\right) = 1000\left(\frac{50}{25}\right) = 2000 \text{ s} = 33.33 \text{ min}$$

Example 5.4. Mixing of two streams of the same pure substance with different temperatures

Two pure water streams of different temperatures are mixed in a tank to produce a stream of desired temperature, as shown in Fig. 5.4. The feeding streams are a hot water stream with mass flowrate of $\dot{M}_{in,H} = 1\frac{kg}{s}$ and a temperature of $T_H = 80$ °C and a cold water stream with mass flowrate of $\dot{M}_{in,C} = 1\frac{kg}{s}$ and a temperature of $T_C = 20$ °C. The startup of the process at a particular occasion begins with an initial temperature of the water in the tank of $T(0) = 35$ °C. Due to the malfunction of the control system, the hot water stream shuts down for 5 min, 1.5 h after start up. Following the loss of the hot water stream, the cold water stream mass flowrate is immediately doubled by the control system to keep the liquid level constant. Find whether the temperature of the product stream falls below 40 °C and becomes off-spec (causing serious disruption in a downstream process operation). The mass of water in the tank is $M = 1000$ kg and is constant. Write the energy balance of the tank and calculate the dynamics of the system (starting up and hot water stream loss) by solving the differential equations in MATLAB using the heat capacity equation for liquid water:

$$c_p = -11.3018 + 0.0846T - 1.7743 \times 10^{-4}T^2 + 1.3735 \times 10^{-7}T^3 + 2.14 \times 10^5 T^{-2} \text{ in kJ/(kg·K)}.$$

Solution

We start with Eq. (5.2.8)

$$\frac{dU(t)}{dt} = \dot{H}_{\text{in}}(t) - \dot{H}_{\text{out}}(t) + \left[\dot{W} + \dot{Q}\right]_{\text{net}}$$

There is no work or heat exchanged between the tank and the environment. There are two incoming streams and one outgoing stream

$$\frac{dU(t)}{dt} = \dot{H}_{\text{in,C}}(t) + \dot{H}_{\text{in,H}}(t) - \dot{H}_{\text{out}}(t)$$

The volume is constant and if a constant density is assumed then the holdup in the vessel is also constant. We also use the equations

$$\dot{H}_{\text{in,C}} = \dot{M}_{\text{in,C}} h(T_{\text{C}})$$

$$\dot{H}_{\text{in,H}} = \dot{M}_{\text{in,H}} h(T_{\text{H}})$$

$$\dot{H}_{\text{out}} = \dot{M}_{\text{out}} h(T) = \left(\dot{M}_{\text{in,C}} + \dot{M}_{\text{in,H}}\right) h(T)$$

Due to the assumptions of constant volume and constant density it follows that $\dot{M}_{\text{out}} = \dot{M}_{\text{in,C}} + \dot{M}_{\text{in,H}}$, which has been used in the last equation. A final observation is that for (nearly) incompressible substances such as solids and liquids we may assume that internal energy is practically equal to enthalpy and using the chain rule we obtain

$$\frac{dU}{dt} = \frac{dH}{dt} = M\frac{dh}{dt} = M\frac{dh}{dT}\cdot\frac{dT}{dt} = Mc_p(T)\frac{dT}{dt}$$

Putting everything together

$$Mc_p(T)\frac{dT}{dt} = \dot{M}_{\text{in,C}} h(T_{\text{C}}) + \dot{M}_{\text{in,H}} h(T_{\text{H}}) - \left(\dot{M}_{\text{in,C}} + \dot{M}_{\text{in,H}}\right) h(T)$$

We solve the problem in MATLAB. We first create the function that calculates the specific enthalpy of liquid water

```
function [h,cp]=SpeEnth(T)
% Calculates the spec. enthalpy and spec. heat capacity
% of liquid water at absolute temperature T
% Properties taken from NIST/Chemistry Webbook
% h is the specific enthalpy of water in kJ/kg
% cp is the specific heat capacity of water in kJ/(kg K)
% T is the absolute temperature in K
% cp = A + BT + CT^2 + DT^3 + ET^(-2)

A    = -11.3018;
B    =   0.0846;
C    =  -1.7743e-04;
D    =   1.3735e-07;
E    =   2.1400e+05;
T_ref = 273.15+25; % reference temperature
% enthalpy of formation of liquid water at 25 degC
Dhf = 285830/18.0153;
h = Dhf + A*(T-T_ref)  + B*(T^2-T_ref^2)/2 +...
                       + C*(T^3-T_ref^3)/3 +...
                       + D*(T^4-T_ref^4)/4 +...
                       - E*(1/T-1/T_ref);
cp = A+B*T+C*T^2+D*T^3+E/T^2;
end
```

Then the function that calculates the right-hand side of the differential equation is created

```
function dTdt=MixingTank(t,T)
% mixing two pure water streams of different temperatures
% starting up at -1.5 h and hot water stream failure at t=0 h

M         = 1000;          % water holdup in the tank in kg
TC        = 273.15 + 20;  % temperature of the cold stream degrees C
hC        = SpeEnth(TC);  % spec.enthalpy of the cold stream in kJ/kg
M_dot_C = 1;               % mass flowrate of the cold stream in kg/s
TH        = 273.15 + 80;  % temperature of the hot stream degrees C
hH        = SpeEnth(TH);  % spec.enthalpy of the hot stream in kJ/kg
M_dot_H = 1;               % mass flowrate of the hot stream in kg/s
[h,cp]    = SpeEnth(T);   % spec.enth. of the product stream in kJ/kg

if ( (t >= 0) && (t <= 300))
    M_dot_H = 0; % hot water stream fails
    M_dot_C = 2; % cold water stream is doubled
end

% mass flowrate of the product stream in kg/s
M_dot    = M_dot_C + M_dot_H;

% differential energy balance
dTdt = (M_dot_H*hH + M_dot_C*hC - M_dot*h)/(M*cp);
end
```

Finally, the problem is solved using the `ode15s` solver

```
clear all
T0 = 273.15 + 35; % initial temperature of the water in the tank
options=odeset('AbsTol',1e-8,'RelTol',1e-8)
t0 = -5400; % initial time in s (start up begins)
tf = +5400; % final time in s
[time,T]=ode15s(@MixingTank,[t0 tf],T0,options);

figure(1)
plot(time/3600,T-273.15,'linewidth',1.2)
xlabel('Time (h)')
ylabel('Temperature (°C)')
grid on
hold on
plot([t0 tf]/3600,[40 40],'r--','LineWidth',1.2)
axis([-0.25 1 35 55])
```

The simulation starts from negative times ($t_0 = -5400$ s) and the system quickly reaches a steady state at which the temperature is 50 °C. Then, from $t = 0$ until $t = 300$ s, the hot water stream is lost and the system temperature is decreasing and violates the minimum temperature constraint, as the water temperature drops below 40 °C, as shown in Fig. 5.5.

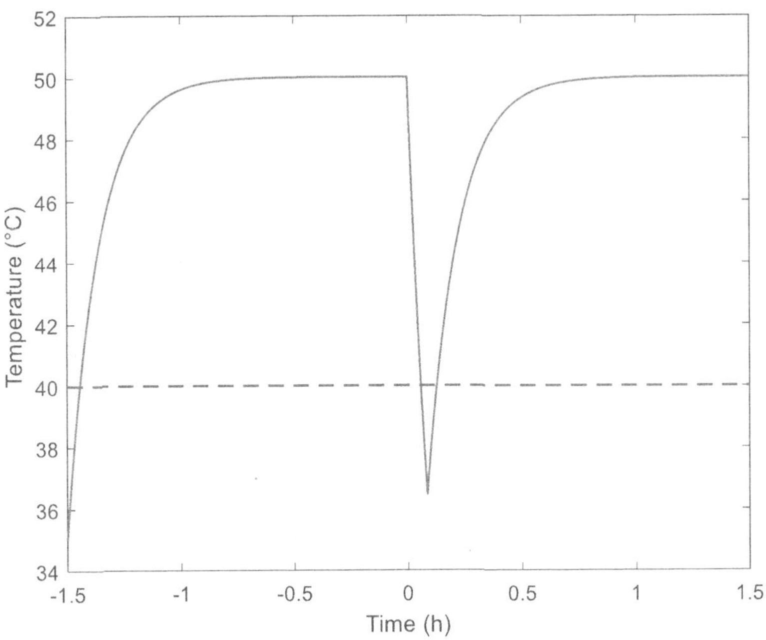

Figure 5.5 Simulation of the mixing of two pure water streams of different temperatures in a tank. The hot water stream fails 1.5 h after starting up for 5 min resulting in an off-spec product stream.

5.4 Enthalpy of reaction

We consider again the following general reaction

$$\sigma_1 R_1 + \sigma_2 R_2 + \ldots + \sigma_r R_r + \sigma_{r+1} P_1 + \sigma_{r+2} P_2 + \ldots + \sigma_{r+p} P_p = 0$$

where R_i is the i-th reactant, P_j is the j-th product, $\sigma_i, i = 1,2\ldots,r$ is the stoichiometric number of the i-th reactant and $\sigma_{r+j}, j = 1,2,\ldots,p$ is the stoichiometric number of the j-th product. We remind ourselves that we have defined the stoichiometric numbers σ_k to be equal to the stoichiometric coefficients in absolute values but are negative for the reactants and positive for the products. If we consider the reaction

$$(-1)CO_2 + (-4)H_2 + (+1)CH_4 + (+2)H_2O = 0 \quad (5.4.1)$$

then $\sigma_{CO_2} = -1$, $\sigma_{H_2} = -4$, $\sigma_{CH_4} = +1$ and $\sigma_{H_2O} = +2$. Note that the conventional way of writing this reaction is

$$CO_2 + 4H_2 \leftrightharpoons CH_4 + 2H_2O \quad (5.4.2)$$

Consider now the hypothetical open system of Fig. 5.6. Stoichiometric amounts of the reactants are fed to a reactor at the reference conditions of 25 °C and 1 bar. The reactants are completely converted to products at the same reference conditions of 25 °C and 1 bar. To keep the temperature constant an amount of heat Q is exchanged with the environment. This amount of heat is of special interest to process systems engineering.

We now apply the energy balance equation, Eq. (5.2.8), at steady state

$$0 = \dot{H}_{in}(t) - \dot{H}_{out}(t) + \dot{Q} \quad (5.4.3)$$

or, after rearrangement and application for a unit period of time

$$\Delta H_{rxn} = H_{out}(t) - H_{in}(t) = Q \quad (5.4.4)$$

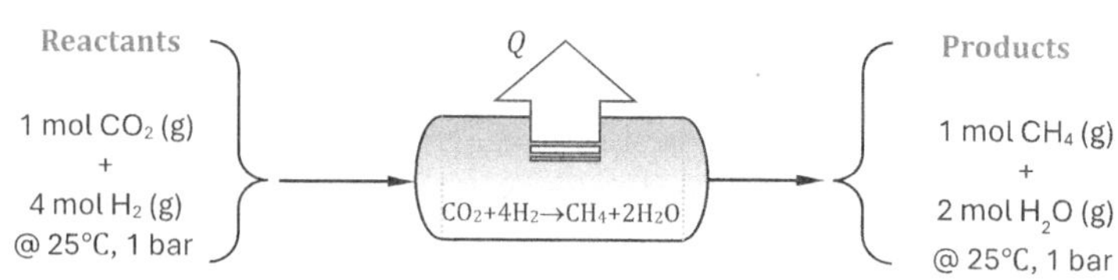

Figure 5.6 Complete conversion of reactants to products at constant reference conditions.

We conclude that the change of the enthalpy during the reaction (rxn) in which stoichiometric amounts of reactants are completely transformed into products under constant temperature and pressure equals the heat exchanged with the environment. This is called the **heat** or **enthalpy of reaction**. If the heat of reaction is negative then the reaction is called **exothermic** while if it is positive then the reaction is called an **endothermic** reaction. In an exothermic reaction heat is released from the system to the environment to keep the temperature constant ($Q<0$). The **standard enthalpy of reaction** (denoted $\Delta h^{\ominus}_{rxn}$) for a chemical reaction is the difference between total product and total reactant molar enthalpies, calculated for substances in their standard states: 25 °C and 1 bar.

According to the way that we have defined enthalpy of pure compounds, which is equal to the standard enthalpy of formation at the standard conditions of temperature and pressure, we have that

$$h_i\left(T_{\text{ref}} = T^{\ominus} = 298.15\ \text{K}, P_{\text{ref}} = P^{\ominus} = 1\ \text{bar}\right) = \Delta h_{f,i}\left(T^{\ominus}, P^{\ominus}\right) = \Delta h^{\ominus}_{f,i} \tag{5.4.5}$$

Using this result, we can write (we remind that $\pi_i = \sigma_i$ are the stoichiometric coefficients for the products and $\rho_j = -\sigma_j$ are the stoichiometric coefficients for the reactants)

$$H_{\text{out}}(t) = \pi_1 \Delta h^{\ominus}_{f,1} + \pi_2 \Delta h^{\ominus}_{f,2} + \ldots + \pi_p \Delta h^{\ominus}_{f,r} = \sum_{i:\text{product}} \pi_i \Delta h^{\ominus}_{f,i} \tag{5.4.6}$$

and

$$H_{\text{in}}(t) = \rho_1 \Delta h^{\ominus}_{f,1} + \rho_2 \Delta h^{\ominus}_{f,2} + \ldots + \rho_r \Delta h^{\ominus}_{f,r} = \sum_{i:\text{reactant}} \rho_i \Delta h^{\ominus}_{f,i} \tag{5.4.7}$$

By replacing (5.4.6) and (5.4.7) into (5.4.4) we obtain

$$\begin{aligned}
\Delta h^{\ominus}_{rxn} = H_{\text{out}}(t) - H_{\text{in}}(t) &= \sum_{i:\text{product}} \pi_i \Delta h^{\ominus}_{f,i} \\
&\quad - \sum_{j:\text{reactant}} \rho_j \Delta h^{\ominus}_{f,j} \\
&= \sum_{i:\text{product}} \sigma_i \Delta h^{\ominus}_{f,i} - \sum_{j:\text{reactant}} (-\sigma_j) \Delta h^{\ominus}_{f,j} \\
&= \sum_{i:\text{product}} \sigma_i \Delta h^{\ominus}_{f,i} + \sum_{j:\text{reactant}} \sigma_j \Delta h^{\ominus}_{f,j}
\end{aligned}$$

or

$$\Delta h^{\ominus}_{\text{rxn}} = \sum_i \sigma_i \Delta h^{\ominus}_{f,i} \tag{5.4.8}$$

Example 5.5. Standard enthalpy of reaction calculation

Calculate the standard enthalpy of reaction for the reaction given by (5.4.2) in two separate forms:

a. $CO_2(g) + 4H_2(g) \leftrightharpoons CH_4(g) + 2H_2O\ (g)$.

and

b. $CO_2(g) + 4H_2(g) \leftrightharpoons CH_4(g) + 2H_2O\ (l)$.

Solution

In Table 5.2 we present the data collected from the NIST Chemistry WebBook and we summarize the calculations involved in Eq. (5.4.8). Note that $\Delta h^{\ominus}_{\text{rxn}} = -164\,750 \frac{\text{kJ}}{\text{kmol}}$ for reaction a (gaseous water) and $\Delta h^{\ominus}_{\text{rxn}} = -252\,750 \frac{\text{kJ}}{\text{kmol}}$ for reaction b (liquid water). The practical meaning is that if CO_2 and H_2 are fed to the reactor of Fig. 5.6 in stoichiometric proportions and converted completely to products (CH_4 and H_2O), then the heat that needs to be removed for each kmol of CO_2 in the feed equals 164 750 kJ (/kmol CO_2 fed) for reaction a and 252 750 kJ (/kmol CO_2 fed) for reaction b.

In most thermodynamics and introductory chemical engineering textbooks, considerable emphasis is placed on the

Table 5.2 Tabular form of the calculation of the standard enthalpy of reaction for the reaction: $CO_2(g) + 4H_2(g) \rightarrow CH_4(g) + H_2O(g\ \text{or}\ l)$.

Species	State	Stoichiometric number σ_i	Standard enthalpy of formation $\Delta h^{\ominus}_{f,i}$ (kJ/kmol)	$\sigma_i \Delta h^{\ominus}_{f,i}$ (reaction a)	$\sigma_i \Delta h^{\ominus}_{f,i}$ (reaction b)
CO_2	(g)	−1	−393 510	393 510	393 510
H_2	(g)	−4	0	0	0
CH_4	(g)	+1	−74 600	−74 600	−74 600
H_2O	(g)	+2	−241830	−483 660	–
H_2O	(l)	+2	−285830	–	−571660
			$\Delta h^{\ominus}_{\text{rxn}} = \sum_i \sigma_i \Delta h^{\ominus}_{f,i} =$	−164 750	−252 750

Data taken from NIST Chemistry WebBook.

standard enthalpy of reaction. However, its practical utility is quite limited. This is primarily because industrial reactions rarely occur under standard conditions. In practice, reactants are typically introduced at elevated temperatures to achieve acceptable reaction rates, and products are often removed at different temperatures. For this reason, we will not rely on the standard enthalpy of reaction ($\Delta h^{\ominus}_{\mathrm{rxn}}$) in this text. Instead, we focus on the more relevant quantity: the **enthalpy of reaction** at arbitrary conditions, $\Delta h_{\mathrm{rxn}}(T, P)$. This term represents the same fundamental concept but without the constraint of standard conditions (25 °C and 1 bar). It applies to any temperature T and pressure P, making it far more applicable in real-world process analysis.

For reactions occurring at constant temperature and pressure, the enthalpy of reaction can be calculated as follows

$$\Delta h_{\mathrm{rxn}}(T, P) = \sum_i \sigma_i h_i(T, P) \tag{5.4.9}$$

In the example that follows we analyze the idea of the enthalpy of reaction further.

Example 5.6. Calculating the enthalpy of reaction at several temperatures

Calculate the enthalpy of reaction for the reaction

$$CO_2(g) + 4H_2(g) \leftrightharpoons CH_4(g) + 2H_2O\ (g)$$

Table 5.3 Data for calculating the standard enthalpy of reaction for the reaction.

Species	Standard enthalpy of formation $\Delta h^{\ominus}_{f,i}$ (kJ/kmol)	A	B×10³	C× 10⁶	D×10⁹	E×10⁻⁶
$CO_2(g)$	−393 510	24.99735	55.186960	−33.691370	7.948387	−0.136638
$H_2(g)$	0	33.066178	−11.363417	11.432816	−2.772874	−0.158558
$CH_4(g)$	−74 600	−0.703029	108.477300	−42.521570	5.862788	0.678565
$H_2O(g)$	−241830	30.092000	6.832514	6.793435	−2.534480	0.082139

$C_p \left[\frac{\mathrm{kJ}}{\mathrm{kmol}}\right] = A + BT + CT^2 + DT^3 + ET^{-2}$

Data taken from NIST Chemistry WebBook: $CO_2(g)+4H_2(g) \rightarrow CH_4(g)+H_2O(g)$.

for temperatures ranging from 300 K to 1000 K at P = 1 bar. Take all necessary data from the NIST Chemistry WebBook (select the most recent data in case more than one value is available).

Solution

Manual calculation of the heat of reaction is impractical, if not impossible, due to its complexity. This is where computational tools become essential. Data collected from NIST Chemistry WebBook are summarized in Table 5.3. The following MATLAB function computes the enthalpy of reaction through a direct application of Eq. (5.4.9)

```
function Dh_rxn=HeatOfReaction(T)
% Calculates the heat of reaction CO2+4H2->CH4+2H2O
% at a given absolute temperature T (row vector).
% All components are in gaseous state.
% Properties taken from NIST/Chemistry Webbook.
data=[-1 -393510 24.997350  55.186960 -33.691370  7.948387 -0.136638
      -4     0    33.066178 -11.363417  11.432816 -2.772874 -0.158558
      +1 -74600  -0.703029 108.477300 -42.521570  5.862788  0.678565
      +2 -241830 30.092000   6.832514   6.793435 -2.534480 .082139];
sigma = data(:,1); % stoichiometric numbers
Dh_f  = data(:,2); % spec.standard enth. of formation in kJ/kmol
A = data(:,3); B = data(:,4)/10^3; C = data(:,5)/10^6;
               D = data(:,6)/10^9; E = data(:,7)*10^6;
T_ref = 273.15+25; % reference temperature
h = Dh_f + A*(T-T_ref) + B*(T.^2-T_ref^2)/2 +...
                        + C*(T.^3-T_ref^3)/3 +...
                        + D*(T.^4-T_ref^4)/4  - E*(1./T-1/T_ref);
Dh_rxn = sigma'*h;
end
```

The calculation of the enthalpy of reaction can now be performed as follows

```
clear all
T=linspace(300,1000,21);
plot(T,HeatOfReaction(T),'-o','Linewidth',1.2)
xlabel('Absolute temperature (K)')
ylabel('Δh_{rxn}(T)')
grid on
text(500,-167500,'CO_{2}+4H_{2}->CH_{4}+2H_{2}O')
```

The results are shown in Fig. 5.7. We note that the heat of reaction varies significantly with temperature.

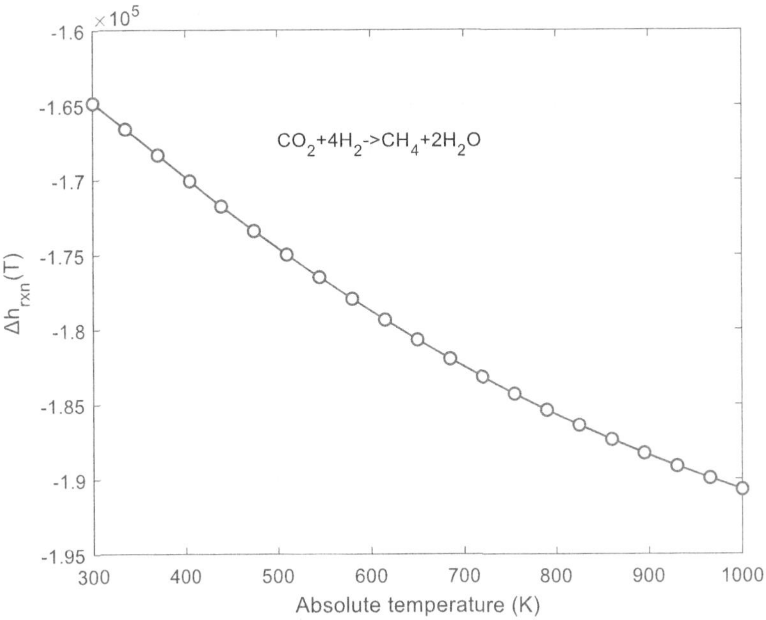

Figure 5.7 Heat of the reaction $CO_2+4H_2 \rightarrow CH_4+2H_2O$ as a function of the absolute temperature.

5.5 Solution of nonlinear algebraic equations

In this section we will consider the problem shown in Fig. 5.8 in which a stream of a pure component with known molar flowrate $\dot{N}$ and initial temperature T_0 is cooled by removing heat at a rate $\dot{Q}$. The process is assumed to be at steady state. The temperature at which the stream is cooled is unknown and needs to be determined. If we apply the general energy balance (Eq. 5.2.8) we obtain

$$\underbrace{\frac{dU(t)}{dt}}_{=0 \text{ at steady-state}} = \dot{H}_{\text{in}}(t) - \dot{H}_{\text{out}}(t) + \left[\underbrace{\dot{W}}_{=0} + \dot{Q}\right]_{\text{net}}$$

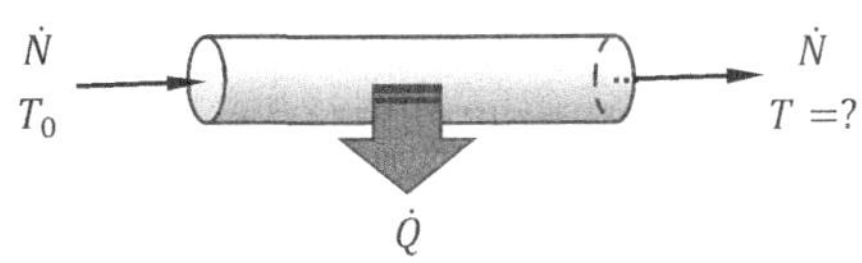

Figure 5.8 A pure component stream is cooled by removing heat at a rate Q.

It then follows that

$$\dot{H}_{\text{out}}(T) = \dot{H}_{\text{in}}(T_0) + \dot{Q} \tag{5.5.1}$$

The right-hand side involves only known quantities while in the left-hand side the unknown exit temperature T appears. Let's assume that a constant mean heat capacity is used, then

$$\dot{H}_{\text{in}}(T_0) = \dot{N}h(T_0) = \dot{N}\left[\Delta h_f(T_{\text{ref}}) + \bar{c}_p(T_0 - T_{\text{ref}})\right] \tag{5.5.2}$$

and

$$\dot{H}_{\text{out}}(T) = \dot{N}h(T) = \dot{N}\left[\Delta h_f(T_{\text{ref}}) + \bar{c}_p(T - T_{\text{ref}})\right] \tag{5.5.3}$$

By replacing Eq. (5.5.2) and (5.5.3) into Eq. (5.5.1) we obtain

$$\dot{N}\left[\Delta h_f(T_{\text{ref}}) + \bar{c}_p(T - T_{\text{ref}})\right] = \dot{N}\left[\Delta h_f(T_{\text{ref}}) + \bar{c}_p(T_0 - T_{\text{ref}})\right] + \dot{Q}$$

or

$$T = T_0 + \frac{\dot{Q}}{\bar{c}_p\dot{N}} \tag{5.5.4}$$

and the solution can be obtained easily. However, the constant specific heat capacity assumption might not be valid and the solution can involve significant error.

If the exact solution is necessary then we must use the exact definition of specific enthalpy (assume that there is no phase change)

$$\dot{H}_{\text{in}}(T_0) = \dot{N}h(T_0) = \dot{N}\left[\Delta h_f(T_{\text{ref}}) + \int_{T_{\text{ref}}}^{T_0} c_p(\theta)d\theta\right] \tag{5.5.5}$$

and

$$\dot{H}_{\text{out}}(T) = \dot{N}h(T) = \dot{N}\left[\Delta h_f(T_{\text{ref}}) + \int_{T_{\text{ref}}}^{T} c_p(\theta)d\theta\right] \tag{5.5.6}$$

By replacing Eq. (5.5.5) and (5.5.6) into Eq. (5.5.1) we obtain

$$\dot{N}\left[\Delta h_f(T_{\text{ref}}) + \int\limits_{T_{\text{ref}}}^{T} c_p(\theta)d\theta\right] = \dot{N}h(T_0) + \dot{Q}$$

or

$$\int\limits_{T_{\text{ref}}}^{T} c_p(\theta)d\theta = \left(h(T_0) - \Delta h_f(T_{\text{ref}})\right) + \frac{\dot{Q}}{\dot{N}} \tag{5.5.7}$$

We assume that the specific heat capacity polynomial is given by Eq. (2.9.11) and then after integration

$$\begin{aligned} A(T - T_{\text{ref}}) &+ \frac{B}{2}\left(T^2 - T_{\text{ref}}^2\right) + \frac{C}{3}\left(T^3 - T_{\text{ref}}^3\right) \\ &+ \frac{D}{4}\left(T^4 - T_{\text{ref}}^4\right) - \text{E}\left(\frac{1}{T} - \frac{1}{T_{\text{ref}}}\right) \\ &= \left(h(T_0) - \Delta h_f(T_{\text{ref}})\right) + \frac{\dot{Q}}{\dot{N}} \end{aligned} \tag{5.5.8}$$

Which can also be written as

$$f(T) = h(T) - \left(h(T_0) + \frac{\dot{Q}}{\dot{N}}\right) = \frac{D}{4}T^4 + \frac{C}{3}T^3 + \frac{B}{4}T^2 + AT - \frac{E}{T} - c_0 = 0 \tag{5.5.9}$$

where

$$c_0 = \frac{D}{4}T_{\text{ref}}^4 + \frac{C}{3}T_{\text{ref}}^3 + \frac{B}{2}T_{\text{ref}}^2 + AT_{\text{ref}} - \frac{E}{T_{\text{ref}}} + \left(h(T_0) - \Delta h_f(T_{\text{ref}})\right) + \frac{\dot{Q}}{\dot{N}} \tag{5.5.10}$$

This is a nonlinear equation in the unknown temperature of the general form

$$f(T) = 0 \tag{5.5.11}$$

Before developing the methodology for solving nonlinear algebraic equations let's see how (5.5.9) looks. We generate the plot of the error function (5.5.9) using the following MATLAB script for the case study with water vapor and

$$\dot{N} = 0.01 \frac{\text{kmol}}{\text{s}}, \dot{Q} = -100 \text{ kW}, T_0 = 450 \text{ °C}$$

```
clear all
% data for water taken from NIST Chemistry Webbook
Dhf = -241830;          % sp. enth. of formation in kJ/kmol
A   = 30.092000;        % cp = A+BT+CT^2+DT^3+ET^(-2)
B   =  6.832514e-3;%        in kJ/(kmol K)
C   =  6.793435e-6;%
D   = -2.534480e-9;%
E   =  0.082139e+6;%
T_ref = 298.15;         % reference temperature in K

% define function for specific enthalpy calculation
h=@(T) Dhf+A*(T.^1-T_ref^1)/1 + B*(T.^2-T_ref^2)/2 +...
           C*(T.^3-T_ref^3)/3 + D*(T.^4-T_ref^4)/4 -...
           E*(1./T-1/T_ref);

T0   =   450+273.15; % feed temperature K
N_dot = 0.01;        % molar flowrate in kmol/s;
Q_dot = -100;        % heat transfer rate in kJ/s=kW

% define the error function
f=@(T) h(T)-(h(T0)+Q_dot/N_dot)

T=273.15+linspace(150,350,101)
plot(T-273.15,f(T),'LineWidth',1.2)
grid on
xlabel('Temperature (°C)')
ylabel('f(T)')
```

The result is shown in Fig. 5.9. The temperature at which the error function $f(T)$ becomes equal to zero (root of the error function) is approximately 173.5 °C. The question that we will answer is how we can determine numerically the temperature that satisfies the error function and is, therefore, the solution to the problem under study. The method that will be developed, known as Newton's method, is repetitive in nature.

We consider now the general nonlinear algebraic Eq. (5.5.11). We start with the definition of the first-order derivative

$$\frac{df(T)}{dT} = \frac{f(T+\delta T) - f(T)}{\delta T} \tag{5.5.12}$$

This equation is then written in a slightly different way

$$f(T+\delta T) = f(T) + \frac{df(T)}{dT}\delta T \tag{5.5.13}$$

The only variable in Eq. (5.5.13) is the variation (step) of the temperature δT. How do we select the step away from an

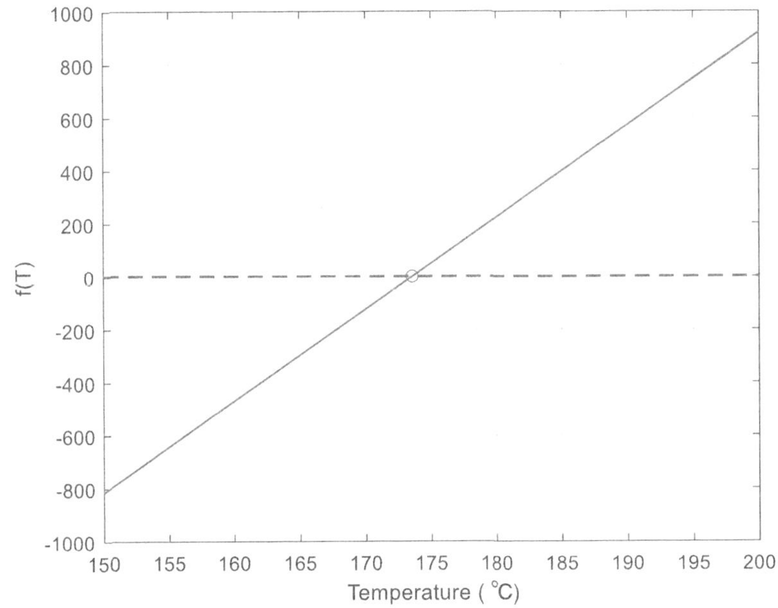

Figure 5.9 The plot of error function given by Eq. (5.5.9) as a function of temperature. The root is approximately 173.5 °C.

estimation of the temperature T that corresponds to an error $f(T)$? The answer is simple: select δT in such a way so that $f(T+\delta T) = 0$. Setting $f(T+\delta T) = 0$ in Eq. (5.5.13) we find the step

$$\delta T = -\frac{f(T)}{\frac{df(T)}{dT}} \tag{5.5.14}$$

This idea suggests a simple strategy to locate the solution of Eq. (5.5.11) (Newton's method):

STEP 1: Select an initial estimate of T^1, set $k = 1$ (the superscript denotes the current iteration).

STEP 2: Calculate $f\left(T^k\right), \frac{df}{dT}\left(T^k\right)$. If $f\left(T^k\right)$ is small enough then the solution has been found; otherwise continue.

STEP 3: Calculate the step from Eq. (5.5.14)

STEP 4: Calculate a new estimate of the solution $T^{k+1} = T^k + a\,\delta T^k$, where $0<a<1$

STEP 5: Set $k = k+1$, and go to STEP 2.

The derivative can be approximated using Eq. (5.5.12).

Example 5.7. Solution of a simple nonlinear algebraic equation using Newton's method

Solve, using Newton's method, the following nonlinear algebraic equation, starting from $x^1 = 1$:

$$f(x) = e^x - 2 = 0$$

Solution

You, most probably, can see immediately that the solution is $x = \log_e 2 = 0.693148$, but pretend that you don't know the solution. Note that

$$\frac{df}{dx} = e^x$$

The steps of the application of Newton's algorithm are the following:

Iteration 1,	Step 1:	$x^1 = 1$
	Step 2:	$f(1) = e^1 - 2 = 0.718282,$
		$df/dx(1) = 2.718282$
	Step 3:	$\delta x^1 = -\frac{0.718282}{2.718282} = -0.264241$
	Step 4:	$x^2 = x^1 + \delta x^1 = 0.735759$
Iteration 2,	Step 2:	$f(0.735759) = 0.087065,$
		$df/dx(0.7357591) = 2.087065$
	Step 3:	$\delta x^2 = -\frac{0.087065}{2.087065} = -0.041716$
	Step 4:	$x^3 = x^2 + \delta x^2 = 0.694043$
Iteration 3,	Step 2:	$f(0.694043) = 0.001792,$
		$df/dx(0.694043) = 2.001792$
	Step 3:	$\delta x^2 = -\frac{0.001792}{2.001792} = -0.000895$
	Step 4:	$x^4 = x^3 + \delta x^3 = 0.693148$
Iteration 4,	Step 2:	$f(0.693148) = 0.000\,001$, The root has been found.

The same steps can be performed in MATLAB

```
>> clear all
>> f=@(x) exp(x)-2

f =  function_handle with value:

    @(x)exp(x)-2

>> df=@(x) exp(x)

df =  function_handle with value:

    @(x)exp(x)

>> x(1)=1;
>> k=1;
>> x(k)=1;
>> f(x(k))
ans =      0.7183

>> dx(k)=-f(x(k))/df(x(k)); % dx  =     -0.2642
>> x(k+1)=x(k)+dx(k)
x =       1.0000     0.7358

>> k=k+1;                       % k   =     2
>> f(x(k))
ans =      0.0871

>> dx(k)=-f(x(k))/df(x(k)); % dx  =    -0.2642    -0.0417
>> x(k+1)=x(k)+dx(k)
x =       1.0000     0.7358     0.6940

>> k=k+1;                       % k   =     3
>> f(x(k))
ans =      0.0018

>> dx(k)=-f(x(k))/df(x(k)); % dx =     -0.2642    -0.0417    -0.0009
>> x(k+1)=x(k)+dx(k)
x =
    1.0000     0.7358     0.6940     0.6931

>> k=k+1;                       % k   =     4
>> f(x(k))
ans =
   8.0100e-07
```

We will now apply the Newton's algorithm for the example under study. We will use MATLAB for the intermediate steps. We select an initial estimation of the temperature $T^1 = 200$ °C, then

```
clear all
% data for water taken from NIST Chemistry Webbook
Dhf = -241830;       % sp. enth. of formation in kJ/kmol
A   = 30.092000;     % cp = A+BT+CT^2+DT^3+ET^(-2)
B   =  6.832514e-3;%      in kJ/(kmol K)
C   =  6.793435e-6;%
D   = -2.534480e-9;%
E   =  0.082139e+6;%
T_ref = 298.15;      % reference temperature in K

% define function for specific enthalpy calculation
h=@(T) Dhf+A*(T.^1-T_ref^1)/1 + B*(T.^2-T_ref^2)/2 +...
           C*(T.^3-T_ref^3)/3 + D*(T.^4-T_ref^4)/4 -...
           E*(1./T-1/T_ref);

T0   =  450+273.15; % feed temperature K
N_dot = 0.01;        % molar flowrate in kmol/s;
Q_dot = -100;        % heat transfer rate in kJ/s=kW

% define the error function
f=@(T) h(T)-(h(T0)+Q_dot/N_dot)

>> dTemp=1e-6;
```

```
>> Temp(1)=200+273.15;        % STEP 1, ITERATION 1
>> k=1;
>> dTemp=1e-6;
>> fun(k)=f(Temp(k))          % STEP 2, ITERATION 1

fun =    920.2991

>> dfun(k)=(f(Temp(k)+dT)-f(Temp(k)))/dT

dfun =    34.9441

>> step(k)=-fun(k)/dfun(k)    % STEP 3, ITERATION 1

step =   -26.3363

>> Temp(k+1)=Temp(k)+step(k)  % STEP 4, ITERATION 1

Temp = 473.1500   446.8137
```

```
>> k=k+1                      % STEP 5, ITERATION 1
```

```
k =       2

>> fun(k)=f(Temp(k))          % STEP 2, ITERATION 2

fun =    920.2991     3.4201

>> dfun(k)=(f(Temp(k)+dT)-f(Temp(k)))/dT

dfun =     34.9441    34.6865

>> step(k)=-fun(k)/dfun(k)    % STEP 3, ITERATION 2
```

```
step =    -26.3363    -0.0986

>> Temp(k+1)=Temp(k)+step(k) % STEP 4, ITERATION 2

Temp =    473.1500  446.8137  446.7151

>> k=k+1;                     % STEP 5, ITERATION 2

k =       3

>> fun(k)=f(Temp(k))          % STEP 2, ITERATION 3

fun =    920.2991     3.4201     0.0000
```

Note that after three iterations we calculate the root $T = 446.72$ K $= 173.57$ °C for which $f < 10^{-4}$.

5.6 Adiabatic flame temperature

Combustion is a high-temperature exothermic chemical reaction between a fuel and an oxidant, usually atmospheric oxygen, that produces oxidized, often gaseous products, in a mixture termed as smoke. For a combustion process that takes place adiabatically with no shaft work, the temperature of the products is referred to as the **adiabatic flame temperature**. This is the maximum temperature that can be achieved for given reactants. The maximum adiabatic flame temperature is achieved when a stoichiometric mixture for a given fuel and oxidizer is used. If more oxidant is used (excess) then the final temperature is lower than the maximum adiabatic flame temperature. The use of excess reactant is a common technique for controlling (lowering) the adiabatic flame temperature.

Calculation of the adiabatic flame temperature is a common and practical calculation that needs to be performed frequently. Let us reconsider the reactor at steady state shown in Fig. 5.10. One mol of a hydrocarbon fuel C_xH_y is mixed with stoichiometric air (79% N_2 and 21% O_2) plus excess air and the mixture is burned. The fuel is completely combusted giving as products x mol of CO_2 and $y/2$ mol of water plus excess air heated up to the adiabatic flame temperature. In the absence of work or heat interaction with the environment the energy "liberated" on combustion is used to heat up the combustion products. If no excess air is used then the maximum adiabatic flame temperature is achieved. The system is considered to operate at steady state and the overall energy balance written for a unit period of time gives

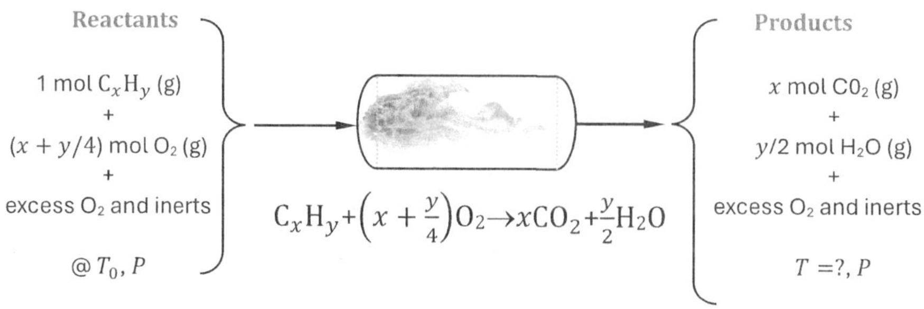

Figure 5.10 Complete combustion of a hydrocarbon fuel with atmospheric air.

$$0 = H_{in}(t) - H_{out}(t) \tag{5.6.1}$$

which can be written as

$$\underbrace{xh_{CO_2}(T) + \frac{y}{2}h_{H_2O}(T)}_{\text{sp.ethalpy of stoichiometric products}} + \underbrace{\left(x + \frac{y}{4}\right)\left(\frac{79}{21}\right)}_{\text{"stoichiometric" } N_2 \text{ in air}} h_{N_2}(T) + \ldots$$
$$= \underbrace{h_{C_xH_y}(T_0) + \left(x + \frac{y}{4}\right)h_{O_2}(T_0)}_{\text{sp.ethalpy of stoichiometric reactants}} \tag{5.6.2}$$
$$+ \underbrace{\left(x + \frac{y}{4}\right)\frac{79}{21}}_{\text{stoichiometricN}_2 \text{ in air}} h_{N_2}(T_0) + \ldots$$

where "+…" denotes all other components present in the feed that are not consumed in the reaction (oxygen and nitrogen in excess air or other inert components).

Let us consider methane as an example fuel (methane is the main constituent in natural gas, usually well over 90%). The combustion reaction with oxygen is

$$CH_4 + 2O_2 \rightarrow CO_2 + 2H_2O \tag{5.6.3}$$

For each mol of methane in the feed we also have 2 mol of oxygen. If air is used then for each mol of oxygen we have (79/21) mol of nitrogen or 100/21 mol of air. Eq. (5.6.3) becomes

$$CH_4 + 2\,O_2 + 2\left(\frac{79}{21}\right)N_2 \rightarrow CO_2 + 2\,H_2O + 2\left(\frac{79}{21}\right)N_2 \tag{5.6.4}$$

Normally the amount of air fed is greater than the stoichiometric amount. The amount of air fed is expressed in terms of the **percent of theoretical or stoichiometric air**. For example, 200% of theoretical air means that the air fed is 2 times the theoretical amount of air. The amount of air supplied can be expressed alternatively as a **percent excess air**. Thus, 200% of theoretical air is equivalent to 100% excess air. If we use the symbol λ% to denote the percent of stoichiometric air used then (5.6.4) can be written as

$$CH_4 + \left(\frac{\lambda\%}{100}\right)\left[2\,O_2 + 2\left(\frac{79}{21}\right)N_2\right] \rightarrow CO_2 + 2\,H_2O + \left(\frac{\lambda\%}{100} - 1\right)2\,O_2 + \left(\frac{\lambda\%}{100}\right)2\left(\frac{79}{21}\right)N_2 \tag{5.6.5}$$

As an example, if 400% of the stoichiometric air is used then

$$CH_4 + 8\,O_2 + 8\left(\frac{79}{21}\right)N_2 \rightarrow CO_2 + 2\,H_2O + 6\,O_2 + 8\left(\frac{79}{21}\right)N_2$$

or

$$\underbrace{CH_4 + 2\,O_2 +}_{\text{stoichiometric reactants}} \left[\underbrace{6\,O_2}_{\text{excess}} + \underbrace{8\left(\frac{79}{21}\right)N_2}_{\text{inert}}\right] \rightarrow \underbrace{CO_2 + 2\,H_2O}_{\text{stoichiometric products}} + \left[\underbrace{6\,O_2}_{\text{excess}} + \underbrace{8\left(\frac{79}{21}\right)N_2}_{\text{inert}}\right] \tag{5.6.6}$$

Example 5.8. Adiabatic flame temperature: combustion of methane with 400% theoretical air

Calculate the adiabatic flame temperature when pure methane is combusted with 400% theoretical air (4 times the stoichiometric air). The fuel and the air are fed at $T_0 = 298.15$ K and 1 bar. Use constant, average values for the heat capacity to obtain an initial estimate and then use MATLAB and the exact equations from Table 2.10 to obtain a better estimate. The average heat capacities for the components involved in the temperature interval 298.15–1000 K are

$$\bar{c}_{p,CH_4} = 54.4\ \frac{\text{kJ}}{\text{kmol K}},$$
$$\bar{c}_{p,O_2} = 32\ \frac{\text{kJ}}{\text{kmol K}},$$
$$\bar{c}_{p,N_2} = 31.6\ \frac{\text{kJ}}{\text{kmol K}},$$
$$\bar{c}_{p,CO_2} = 47.6\ \frac{\text{kJ}}{\text{kmol K}},$$
$$\bar{c}_{p,H_2O} = 37.05\ \frac{\text{kJ}}{\text{kmol K}}$$

Solution

The stoichiometric reaction is given by Eq. (5.6.3), and Eq. (5.6.2) can be written as

$$h_{CO_2}(T) + 2h_{H_2O}(T) + 6\,h_{O_2}(T) + 8\left(\frac{79}{21}\right)h_{N_2}(T) - \left[h_{CH_4}(T_0) + 8h_{O_2}(T_0) + 8\left(\frac{79}{21}\right)h_{N_2}(T_0)\right] = 0 \tag{a}$$

Note that the specific enthalpy of formation of oxygen and nitrogen is zero as the feed is at standard conditions, so

$$h_{CH_4}(T_0) + 8h_{O_2}(T_0) + 8\left(\frac{79}{21}\right)h_{N_2}(T_0) = \Delta h_{f,CH_4}(T_{ref})$$

where $T^{\ominus} = T_{ref} = T_0 = 298.15$ K. We replace into Equation (a) and express the specific enthalpies through the usual definition we obtain

$$\left(\Delta h_{f,CO_2}(T_{ref}) + \int_{T_{ref}}^{T} c_{p,CO_2}(\theta)d\theta\right) + 2\left(\Delta h_{f,H_2O}(T_{ref}) + \int_{T_{ref}}^{T} c_{p,H_2O}(\theta)d\theta\right) + 6\left(\underbrace{\Delta h_{f,O_2}(T_{ref})}_{=0} + \int_{T_{ref}}^{T} c_{p,O_2}(\theta)d\theta\right)$$
$$+ 8\left(\frac{79}{21}\right)\left(\underbrace{\Delta h_{f,N_2}(T_{ref})}_{=0} + \int_{T_{ref}}^{T} c_{p,N_2}(\theta)d\theta\right) - \Delta h_{f,CH_4}(T_{ref}) = 0$$

or

$$\int_{T_{ref}}^{T}\left(c_{p,CO_2}(\theta) + 2c_{p,H_2O}(\theta) + 6c_{p,O_2}(\theta) + 8\left(\frac{79}{21}\right)c_{p,N_2}(\theta)\right)d\theta - \left(\Delta h_{f,CH_4}(T_{ref}) - \Delta h_{f,CO_2}(T_{ref}) - 2\Delta h_{f,H_2O}(T_{ref})\right) = 0 \tag{b}$$

For constant heat capacities

$$\left(\bar{c}_{p,CO_2} + 2\bar{c}_{p,H_2O} + 6\bar{c}_{p,O_2} + 8\left(\frac{79}{21}\right)\bar{c}_{p,N_2}\right)(T - T_{ref})$$
$$-\left(\Delta h_{f,CH_4}(T_{ref}) - \Delta h_{f,CO_2}(T_{ref}) - 2\Delta h_{f,H_2O}(T_{ref})\right) = 0$$

The following specific enthalpy of formation at standard conditions are obtained from NIST Chemistry WebBook

$$\Delta h_{f,CH_4}(T_{ref}, \text{gas}) = -74\,600\ \frac{\text{kJ}}{\text{kmol}}, \Delta h_{f,CO_2}(T_{ref}, \text{gas}) = -393\,520\ \frac{\text{kJ}}{\text{kmol}}, \Delta h_{f,H_2O}(T_{ref}, \text{gas}) = -241\,826\ \frac{\text{kJ}}{\text{kmol}}$$

We replace the numerical values

$$\left(47.6 + 2 \times 37.05 + 6 \times 32 + 8\left(\frac{79}{21}\right) \times 31.6\right)(T - 298.15) - (-74600 - (-393520) - 2(-241826)) = 0$$

or

$$T = 932.74\ \text{K} = 659.6\ ^\circ\text{C}$$

The following script can be used to estimate the enthalpy of any mixture of the five components involved in this case study

```
function h=AdFlameTemp(n,T)
% Calculates the enthalpy of gaseous mixtures of
% 1 : CH4     2: O2      3:N2     4:CO2     5:H2O
% at a given absolute temperature T (row vector)
% and n(i) mole of component i in kmol
% All components are in gaseous state.
% Properties taken from NIST/Chemistry Webbook.

%       Δh_f      A           B           C           D          E
%                          x10^3       x10^6        x10^9     x10^-3
data=[-74600   -0.703029 108.477300 -42.521570    5.862788  678.565
        0       31.322340 -20.235310  57.866440 -36.506240   -7.374
        0       28.986410   1.853980  -9.647459  16.635370    0.117
     -393520 24.997350  55.186960 -33.691370    7.948387 -136.638
     -241830 30.092000   6.832514   6.793435  -2.534480   82.139];
Dh_f  = data(:,1); % spec.standard enth. of formation in kJ/kmol
% cp = A + BT + CT^2 + DT^3 + ET^(-2) in kJ/kmol
A = data(:,2); B = data(:,3)/10^3; C = data(:,4)/10^6;
               D = data(:,5)/10^9; E = data(:,6)*10^3;
T_ref = 273.15+25; % reference temperature
h = Dh_f + A*(T-T_ref) + B*(T.^2-T_ref^2)/2 +...
                       + C*(T.^3-T_ref^3)/3 +...
                       + D*(T.^4-T_ref^4)/4 +...
                       - E*(1./T-1/T_ref);

h = n(:)'*h;
end
```

Then, the deviation of the energy balance from the zero value is plotted

```
clear all
n0 = [1; 4*2; 4*2*(79/21); 0; 0];
T0 = 298.15;
n  = [0; 3*2; 4*2*(79/21); 1; 2];

f=@(T) AdFlameTemp(n,T)-AdFlameTemp(n0,T0)

T=273.15+linspace(660,675,11)
plot(T-273.15,f(T),'-o','linewidth',1.2)
grid on
xlabel('Temperature (℃)')
ylabel('H_{out}(T)-H_{in}(T_{0})')
```

The results are pictured in Fig. 5.11. The adiabatic flame temperature is 667.75 °C.

5.7 MATLAB and Python for solving nonlinear algebraic equations

5.7.1 Solving polynomial equations and nonlinear algebraic equations in MATLAB

MATLAB has three special functions for solving nonlinear and polynomial equations. The first one can be used for solving polynomial equations of the form

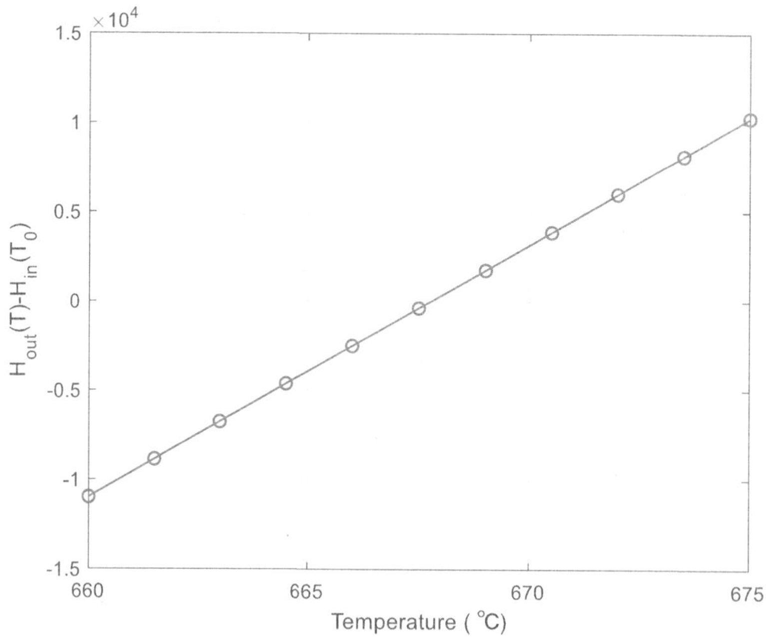

Figure 5.11 The error in the enthalpy balance as a function of temperature.

$$p(x) = c(1)x^n + c(2)x^{n-1} + c(3)x^{n-2} + c(n)x^1 + c(n+1)x^0 = 0$$

To find the roots of this polynomial you need to define the vector c and then type `roots(c)`, as it is shown in the examples below where we calculate the roots of the polynomials:

$$p_1(x) = x - 2$$

$$p_2 = x^2 - 3x + 2$$

$$p_3(x) = x^4 - 10x^3 + 35x^2 - 50x + 24$$

```
>> roots([1 -2])

ans =
     2

>> roots([1 -3 2])

ans =
     2
     1

>> c=[1 -10 35 -50 24];
>> roots(c)

ans =

    4.0000
    3.0000
    2.0000
    1.0000
```

We are mainly interested here in the solution of general nonlinear algebraic equations of the form

$$f(x) = 0, x \in R, f: R \to R \tag{5.7.1}$$

We have presented the steps of Newton's method. The implementation in MATLAB is the `fzero` function which has the following structure

```
[x_solution, fval_solution, exitflag] = fzero(fun, x0, options)
```

where `fun` is the function to solve, specified as a handle to a scalar-valued function or the name of such a function. `fun` accepts a scalar x and returns a scalar `fun(x)`.

`x0` is an initial guess of the solution if it is a single variable. If `x0` is a two-element vector, then the algorithm seeks a root in the interval `[x0(1) x0(2)]`. For this to be possible it must hold that `fun(x0(1))` has a different sign than `fun(x0(2))`. This is due to the fact that `fzero` uses a bisection algorithm: the middle of the interval `[x0(1) x0(2)]` is chosen `xm=((x0(1)+x0(2))/2`, then `fun(xm)` is calculated and `xm` replaces either `x0(1)`, if they have the same sign, or `x0(2)`, if they have the same sign. Thus, in each iteration the length of the interval is halved and convergence is fast. This, however, works well only when there is a single root in the interval `[x0(1) x0(2)]`.

`options` is an optional argument (structure set using the `optimset` command) that is used to interact with the `fzero` function.

Following convergence, the `fzero` function returns the solution in `x_solution`, the value of the function at the solution `fval_solution`, and an integer `exitflag` to flag convergence (if `exitflag=1`) or failure (`exitflag<0`).

To solve the problem of finding the root of $f(x) = e^x - 2 = 0$ we type the following in the command window

```
>> clear all
>> f=@(x) exp(x)-2
f =
  function_handle with value:
    @(x)exp(x)-2

>> options=optimset('Display','Iter');
>> x0=[0 2];
>> [x_solution,fval_solution,flag]=fzero(f,x0,options)

 Func-count           x          f(x)          Procedure
    2                 0            -1          initial
    3          0.313035      -0.63243          interpolation
    4          0.795066      0.214588          interpolation
    5          0.672946    -0.0399968          interpolation
    6          0.692132   -0.00202946          interpolation
    7          0.693148   1.36536e-06          interpolation
    8          0.693147  -6.93206e-10          interpolation
    9          0.693147  -2.22045e-16          interpolation
   10          0.693147  -2.22045e-16          interpolation

Zero found in the interval [0, 2]

x_solution =    0.6931

fval_solution =   -2.2204e-16

flag =      1
```

We may use fzero to find the adiabatic flame temperature of methane when combusted with 400% of theoretical air. The solution is pictured in Fig. 5.11. The solution can be found using fzero as follows

```
clear all
n0 = [1; 4*2; 4*2*(79/21); 0; 0];
T0 = 298.15;
n  = [0; 3*2; 4*2*(79/21); 1; 2];
f=@(T) AdFlameTemp(n,T)-AdFlameTemp(n0,T0)

AFT=fzero(f,273.15+[600 800],optimset('Display','Iter'))
```

```
>> runAdFlameTemp

f =  function_handle with value:

    @(T)AdFlameTemp(n,T)-AdFlameTemp(n0,T0)

 Func-count     x          f(x)             Procedure
    2         873.15     -94267.8           initial
    3         938.23     -3795.19           interpolation
    4        940.908     -0.68971           interpolation
    5        940.909   -3.1892e-06          interpolation
    6        940.909            0           interpolation

Zero found in the interval [873.15, 1073.15]

AFT =   940.9089

>> AFT-273.15

ans =    667.7589
```

The third function available in MATLAB is called fsolve and solves the more general problem of n equations in n unknowns

$$\boldsymbol{f}(\boldsymbol{x}) = \boldsymbol{0}, \boldsymbol{x} \in R^n, \boldsymbol{f}: \mathbb{R}^n \rightarrow \mathbb{R}^n \qquad (5.7.2)$$

We will present this function in a later chapter. For the time being it suffices to say that the syntax of the function is the same as that of fzero.

5.7.2 Solving polynomial equations and nonlinear algebraic equations in Python

Polynomial equations can be solved in Python using the numpy.roots command as shown below

```
In [3]: import numpy as np

In [4]: coeff = [1, -2]

In [6]: print(np.roots(coeff))
        [2.]

In [7]: coeff = [1, -3, 2]

In [8]: print(np.roots(coeff))
        [2. 1.]

In [9]: coeff = [1, -10, 35, -50, 24]

In [10]: print(np.roots(coeff))
        [4. 3. 2. 1.]
```

Solution of nonlinear algebraic equations can be performed using `scipy.optimize`, using bisection or Newton's method

```
In [11]: from scipy import optimize

In [12]: def f(x): return (np.exp(x) - 2)

In [13]: root = optimize.bisect(f, 0, 2)

In [14]: print(root)
        0.6931471805601177

In [15]: root = optimize.newton(f, 1.5)

In [16]: print(root)
        0.693147180559946
```

Chapter summary

The energy balance equation can be written as

$$\frac{dU(t)}{dt} = \dot{H}_{\text{in}}(t) - \dot{H}_{\text{out}}(t) + \left[\dot{W} + \dot{Q}\right]_{\text{net}}$$

where each term on the right-hand side is repeated for each material, work, or heat stream crossing the system boundaries. To solve the energy balance equation a model for calculating the specific internal energy and enthalpy is necessary. For pure components the specific enthalpy $h(T, P)$ at a particular temperature T and pressure P can be calculated by the equation

$$h(T, P) = h^{ig}(T) + \Delta h^R(T, P)$$

where $h^{ig}(T)$ is the specific enthalpy of the ideal gas and $\Delta h^R(T, P)$ is the residual specific enthalpy. A simplified model is the following

- *if the pure fluid is in the gas state*

$$h(T, P, \text{state} = \text{gas}) = h^{ig}(T) = \Delta h_f^{ig}(298.15\ \text{K}) + \int_{298.15\ \text{K}}^{T} c_p^{ig}(T)dT$$

- *if the pure fluid is in the liquid state*

$$h(T, P, \text{state} = \text{liquid}) = h^{\text{ideal gas}}(T) - \Delta h_{\text{vap}}(T)$$

$$= \Delta h_f^{ig}(298.15\ \text{K}) + \int_{298.15\ \text{K}}^{T} c_p^{ig}(T)dT - \Delta h_{\text{vap}}(T)$$

Problems

5.1 For the data of Table P5.1 calculate the enthalpy and internal energy of methane at two conditions:
a. $T = -180\ °C$, $P = 1$ bar, liquid.
b. $T = 0.0\ °C$, $P = 1$ bar, vapor
Compare the results with the data shown in Table P5.1. Note that the reference conditions for calculating enthalpy is $h = 0$ at the normal boiling point and at the state of saturated liquid. Explain the reason(s) for the significant deviations.

5.2 For the data of Table P5.1 calculate the enthalpy and internal energy of methane at two conditions:
a. $T = -160\ °C$, $P = 100$ bar, liquid
b. $T = 0.0\ °C$, $P = 100$ bar, supercritical fluid
Compare the results with the data shown in Table P5.1. Note that the reference conditions for calculating enthalpy is $h = 0$ at the normal boiling point and at the state of saturated liquid. Explain the reason(s) for the significant deviations.

5.3 In Example 5.2, where 1000 kg of water are heated using an electrical resistor, the vessel used to store the water is a cylindrical vessel of diameter $D = 1$ m with flat ends. During operation it was observed that there is a significant heat

Table P5.1 Properties of methane at different temperatures at 1 bar and 100 bar.

Temperature (C)	Pressure (bar)	Density (kg/m³)	Internal Energy (kJ/kg)	Enthalpy (kJ/kg)	Phase
Isobaric Data for P = 1 bar					
−180.00		448.25	−63.620	−63.396	
−161.64		422.59	−0.79398	−0.55734	liquid
−161.64		1.7946	454.84	510.56	
−160.00		1.7656	457.55	514.19	
−140.00		1.4799	489.95	557.52	
−120.00	1.0000	1.2775	521.76	600.03	
−100.00		1.1252	553.35	642.22	Vapor
−80.000		1.0060	584.90	684.30	
−60.000		0.90997	616.55	726.44	
−40.000		0.83086	648.45	768.80	
−20.000		0.76452	680.74	811.55	
0.0000		0.70805	713.60	854.83	
Reference States, Normal Boiling Point Convention. Enthalpy H = 0 at the normal boiling point for saturated liquid.					
Isobaric Data for P = 100 bar					
−160.00		428.99	-3.5123	19.798	
−140.00		401.35	63.731	88.647	liquid
−120.00		370.48	133.31	160.31	
−100.00		334.08	207.12	237.05	
−80.000	100.00	286.88	289.49	324.35	
−60.000		216.45	390.70	436.90	
−40.000		143.62	494.57	564.20	Supercritical
−20.000		108.45	565.13	657.35	
0.0000		90.040	619.23	730.29	

Reference States, Normal Boiling Point Convention.
Enthalpy H = 0 at the normal boiling point for saturated liquid.

Data for methane from https://WebBook.nist.gov/chemistry/fluid/.

loss to the environment from the bottom of the vessel and the side that is covered by water which can be described by the equation

$$\dot{Q}_{loss} = AU(T - T_{air})$$

where $\dot{Q}_{loss}$ is the rate of heat loss to the environment, A is the heat transfer area (bottom of the vessel and the side area covered by the liquid), T is the temperature of the water and $T_{air} = 15\ °C$ is the temperature of the surrounding air. Currently there is no insulation used and the overall heat transfer coefficient is $U = 0.2\ kJ/(m^2 s\ K)$. For the data given in Example 5.2 determine the time necessary to reach the temperature of 75 °C. What is the cost of heat losses if 1 kWh of electrical energy costs 0.3 $? Assume a constant heat capacity.

State	A	B×10^3	C×10^6	D×10^9	E×10^{-6}	A_{VAP}	a	β
gas	−0.703029	108.4773	-42.52157	5.862788	0.678565	10110	−0.22	0.388

Additional fluid properties

Critical temperature (T_c): −82.586 °C

Critical pressure (P_c): 45.992 bar

Normal boiling point: −161.483 °C

Standard heat of formation: −74600 kJ/kmol

$$C_p^o = A + B\,t + Ct^2 + Dt^3 + E/t^2, \quad t = T \text{ in K}/1000$$

$$\Delta_{vap}H = A_{VAP}\exp(-\alpha T_r)\,(1 - T_r)^{\beta}$$

5.4 One thousand metal spheres consisting of a stainless steel alloy with diameter $d = 3$ cm, density $\rho_s = 7800\ kg/m^3$ and initial temperature $T_s(0) = 600\ °C$ are immersed in $M_w = 100$ kg of water to cool them down quickly as part of their manufacturing process. Calculate the temperature evolution of the water bath and of the metal spheres.

The heat transferred from the solid spheres to the water bath is described by the equation

$$\dot{Q} = A_s U\left(T_s(t) - T_w(t)\right)$$

where $\dot{Q}$ is the rate of heat transfer from the solid spheres to the water bath, A_s is the heat transfer area (external area of the 1000 spherical, solid spheres), $T_s(t)$ is the temperature of the solid spheres, $T_w(t)$ is the temperature of the water

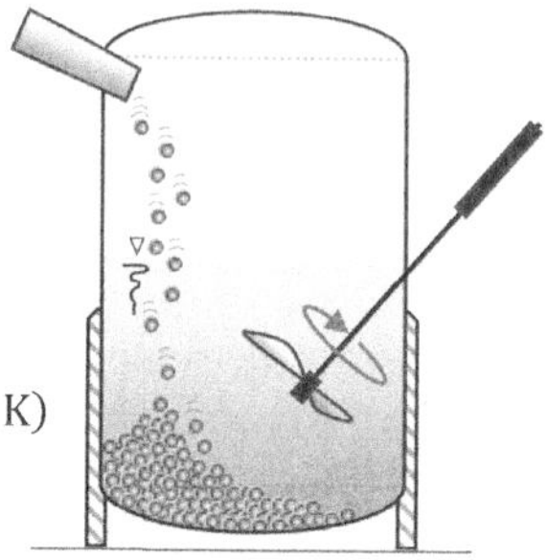

Figure P5.4 Cooling of metal spheres in water using a well-mixed tank.

bath, and $U = 5\ \mathrm{kJ}/(\mathrm{m}^2\mathrm{s\ K})$ is the heat transfer coefficient between the solids and the water. Other model parameters are given in Fig. P5.4.

5.5 Liquid water with an initial temperature of $T(0) = 25\ ^\circ C$ is fed to a tube with mass flowrate of $\dot{M} = 1\ \mathrm{kg/s}$ (see Fig. P5.5). On the outer surface of the tube, low pressure, saturated steam is condensed at a constant temperature of $T_S = 120\ ^\circ C$. Assume that the heat capacity of water is constant $c_{p,w} = 4.2\ \mathrm{kJ}/(\mathrm{kg\ K})$ and that the systems is at steady state. If the tube has a length of 100 m, determine the profile of the temperature of the water in the tube.
The heat transferred from the condensing steam to the water is described by the equation

$$\dot{Q} = AU(T_s - T(x))$$

where $\dot{Q}$ is the rate of heat transfer from the condensing steam to the water, A is the heat transfer area (external area of the tube which has a diameter of $d_t = 0.0357$ m) and $U = 0.5\ \mathrm{kJ}/(\mathrm{m}^2\mathrm{s\ K})$ is the heat transfer coefficient between the steam and the water.

5.6 In Fig. P5.6a, a simple solar water heater is shown. The basic idea is relatively simple. A black, horizontal, metal surface is placed in an enclosure with the water flowing slowly above it. Solar irradiation is absorbed from the black, metal surface and is heated up during the hours of the day when sunlight is intense. Then, the hot metal surface heats up the water. This is a very simple and cheap principle that can be explored to produce hot water but, more importantly, portable water if evaporation of wastewater is achieved.
In the particular case study of Fig. P5.6a the solar heater surface area is $A = 1\ \mathrm{m}^2$. The metal surface absorbs a fraction $a = 0.95$ of the incident solar radiation $\dot{G}_s''$ (expressed in J/(s

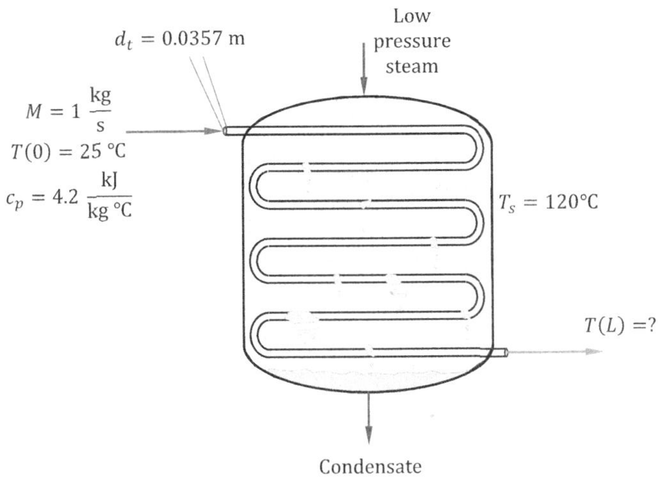

Figure P5.5 Water heating from condensing steam.

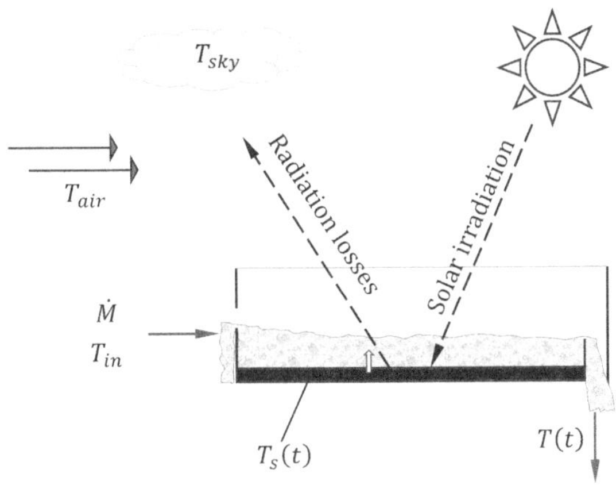

Figure P5.6a A simple solar heater for water purification.

m²)), which is a day with clear skies at Rio, Patras, Greece is described by the equation

$$\dot{G}_s''\left[\frac{\text{J}}{\text{s m}^2}\right] = \max\{0, -5830 + 1260t - 73\,t^2 + 1.18t^3\},\ t \text{ in h}$$

where t = 0 at midnight (00:00). Figure P5.6.b shows the solar irradiation for the specific day.

The heat transfer between the hot metal surface and the water is given by the equation

$$\dot{Q} = AU(T_s(t) - T(t))$$

where $\dot{Q}$ is the rate of heat transfer from the condensing steam to the water, A is the heat transfer area (area of the black surface $A = 1\ \text{m}^2$), $U = 0.05\ \text{kJ}/(\text{m}^2\text{s K})$ is the heat transfer coefficient between the metal surface (temperature $T_s(t)$) and the water (temperature $T(t)$) Fig. P5.6.b.
Heat is lost from the solar heater due to radiation $\dot{Q}''_{\text{rad}}$, which is described by

$$\dot{Q}'' = \varepsilon\sigma\left(T_s(t)^4 - T_{\text{sky}}^4\right)$$

where $\varepsilon = 0.98$ is the emissivity of the water, $\sigma = 5.67 \times 10^{-8}\ \text{J}/(\text{m}^2\ \text{s K}^4)$ is the Maxwell–Stefan constant and $T_{\text{sky}} = 10\ °C$ is the temperature of the sky.
The metal surface has a thickness of $H_s = 0.01$ m, density of $\rho_s = 7800\ \text{kg/m}^3$ and a heat capacity of $c_{p,s} = 0.5$ kJ/(kg K). The heat capacity of water is $c_p = 4.2$ kJ/(kg K) and the mass flowrate of water is $\dot{M} = 20$ kg/h. The level of the water in the solar heater is $H = 0.02$ m and the density $\rho = 1000\ \text{kg/m}^3$.

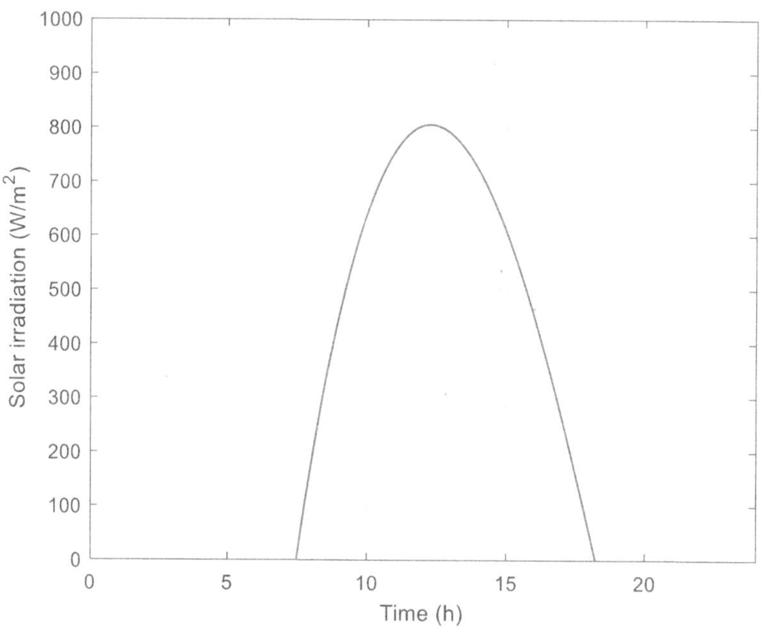

Figure P5.6.b Solar irradiation on the solar heater.

Assume that the temperature of the air is constant $T_{air} = 25\ °C$ and at $t = 0$ both the temperature of the solid surface and of the water are equal to T_{air}.
Calculate the temperature of the solid surface and of the water from $t = 0$ to $t = 24$ h.

5.7 Consider the catalytic production of ethanol by catalytic hydration of ethylene

$$C_2H_4 + H_2O \rightleftharpoons C_2H_5OH$$

Calculate the enthalpy of reaction for temperatures in the range 100–500 °C. What do you observe?

5.8 Consider the reaction for the ammonia production from nitrogen and hydrogen

$$N_2 + 3\ H_2 \rightleftharpoons 2\ NH_3$$

Calculate the enthalpy of reaction for temperatures in the range 0–1000 °C. What do you observe?

5.9 Consider Example 5.8. Repeat the calculations with theoretical air in the range 100%–1000% and plot the adiabatic flame temperature as a function of theoretical air.

5.10 One kg of water is heated up by adding $Q = 1200$ kJ. The initial temperature of water is $T = 40\ °C$ and the heating is performed under constant pressure of 101.325 kPa. What is the final temperature of water? Important note: Water boils at 100 °C when the pressure is 101.325 kPa. Use any data necessary from NIST Chemistry WebBook.

5.11 One kmol/s of carbon dioxide is heated up by adding $\dot{Q} = 10^4$ kW. The initial temperature of CO_2 is $T_{in} = 0\ °C$ and the heating is performed under constant pressure of 101.325 kPa. At these conditions CO_2 is in the gaseous state. What is the final temperature of carbon dioxide?
Use the following equation for the heat capacity of carbon dioxide and perform all calculations "manually"

$$c_p(T)\left[\frac{\text{kJ}}{\text{kmol K}}\right] = 28.855 + 3.6024 \times 10^{-2}T - 1.09248 \times 10^{-5}T^2,\ T \text{ in K}$$

Use an average heat capacity in the temperature range 273 K–500 K, to get an initial estimate of the final temperature.

5.12 Consider the reaction of hydrogen combustion with air

$$H_2 + \frac{1}{2}O_2 \rightarrow H_2O$$

Repeat the calculations with theoretical air in the range 100%–1000% and plot the adiabatic flame temperature as a function of theoretical air. The feed is available at 25 °C.

5.13 Consider the reaction of hydrogen combustion with air

$$H_2 + \frac{1}{2}O_2 \rightarrow H_2O$$

The reaction is performed in a furnace (see Fig. P5.13) and the exit temperature of the flue gas is 400 °C when a theoretical air of 500% is used. The feed to the furnace is at 25 °C and has molar flowrate of 1 kmol H_2/s, 2.5 kmol O_2/s, and the corresponding nitrogen (2.5×79/21 kmol N_2/s). What is the amount of heat recovered by a boiler that produces high-pressure steam? Assume that the losses are negligible.

5.14 (This can be a group project and is a modification of an example in U. Hauptmanns, *Process and Plant Safety*, Springer, Heidelberg, 2015, p. 120) RDX (hexogen) is produced via the continuous SH nitration process, in which hexamine is reacted with a large excess of concentrated nitric acid (see Fig. P5.14). The reaction is highly exothermic and chemically unstable, requiring strict temperature control to prevent runaway or explosion. The reaction temperature must not exceed 23 °C, with optimal operation around 12 °C for maximum yield and safety. The reaction is the following

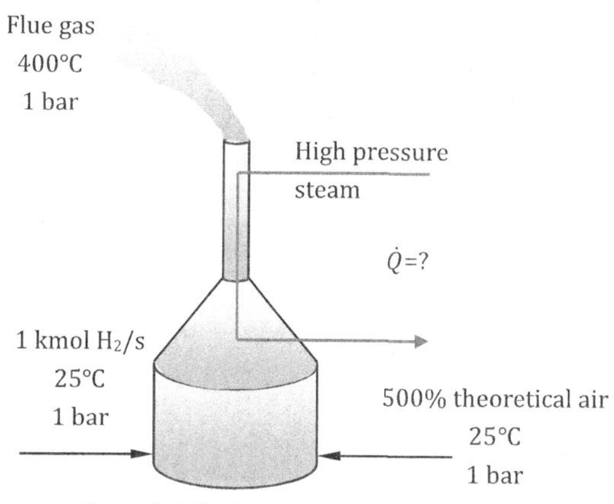

Figure P5.13 A simple hydrogen furnace.

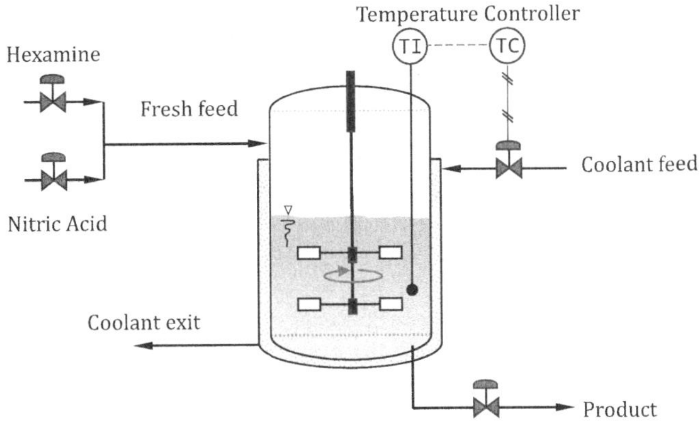

Figure P5.14 A continuous stirred tank reactor with an exothermic reaction.

$$HA + 10\ NA \rightarrow HXG + 3\ NDDN + AMN + 3\ H_2O$$

Details of the specific components are given in Table P5.14. The startup of the process is taking place by first filling up the reactor ($V = 0.63\ m^3$) with nitric acid with initial concentration $C_{NA}(0) = 24.44\ kmol/m^3$ and initial temperature $T(0) = 4\ ^\circ C$ (277.15 K). Then, fresh feed in fed to the reactor with volumetric flowrate $\dot{V} = 2.5 \times 10^{-4}\ m^3/s$ that contains only the two reactants with concentrations $C_{HA,\text{in}} = 0.9851\ kmol/m^3$ and $C_{NA,\text{in}} = 20.9087\ kmol/m^3$ and temperatures $T_{HA,\text{in}} = 20\ ^\circ C$ (293.15 K) and $T_{NA,\text{in}} = 4\ ^\circ C$ (297.15 K).
The reaction rate can be expressed as:

$$\dot{R}''' = k_0 \exp\left(-\frac{E}{RT}\right) C_{HA}^n C_{NA}, \quad n = 9.958$$

Table P5.14 Thermodynamic data for Problem 5.14.

Component	HA	NA	HXG	NDDN	AMN	W
Chemical formula	$C_6H_{12}N_4$	HNO_3	$C_3H_6N_6O_6$	$CH_2(ONO_2)_2$	NH_4NO_3	H_2O
Name	Hexamine	Nitric acid	Hexogen	Methanediol dinitrate	Ammonium nitrate	Water
Molar mass (MM_i)	140.19	63.01	222.12	138.04	80.05	18
Heat capacity (kJ/(kg K)) ($c_{p,i}$)	1.256	1.989	1.190	1.926	1.759	4.187

where the preexponential constant is $k_0 = 4.75048 \times 10^7$ $(\mathrm{m}^3/\mathrm{kmol})^n(1/\mathrm{s})$, $n = 9.958$, the activation energy is $E = 34390.9$ kJ/kmol and the heat of the reaction is $\Delta h_{\mathrm{rxn}} = -274730$ kJ/kmol.

The temperature of the reactor is measured through a thermometer with first-order dynamics with a time constant of $\tau_m = 5$ s

$$\tau_m \frac{dT_m}{dt} + T_m = T, T_m(0) = T(0)$$

A proportional-integral (PI) controller is used to control the temperature in the reactor by varying the volumetric flowrate of the coolant that is fed to the jacket of the reactor. The equations that describe the operation of the PI are the following

$$F_c = F_{c,0} + k_c\left[\left(T^{SP} - T_m\right) + \frac{e_I}{\tau_I}\right]$$

$$\frac{de_I}{dt} = T^{SP} - T_m, e_I(0) = 0$$

where the controller gain is $k_c = -0.05\ \frac{\mathrm{m}^3}{\mathrm{s}}/\mathrm{K}$, the integral time is $\tau_I = 36000$ s and the bias is $F_{c,0} = 0.001\frac{\mathrm{m}^3}{\mathrm{s}}$. The desired reaction temperature is $T^{SP} = 12\ ^\circ C$.

The coolant is fed to the jacket of the reactor which has a volume of $V_J = 0.11\ \mathrm{m}^3$ and is a mixture of methanol and water with incoming temperature of $T_{c,\mathrm{in}} = -5\ ^\circ C$ and a heat capacity of $c_{p,c} = 3.6\frac{\mathrm{kJ}}{\mathrm{kg\ K}}$ and a density of $\rho_c = 938.4\ \frac{kg}{m^3}$. The energy balance of the cooling medium can be written as

$$V_J\rho_c c_{p,c}\frac{dT_C}{dt} = F_c\rho_c c_{p,c}(T_{c,\mathrm{in}} - T_c) + \dot{Q}_c$$

where Q_c is the cooling rate that depends on the temperature difference between the temperature of the reacting mixture and the temperature of the coolant:

$$Q_c = AU(T - T_c)$$

A is the heat transfer area ($A = 2.15\ m^2$) and U is the overall heat transfer coefficient $\left(U = 1.4\ \frac{\mathrm{kJ}}{\mathrm{s\ m^2\ K}}\right)$.

The mass balances can be written as (we assume constant volume of the reacting mixture)

$$\frac{dC_i}{dt} = \sigma_i \dot{R}'''$$

where σ_i is the stoichiometric number of component i, $i = \{\text{HA, NA, HXG, NDDN, AMN, W}\}$. The energy balance of the reactor can be written as

$$\frac{dH}{dt} = \dot{V} \sum_{i=\{\text{HA,NA}\}} MM_i\, c_{p,i}(C_{i,\text{in}}\, T_{i,\text{in}} - C_i T) + V(-\Delta h_{\text{rxn}})\dot{R}''' - \dot{Q}_c$$

where

$$H = V\left(\sum_{i=\{\text{HA,NA,HXG,NDDN,AMN,W}\}} MM_i\, c_{p,i} C_i\right) T$$

a. Simulate the startup process in MATLAB or Python for the first 3 h of operation and validate that steady-state operation is achieved half an hour after the start-up.
b. If there is a failure in the coolant medium feed line and cooling is lost, validate that the temperature of the reacting mixture is increasing unwieldily.
c. Propose and validate a strategy to shut down safely the process in case of loss of coolant flowrate.

5.15 In the batch reactor of Fig. P5.15 the following reactions are taking place

$$\text{R} \rightarrow \text{P} \rightarrow \text{B}$$

where R denotes the reactant, P denotes the desired product, and B an undesired byproduct. The reaction R→P is second-order with respect to the concentration of the reactant R while the reaction P→B is first-order with respect to the concentration of the product P. The heat released by the reactions is removed by using a coolant medium that is fed to the jacket of the reactor. Details about the process are given in Table P5.15.

a. Develop the mathematical model of the batch reactor that involves the mole balances of the three components, the energy balance of the reacting mixture, and the energy balance of the cooling medium.
b. Assume that the coolant volumetric flowrate can be varied between 0.0001 m^3/s and 0.001 m^3/s. Simulate the process to select the coolant volumetric flowrate that maximizes the production of product P.

5.16 A shell-and-tube heat exchanger is one of the most common types of heat exchangers used in industry to transfer heat

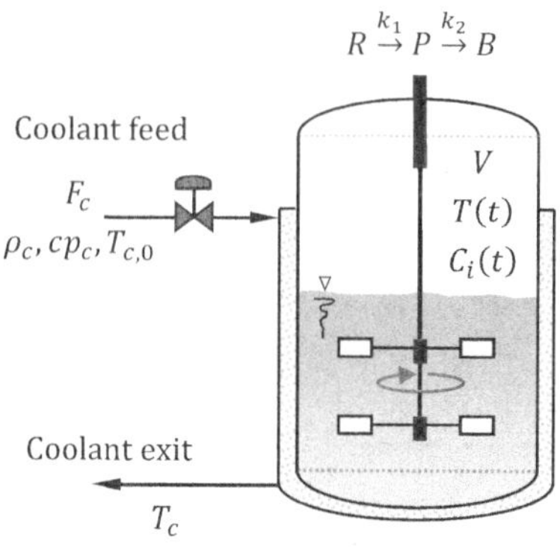

Figure P5.15 A batch reactor.

between two fluids—one flowing through tubes and the other flowing around the tubes within an outer shell. A typical arrangement of a 1-2 shell-and-tube heat exchanger is shown in Fig. P5.16. The tube bundle is divided into two

Table P5.15 Numerical data for Problem P5.15.

Symbol	Variable	Numerical value	Units
$C_R(0)$	Initial concentration of the reactant	1	kmol/m^3
$T(0)$	Initial temperature of the reacting mixture	298.15	K
V_L	Volume of the reacting mixture	1	m^3
D	Diameter of the reactor	$\sqrt[3]{\frac{4V_L}{\pi}}$	m
A	Heat transfer area	$\frac{\pi D^2}{4} + \pi D \cdot D$	m
U	Overall heat transfer coefficient	0.4	kJ/(s m^2 K)
$k_{0,1}$	Preexponential factor of the reaction rate constant of the R→P reaction	1	(1/s) (m^3/kmol)
E_1	Activation energy for the R→P reaction	20 000	kJ/kmol
$k_{0,2}$	Preexponential factor of the reaction rate constant of the P →B reaction	100	(1/s)
E_2	Activation energy for the P→B reaction	40 000	kJ/kmol
c_p	Heat capacity of the reacting mixture	100	kJ/(kmol K)
$\Delta h^{ref}_{f,R}$	Heat of formation of liquid R at reference temperature	−100 000	kJ/kmol
$\Delta h^{ref}_{f,P}$	Heat of formation of liquid P at reference temperature	−200 000	kJ/kmol
$\Delta h^{ref}_{f,B}$	Heat of formation of liquid B at reference temperature	−300 000	kJ/kmol
T_{ref}	Reference temperature for specific enthalpy calculation	298.15	K
V_J	Volume of the reactor jacket (thickness of 2 in)	$0.0508(\frac{\pi D^2}{4} + \pi D \cdot 2D)$	m^3
ρ_c	Density of coolant	1000	kg/m^3
$c_{p,c}$	Heat capacity of coolant	4.187	kJ/(kg K)
$T_{c,0}$	Temperature of the coolant at its feed	298.15	K

sections using a partition (called a partition plate or pass divider) in the tube sheet or channel head. The tube-side fluid enters one side, flows through half the tubes to the far end, reverses direction through the return header (or U-bend in a U-tube design), and then comes back through the other half of the tubes. The shell-side fluid flows only once across the tube bundle—typically perpendicular to the tubes—before exiting the shell (see Fig. P5.16a). Modeling of the temperature dynamics of shell-and-tube heat exchangers can be complicated, and a relatively efficient way is proposed in Fig. P5.16.b. In this method the tube side is divided into two compartments per pass and the shell is also divided into two compartments.

The mathematical model of the heat exchanger is the following (subscript t denotes the tube side, subscript s denotes the shell side, T denotes temperature, ρ the density, c_p denotes the heat capacity, V the volume, A the heat transfer area, and U the overall heat transfer coefficient:

$$\frac{(V\rho c_p)_t}{4}\frac{dT_1}{dt} = F_t(\rho c_p)_t(T_{t,\mathrm{in}} - T_1) - \frac{AU}{4}(T_1 - T_5)$$

$$\frac{(V\rho c_p)_t}{4}\frac{dT_2}{dt} = F_t(\rho c_p)_t(T_1 - T_2) - \frac{AU}{4}(T_2 - T_6)$$

$$\frac{(V\rho c_p)_t}{4}\frac{dT_3}{dt} = F_t(\rho c_p)_t(T_2 - T_3) - \frac{AU}{4}(T_3 - T_6)$$

$$\frac{(V\rho c_p)_t}{4}\frac{dT_4}{dt} = F_t(\rho c_p)_t(T_3 - T_4) - \frac{AU}{4}(T_4 - T_5)$$

$$\frac{(V\rho c_p)_s}{2}\frac{dT_5}{dt} = F_s(\rho c_p)_s(T_6 - T_5) + \frac{AU}{4}(T_1 - T_5) + \frac{AU}{4}(T_4 - T_5)$$

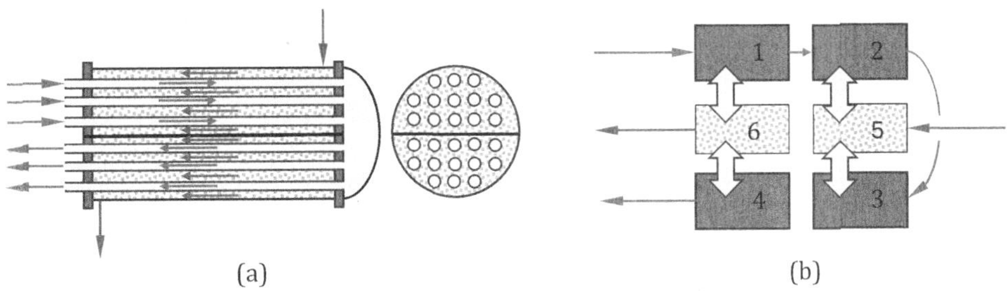

Figure P5.16 A shell-and-tube heat exchanger.

$$\frac{(V\rho c_p)_s}{2}\frac{dT_6}{dt} = F_s(\rho c_p)_s(T_{s,\text{in}} - T_6) + \frac{AU}{4}(T_2 - T_6) + \frac{AU}{4}(T_3 - T_6)$$

a. Validate the mathematical model proposed above.

b. Show that at steady state the model can be written as a linear system of equations:

$$\begin{bmatrix} -F_t(\rho c_p)_t - \frac{AU}{4} & 0 & 0 & 0 & \frac{AU}{4} & 0 \\ F_t(\rho c_p)_t & -F_t(\rho c_p)_t - \frac{AU}{4} & 0 & 0 & 0 & \frac{AU}{4} \\ 0 & F_t(\rho c_p)_t & -F_t(\rho c_p)_t - \frac{AU}{4} & 0 & 0 & \frac{AU}{4} \\ 0 & 0 & F_t(\rho c_p)_t & -F_t(\rho c_p)_t - \frac{AU}{4} & \frac{AU}{4} & 0 \\ \frac{AU}{4} & 0 & 0 & \frac{AU}{4} & -F_s(\rho c_p)_s - 2\frac{AU}{4} & F_s(\rho c_p)_s \\ 0 & \frac{AU}{4} & \frac{AU}{4} & 0 & 0 & -F_s(\rho c_p)_s - 2\frac{AU}{4} \end{bmatrix}$$

$$\begin{bmatrix} T_1 \\ T_2 \\ T_3 \\ T_4 \\ T_5 \\ T_6 \end{bmatrix} = \begin{bmatrix} -F_t(\rho c_p)_t T_{t,\text{in}} \\ 0 \\ 0 \\ 0 \\ 0 \\ -F_s(\rho c_p)_s T_{s,\text{in}} \end{bmatrix}$$

c. In a case of a 1-2 shell-and-tube heat exchanger with 13 tubes of 0.0254 m diameter and 1 m length, a shell of diameter of 0.254 m, $\rho c_p = 4180\ \text{kJ}/(m^3\ °\text{C})$ for both fluids, $U = 1\ \text{kJ}/(\text{s m}^2\ °\text{C})$, and

$$F_t = 0.0010\ \text{m}^3/\text{s},\ T_{t,\text{in}} = 20\ °\text{C},\ F_s = 0.0005\ \text{m}^3/\text{s},\ T_{s,\text{in}} = 160\ °\text{C}$$

find the steady-state solution.

d. A proportional-integral controller is used with $k_c = 5 \times 10^{-5}$, $\tau_I = 30$ s, and

$$\frac{de_I(t)}{dt} = T^{SP} - T_m(t),\ e_I(0) = 0$$

$$F_s(t) = 0.0005 + k_c\left[\left(T^{SP} - T_m(t)\right) + e_I(t)\right]$$

$$\tau_m\frac{dT_m(t)}{dt} = T_4(4) - T_m(4), \tau_m = 3\ \text{s}, T_m(0) = T_4(0)$$

Simulate the dynamics of the system for 2000 s where F_t changes from 0.00100 m^3/s to 0.00075 m^3/s at 500 s, from

0.00075 m^3/s to 0.00100 m^3/s at 1000 s, and finally to 0.00125 m^3/s at 1500 s.

Further reading

Many definitions have been retrieved from the Wikipedia, the free encyclopedia.

Books on material and energy balances

Andersen, L.B., Wenzel, L.A., 1961. Introduction to Chemical Engineering. McGraw Hill, NY.

Duncan, T.M., Reimer, J.A., 1998. Chemical Engineering Design and Analysis. Cambridge University Press, NY.

Foust, A.S., Wenzel, L.A., Clump, C.W., Maus, L., Andersen, L.B., 1980. Principles of Unit Operations, second ed. John Wiley & Sons, NY.

Henley, E.J., Bieber, H., 1959. Chemical Engineering Calculations. McGraw Hill, NY.

Himmelblau, D.M., Bischoff, K.B., 1968. Process analysis and simulation. In: Deterministic Systems. John Wiley & Sons, NY.

Hougen, O.A., Watson, k.m., Ragatz, R.A., 1943. Chemical Process Principles, Part I, Material and Energy Balances. John Wiley and Sons, NY.

Myers, A.L., Seider, W.D., 1976. Introduction to Chemical Engineering and Computer Calculations. Prentice Hall, Inc., Englewood Cliffs, NJ.

Reklaitis, G.V., 1983. Introduction to Material and Energy Balances. John Wiley and Sons, NY.

Russell, T.W.F., Denn, M.M., 1972. Introduction to Chemical Engineering Analysis. John Wiley and Sons, NY.

Thermodynamics books

Abbott, M.M., Van Ness, H.C., 1972. Theory and Problems of Thermodynamics, Schaums Outline Series. McGraw-Hill Book Company, NY.

Elliot, J.R., Lira, C.T., 2012. Introductory Chemical Engineering Thermodynamics, second ed. Prentice Hall, Upper Saddle River, NJ.

Matsoukas, T., 2012. Fundamentals of Chemical Engineering Thermodynamics. Prentice Hall, Upper Saddle River, NJ.

Moran, M.J., Shapiro, H.N., 1988. Fundamentals of Engineering Thermodynamics. John Wiley and Sons, NY.

Sandler, S.I., 1977. Chemical and Engineering Thermodynamics. John Wiley and Sons, NY.

Books on numerical methods and ordinary differential equations

Borrelli, R.L., Coleman, C.S., 1998. Differential Equations, A Modelling Perspective. John Wiley & Sons, NY.

Clough, D.E., Chapra, S.C., 2023. Introduction to Engineering and Scientific Computing with Python. CRC Press, Boca Raton, FL.

Dorfman, K.D., Daoutidis, P., 2017. Numerical Methods with Chemical Engineering Applications. Cambridge University Press.

Mostoufi, N., Constantinides, A., 2023. Applied Numerical Methos for Chemical Engineers. Academic Press, London.

Mitsotakis, D., 2023. Computational Mathematics: An Introduction to Numerical Analysis and Scientific Computing with Python. CRC Press, Boca Raton, FL.

6

Mass balances at steady state

6.1 Introduction

In Chapter 4 we developed the general mole balance equations for cases where there is no chemical reaction, which are summarized as follows (we will use an extended notation, given in Table 6.1, to account explicitly for the fact that there might be *more than one entering and/or exiting streams,* as it is pictured in Fig. 6.1).

Table 6.1 Notation used in the mole and material balance equation.

Symbol	Definition	SI units
Components : $i = 1, 2, \ldots, \mathrm{NC}$		
Feed streams : $k = 1, 2, \ldots, f$		
$\dot{N}_k^{in} / \dot{M}_k^{in}$	Total molar/mass flowrate of the k-th entering (feed) stream	mol/s or kg/s
$\dot{n}_{k,i}^{in} / \dot{m}_{k,i}^{in}$	Component i molar/mass flowrate in the k-th entering stream	mol/s or kg/s
$\dot{F}_k^{in}$	Volumetric flowrate of the k-th entering stream	m^3/s
$C_{k,i}^{in}$	Concentration of component i in the k-th entering stream	mol/m^3 or kg/m^3
$x_{k,i}^{in} / w_{k,i}^{in}$	Mole/mass fraction of component i in the k-th entering stream	–
$k = 1, 2, \ldots, p$		
$\dot{N}_k^{out} / \dot{M}_k^{out}$	Total molar/mass flowrate of the k-th exiting (product) stream	mol/s or kg/s
$\dot{n}_{k,i}^{out} / \dot{m}_{k,i}^{out}$	Component i molar/mass flowrate in the k-th exiting stream	mol/s or kg/s
$\dot{F}_k^{out}$	Volumetric flowrate of the k-th exiting stream	m^3/s
$C_{k,i}^{out}$	Concentration of component i in the k-th exiting stream	mol/m^3 or kg/m^3
$x_{k,i}^{out} / w_{k,i}^{out}$	Mole/mass fraction of component i in the k-th exiting stream	–
Other		
$\dot{R}$	Reaction rate	mol/s
σ_i	Stoichiometric number of component i	–

Introduction to Process Systems Engineering. https://doi.org/10.1016/B978-0-443-45444-8.00006-3

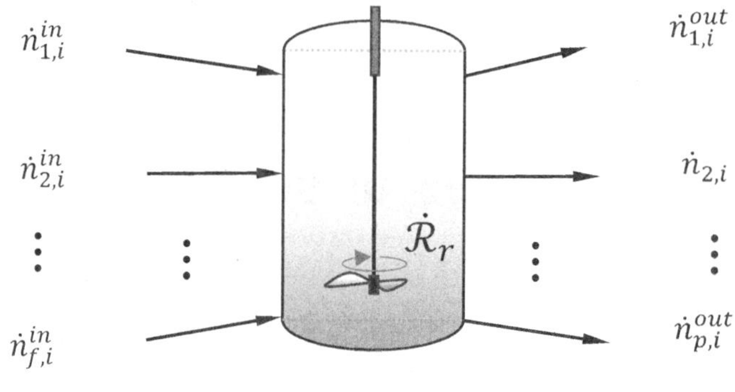

Figure 6.1 Mole balance of a chemical component at steady state, notation used.

- *Total mole balance*

$$\frac{dN}{dt} = \sum_{k=1}^{f} \dot{N}_k^{\text{in}} - \sum_{k=1}^{p} \dot{N}_k^{\text{out}} \tag{6.1.1}$$

where the superscript in denotes an entering stream and the superscript out denotes an exiting stream. There are $k = 1, 2, \ldots, f$ entering or **f**eed streams and $k = 1, 2, \ldots, p$ exiting or **p**roduct streams.

- *Component mole balances for component* $i = 1, 2, \ldots NC$.

$$\frac{dn_i}{dt} = \sum_{k=1}^{f} \dot{n}_{k,i}^{\text{in}} - \sum_{k=1}^{p} \dot{n}_{k,i}^{\text{out}} \tag{6.1.2}$$

where NC denotes the **n**umber of **c**omponents. At steady state none of the variables appearing in the balance equations depends on time, and all time derivatives are, therefore, zero. The mole balances become at steady state

$$\sum_{k=1}^{f} \dot{N}_k^{\text{in}} = \sum_{k=1}^{p} \dot{N}_k^{\text{out}} \tag{6.1.3}$$

$$\sum_{k=1}^{f} \dot{n}_{k,i}^{\text{in}} = \sum_{k=1}^{p} \dot{n}_{k,i}^{\text{out}} \tag{6.1.4}$$

These can also be written in mass units

$$\sum_{k=1}^{f} \dot{M}_k^{\text{in}} = \sum_{k=1}^{p} \dot{M}_k^{\text{out}} \tag{6.1.5}$$

$$\sum_{k=1}^{f} \dot{m}_{k,i}^{\text{in}} = \sum_{k=1}^{p} \dot{m}_{k,i}^{\text{out}} \tag{6.1.6}$$

In cases with chemical reaction(s) we use the balance equations in molar units and add the chemical reaction rate term(s)

$$\sum_{k=1}^{p} \dot{N}_k^{\text{out}} = \sum_{k=1}^{f} \dot{N}_k^{\text{in}} + \sum_{r=1}^{NR} \Sigma_r \dot{R}_r \tag{6.1.7}$$

$$\sum_{k=1}^{p} \dot{n}_{k,i}^{\text{out}} = \sum_{k=1}^{f} \dot{n}_{k,i}^{\text{in}} + \sum_{r=1}^{NR} \sigma_{r,i} \dot{R}_r, i = 1, 2, \ldots, NC \tag{6.1.8}$$

where $\dot{R}_r$ is the rate of the r-th reaction $r = 1, 2, \ldots, NR$, NR is the **n**umber of **r**eactions and

$$\Sigma_r = \sum_{i=1}^{NC} \sigma_{r,i}$$

is the total change in the number of moles due to the r-th reaction.

If we only have one entering stream, one exiting stream, and one reaction we obtain

$$\dot{N}^{\text{out}} = \dot{N}^{\text{in}} + \Sigma \dot{R} \tag{6.1.9}$$

$$\dot{n}_i^{\text{out}} = \dot{n}_i^{\text{in}} + \sigma_i \dot{R}, i = 1, 2, \ldots, NC \tag{6.1.10}$$

The last equation can be written in two additional equivalent forms.

- *In terms of volumetric flowrates and concentrations*

$$\dot{F}^{\text{out}} C_i^{\text{out}} = \dot{F}^{\text{in}} C_i^{\text{in}} + \sigma_i \dot{R} \tag{6.1.11}$$

- *In terms of molar flowrates and mole fractions*

$$\dot{N}^{\text{out}} x_i^{\text{out}} = \dot{N}^{\text{in}} x_i^{\text{in}} + \sigma_i \dot{R} \tag{6.1.12}$$

At steady state, without chemical reactions, the principle holds that the mass (or quantity) exiting the system equals the mass entering it:

$$\dot{n}_i^{\text{out}} = \dot{n}_i^{\text{in}} \tag{6.1.13}$$

- *In terms of volumetric flowrates and concentrations*

$$\dot{F}^{\text{out}} C_i^{\text{out}} = \dot{F}^{\text{in}} C_i^{\text{in}} \tag{6.1.14}$$

- *In terms of molar flowrates and mole fractions*

$$\dot{N}^{\text{out}} x_i^{\text{out}} = \dot{N}^{\text{in}} x_i^{\text{in}} \tag{6.1.15}$$

It is also reminded that if we add all Eq. (6.1.15) we obtain

$$\sum_{i=1}^{NC} \dot{N}^{\text{out}} x_i^{\text{out}} = \sum_{i=1}^{NC} \dot{N}^{\text{in}} x_i^{\text{in}}$$

or

$$\dot{N}^{\text{out}} \sum_{i=1}^{NC} x_i^{\text{out}} = \dot{N}^{\text{in}} \sum_{i=1}^{NC} x_i^{\text{in}}$$

From the definition of the mole fractions it follows that $\sum_{i=1}^{NC} x_i^{\text{out}} = \sum_{i=1}^{NC} x_i^{\text{in}} = 1$, and the total mole balance is obtained

$$\dot{N}^{\text{out}} = \dot{N}^{\text{in}} \tag{6.1.16}$$

i.e., from the set of the $NC + 1$ equations involving the total mole balance and the NC component mole balances, only NC are independent (1 mole balance is always redundant).

The mole balance equations can also be transformed into mass balance equations (change $\dot{n}_i$ to $\dot{m}_i$, $\dot{N}$ to $\dot{M}$, concentrations in mass units, and mole fractions x_i to mass fractions w_i). All forms are completely equivalent, and the selection needs to be made on a case-by-case basis.

6.2 Steady state mass balances without chemical reactions

6.2.1 Independent steady state mass balances

In the previous section, we presented the general set of equations used to describe mass balances at steady state for both reactive and nonreactive systems. In this section, we will focus specifically on non-reactive systems, with the goal of developing the set of independent equations needed for analysis.

To illustrate these concepts, let us consider a specific example. Two streams, Stream 1 and Stream 2, are fed into a mixing vessel, as shown in Fig. 6.2. Stream 1 is a mixture of water and methanol with a molar flow rate of $\dot{N}_1^{\text{in}} = 1\ \text{kmol}/\text{s}$ and contains 90 mol% methanol (throughout this text, we use the

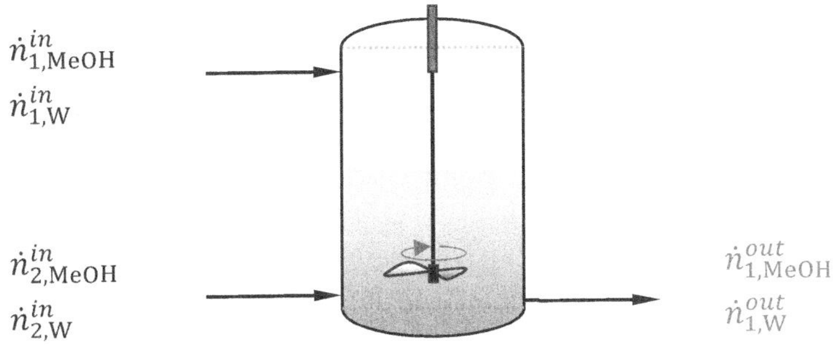

Figure 6.2 Mixing of two streams.

notation wt%, mol%, and vol% to denote weight percent, mole percent, and volume percent, respectively). Stream 2 is a pure water stream with a molar flow rate of $\dot{N}_2^{in} = 2\ \text{kmol}/\text{s}$. There is a single product stream exiting the vessel. The objective is to determine the molar flow rates of water and methanol in the outlet stream.

For the system of Fig. 6.2 we can write one total mole balance and two component mole balances (subscript MeOH denotes methanol and W denotes water)

$$\begin{aligned} \dot{N}_1^{out} &= \dot{N}_1^{in} + \dot{N}_2^{in} \\ \dot{n}^{out}_{1,MeOH} &= \dot{n}^{in}_{1,\mathrm{MeOH}} + \dot{n}^{in}_{2,\mathrm{MeOH}} \\ \dot{n}^{out}_{1,\mathrm{W}} &= \dot{n}^{in}_{1,\mathrm{W}} + \dot{n}^{in}_{2,\mathrm{W}} \end{aligned}$$

In the data, the total molar flowrate of the entering streams as well as the mole fractions are given, and we can calculate the molar flowrates of the two components in the two incoming streams:

$$\dot{n}^{in}_{1,\mathrm{MeOH}} = \dot{N}_1^{in} x^{in}_{1,\mathrm{MeOH}} = \left(1\ \frac{\mathrm{kmol}}{\mathrm{s}}\right) \times (0.9) = 0.9\ \frac{\mathrm{kmol}}{\mathrm{s}}$$

$$\dot{n}^{in}_{1,\mathrm{W}} = \dot{N}_1^{in} x^{in}_{1,\mathrm{W}} = \left(1\ \frac{\mathrm{kmol}}{\mathrm{s}}\right) \times (0.1) = 0.1\ \frac{\mathrm{kmol}}{\mathrm{s}}$$

$$\dot{n}^{in}_{2,\mathrm{MeOH}} = \dot{N}_2^{in} x^{in}_{2,\mathrm{MeOH}} = \left(2\ \frac{\mathrm{kmol}}{\mathrm{s}}\right) \times (0) = 0\ \frac{\mathrm{kmol}}{\mathrm{s}}$$

$$\dot{n}^{in}_{2,\mathrm{W}} = \dot{N}_2^{in} x^{in}_{2,\mathrm{W}} = \left(2\ \frac{\mathrm{kmol}}{\mathrm{s}}\right) \times (1.0) = 2.0\ \frac{\mathrm{kmol}}{\mathrm{s}}$$

We can now replace the numerical values in the mole balances and obtain the molar flowrates of the components in the exiting stream

$$\begin{aligned} \dot{N}_1^{out} &= 1.0 + 2.0 = 3.0\ \frac{kmol}{s} \\ \dot{n}_{1,MeOH}^{out} &= 0.9 + 0 = 0.9\ \frac{kmol}{s} \\ \dot{n}_{1,W}^{out} &= 0.1 + 2.0 = 2.1\ \frac{kmol}{s} \end{aligned}$$

Had we considered the fact that the total molar flowrate of any stream is the sum of the individual molar flowrates of all components, then

$$\dot{N}_1^{out} = \sum_{k \in \{MeOH,W\}} \dot{n}_{1,k}^{out} = \dot{n}_{1,MeOH}^{out} + \dot{n}_{1,W}^{out} = 0.9 + 2.1 = 3.0\ \frac{kmol}{s}$$

It then follows that the total material balance is redundant.

Let us now use the total molar flowrates and the mole fractions

$$\begin{aligned} \dot{N}_1^{out} &= \dot{N}_1^{in} + \dot{N}_2^{in} \\ \dot{N}_1^{out} x_{1,MeOH}^{out} &= \dot{N}_1^{in} x_{1,MeOH}^{in} + \dot{N}_2^{in} x_{2,MeOH}^{in} \\ \dot{N}_1^{out} x_{1,W}^{out} &= \dot{N}_1^{in} x_{1,W}^{in} + \dot{N}_2^{in} x_{2,W}^{in} \end{aligned}$$

Our data is in exactly this form and we can replace directly with the numerical values to obtain

$$\begin{aligned} \dot{N}_1^{out} &= 1.0 + 2.0 = 3.0\ \frac{kmol}{s} \\ \dot{N}_1^{out} x_{1,MeOH}^{out} &= \left(1\ \frac{kmol}{s}\right) \times (0.9) + \left(2\ \frac{kmol}{s}\right) \times (0.0) = 0.9\ \frac{kmol}{s} \\ \dot{N}_1^{out} x_{1,W}^{out} &= \left(1\ \frac{kmol}{s}\right) \times (0.1) + \left(2\ \frac{kmol}{s}\right) \times (1.0) = 2.1\ \frac{kmol}{s} \end{aligned}$$

To find the mole fraction of methanol in the product stream we use the second equation and replace the value of $\dot{N}_1^{out}$ from the first

$$\left(3.0\ \frac{kmol}{s}\right) x_{1,MeOH}^{out} = 0.9\ \frac{kmol}{s} \quad \Rightarrow x_{1,MeOH}^{out} = \frac{0.9}{3.0} = 0.3$$

To calculate $x_{1,W}^{out}$ we can follow the same approach, or we can use the definition of mole fractions and the fact that their sum is always equal to 1:

$$x_{1,\mathrm{MeOH}}^{\mathrm{out}} + x_{1,\mathrm{W}}^{\mathrm{out}} = 1 \Rightarrow x_{1,\mathrm{W}}^{\mathrm{out}} = 1 - x_{1,\mathrm{MeOH}}^{\mathrm{out}} = 1 - 0.3 = 0.7$$

We conclude that the mole balance of water is redundant.

To avoid redundancy in our formulations, it is essential to define **an independent set of mole or mass balances** in a systematic manner. We propose selecting any subset of *NC* balances from the total of $NC + 1$ possible equations that can be formulated. Since most practical and industrial formulations are based on total molar (or mass) flow rates and mole (or mass) fractions, we adopt a common approach: using the total mole balance along with $NC - 1$ component mole balances. The mole fraction of the remaining component (assumed to be component *NC*, the last in the list) can then be determined using the mole fraction summation constraint, which requires that the sum of all mole fractions equals one

$$x_{NC} = 1 - \sum_{i=1}^{NC-1} x_i \tag{6.2.1}$$

The following set of *NC* independent equations will be used in this text in *molar units:*

$$\sum_{k=1}^{p} \dot{N}_k^{\mathrm{out}} = \sum_{k=1}^{f} \dot{N}_k^{\mathrm{in}} \tag{6.2.2}$$

$$\sum_{k=1}^{p} \dot{N}_k^{\mathrm{out}} x_{k,i}^{\mathrm{out}} = \sum_{k=1}^{f} \dot{N}_k^{\mathrm{in}} x_{k,i}^{\mathrm{in}}, \; i = 1, 2, \ldots, NC - 1 \tag{6.2.3}$$

or in mass units:

$$\sum_{k=1}^{p} \dot{M}_k^{\mathrm{out}} = \sum_{k=1}^{f} \dot{M}_k^{\mathrm{in}} \tag{6.2.4}$$

$$\sum_{k=1}^{p} \dot{M}_k^{\mathrm{out}} w_{k,i}^{\mathrm{out}} = \sum_{k=1}^{f} \dot{M}_k^{\mathrm{in}} w_{k,i}^{\mathrm{in}}, \; i = 1, 2, \ldots, NC - 1 \tag{6.2.5}$$

where *w* denotes weight fraction. The mole or mass fraction of the last component is calculated from Eq. (6.2.1).

6.2.2 Number of degrees of freedom and independent specifications

An important question to consider is: **how many variables are needed to fully describe a chemical process based on all entering and exiting streams**? Each stream contributes a number of variables equal to the number of components it contains. These variables can be expressed, for example, as the molar flow rates of the individual components within each stream

$$\dot{n}_{k,i}, \quad i = 1, 2, \ldots, NC$$

or as the total molar (or mass) flowrate and the $NC - 1$ mole (or mass) fractions

$$N_k, \; x_{k,1}, \; x_{k,2}, \ldots, \; x_{k,NC-1}$$

There are f entering and p exiting streams and $NS = f + p$ in total, where NS denotes the **n**umber of **s**treams. In the most general case, all NC components appear in all streams. Thus, the maximum number of variables is

$$\text{maximum number of variables} = NS \cdot NC$$

We have already discussed that

$$\text{number of independent molar balances} = NC$$

For our analysis to result in a well-defined system of equations that can have a unique solution, **we must have as many equations as there are variables**. The mole balances give NC independent equations only and we need to have

$$\text{number of variables} - \text{number of independent molar balances}$$

additional independent specifications for the system of equations to have a unique solution. This is called the **number of degrees of freedom** (N_{DOF}) and is given by

$$N_{\text{DOF}} = N_{\text{VAR}} - N_{EQ} = NS \cdot NC - NC = (NS - 1)NC \tag{6.2.6}$$

This represents the number of **independent pieces of information that must be specified in order to fully define the system**. Combined with the set of independent molar balances, these specifications allow us to solve the mathematical model and determine all relevant process variables. Note that in the classical chemical engineering literature the number of degrees of freedom is defined in a different, more restricted way (see literature at the end of the chapter).

Consider again the case study of Fig. 6.2. How many variables are there? The answer follows by noting that $NC = 2$, in all streams, and $NS = 3$ (as $f = 2, p = 1$). There are $NS{\cdot}NC = 3 \times 2 = 6$ variables, namely

$$\dot{n}_{1,\mathrm{MeOH}}^{\mathrm{in}},\ \dot{n}_{1,\mathrm{W}}^{\mathrm{in}},\ \dot{n}_{2,\mathrm{MeOH}}^{\mathrm{in}},\ \dot{n}_{2,\mathrm{W}}^{\mathrm{in}},\ \dot{n}_{1,\mathrm{MeOH}}^{\mathrm{out}},\ \dot{n}_{1,\mathrm{W}}^{\mathrm{out}}$$

or

$$\dot{N}_1^{\mathrm{in}},\ x_{1,\mathrm{MeOH}}^{\mathrm{in}},\ \dot{N}_2^{\mathrm{in}},\ x_{2,\mathrm{MeOH}}^{\mathrm{in}},\ \dot{N}_1^{\mathrm{out}},\ x_{1,\mathrm{MeOH}}^{\mathrm{out}}$$

Both entering streams are fully specified as the values of their total molar flowrates and methanol mole fractions are given, four in total. The number of specifications is equal to N_{DOF}

$$N_{DOF} = (NS - 1)NC = (3 - 1) \times 2 = 4$$

This is why (number of specifications = number of degrees of freedom) we have been able to solve the 2 molar balances for the two unknowns.

Consider now the alternative specifications for the two-stream mixing process shown in Fig. 6.3. The product stream is now fully specified. In addition, the composition of the feed streams, which are again binary mixtures of methanol and water, is known. In overall we have four specifications:

$$x_{1,\mathrm{MeOH}}^{\mathrm{in}},\ x_{2,\mathrm{MeOH}}^{\mathrm{in}},\ \dot{N}_1^{\mathrm{out}},\ x_{1,\mathrm{MeOH}}^{\mathrm{out}}$$

and two unknown variables

$$\dot{N}_1^{\mathrm{in}}, \dot{N}_2^{\mathrm{in}}$$

The specifications are equal to the N_{DOF}, which are equal to the previous case study, and as a direct result, the number of unknowns equals the number of equations (mole balances). This is not accidental but the result of a systematic approach. We write down the overall mole balance and the mole balance of methanol

$$\begin{aligned} \dot{N}_1^{\mathrm{out}} &= \dot{N}_1^{\mathrm{in}} + \dot{N}_2^{\mathrm{in}} \\ \dot{N}_1^{\mathrm{out}} x_{1,\mathrm{MeOH}}^{\mathrm{out}} &= \dot{N}_1^{\mathrm{in}} x_{1,\mathrm{MeOH}}^{\mathrm{in}} + \dot{N}_2^{\mathrm{in}} x_{2,\mathrm{MeOH}}^{\mathrm{in}} \end{aligned}$$

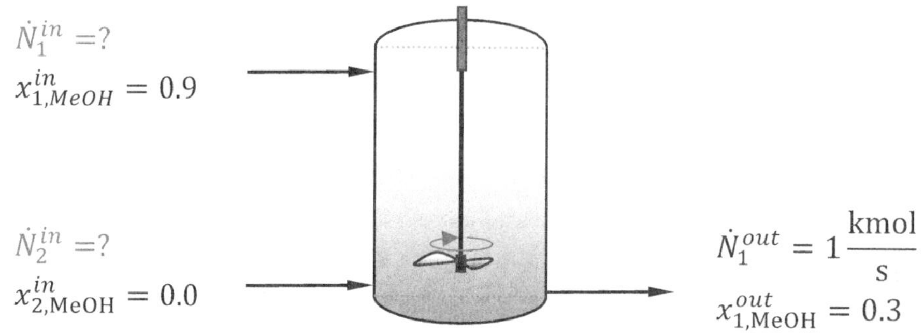

Figure 6.3 Alternative specifications in the mixing of two streams process.

As $x_{2,\text{MeOH}}^{\text{in}} = 0$, the equations are easy to solve.

Let us now consider a slightly modified process, illustrated in Fig. 6.4. As before, there are two inlet streams and one outlet stream. The first inlet stream is a methanol/water mixture containing 80 mol% methanol, while the second is an ethanol/water mixture with 50 mol% ethanol. The product stream contains three components—methanol, ethanol, and water—and both its total flow rate and component mole fractions are provided.

Please copy the process description into your notebook and attempt the analysis independently before proceeding. It is strongly recommended that you invest the time to perform a thorough evaluation of the system and reach a clear conclusion regarding whether the mathematical model is solvable and, if so, what the unique, physically meaningful solution would be. Do not turn to the next page until you are confident in your analysis!

Welcome back! So, what is the solution to this problem? Let us follow the steps:

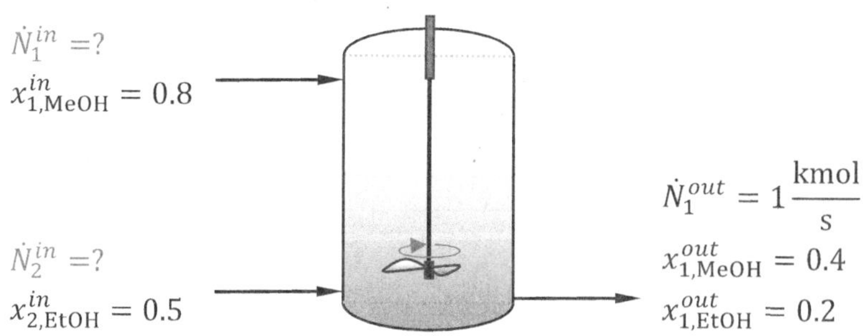

Figure 6.4 Mixing of two streams process with three components.

STEP 1: For each stream—whether entering or exiting—count the number of components it contains. This number corresponds to the number of variables required to fully describe that stream (e.g., component molar flow rates).

STEP 2: Add up all variables from every stream to determine the total number of variables, denoted as N_{VAR}. This represents the total number of unknowns in the problem. Clearly list these variables and define your notation.

STEP 3: Determine the total number of distinct components present in the overall system. This is the maximum number of independent mole or mass balances that can be written and is denoted as N_{EQ}. Now calculate the degrees of freedom using: $N_{DOF} = N_{VAR} - N_{EQ}$.

STEP 4: Count the number of independent specifications provided in the problem data, denoted as N_{SPEC}.

- If $N_{SPEC} = N_{DOF}$, the problem is **well-defined** and a unique solution may exist.
- If $N_{SPEC} > N_{DOF}$, there are too many specifications; the problem may be **overspecified** or inconsistent.
- If $N_{SPEC} < N_{DOF}$ **additional information is needed** to fully define the problem.

Let us follow these steps for the process of Fig. 6.4:

STEP 1, feed streams, stream 1: there are 2 components (MeOH, W) and two variables

stream 2: there are 2 components (EtOH, W) and two variables

product streams, stream 1: there are three components (MeOH, EtOH, W) and three variables.

STEP 2, 2 + 2+3 = 7 variables: $\dot{N}_1^{in}$, $x_{1,MeOH}^{in}$, $\dot{N}_2^{in}$, $x_{2,EtOH}^{in}$, $\dot{N}_1^{out}$, $x_{1,MeOH}^{out}$, $x_{1,EtOH}^{out}$

STEP 3, there are three components appearing in the process and $N_{EQ} = 3$, it follows that

$$N_{DOF} = 7 - 3 = 4$$

STEP 4, we count the specifications.

1. $x_{1,MeOH}^{in} = 0.8$.
2. $x_{2,EtOH}^{in} = 0.5$.
3. $\dot{N}_1^{out} = 1\frac{\text{kmol}}{\text{s}}$.
4. $x_{1,MeOH}^{out} = 0.4$.
5. $x_{1,EtOH}^{out} = 0.2$.

We note that $N_{SPEC} = 5 > N_{DOF} = 4$ and the system is overspecified!

The same conclusion follows if we write down the 3 molar balances:

$$\begin{aligned}
\dot{N}_1^{\text{out}} &= \dot{N}_1^{\text{in}} + \dot{N}_2^{\text{in}} \\
\dot{N}_1^{\text{out}} x_{1,\text{MeOH}}^{\text{out}} &= \dot{N}_1^{\text{in}} x_{1,\text{MeOH}}^{\text{in}} + \underbrace{\dot{N}_2^{\text{in}} x_{2,\text{MeOH}}^{\text{in}}}_{=0} \\
\dot{N}_1^{\text{out}} x_{1,\text{EtOH}}^{\text{out}} &= \underbrace{\dot{N}_1^{\text{in}} x_{1,\text{EtOH}}^{\text{in}}}_{=0} + \dot{N}_2^{\text{in}} x_{2,\text{EtOH}}^{\text{in}}
\end{aligned}$$

Note that there are three equations but only two unknown variables ($\dot{N}_1^{\text{in}}$ and $\dot{N}_2^{\text{in}}$).

Many assume there is a straightforward solution to this problem: since there are two unknowns, one might simply select any two of the three available equations and disregard the third. However, what are the consequences of this approach? To explore this, let us consider the case where we retain the two component balance equations and omit the overall balance

$$\begin{aligned}
\dot{N}_1^{\text{out}} x_{1,\text{MeOH}}^{\text{out}} &= \dot{N}_1^{\text{in}} x_{1,\text{MeOH}}^{\text{in}} + \underbrace{\dot{N}_2^{\text{in}} x_{2,\text{MeOH}}^{\text{in}}}_{=0} \\
\dot{N}_1^{\text{out}} x_{1,\text{EtOH}}^{\text{out}} &= \underbrace{\dot{N}_1^{\text{in}} x_{1,\text{EtOH}}^{\text{in}}}_{=0} + \dot{N}_2^{\text{in}} x_{2,\text{EtOH}}^{\text{in}}
\end{aligned}$$

The solution is

$$\dot{N}_1^{\text{in}} = \dot{N}_1^{\text{out}} \frac{x_{1,\text{MeOH}}^{\text{out}}}{x_{1,\text{MeOH}}^{\text{in}}} = 1 \frac{\text{kmol}}{\text{s}} \times \frac{0.4}{0.8} = 0.5 \frac{\text{kmol}}{\text{s}}$$

$$\dot{N}_2^{\text{in}} = \dot{N}_1^{\text{out}} \frac{x_{1,\text{EtOH}}^{\text{out}}}{x_{2,\text{EtOH}}^{\text{in}}} = 1 \frac{\text{kmol}}{\text{s}} \times \frac{0.2}{0.5} = 0.4 \frac{\text{kmol}}{\text{s}}$$

The equation for the overall molar balance gives

$$N_1^{\text{out}} = \dot{N}_1^{\text{in}} + \dot{N}_2^{\text{in}} = 0.5 \frac{\text{kmol}}{\text{s}} + 0.4 \frac{\text{kmol}}{\text{s}} = 0.9 \frac{\text{kmol}}{\text{s}} \neq 1 \frac{\text{kmol}}{\text{s}}$$

and the specification is not satisfied! The quick fix does not work, as expected.

The only feasible solution is to relax one of the three specifications for the product stream or, alternatively, to increase the number of degrees of freedom so that it matches the number of specifications ($N_{DOF} = N_{SPEC}$). One way to achieve this is by introducing an additional input stream—specifically, a pure water feed, as illustrated in Fig. 6.5. This increases the degrees of freedom by one, bringing them into alignment with the number of specified values. With this modification, the mathematical model we need to solve is as follows (be sure to validate each step)

$$\begin{aligned}
\dot{N}_1^{out} &= \dot{N}_1^{in} + \dot{N}_2^{in} + \dot{N}_3^{in} \\
\dot{N}_1^{out} x_{1,MeOH}^{out} &= \dot{N}_1^{in} x_{1,MeOH}^{in} + \underbrace{\dot{N}_2^{in} x_{2,MeOH}^{in}}_{=0} + \underbrace{\dot{N}_3^{in} x_{3,MeOH}^{in}}_{=0} \\
\dot{N}_1^{out} x_{1,EtOH}^{out} &= \underbrace{\dot{N}_1^{in} x_{1,EtOH}^{in}}_{=0} + \dot{N}_2^{in} x_{2,EtOH}^{in} + \underbrace{\dot{N}_3^{in} x_{3,EtOH}^{in}}_{=0}
\end{aligned}$$

and the solution is

$$\dot{N}_1^{in} = \dot{N}_1^{out} \frac{x_{1,MeOH}^{out}}{x_{1,MeOH}^{in}} = 1 \frac{\text{kmol}}{\text{s}} \times \frac{0.4}{0.8} = 0.5 \frac{\text{kmol}}{\text{s}}$$

$$\dot{N}_2^{in} = \dot{N}_1^{out} \frac{x_{1,EtOH}^{out}}{x_{2,EtOH}^{in}} = 1 \frac{\text{kmol}}{\text{s}} \times \frac{0.2}{0.5} = 0.4 \frac{\text{kmol}}{\text{s}}$$

$$\begin{aligned}
\dot{N}_3^{in} = N_1^{out} - \dot{N}_1^{in} - \dot{N}_2^{in} &= 1.0 \frac{\text{kmol}}{\text{s}} - 0.5 \frac{\text{kmol}}{\text{s}} - 0.4 \frac{\text{kmol}}{\text{s}} \\
&= 0.1 \frac{\text{kmol}}{\text{s}}
\end{aligned}$$

You may have noticed a new symbol introduced in Fig. 6.5: a hexagon containing a number appears alongside each stream.

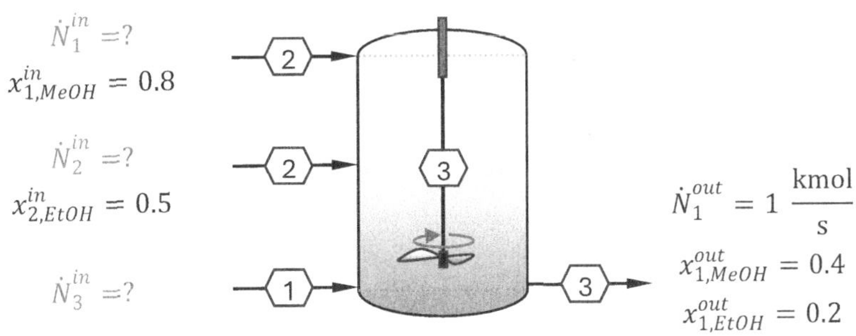

Figure 6.5 Mixing of three streams process with three components, degrees of freedom analysis.

This number answers the question: **How many components are present in this stream?** Please validate the values shown in each hexagon. As previously discussed, **the number of components in a stream corresponds to the number of variables required to fully describe that stream**. Therefore, **the sum of all hexagon values across the streams represents the total number of variables needed to describe the entire process**. In this case, the sum is: 2 + 2+1 + 3 = 8 variables. You will also see a hexagon inside the process boundary, which contains the number of distinct components appearing in the incoming and outgoing streams, that is, the components that cross the system boundaries. Validate the number shown in Fig. 6.5: it should be 3, representing methanol (MeOH), ethanol (EtOH), and water (W). The difference between the total number of variables and the number of independent component balances gives the number of independent specifications required for a well-defined system:

$$N_{SPEC} = N_{VAR} - N_{EQ} = 8 - 3 = 5$$

As shown in Fig. 6.5, five independent specifications have been provided, making the system mathematically well-defined.

1. $x^{in}_{1,\mathrm{MeOH}} = 0.8$.
2. $x^{in}_{2,\mathrm{EtOH}} = 0.5$.
3. $\dot{N}^{out}_{1} = 1\frac{\mathrm{kmol}}{\mathrm{s}}$.
4. $x^{out}_{1,\mathrm{MeOH}} = 0.4$.
5. $x^{out}_{1,\mathrm{EtOH}} = 0.2$.

It follows that we have a well-defined system.

Example 6.1. A distillation separation unit

An equimolar mixture of benzene (B), toluene (T), and xylene (X), known as BTX mixture, with molar flowrate 100 kmol/h is fed to a distillation unit, as shown in Fig. 6.6. A distillation column is a separating unit that has usually two product streams: a top product stream that is dominated by the light (low boiling point) components and a bottoms product stream that is dominated by the heavy (high boiling point) components. The top product has a molar flowrate of 50 kmol/h. It has been specified that the top product is 66 mol% benzene while the bottom product is 33 mol% toluene. Determine all variables that are relevant to the system.

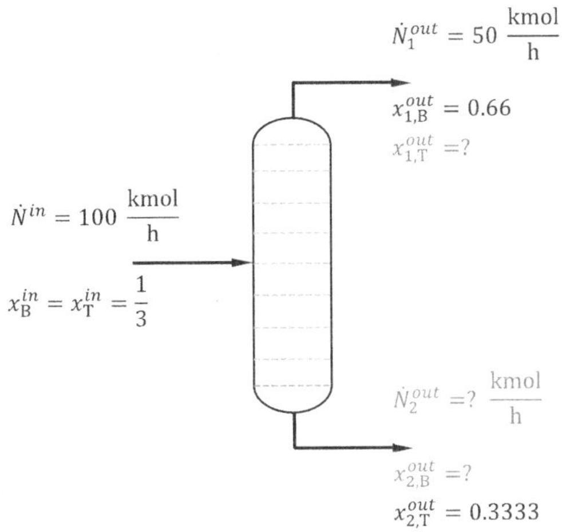

Figure 6.6 Separation of a BTX mixture in a distillation column.

Solution.

We follow the systematic steps:

STEP 1, feed stream, stream 1: there are three components (B, T, X) and three variables

product streams, stream 1 (top product): there are three components and three variables

stream 2 (bottom product): there are three components and three variables.

STEP 2, 3 + 3+3 = 9 variables: $\dot{N}^{in}$, x_B^{in}, x_T^{in}, $\dot{N}_1^{out}$, $x_{1,B}^{out}$, $x_{1,T}^{out}$, $\dot{N}_2^{out}$, $x_{2,B}^{out}$, $x_{2,T}^{out}$

STEP 3, there are three components appearing in the process and $N_{EQ} = 3$, it follows that $N_{DOF} = 9 - 3 = 6$

STEP 4, we count the specifications.

1. $\dot{N}^{in} = 100 \frac{kmol}{h}$.
2. $x_B^{in} = 1/3$.
3. $x_T^{in} = 1/3$.
4. $\dot{N}_1^{out} = 50 \frac{kmol}{h}$.
5. $x_{1,B}^{out} = 0.66$.
6. $x_{2,T}^{out} = 0.3333$.

We note that $N_{SPEC} = 6 = N_{DOF}$ and the system is well defined.

The same result is obtained from Fig. 6.7, from which it follows that there are (3 + 3+3 =) nine variables and three independent equations. For a well-defined system we need six independent specifications.

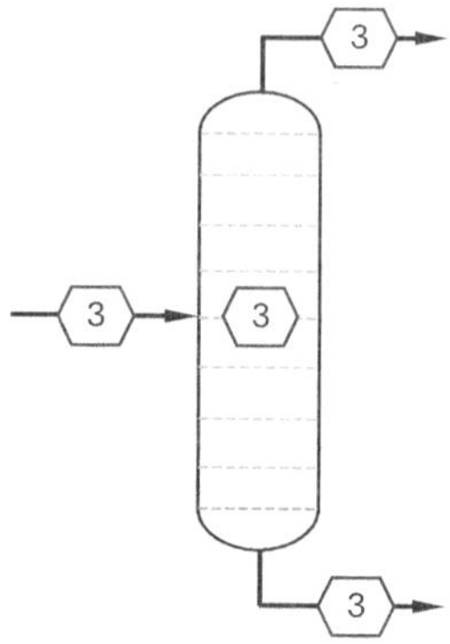

Figure 6.7 Separation of a BTX mixture in a distillation column, degrees of freedom analysis.

We write the 3 molar balances

$$\begin{aligned}
\dot{N}^{in} &= \dot{N}_1^{out} + \dot{N}_2^{out} \\
\dot{N}^{in} x_B^{in} &= \dot{N}_1^{out} x_{1,B}^{out} + \dot{N}_2^{out} x_{2,B}^{out} \\
\dot{N}^{in} x_T^{in} &= \dot{N}_1^{out} x_{1,T}^{out} + \dot{N}_2^{out} x_{2,T}^{out}
\end{aligned}$$

The solution is

$$\dot{N}_2^{out} = 100 - 50 = 50 \text{ kmol/h}$$

$$x_{2,B}^{out} = \left(\dot{N}^{in} x_B^{in} - \dot{N}_1^{out} x_{1,B}^{out}\right) / \dot{N}_2^{out} = (33.333 - 50 \cdot 0.66) / 50 = 0.0067$$

$$x_{1,T}^{out} = \left(\dot{N}^{in} x_T^{in} - \dot{N}_2^{out} x_{2,T}^{out}\right) / \dot{N}_1^{out} = (33.333 - 50 \cdot 0.3333) / 50 = 0.3333$$

Example 6.2. Acetone recovery in an absorption unit

In Fig. 6.8, an absorption unit is used to remove acetone (Ac) from an 100 kmol/h air stream containing 10 mol% acetone. The air enters the bottom of the absorption column, while a liquid solvent (S)—initially free of acetone—is introduced at the top. The air flows upward, countercurrent to the downward flow of the solvent. The solvent selectively absorbs only acetone from the air; it does not dissolve any air, and it is assumed to be nonvolatile. The objective is to reduce the acetone concentration in the exiting air stream to 0.1 mol%. The molar flow rate of the solvent is twice that of the air. Given these conditions, determine the mole fraction of acetone in the solvent exiting the absorption unit.

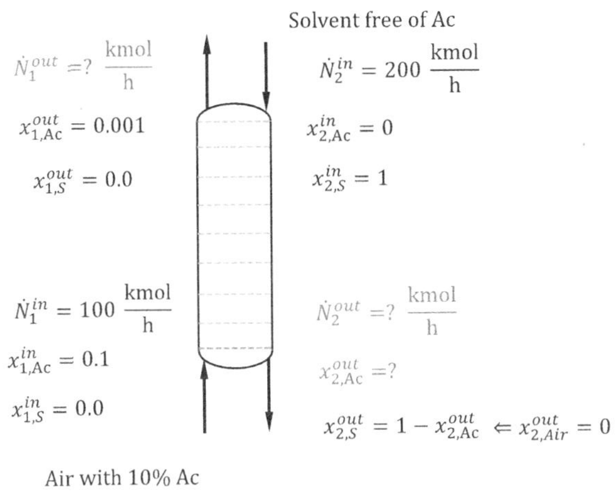

Figure 6.8 Acetone removed from an air stream by a liquid solvent.

Solution.

In Fig. 6.9 we present the analysis of the degrees of freedom. We consider all components acetone, solvent, and air even if some of them do not appear in a stream. We assume that all three components are present in all streams and add a specification for the component that is not present (air in the solvent-rich stream and solvent in the air stream). Study the specifications in Fig. 6.8 carefully. The number of variables is

$$N_{VAR} = 3 + 3 + 3 + 3 = 12$$

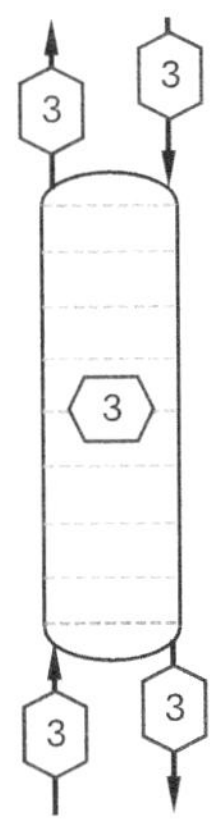

Figure 6.9 Acetone removed from an air stream by a liquid solvent, degrees of freedom analysis.

The number of equations is

$$N_{EQ} = 3$$

and the number of degrees of freedom

$$N_{\text{DOF}} = N_{\text{VAR}} - N_{EQ} = 12 - 3 = 9$$

We count the specifications given

1. $\dot{N}_1^{\text{in}} = 100 \dfrac{\text{kmol}}{\text{h}}$.
2. $x_{1,\text{Ac}}^{\text{in}} = 0.1$.
3. $x_{1,\text{S}}^{\text{in}} = 0.0$.
4. $x_{1,\text{Ac}}^{\text{out}} = 0.001$.
5. $x_{1,\text{S}}^{\text{out}} = 0$.
6. $x_{2,\text{Ac}}^{\text{in}} = 0$.
7. $\dot{N}_2^{\text{in}} = 200 \dfrac{\text{kmol}}{\text{h}}$.
8. $x_{2,\text{S}}^{\text{in}} = 1$.
9. $x_{2,\text{S}}^{\text{out}} = 1 - x_{2,\text{Ac}}^{\text{out}}$ (i.e., no air present)

The number of specifications is equal to the number of degrees of freedom, and the problem is well defined. We write the three mole balances

$$\text{Total molar balance:}\ \dot{N}_1^{\text{in}} + \dot{N}_2^{\text{in}} = \dot{N}_1^{\text{out}} + \dot{N}_2^{\text{out}}$$

$$\text{Acetone mole balance:}\ \dot{N}_1^{\text{in}} x_{1,\text{Ac}}^{\text{in}} + \underbrace{\dot{N}_2^{\text{in}} x_{2,\text{Ac}}^{\text{in}}}_{=0} = \dot{N}_1^{\text{out}} x_{1,\text{Ac}}^{\text{out}} + \dot{N}_2^{\text{out}} x_{2,\text{Ac}}^{\text{out}}$$

$$\text{Solvent mole balance:}\ \dot{N}_2^{\text{in}} = \dot{N}_2^{\text{out}}\left(1 - x_{2,\text{Ac}}^{\text{out}}\right)$$

We replace the numerical values:

$$300 = \dot{N}_1^{\text{out}} + \dot{N}_2^{\text{out}}$$

$$10 = \dot{N}_1^{\text{out}} 0.001 + \dot{N}_2^{\text{out}} x_{2,\text{Ac}}^{\text{out}}$$

$$200 = \dot{N}_2^{\text{out}} - \dot{N}_2^{\text{out}} x_{2,\text{Ac}}^{\text{out}}$$

We add the second and the third equation

$$210 = \dot{N}_1^{\text{out}} 0.001 + \dot{N}_2^{\text{out}}$$

We subtract from the first

$$90 = 0.999\dot{N}_1^{\text{out}} \quad \Rightarrow \quad \dot{N}_1^{\text{out}} = \frac{90}{0.999} = 90.09\ \frac{\text{kmol}}{\text{h}}$$

$$\dot{N}_2^{\text{out}} = 300 - \dot{N}_1^{\text{out}} = 209.91\ \frac{\text{kmol}}{\text{h}}$$

$$x_{2,\text{Ac}}^{\text{out}} = 1 - \frac{200}{\dot{N}_2^{\text{out}}} = 1 - \frac{200}{209.91} = 0.047$$

Example 6.3. Fresh water from seawater

Desalination is an artificial process by which saline water is converted to fresh water. The most common desalination processes are distillation and reverse osmosis. The traditional process of desalination is distillation (i.e., boiling and re-condensation of seawater to leave salt and impurities behind). An evaporator for producing fresh water is shown in Fig. 6.10. One thousand kg/h of seawater is fed to an evaporator in which, using thermal energy, the water is evaporated to produce potable water and a concentrated salt solution is received as a byproduct. The concentration of salt in seawater is 3.5 wt% and is concentrated to 10 wt%. Can you determine the mass flowrates of the two product streams?

Solution.

Perform a degree of freedom analysis to prove that this is a well-defined problem. There are two components appearing in the feed stream (water and salt), two components in the liquid byproduct stream, while the vapor stream is pure water. We need, therefore, five variables for the complete description of the process

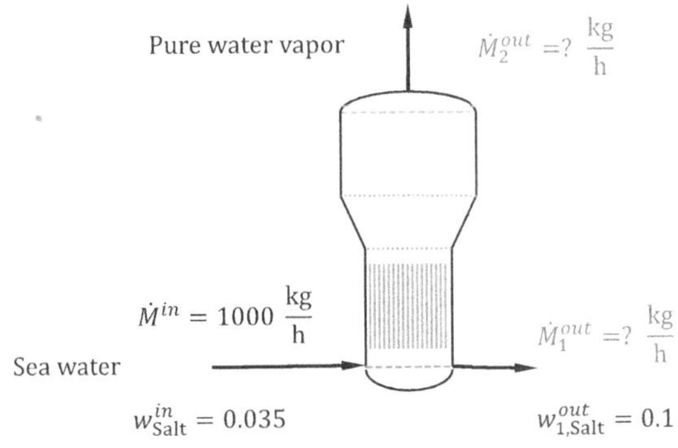

Figure 6.10 An evaporation process for producing fresh water from seawater.

$$\dot{M}^{\text{in}}, w_{\text{Salt}}^{\text{in}}, \dot{M}_1^{\text{out}}, w_{1,\text{Salt}}^{\text{out}}, \dot{M}_2^{\text{out}}$$

Two components are present in the streams and we can write two mass balances:

Total mass balance

$$\dot{M}^{\text{in}} = \dot{M}_1^{\text{out}} + \dot{M}_2^{\text{out}}$$

Salt mass balance

$$\dot{M}^{\text{in}} w_{\text{Salt}}^{\text{in}} = \dot{M}_1^{\text{out}} w_{1,\text{Salt}}^{\text{out}} + \dot{M}_2^{\text{out}} \underbrace{w_{2,\text{Salt}}^{\text{out}}}_{=0}$$

The solution is straightforward. We first solve the salt mass balance for the mass flowrate of the liquid product

$$\dot{M}_1^{\text{out}} = \dot{M}^{\text{in}} \frac{w_{\text{Salt}}^{\text{in}}}{w_{1,\text{Salt}}^{\text{out}}} = \left(1000 \frac{\text{kg}}{\text{h}}\right) \times \frac{0.035}{0.100} = 350 \frac{\text{kg}}{\text{h}}$$

and we then solve the total mass balance for the mass flowrate of the (pure water) vapor stream

$$\dot{M}_2^{\text{out}} = \dot{M}^{\text{in}} - \dot{M}_1^{\text{out}} = 1000 \frac{\text{kg}}{\text{h}} - 350 \frac{\text{kg}}{\text{h}} = 650 \frac{\text{kg}}{\text{h}}$$

Example 6.4. Bicarbonate soda recovery

Sodium bicarbonate, commonly known as baking soda or bicarbonate of soda, is a chemical compound with the formula $NaHCO_3$, used in baking as a leavening agent (carbon dioxide is released, which causes expansion of the batter and forms the characteristic texture and grain in cakes). Soda forms a saturated solution in water at 25 °C with 10 wt% soda. A stream of 25 wt% soda and unknown mass flowrate is fed to a cooling crystallizer, in which it is cooled into 25 °C, as shown in Fig. 6.11. The liquid becomes saturated and exits the crystallizer with 10 wt% soda while a solid soda (crystals) with 10 wt% moisture is also removed from the crystallizer. If 1000 kg/h of wet crystals are produced, find the mass flowrate of all streams in the process.

Solution.

Perform a degree of freedom analysis to prove that this is a well-defined problem. There are two components appearing in the feed stream (water and soda), two components in the saturated liquid byproduct stream, and two components in the solids stream. We need, therefore, six variables for the complete description of the process

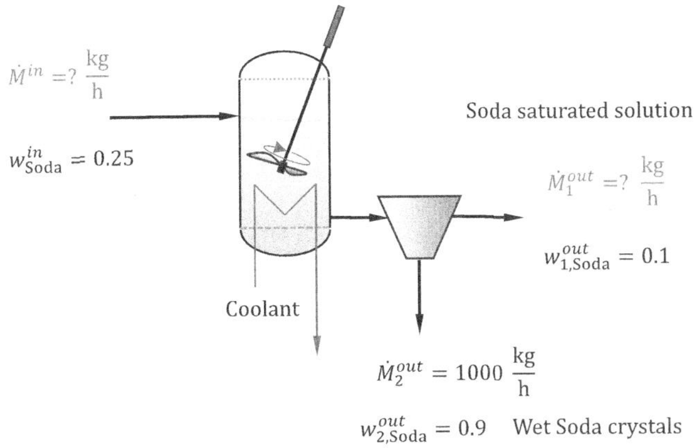

Figure 6.11 A cooling crystallizer for bicarbonate soda recovery.

$$\dot{M}^{in}, w^{in}_{Soda}, \dot{M}^{out}_1, w^{out}_{1,Soda}, \dot{M}^{out}_2, w^{out}_{2,Soda}$$

Two components are present in the streams and we can write two mass balances:

Total mass balance

$$\dot{M}^{in} = \dot{M}^{out}_1 + \dot{M}^{out}_2$$

Soda mass balance

$$\dot{M}^{in} w^{in}_{Soda} = \dot{M}^{out}_1 w^{out}_{1,Soda} + \dot{M}^{out}_2 w^{out}_{2,Soda}$$

We replace the numerical values

$$\dot{M}^{in} - \dot{M}^{out}_1 = 1000$$

$$0.25\,\dot{M}^{in} - 0.1\dot{M}^{out}_1 = 900$$

To calculate the solution, we multiply the second equation by −4 to obtain

$$\dot{M}^{in} - \dot{M}^{out}_1 = 1000$$

$$-\dot{M}^{in} + 0.4\dot{M}^{out}_1 = -3600$$

We add to the overall mass balance equation to obtain

$$-0.6\dot{M}^{out}_1 = -2600$$

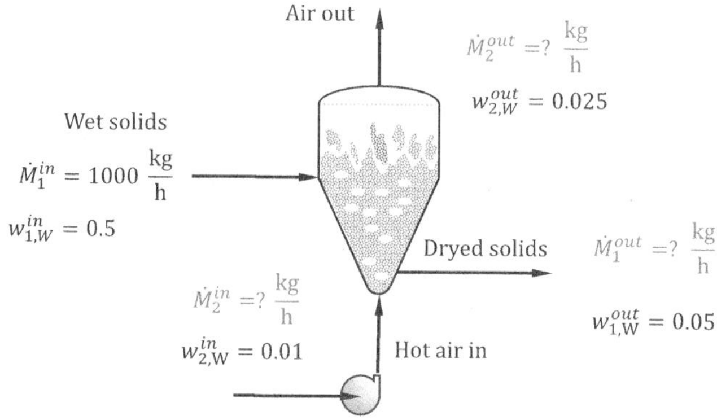

Figure 6.12 A fluidized bed dryer problem.

and

$$\dot{M}_1^{\text{out}} = \frac{-2600}{-0.6} = 4333.33 \ \frac{\text{kg}}{\text{h}}$$

Finally

$$\dot{M}^{\text{in}} = \dot{M}_1^{\text{out}} + \dot{M}_2^{\text{out}} = 4333.33 \ \frac{\text{kg}}{\text{h}} + 1000 \ \frac{\text{kg}}{\text{h}} = 5333.3 \ \frac{\text{kg}}{\text{h}}$$

Example 6.5. Solids drying in a fluidized bed dryer

One thousand kg/h of wet solids with 50 wt% moisture is fed to a (fluidized bed) dryer to reduce the moisture to 5 wt% (see Fig. 6.12). Hot air with initial moisture of 1% is used to remove the moisture from the solids. The moisture in the air stream at the exit is increasing to 2.5 wt%. Find the mass flowrate of all streams.

Solution.

The two air streams involve only air and moisture and the two solid streams involve only solids and moisture. There are 2 variables necessary in each one of the four streams and eight in total:

$$\dot{M}_1^{\text{in}}, w_{1,W}^{\text{in}}, \dot{M}_2^{\text{in}}, w_{2,W}^{\text{in}}, \dot{M}_1^{\text{out}}, w_{1,W}^{\text{out}}, \dot{M}_2^{\text{out}}, w_{2,W}^{\text{out}}$$

There are three components (air, water, solid) and we can write an equal number of mass balances. The number of degrees of freedom are $N_{\text{DOF}} = N_{\text{VAR}} - N_{EQ} = 8 - 3 = 5$, which is equal to the number of specifications:

1. $\dot{M}_1^{\text{in}} = 1000 \ \frac{\text{kg}}{\text{h}}$.
2. $w_{1,W}^{\text{in}} = 0.5$.
3. $w_{2,W}^{\text{in}} = 0.01$.

4. $w_{1,W}^{out} = 0.05$.

5. $w_{2,W}^{out} = 0.025$.

The three mass balances are the following.

Total mass balance

$$\dot{M}_1^{in} + \dot{M}_2^{in} = \dot{M}_1^{out} + \dot{M}_2^{out}$$

Water mass balance

$$\dot{M}_1^{in} w_{1,W}^{in} + \dot{M}_2^{in} w_{2,W}^{in} = \dot{M}_1^{out} w_{1,W}^{out} + \dot{M}_2^{out} w_{2,W}^{out}$$

Solids mass balance

$$\dot{M}_1^{in}\left(1 - w_{1,W}^{in}\right) = \dot{M}_1^{out}\left(1 - w_{1,W}^{out}\right)$$

We can also replace the solids mass balance with the air mass balance

$$\dot{M}_2^{in}\left(1 - w_{2,W}^{in}\right) = \dot{M}_2^{out}\left(1 - w_{2,W}^{out}\right)$$

but the solution is more involved.

From the solids mass balance, we calculate the mass flowrate of the solids exiting stream

$$\dot{M}_1^{out} = \dot{M}_1^{in} \frac{\left(1 - w_{1,W}^{in}\right)}{\left(1 - w_{1,W}^{out}\right)} = 1000 \, \frac{\text{kg}}{\text{h}} \times \frac{0.5}{0.95} = 526.3 \, \frac{\text{kg}}{\text{h}}$$

We replace the numerical values to the total mass balance and the water material balance

$$-\dot{M}_2^{in} + \dot{M}_2^{out} = 473.7$$

$$-0.01\dot{M}_2^{in} + 0.025\dot{M}_2^{out} = 473.7$$

We multiply the second equation by −100 and we add the result to the first

$$-1.5\dot{M}_2^{out} = -46897$$

or

$$\dot{M}_2^{out} = 31\,264.7 \, \frac{\text{kg}}{\text{h}}$$

From the total mass balance

$$\dot{M}_2^{in} = \dot{M}_1^{out} + \dot{M}_2^{out} - \dot{M}_1^{in} = 526.3 + 31\,264.7 - 1000 = 30\,791 \frac{kg}{h}$$

Example 6.6. Soybean oil extraction using hexane as solvent

One thousand kg/h of soybeans that contain 30 wt% soybean oil, 10 wt% water, and is free of hexane, are fed to an extractor/separator unit together with 1.5 t/h pure hexane (see Fig. 6.13). The cake in the exit of the extractor/separator unit contains 5 wt% oil, 8 wt% water, and 1% hexane. The solvent-rich product contains 4 wt% water. All soybeans are obtained in the cake. Find the mass flowrate and mass fractions of all streams.

Solution.

The soybeans stream contains oil (Oil), water (W), and beans (B) and we need three variables to fully describe the stream. The fresh solvent stream contains pure hexane and we only need one variable to describe the stream. The cake contains all components, including the solvent (S) and we need four variables. The solvent-rich product contains solvent, oil, and water and we need three variables to fully describe the stream. The variables are 3 + 1+4 + 3 = 11 variables in total

$$\dot{M}_1^{in}, w_{1,Oil}^{in}, w_{1,W}^{in}, \dot{M}_2^{in}, \dot{M}_1^{out}, w_{1,Oil}^{out}, w_{1,W}^{out}, w_{1,S}^{out}, \dot{M}_2^{out}, w_{2,Oil}^{out}, w_{2,W}^{out}$$

There are four components appearing in the process (oil, water, solvent, and beans) and we can write four mass balances.

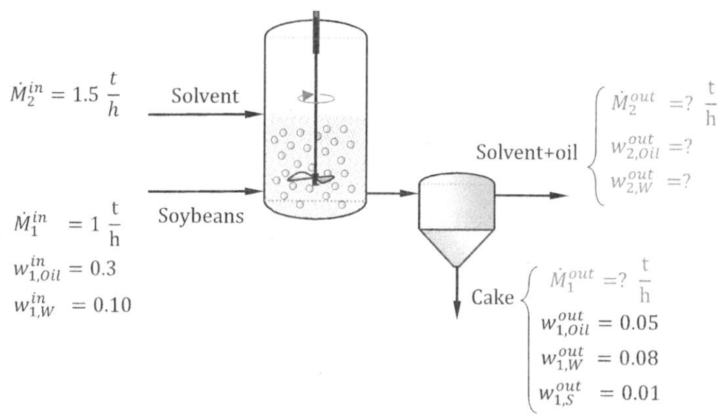

Figure 6.13 Mixing of three streams process with three components, degrees of freedom analysis.

Total mass balance

$$\dot{M}_1^{\text{in}} + \dot{M}_2^{\text{in}} = \dot{M}_1^{\text{out}} + \dot{M}_2^{\text{out}}$$

Oil mass balance

$$\dot{M}_1^{\text{in}} w_{1,Oil}^{\text{in}} = \dot{M}_1^{\text{out}} w_{1,Oil}^{\text{out}} + \dot{M}_2^{\text{out}} w_{2,\text{Oil}}^{\text{out}}$$

Water mass balance

$$\dot{M}_1^{\text{in}} w_{1,W}^{\text{in}} = \dot{M}_1^{\text{out}} w_{1,W}^{\text{out}} + \dot{M}_2^{\text{out}} w_{2,W}^{\text{out}}$$

Solvent mass balance

$$\dot{M}_2^{\text{in}} = \dot{M}_1^{\text{out}} w_{1,S}^{\text{out}} + \dot{M}_2^{\text{out}} \left(1 - w_{2,Oil}^{\text{out}} - w_{2,W}^{\text{out}}\right)$$

We replace the numerical values

$$\dot{M}_1^{\text{out}} + \dot{M}_2^{\text{out}} = 2.5$$

$$0.05\dot{M}_1^{\text{out}} + \dot{M}_2^{\text{out}} w_{2,Oil}^{\text{out}} = 0.3$$

$$0.08\dot{M}_1^{\text{out}} + \dot{M}_2^{\text{out}} w_{2,W}^{\text{out}} = 0.1$$

$$0.01\dot{M}_1^{\text{out}} + \dot{M}_2^{\text{out}} \left(1 - w_{2,Oil}^{\text{out}} - w_{2,W}^{\text{out}}\right) = 1.5$$

The solution is not obvious and a numerical solution is necessary. To this end, we define the vector

$$\boldsymbol{x} = \begin{bmatrix} \dot{M}_1^{\text{out}} \\ \dot{M}_2^{\text{out}} \\ w_{2,\text{Oil}}^{\text{out}} \\ w_{2,W}^{\text{out}} \end{bmatrix}$$

and we solve the system of nonlinear algebraic equations

$$\boldsymbol{f}(\boldsymbol{x}) = \begin{bmatrix} f_1(\boldsymbol{x}) \\ f_2(\boldsymbol{x}) \\ f_3(\boldsymbol{x}) \\ f_4(\boldsymbol{x}) \end{bmatrix} = \begin{bmatrix} x_1 + x_2 - 2.5 \\ 0.05x_1 + x_2x_3 - 0.3 \\ 0.08x_1 + x_2x_4 - 0.1 \\ 0.01x_1 + x_2(1 - x_3 - x_4) - 1.5 \end{bmatrix}$$

The MATLAB implementation is as follows.
and

```
clear all
x0 =[1; 1; 0.1; 0.1]
x=fsolve(@SolvedExample6_6,x0,optimset('Display','iter'))

function f=SolvedExample6_6(x)
f(1) = x(1)+x(2)-2.5;
f(2) = 0.05*x(1)+x(2)*x(3)-0.3;
f(3) = 0.08*x(1)+x(2)*x(4)-0.1;
f(4) = 0.01*x(1)+x(2)*(1-x(3)-x(4))-1.5;
f=f(:);
end
```

```
                                          Norm of      First-order   Trust-region
 Iteration  Func-count     f(x)           step         optimality    radius
     0          5           0.755                          1.06               1
     1         10       0.0182378       0.872262          0.271               1
     2         15     4.88616e-19      0.0713841        1.41e-09           2.18
Equation solved.
x =
       0.6977
       1.8023
       0.1471
       0.0245
```

$$
\boldsymbol{x} = \begin{bmatrix} \dot{M}_1^{out} \\ \dot{M}_2^{out} \\ w_{2,Oil}^{out} \\ w_{2,W}^{out} \end{bmatrix} = \begin{bmatrix} 0.6977\ \dfrac{kg}{h} \\ 1.8023\ \dfrac{kg}{h} \\ 0.1471 \\ 0.0245 \end{bmatrix}
$$

6.3 Steady state mass balances without chemical reactions in multiple-unit processes

While the level of analysis presented may appear excessive for the simple case studies discussed so far, it is, in fact, essential. The real challenge in developing mass balances for chemical processes lies in formulating a well-posed problem and being able to identify ill-defined or problematic cases. This challenge becomes especially apparent in multi-unit chemical processes, where the number of variables and equations increases rapidly, making the analysis significantly more complex and difficult to manage.

In industrial applications, it is not uncommon to encounter systems involving 10 to 20 process units, with several hundred to several thousand variables. The presence of **recycle streams** further complicates the situation, often obstructing straightforward solution procedures and making sequential methods for solving mass balances across units ineffective.

Although we will focus on case studies involving a limited number of units, the systematic approach developed here is fully applicable to large-scale industrial problems. As problem size grows, such an approach becomes not only beneficial but necessary—especially to avoid incorrect specifications and modeling errors.

In this section, we will examine a series of multi-unit processes of increasing complexity to demonstrate how the core principles introduced earlier can be extended and effectively applied to more complex systems.

Let us consider the hypothetical process of Fig. 6.14. We have available two specific aqueous solutions of A and B. The first one is 40 wt% A and 10 wt% B and the second one is 10 wt% A and 40 wt% B. We have been asked to prepare 1 t/h of an aqueous solution 10 wt% in both A and B. Mixing the two available solutions will not achieve the given specifications. A colleague of yours had the idea to first mix the two solutions and then use pure water (W) to achieve the given specifications. You have been asked to determine the mass flowrates of all streams.

The first thing you realize is that the notation that we have used so far:

$$\text{variable}^{\text{in/out}}_{\text{stream,component}}$$

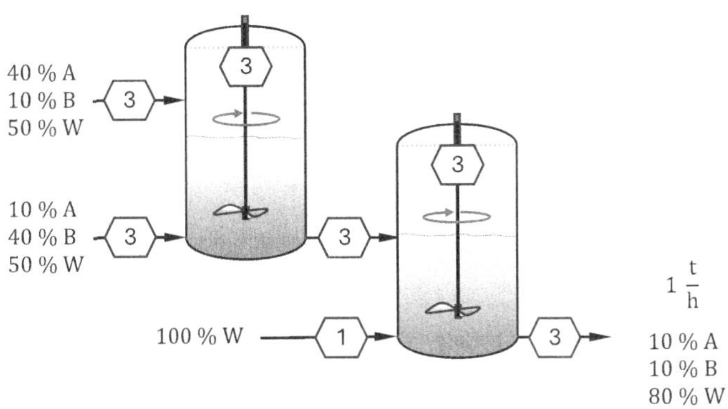

Figure 6.14 Mixing of three streams in two process units to produce a product stream with prespecified composition.

is not flexible enough for multi-unit processes where a specific stream can be an exiting stream for one unit and an entering stream for another unit. We are forced to abandon this notation for a less systematic but more flexible one. We assign numbers to all streams and we use the notation

$$\text{variable}_{\text{stream,component}}$$

By applying this new stream numbering and notation scheme we obtain Fig. 6.15.

We now proceed by applying the methodology of counting variables, equations, and specifications. In this case, however, the system comprises multiple units. This raises key questions.

- *how many independent mass balance equations can be formulated*? The answer is simple: **for each unit we can write as many independent mass balance equations as there are components appearing in the streams that cross the system boundaries**, i.e., we examine each process unit independently and we then add the numbers to get the total number of independent equations.
- *how many variables need to be defined for the full description of the system*? The answer is again simple, **for each stream we count the number of variables (that are equal to the number of components) and we add all numbers together to get the overall number of variables.**

To conclude, our methodology remains valid for the case of multi-unit processes.

There are 3 + 3+3 + 1+3 = 13 variables necessary to describe the process, as shown in the analysis of Fig. 6.14 (see numbers inside the hexagons in Fig. 6.14). For each one of the two units we

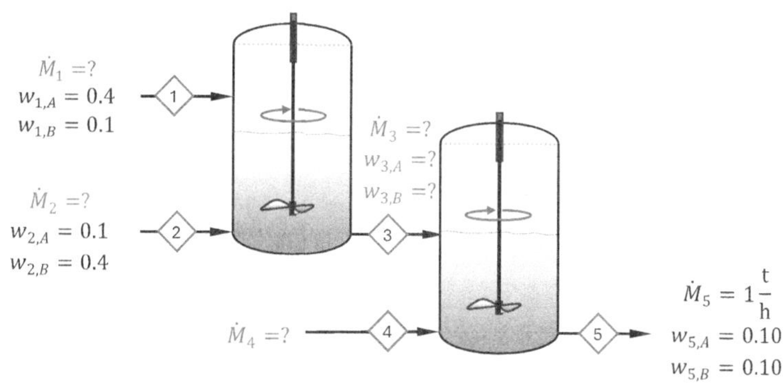

Figure 6.15 Mixing of three streams in two process units to produce a product stream with prespecified composition: stream numbering and notation used.

can formulate 3 ($= NC$) mass balances and six in total. It then follows that

$$N_{DOF} = N_{VAR} - N_{EQ} = 13 - 6 = 7$$

There are seven specifications that are summarized in Figs. 6.14 and 6.15. We conclude that our problem is well defined and can have a unique solution.

We write down the 6 mass balances, three for the first tank and three for the second (we replace all known numerical values to save space)

Total mass balance for the first tank

$$\dot{M}_1 + \dot{M}_2 = \dot{M}_3$$

Mass balance for A in the first tank

$$0.4\dot{M}_1 + 0.1\dot{M}_2 = w_{3,A}\dot{M}_3$$

Mass balance for B in the first tank

$$0.1\dot{M}_1 + 0.4\dot{M}_2 = w_{3,B}\dot{M}_3$$

Total mass balance for the second tank

$$\dot{M}_3 + \dot{M}_4 = 1$$

Mass balance for A in the second tank

$$w_{3,A}\dot{M}_3 = 0.1$$

Mass balance for B in the second tank

$$w_{3,B}\dot{M}_3 = 0.1$$

This is a system of nonlinear equations and we need to apply a numerical method to obtain the solution. However, there is a way to simplify the system by forming a system of linear equations for the total mass flowrates. We substitute the mass balances of the two components in the second tank to the mass balances of the two components in the first tank to obtain

$$0.4\dot{M}_1 + 0.1\dot{M}_2 = 0.1$$

$$0.1\dot{M}_1 + 0.4\dot{M}_2 = 0.1$$

We multiply the second equation by (-4) and we add the result to the first

$$-1.5\dot{M}_2 = -0.3 \quad \Rightarrow \dot{M}_2 = 0.2$$

and $\dot{M}_1 = 0.2$. It then follows that $\dot{M}_3 = 0.4$ and $\dot{M}_4 = 0.6$. We can now calculate the mass fractions in stream three

$$w_{3,A} = \frac{0.1}{\dot{M}_3} = \frac{0.1}{0.4} = 0.25$$

$$w_{3,B} = \frac{0.1}{\dot{M}_3} = \frac{0.1}{0.4} = 0.25$$

A second multi-unit process is shown in Fig. 6.16. We have been asked to design a process to produce 1 t/h of dried powder with 3 wt% moisture from an initial wet powder that contains 90 wt% moisture. To achieve that we first use an evaporator in which part of the water in the wet powder is evaporated and the concentrated powder is fed to a fluidized bed dryer. Hot air with 1% moisture is used in the dryer to remove most of the remaining moisture to produce the dried powder.

The analysis of degrees of freedom is pictured in Fig. 6.17. From this figure it follows that there are 2 + 1+2 + 2+2 + 2 = 11 variables, 2 + 3 = 5 independent mass balances and 11-5 = 6 degrees of freedom. We were given two specifications for the dry powder, one specification for the fresh wet powder, and one specification for the incoming air, i.e., we were given four specifications. The problem is not well defined and we need two (=6–4) additional specifications. We select the following specifications

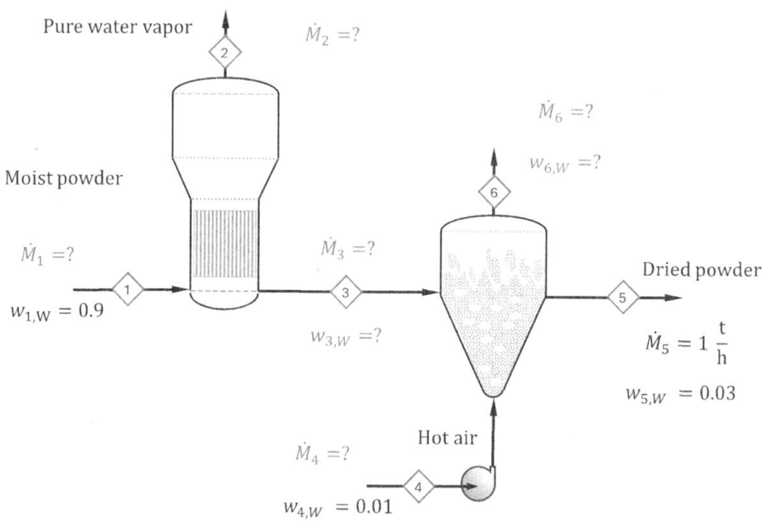

Figure 6.16 A process to dry moist powder using a combination of evaporation and a bed dryer.

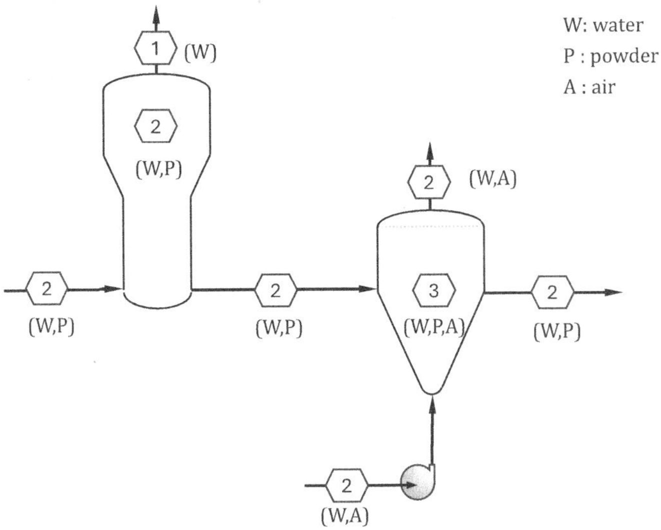

Figure 6.17 A process to dry moist powder using a combination of evaporation and a bed dryer: Analysis of degrees of freedom.

$$w_{3,W} = 0.3$$

$$w_{6,W} = 0.025$$

We write the material balances.
Total mass balance in the evaporator

$$\dot{M}_1 = \dot{M}_2 + \dot{M}_3$$

Water mass balance in the evaporator

$$0.9\dot{M}_1 = \dot{M}_2 + 0.3\dot{M}_3$$

Total mass balance in the dryer

$$\dot{M}_3 + \dot{M}_4 = 1 + \dot{M}_6$$

Water mass balance in the dryer

$$0.3\dot{M}_3 + 0.01\dot{M}_4 = 0.025\dot{M}_6 + 0.03$$

Air balance in the dryer

$$0.99\dot{M}_4 = 0.975\dot{M}_6$$

This is a system of linear equations that can be written in matrix form.

$$\begin{bmatrix} +1 & -1 & -1 & 0 & 0 \\ 0.9 & -1 & -0.3 & 0 & 0 \\ 0 & 0 & +1 & +1 & -1 \\ 0 & 0 & 0.3 & 0.01 & -0.025 \\ 0 & 0 & 0 & 0.99 & -0.975 \end{bmatrix} \begin{bmatrix} \dot{M}_1 \\ \dot{M}_2 \\ \dot{M}_3 \\ \dot{M}_4 \\ \dot{M}_6 \end{bmatrix} = \begin{bmatrix} 0 \\ 0 \\ 1 \\ 0.03 \\ 0 \end{bmatrix}$$

The solution is obtained in MATLAB.

```
>> clear all
>> A=[1 -1 -1 0 0;0.9 -1 -0.3 0 0;0 0 1 1 -1;0 0 0.3 0.01 -0.025; 0 0 0 0.99
-0.975]

A = 1.0000   -1.0000   -1.0000         0         0
    0.9000   -1.0000   -0.3000         0         0
         0         0    1.0000    1.0000   -1.0000
         0         0    0.3000    0.0100   -0.0250
         0         0         0    0.9900   -0.9750

>> b=[0;0;1;0.03;0]

b =      0
         0
    1.0000
    0.0300
         0
>> A\b

ans = 9.7000
      8.3143
      1.3857
     25.0714
     25.4571
```

Let us now consider the process of heterogeneous azeotropic distillation of Fig. 6.18. The feed to the distillation column is a mixture of 79.81 mol% water (W), 19.96 mol% acetic acid (A) and the rest is n-propyl acetate. The feed molar flowrate is 1000 kmol/h. The bottom product is 1 mol% water and 99 mol% acetic acid and has molar flowrate 200 kmol/h. The top product of the column is of unknown molar flowrate and composition and is fed to a horizontal tank in which two immiscible liquid phases are formed (this unit is called a decanter). The lower phase is fed back to the top of the column and has composition 15.3 mol % water and 0.35 mol% acetic acid. The upper phase is the second product stream of the process, has molar flowrate 800 kmol/h, and has unknown composition.

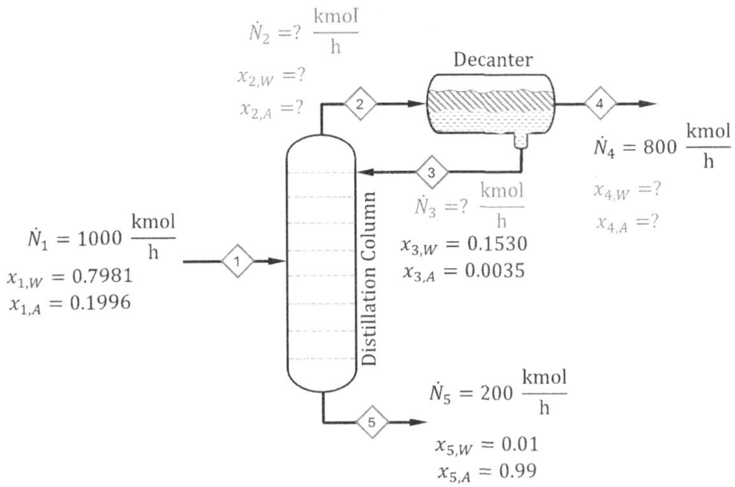

Figure 6.18 Separation of water and acetic acid in a heterogeneous azeotropic distillation.

The process has five streams, two units, and three components resulting in 15 variables and 6 mole balances. The number of degrees of freedom is 15 − 6 = 9 and nine specifications are necessary to have a well-posed problem. There are nine specifications given

1. $\dot{N}_1 = 1000 \frac{kmol}{h}$.
2. $x_{1,W} = 0.7981$.
3. $x_{1,A} = 0.1996$.
4. $x_{3,W} = 0.1530$.
5. $x_{3,A} = 0.0035$.
6. $\dot{N}_4 = 800 \frac{kmol}{h}$.
7. $\dot{N}_5 = 200 \frac{kmol}{h}$.
8. $x_{5,W} = 0.01$.
9. $x_{5,A} = 0.99$.

We write the 6 mole balances.

Distillation column

$$\dot{N}_1 + \dot{N}_3 = \dot{N}_2 + \dot{N}_5$$

$$x_{1,W}\dot{N}_1 + x_{3,W}\dot{N}_3 = x_{2,W}\dot{N}_2 + x_{5,W}\dot{N}_5$$

$$x_{1,A}\dot{N}_1 + x_{3,A}\dot{N}_3 = x_{2,A}\dot{N}_2 + x_{5,A}\dot{N}_5$$

Decanter

$$\dot{N}_2 = \dot{N}_3 + \dot{N}_4$$

$$x_{2,W}\dot{N}_2 = x_{3,W}\dot{N}_3 + x_{4,W}\dot{N}_4$$

$$x_{2,A}\dot{N}_2 = x_{3,A}\dot{N}_3 + x_{4,A}\dot{N}_4$$

Everything seems to be normal and we then try to solve the system of nonlinear algebraic equations. We note that two overall mole balances are linear and involve only two variables: $\dot{N}_2$ and $\dot{N}_3$, we write the total mole balances in a slightly different form:

$$\dot{N}_2 - \dot{N}_3 = \dot{N}_1 - \dot{N}_5$$

$$\dot{N}_2 - \dot{N}_3 = \dot{N}_4$$

We can solve these two equations to determine the 2 molar flowrates and then we can determine all mole fractions as the remaining equations, after replacing the known values of the total molar flowrates, will be linear equations as well.

$$x_{1,W}\dot{N}_1 + x_{3,W}\dot{N}_3 = x_{2,W}\dot{N}_2 + x_{5,W}\dot{N}_5$$

$$x_{1,A}\dot{N}_1 + x_{3,A}\dot{N}_3 = x_{2,A}\dot{N}_2 + x_{5,A}\dot{N}_5$$

$$x_{2,W}\dot{N}_2 = x_{3,W}\dot{N}_3 + x_{4,W}\dot{N}_4$$

$$x_{2,A}\dot{N}_2 = x_{3,A}\dot{N}_3 + x_{4,A}\dot{N}_4$$

We replace the numerical values $\dot{N}_1 = 1000$, $\dot{N}_4 = 800$ and $\dot{N}_5 = 200$ to the system of the overall mole balances:

$$\dot{N}_2 - \dot{N}_3 = 1000 - 200 = 800$$

$$\dot{N}_2 - \dot{N}_3 = 800$$

To our surprise the equations are the same and as a result we must calculate two unknown variables from a single equation, which is impossible. So, what went wrong?

The answer is the following: when we refer to specifications, we always refer to **independent specifications**. In practical terms this means that each additional specification must add information that is not already available. In Fig. 6.19 we consider a control volume that includes the whole process. For this control volume we can formulate the overall mole balance:

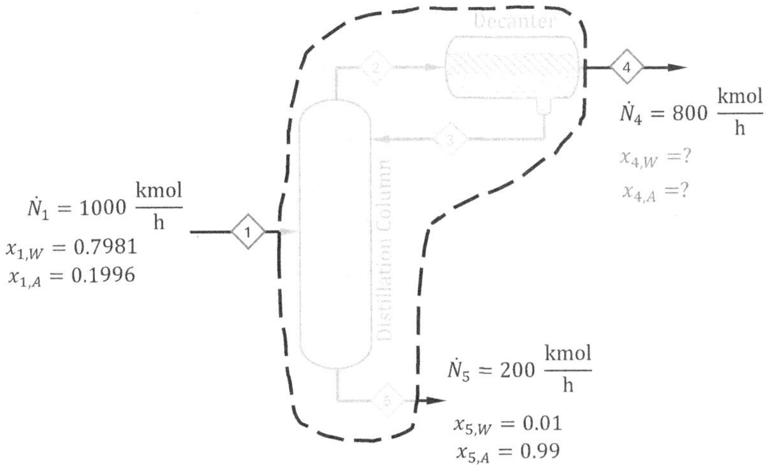

Figure 6.19 Separation of water and acetic acid in a heterogeneous azeotropic distillation: The external molar flowrates are constrained by the overall mole balance.

$$\dot{N}_1 = \dot{N}_4 + \dot{N}_5$$

The molar flow rates of the three external streams are, therefore, interrelated (constrained) through the overall mole balance constraint of the process. Consequently, specification of any two flow rates uniquely determines the third. Imposing an additional specification on the third stream introduces redundancy, as it does not constitute an independent piece of information.

Things could have been even worse. Imagine that we have specified the following values for the three external molar flowrates

1. $\dot{N}_1 = 1000\ \frac{\text{kmol}}{\text{h}}$.
2. $\dot{N}_4 = 800\ \frac{\text{kmol}}{\text{h}}$.
3. $\dot{N}_5 = 500\ \frac{\text{kmol}}{\text{h}}$.

Then the overall mole balance equation is not only redundant but also inconsistent.

So, how should we proceed to ensure that our specifications are independent? Unfortunately, there is no universally simple answer. While experience plays a valuable role, a definitive approach involves performing a structural analysis of the material balances and model specifications. Commercial process simulators implement sophisticated algorithms to detect inconsistencies or redundancies in user-defined specifications. These methods typically involve identifying subsets of equations where the number of equations exceeds the number of unknowns, thereby revealing overspecified sub-systems. Most simulators can then isolate and flag the redundant or conflicting specifications for the user.

The previously discussed case of inconsistent specifications pertains to structural issues within the mathematical formulation of the problem—specifically, conflicts arising from the relationships between equations and variables. However, a more subtle and often more challenging form of inconsistency can arise from physically unattainable specifications. These occur when constraints are imposed on the system that, while mathematically definable, cannot be realized by the actual process hardware or physics. This is analogous to specifying that a family sedan should perform like a race car—while such a target can be set, the underlying system is fundamentally incapable of meeting it.

Let us return to the case study presented in Fig. 6.15. Consider the first tank in which the two streams with different compositions of components A and B are mixed. What are the possible compositions of stream 3? Let us assume that we mix stream A (40 wt% A, 10 wt% B) and stream B (10 wt% A, 40 wt% B). Then, for all possible relative proportions, the composition of stream three lies on the line segment connecting point A and B in Fig. 6.20 (can you prove this?). Stream three is then mixed with stream four (point O in Fig. 6.20) and, depending on their relative proportions, the final stream lies on the linear segment that connects a point on the line segment AB with point O. All possible points that can be attained by the process are within the triangle shown in Fig. 6.20. The solution to the case study of Fig. 6.15 is shown in Fig. 6.20 with two circles, the first one is the point

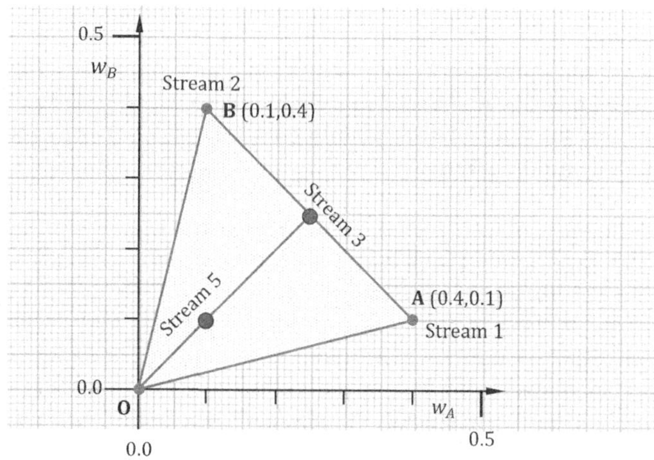

Figure 6.20 Graphical analysis of the three-stream mixing in two mixing tank process shown in Figure 6.15.

(0.25, 0.25), which represents stream 3, and the second one is point (0.1, 0.1), which represents stream 5.

However, what if the specifications chosen for stream four lie outside the attainable region denoted by the triangle in Fig. 6.20? In that case, the simulation (mathematical) problem—although well-defined and mathematically well-posed—will have no feasible (realizable) solution, as the specified conditions fall outside the physically realizable operating envelope of the process! This type of inconsistency often goes undetected until the solver is executed and returns a result that, while mathematically valid, violates physical constraints. In other words, the numerical solution process may converge successfully, but the resulting solution is not physically meaningful or implementable.

We will not delve further into the analysis of inconsistencies commonly encountered in the formulation of mass balances for multi-unit chemical processes. With experience, you will become more proficient in defining consistent process specifications, and the use of commercial process simulators will serve as an invaluable aid in developing this expertise. However, you must always be aware of that possibility, and this is the reason why, in this text, we have followed a systematic and structured way in analyzing and developing the mathematical model of a multi-unit process that may help you to avoid common pitfalls. We now proceed and include the possibility of chemical reaction(s) taking place in the process under study.

6.4 Steady state mass balances with chemical reactions

The first question we must address in this section is: why do we need a dedicated discussion on mass balances in the presence of chemical reactions? The answer is straightforward. Up to this point, our focus has been primarily on the properties of the process streams, with minimal consideration of the role played by the process unit itself. This may appear counterintuitive, as it suggests that the type of process unit has no influence on the formulation of mass balances. In fact, the mathematical structure of the mass balance equations has so far remained independent of the nature of the process—implying that, in the absence of reaction, the process unit is largely irrelevant to the balance formulation. This highlights a fundamental limitation of the discussion thus far. The occurrence of chemical reactions marks the first instance where it becomes essential to consider the internal behavior of the process unit, rather than relying solely on the properties of the inlet and outlet streams. The two additional cases we will examine shortly are chemical and thermodynamic equilibrium, and rate-based processes involving mass and energy transfer. This will offer us the opportunity to face problems relevant to industrial practice and the chemical engineering profession.

We have shown that for the case of a single-input-single-output process in which a single chemical reaction is taking place with reaction rate $\dot{R}$ the overall mole balance and the *i-th* component mole balance can be written as

$$\dot{N}^{\text{out}} = \dot{N}^{\text{in}} + \Sigma\dot{R} \tag{6.4.1}$$

$$\dot{n}_i^{out} = \dot{n}_i^{\text{in}} + \sigma_i\dot{R}, \quad i = 1, 2, \ldots, NC - 1 \tag{6.4.2}$$

where

$$\Sigma = \sum_{i=1}^{NC} \sigma_i$$

is the total change in the number of moles due to the reaction. The component mole balance equation can be written in two additional equivalent forms.

- *In terms of volumetric flowrates and concentrations*

$$\dot{F}^{\text{out}} C_i^{\text{out}} = \dot{F}^{\text{in}} C_i^{\text{in}} + \sigma_i\dot{R} \tag{6.4.3}$$

- *In terms of molar flowrates and mole fractions*

$$\dot{N}^{\text{out}} x_i^{\text{out}} = \dot{N}^{\text{in}} x_i^{\text{in}} + \sigma_i\dot{R} \tag{6.4.4}$$

The number of independent mole balance equations does not change due to the presence of the chemical reaction(s). However, the number of variables increases by the number of independent chemical reactions ($\dot{R}_r, r = 1, 2, \ldots, NR$). Consequently, a process that is well-defined under the criteria established for nonreactive systems will require additional independent specifications when chemical reactions are present. Specifically, **the number of additional independent pieces of information must equal the number of independent reactions** to fully define the system.

The additional specification needed to fully specify a process that involves chemical reaction(s) can have the form of chemical reaction rate expressions (including equilibrium conditions)

$$\dot{R}''' = k(T)f(C_1, C_2, \ldots, C_n) \tag{6.4.5}$$

or the form of reaction conversions

$$X = \frac{\text{mole of limiting reactant that react}}{\text{mole of limiting reactant in the feed}} \tag{6.4.6}$$

or the form of extent of reaction

$$\dot{\xi} = V\dot{R}''' \tag{6.4.7}$$

We will now examine several case studies to illustrate how information about chemical reactions can be integrated with the mole balances of a reacting system, enabling the calculation of all relevant process variables.

In a chemical reactor the following reactions are taking place

$$E + \frac{1}{2}O_2 \longrightarrow A + W$$

$$E + 3O_2 \longrightarrow 2CO_2 + 3W$$

$$2A + O_2 \longrightarrow 2Ac$$

where E: ethanol (CH_3CH_2OH), A: Acetaldehyde (CH_3CHO), W: water, Ac: acetic acid (CH_3COOH). O_2 is supplied using atmospheric air and when the ratio of the molar flowrate of E to the molar flowrate of O_2 is 5.7, then oxygen is completely consumed. If the feed molar flowrate of ethanol is 570 kmol/h, what is the product composition if A, CO_2, and Ac are produced in equal amounts?

The system under study is pictured in Fig. 6.21. There are seven components appearing in two streams giving rise to 14 variables pertinent to process streams. If we take into consideration the fact that three reactions are taking place, we finally obtain 17 variables. We can formulate 7 mol balances, and we therefore need $17 - 7 = 10$ independent specifications. All molar flowrates in the feed are known (7 specifications), oxygen is not present in the exiting stream (1 specification), while the molar flowrates of A, CO_2, and Ac are equal (2 specifications). We therefore have 10 specifications, and the system is well posed. The mathematical model is as follows (10 equations in 10 unknowns)

$$\dot{n}_i^{\text{out}} = \dot{n}_i^{\text{in}} + \sigma_{i,1}\dot{\xi}_1 + \sigma_{i,2}\dot{\xi}_2 + \sigma_{i,3}\dot{\xi}_3,\ i = \text{E}, \text{O}_2, \text{N}_2, \text{A}, \text{W}, \text{CO}_2, \text{Ac}$$

$$n_{O_2}^{\text{out}} = 0$$

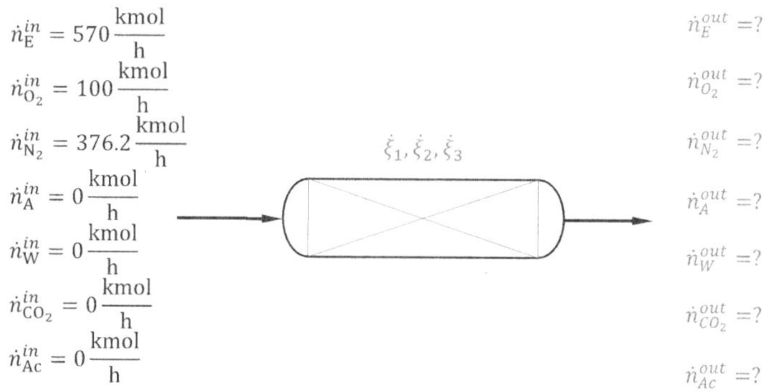

Figure 6.21 Three simultaneous chemical reactions that take place in a chemical reactor.

$$n_A^{out} = n_{CO_2}^{out}$$

$$n_{Ac}^{out} = n_{CO_2}^{out}$$

Let us write all equations explicitly

$$\begin{array}{lclllll}
\dot{n}_E^{out} & = & 570 & -\dot{\xi}_1 & -\dot{\xi}_2 & \\
\dot{n}_{O_2}^{out} & = & 100 & -0.5\dot{\xi}_1 & -3\dot{\xi}_2 & -\dot{\xi}_3 \\
\dot{n}_{N_2}^{out} & = & 376.2 & & & \\
\dot{n}_A^{out} & = & & +\dot{\xi}_1 & & -2\dot{\xi}_3 \\
\dot{n}_W^{out} & = & & +\dot{\xi}_1 & +3\dot{\xi}_2 & \\
\dot{n}_{CO_2}^{out} & = & & & +2\dot{\xi}_2 & \\
\dot{n}_{Ac}^{out} & = & & & & +2\dot{\xi}_3 \\
\dot{n}_{O_2}^{out} & = & 0 & & & \\
\dot{n}_A^{out} & = & \dot{n}_{CO_2}^{out} & & & \\
\dot{n}_{Ac}^{out} & = & \dot{n}_{CO_2}^{out} & & &
\end{array}$$

Combining the eighth equation with the second equation

$$0.5\dot{\xi}_1 + 3\dot{\xi}_2 + \dot{\xi}_3 = 100$$

The fourth and sixth equations with the ninth

$$\dot{\xi}_1 - 2\dot{\xi}_3 - 2\dot{\xi}_2 = 0$$

and the sixth and seventh with the last one

$$-\dot{\xi}_3 + \dot{\xi}_2 = 0$$

It immediately follows that $\dot{\xi}_2 = \dot{\xi}_3$, $\dot{\xi}_1 = 4\dot{\xi}_3$ and then $\dot{\xi}_3 = \dot{\xi}_2 = \frac{100}{6}\frac{\text{kmol}}{\text{h}}$, $\dot{\xi}_1 = \frac{100}{1.5}\frac{\text{kmol}}{\text{h}}$ and

$$
\begin{aligned}
\dot{n}_{E}^{out} &= 570 - \dot{\xi}_1 - \dot{\xi}_2 &&= 486.7\\
\dot{n}_{O_2}^{out} &= 100 - 0.5\dot{\xi}_1 - 3\dot{\xi}_2 - \dot{\xi}_3 &&= 0\\
\dot{n}_{N_2}^{out} &= 376.2 &&= 376.2\\
\dot{n}_{A}^{out} &= +\dot{\xi}_1 - 2\dot{\xi}_3 &&= 33.333\\
\dot{n}_{W}^{out} &= +\dot{\xi}_1 + 3\dot{\xi}_2 &&= 116.67\\
\dot{n}_{CO_2}^{out} &= +2\dot{\xi}_2 &&= 33.333\\
\dot{n}_{Ac}^{out} &= +2\dot{\xi}_3 &&= 33.333
\end{aligned}
$$

We now consider the simplified ammonia production process illustrated in Fig. 6.22. Stoichiometric amounts of the reactants are fed into the system: 100 kmol/h of nitrogen and 300 kmol/h of hydrogen. A fraction X of the reactants is converted into the reactor. The reactor effluent is then sent to an ideal separation unit, where all ammonia is recovered and exits the process. **The unreacted reactants from the ideal separation unit are recycled**, mixed with fresh feed, and returned to the reactor to fully consume them. Our aim is to determine the molar flowrate of the recycle stream as a function of conversion.

We assume that all components appear in all streams and

- add a specification in streams 1 and 4 that makes the molar flowrate of ammonia equal to zero, and

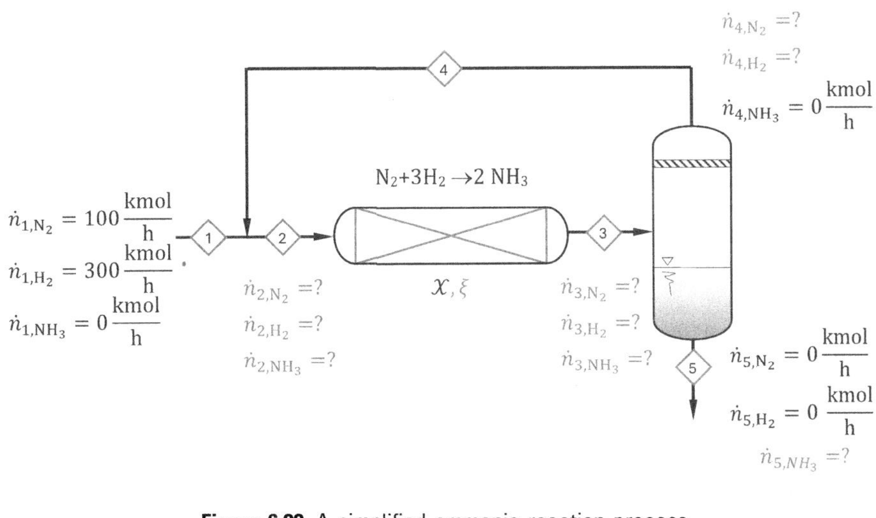

Figure 6.22 A simplified ammonia reaction process.

- add two specifications in product stream to make nitrogen and hydrogen molar flowrates equal to zero (see Fig. 6.22).

The overall number of variables associated with the streams is $(5\times3=)$ 15 and we add one variable (X) for the reaction. There are three units: mixing point, reactor and separation unit and we can formulate 9 mole balances. Then, the number of degrees of freedom is $(15+1–9=)$ 7. There are six specifications marked clearly in Fig. 6.22. An additional specification is the specific value of the fraction conversion used to determine the molar flowrate of the recycle stream. We will also be using the extent of reaction to build the mole balances but we will add an additional equation that relates the extent of reaction to fraction conversion. We write the 9 mole balances (we also replace the known numerical values to save space):

Mixing point

$$\dot{n}_{2,N_2} - \dot{n}_{4,N_2} = 100 \tag{6.4.8}$$

$$\dot{n}_{2,H_2} - \dot{n}_{4,H_2} = 300 \tag{6.4.9}$$

$$\dot{n}_{2,NH_3} = 0 \tag{6.4.10}$$

Reactor

$$\dot{n}_{3,N_2} - \dot{n}_{2,N_2} + \dot{\xi} = 0 \tag{6.4.11}$$

$$\dot{n}_{3,H_2} - \dot{n}_{2,H_2} + 3\dot{\xi} = 0 \tag{6.4.12}$$

$$\dot{n}_{3,NH_3} - 2\dot{\xi} = 0 \tag{6.4.13}$$

Ideal separation unit

$$\dot{n}_{4,N_2} - \dot{n}_{3,N_2} = 0 \tag{6.4.14}$$

$$\dot{n}_{4,H_2} - \dot{n}_{3,H_2} = 0 \tag{6.4.15}$$

$$\dot{n}_{5,NH_3} - \dot{n}_{3,NH_3} = 0 \tag{6.4.16}$$

The equation for fraction conversion is

$$X = \frac{\dot{n}_{2,N_2} - \dot{n}_{3,N_2}}{\dot{n}_{2,N_2}}$$

or

$$(1-X)\dot{n}_{2,N_2} - \dot{n}_{3,N_2} = 0 \tag{6.4.17}$$

To solve the equations, we use Eq. (6.4.14) to eliminate $\dot{n}_{4,N_2}$ in Eq. (6.4.8′)

$$\dot{n}_{2,N_2} - \dot{n}_{3,N_2} = 100 \tag{6.4.8′}$$

Note that Eq. (6.4.11′) can be written as

$$\dot{n}_{2,N_2} - \dot{n}_{3,N_2} = \dot{\xi} \tag{6.4.11′}$$

From Eq. (6.4.8′) and (6.4.11′) it follows that

$$\dot{\xi} = 100\ \frac{\text{kmol}}{\text{h}} \tag{6.4.18}$$

We then subtract Eq. (6.4.8′) from Eq. (6.4.17) to obtain

$$\dot{n}_{2,N_2} = \frac{100}{X} \tag{6.4.19}$$

Substitution of Eq. (6.4.19) into (6.4.8) gives

$$\dot{n}_{4,N_2} = 100\left(\frac{1}{X} - 1\right) \tag{6.4.20}$$

Following a similar approach, we obtain

$$\dot{n}_{4,H_2} = 300\left(\frac{1}{X} - 1\right) \tag{6.4.21}$$

Adding Eq. (6.4.20) and (6.4.21) we obtain

$$\dot{N}_4 = \dot{n}_{4,N_2} + \dot{n}_{4,H_2} = 400\left(\frac{1}{X} - 1\right) \tag{6.4.22}$$

Fig. 6.23 illustrates how the molar flow rate of the recycle stream varies with conversion. Note that the vertical axis uses a logarithmic scale. As conversion decreases, the molar flow rate of the recycle stream increases sharply—eventually becoming unbounded at low conversions. At 50% conversion, the recycle flow rate equals the molar flow rate of the fresh feed. As the conversion increases, the recycle flow rate steadily decreases, approaching zero as conversion nears 100%. This trend is not unique to the specific case shown but is a general characteristic of systems involving recycle loops. In such systems, **unreacted reactants are recycled to improve overall conversion in reactions that are not complete in a single pass**. Before closing this section, we need to make a very important observation related to the definition of conversion.

The key observation is that the extent of reaction remains constant at $\dot{\xi} = 100$ kmol/h, regardless of the per-pass conversion. To

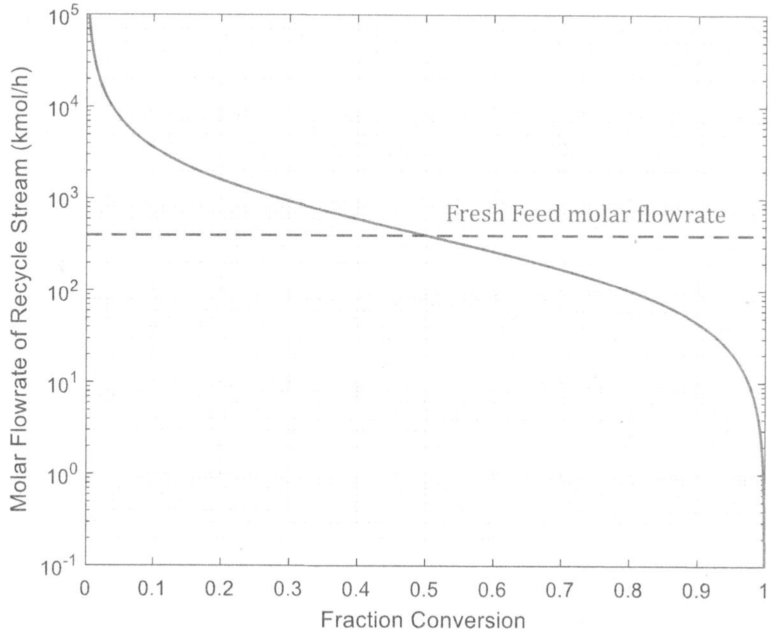

Figure 6.23 Variation of the recycle stream molar flowrate as a function of the per-pass fraction conversion.

understand why, consider the overall conversion based on the fresh feed and the ammonia product stream. Using the standard definition of conversion with respect to the limiting reactant:

$$\text{fraction conversion} = \frac{\text{mole reacted}}{\text{mole in the feed}}$$

Assuming nitrogen is the limiting reactant, and comparing the fresh feed to the product stream:

$$\text{fraction conversion} = \frac{\text{mole in the fresh feed} - \text{mole in the product stream}}{\text{mole in the fresh feed}}$$

$$= \frac{100 - 0}{100} = 1$$

This corresponds to 100% overall conversion, independent of the per-pass conversion X used in intermediate calculations. This outcome reflects the effect of complete recycle of unreacted materials, which ensures all reactants are eventually converted. This also explains why, in all cases, the extent of reaction remains constant at $\xi = 100$ kmol/h. It is therefore essential to clearly distinguish between **overall conversion**, which is based on the fresh feed and final product streams, and **per-pass conversion**, which refers specifically to the conversion occurring between the reactor's inlet and outlet streams.

6.5 Purge streams

A purge stream is a controlled flow of material that is removed—or "purged"—from a process system to **prevent the buildup of unwanted or inert components** that could otherwise accumulate and disrupt the operation.

In continuous chemical processes, especially in systems with recycle loops (see Fig. 6.22), certain components may not react or be fully separated. These could include:

- Inert gases (like argon)
- Impurities in feedstock
- By-products or side reactions

Over time, these non-reactive or low-reactivity substances can accumulate in the recycle stream, potentially leading to reduced reaction efficiency, equipment fouling or corrosion, and safety hazards (e.g., pressure buildup, flammable mixtures). Achieving a steady-state operation is impossible in the absence of a purge stream or, in case a steady state is achieved, the product purity is severely degraded as the product is contaminated by the unwanted or inert component. To avoid these issues, a small portion of the recycle stream is purged from the system, helping maintain steady-state operation and safe conditions.

In Fig. 6.24, the simplified ammonia reaction process of Fig. 6.22 is presented with one important difference: the fresh feed contains an inert (I) component, which has a molar flowrate of 1 kmol/h. What will be the consequences of this inert component in the operation of the process? Let us assume that the inert component is a volatile component like argon. To answer this

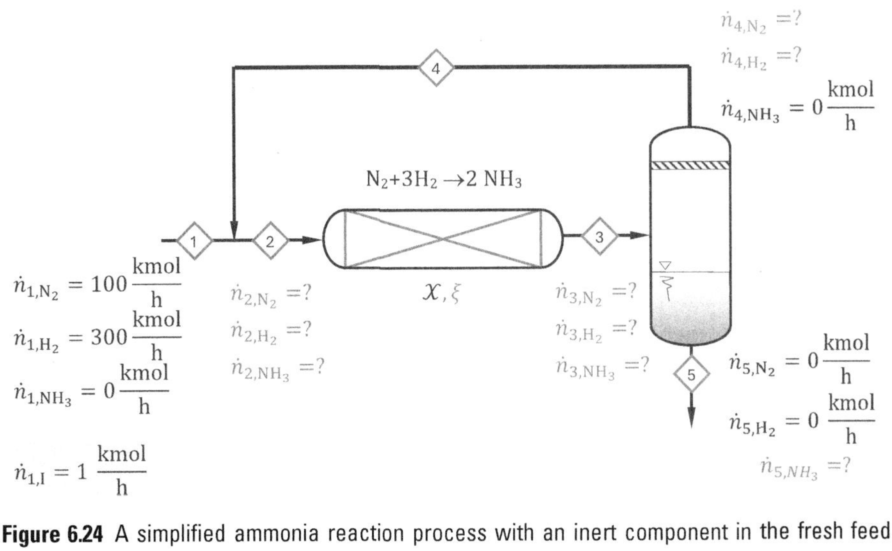

Figure 6.24 A simplified ammonia reaction process with an inert component in the fresh feed.

question, think about the total mole balance of the inert component. We may write the following equation

$$\frac{dn_I(t)}{dt} = \dot{n}_{1,I} - \dot{n}_{5,I}$$

where $n_I(t)$ is the molar holdup of the inert component in the process, $\dot{n}_{1,I} = 1$ kmol/h is the molar flowrate of the inert component in the fresh feed and $\dot{n}_{5,I}$ is the molar flowrate of the inert component in the product stream. But we have assumed that the product stream is pure ammonia, i.e., $\dot{n}_{5,I} = 0$. We therefore have that

$$\frac{dn_I(t)}{dt} = \dot{n}_{1,I}$$

or, after integration

$$n_I(t) = n_I(0) + \dot{n}_{1,I}t$$

This is a rather unwanted situation as the molar holdup of the inert component will increase linearly with time and will cause the pressure to increase, posing a serious safety hazard. If the equipment can withstand the increase in the pressure, then due to increased pressure the inert component (and possibly the reactants) will start to condense and contaminate the product. The process will reach a steady state but the product stream will, in the best case, consist of 100 kmol/h ammonia and 1 kmol/h inert component, reducing the ammonia purity to $100/101 \approx 99\%$ ammonia.

You typically need a purge stream when:

- Recycle loops are used – to prevent accumulation of inert or undesired components.
- The feed contains impurities – and there's no perfect separation process to remove them.
- Maintaining product purity is critical.
- Controlling pressure or composition – in gas-phase processes or reactors.

In short, a purge stream is a vital part of many chemical systems, helping maintain stability, safety, and efficiency by managing components that can't be removed or reacted away in the process.

The process of Fig. 6.24 is modified by introducing a purge stream as shown in Fig. 6.25. In the same figure, the variables necessary to completely describe the process are indicated. There are 17 variables

$$\dot{N}_1, \dot{N}_2, \dot{N}_3, \dot{N}_4, \dot{N}_5, \dot{N}_6, \dot{N}_7, x_{1,N_2}, x_{2,N_2}, x_{3,N_2}, x_{4,N_2}, x_{1,H_2}, x_{2,H_2}, x_{3,H_2}, x_{4,H_2}, x_{3,NH_3}, \dot{\xi}$$

The independent mole balances that we can formulate are:

- 3 mol balances for the mixing point
- 4 mol balances for the reactor
- 4 mol balances for the separation unit
- 1 mol balance for the splitting point (we will explain this shortly)

There are 12 independent mole balances in total. The degrees of freedom are (17–12 =) 5.

We are given the total flowrate and the composition of the fresh feed (3 specifications). To be able to solve the model we need two additional specifications: we select

- the per-pass fraction conversion X, and
- the ratio of the purge stream and the recycle stream ($\dot{N}_6/\dot{N}_7$).

The complete set of equations is the following:

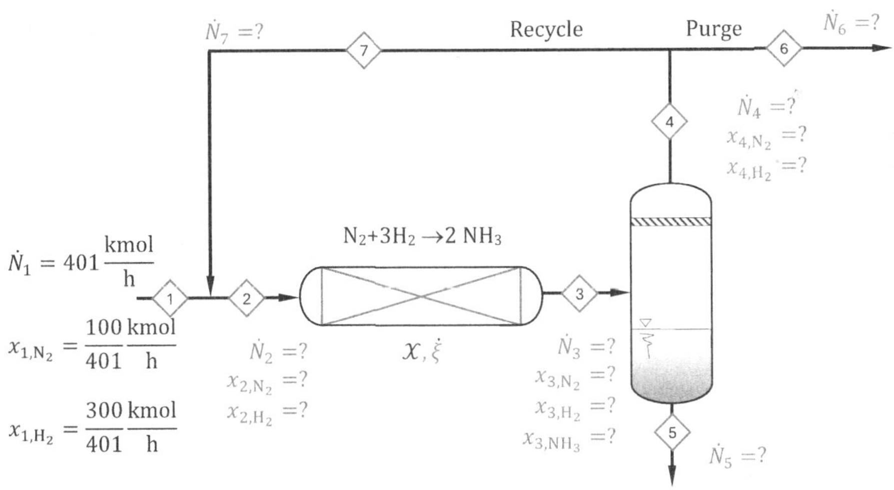

Figure 6.25 A simplified ammonia reaction process with an inert component in the fresh feed, a recycle stream for complete conversion of the reactants, and a purge stream for removing an inert component present in the feed stream.

Mixing point

$$\dot{N}_2 = 401 + \dot{N}_7 \tag{6.5.1}$$

$$\dot{N}_2 x_{2,N_2} = 100 + \dot{N}_7 x_{4,N_2} \tag{6.5.2}$$

$$\dot{N}_2 x_{2,H_2} = 300 + \dot{N}_7 x_{4,H_2} \tag{6.5.3}$$

Reactor

$$\dot{N}_3 = \dot{N}_2 - 2\dot{\xi} \tag{6.5.4}$$

$$\dot{N}_3 x_{3,N_2} = \dot{N}_2 x_{2,N_2} - \dot{\xi} \tag{6.5.5}$$

$$\dot{N}_3 x_{3,H_2} = \dot{N}_2 x_{2,H_2} - 3\dot{\xi} \tag{6.5.6}$$

$$\dot{N}_3 x_{3,NH_3} = +2\dot{\xi} \tag{6.5.7}$$

Ideal separation unit

$$\dot{N}_3 = \dot{N}_5 + \dot{N}_4 \tag{6.5.8}$$

$$\dot{N}_3 x_{3,N_2} = \dot{N}_4 x_{4,N_2} \tag{6.5.9}$$

$$\dot{N}_3 x_{3,H_2} = \dot{N}_4 x_{4,H_2} \tag{6.5.10}$$

$$\dot{N}_3 x_{3,NH_3} = \dot{N}_5 \tag{6.5.11}$$

Splitting point

$$\dot{N}_4 = \dot{N}_6 + \dot{N}_7 \tag{6.5.12}$$

The equation for the per-pass fraction conversion is

$$X = \frac{\dot{N}_2 x_{2,N_2} - \dot{N}_3 x_{3,N_2}}{\dot{N}_2 x_{2,N_2}}$$

or

$$(1 - X)\dot{N}_2 x_{2,N_2} = \dot{N}_3 x_{3,N_2} \tag{6.5.13}$$

Finally, the specification on the ratio of the purge stream to the recycle stream is as follows

$$\left(\frac{P}{R}\right) = \frac{\dot{N}_6}{\dot{N}_7} \tag{6.5.14}$$

The conclusion is that there are 14 equations in the 14 unknowns of the problem. By selecting specific values for the fraction, per-pass conversion X, and the ratio $\left(\frac{P}{R}\right)$ the system of equations can be solved to obtain all molar flowrates and compositions.

A point that needs further clarification is the number of independent equations that we can formulate for a splitting point. The general case is pictured in Fig. 6.26. The important note is that all outgoing streams share the same composition as the incoming stream. There is no need to formulate component mole balances as they all simplify to the total mole balance:

$$\dot{N}^{\text{in}} = \dot{N}_1^{\text{out}} + \dot{N}_2^{\text{out}} + \cdots + \dot{N}_p^{\text{out}} \tag{6.5.15}$$

We can only formulate one total mole balance irrespective of the number of outgoing streams.

We now return to the ammonia production case study with an inert present in the fresh feed. The mathematical model that consists of Eqs. (6.5.1) to (6.5.14) is a nonlinear model, and its solution necessitates the use of software. We will solve the model for a representative set of specifications:

$$X = 0.5, \left(\frac{P}{R}\right) = \frac{1}{19}$$

To solve the nonlinear equations in MATLAB, we first create the following function that implements Eqs. (6.5.1) to (6.5.14).

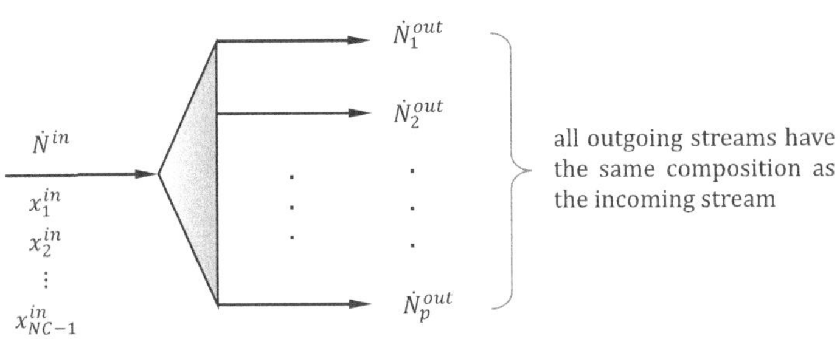

Figure 6.26 A stream splitting unit: all outgoing streams have the same composition as the incoming stream, and there is only one total mole balance that we can formulate for any splitting unit irrespective of the number of outgoing streams.

```
function f=AmmoniaProductionWithPurge(x,X,P_R)
% steady-state model of simplified ammonia production with purge
N(2)     = x(1); % molar flowrate of stream 2
N(3)     = x(2); % molar flowrate of stream 3
N(4)     = x(3); % molar flowrate of stream 4
N(5)     = x(4); % molar flowrate of stream 5
N(6)     = x(5); % molar flowrate of stream 6
N(7)     = x(6); % molar flowrate of stream 7
xN2(2)   = x(7); % mole fraction of N2 in of stream 2
xN2(3)   = x(8); % mole fraction of N2 in of stream 3
xN2(4)   = x(9); % mole fraction of N2 in of stream 4
xH2(2)   = x(10);% mole fraction of H2 in of stream 2
xH2(3)   = x(11);% mole fraction of H2 in of stream 3
xH2(4)   = x(12);% mole fraction of H2 in of stream 4
xNH3(3)  = x(13);% mole fraction of NH3 in of stream 3
ksi      = x(14);% extent of reaction

%% MIXING POINT
f(1) =401 + N(7)         - N(2);          % total mole balance
f(2) =100 + N(7)*xN2(4) - N(2)*xN2(2);   % N2    mole balance
f(3) =300 + N(7)*xH2(4) - N(2)*xH2(2);   % H2    mole balance
%% REACTOR
f(4) = N(2)         - N(3)        -2*ksi; % total mole balance
f(5) = N(2)*xN2(2) - N(3)*xN2(3) -1*ksi; % N2    mole balance
f(6) = N(2)*xH2(2) - N(3)*xH2(3) -3*ksi; % H2    mole balance
f(7) =              - N(3)*xNH3(3)+2*ksi; % NH3   mole balance
%% SEPARATOR
f(8) = N(3)          - N(4)         - N(5);% total mole balance
f(9) = N(3)*xN2(3)   - N(4)*xN2(4)       ;% N2    mole balance
f(10)= N(3)*xH2(3)   - N(4)*xH2(4)       ;% H2    mole balance
f(11)= N(3)*xNH3(3)                 - N(5);% NH3   mole balance
%% SPLITTING POINT
f(12)= N(4) - N(6) - N(7);                % total mole balance
%% SPECS
f(13)= (1-X)*N(2)*xN2(2) - N(3)*xN2(3);  % spec on conversion
f(14)= N(6)-P_R*N(7);                     % spec on P over R ratio

f=f(:);
end
```

The model is then solved using the fsolve command in MATLAB.

```
clear all

x0(1)  = 1000;
x0(2)  = 800;
x0(3)  = 600;
x0(4)  = 200;
x0(5)  = 20;
x0(6)  = 600;
x0(7)  = 0.25;
x0(8)  = 0.15;
x0(9)  = 0.25;
x0(10)= 0.70;
x0(11)= 0.50;
x0(12)= 0.60;
x0(13)= 0.10;
x0(14)= 75;

opts = optimset('Display','iter');
x0 = fsolve(@AmmoniaProductionWithPurge,x0,opts,0.5,1/19);

%
fprintf('molar flowrate  of stream 2 %12.4f kmol/h,  \n',x0(1))
fprintf('molar flowrate  of stream 3 %12.4f kmol/h,  \n',x0(2))
fprintf('molar flowrate  of stream 4 %12.4f kmol/h,  \n',x0(3))
fprintf('molar flowrate  of stream 5 %12.4f kmol/h,  \n',x0(4))
fprintf('molar flowrate  of stream 6 %12.4f kmol/h,  \n',x0(5))
fprintf('molar flowrate  of stream 7 %12.4f kmol/h,  \n',x0(6))
fprintf('m. f. of N_{2}  in stream 2 %12.4f -, \n',x0(7))
fprintf('m. f. of N_{2}  in stream 3 %12.4f -, \n',x0(8))
fprintf('m. f. of N_{2}  in stream 4 %12.4f -, \n',x0(9))
fprintf('m. f. of H_{2}  in stream 2 %12.4f -, \n',x0(10))
fprintf('m. f. of H_{2}  in stream 3 %12.4f -, \n',x0(11))
fprintf('m. f. of H_{2}  in stream 4 %12.4f -, \n',x0(12))
fprintf('m. f. of NH_{3} in stream 3 %12.4f -, \n',x0(13))
fprintf('extent of reaction ξ        %12.4f kmol/h. \n',x0(14))

>> RUNAmmoniaProductionWithPurge

                                                Norm of      First-order   Trust-region
 Iteration  Func-count      f(x)                step         optimality    radius
     0          15          35110.1                          1.52e+05            1
     1          30          11062.4                   1       5.5e+04            1
...
    16         241      3.25389e-15       0.0110938          2.19e-05          381
    17         256      2.64552e-26     9.60153e-08          8.89e-11          381

Equation solved.
<stopping criteria details>

molar flowrate  of stream 2     781.9048 kmol/h,
molar flowrate  of stream 3     591.4286 kmol/h,
molar flowrate  of stream 4     400.9524 kmol/h,
```

```
molar flowrate  of stream 5      190.4762 kmol/h,
molar flowrate  of stream 6       20.0476 kmol/h,
molar flowrate  of stream 7      380.9048 kmol/h,
m. f. of N_{2}  in stream 2        0.2436 -,
m. f. of N_{2}  in stream 3        0.1610 -,
m. f. of N_{2}  in stream 4        0.2375 -,
m. f. of H_{2}  in stream 2        0.7308 -,
m. f. of H_{2}  in stream 3        0.4831 -,
m. f. of H_{2}  in stream 4        0.7126 -,
m. f. of NH_{3} in stream 3        0.3221 -,
extent of reaction ξ              95.2381 kmol/h.
```

We will close this chapter by posing a final question. Given a process flowsheet of a multi-unit process, like the one shown in Fig. 6.25, what is the maximum number of independent control volumes that can be defined? The meaning of independent control volumes is that the equations we can write (mass balances) for these control volumes are indeed independent from each other. For the process of Fig. 6.25 the answer is given in Fig. 6.27: there are four independent control volumes, which is equal to the number of independent process units appearing in the flowsheet (mixing point, reactor, flash distillation unit, and the splitting point). Several alternatives involving fewer independent control volumes can be defined, as shown in Fig. 6.28. The reason for defining control volumes that involve more than one unit is simply to eliminate intermediate streams for which we do not have enough specifications to determine. However, we

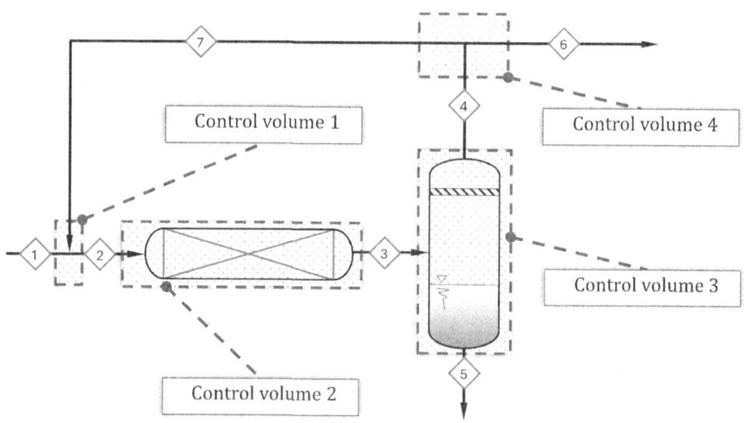

Figure 6.27 The maximum number of independent control volumes that can be defined for the process of Fig. 6.25.

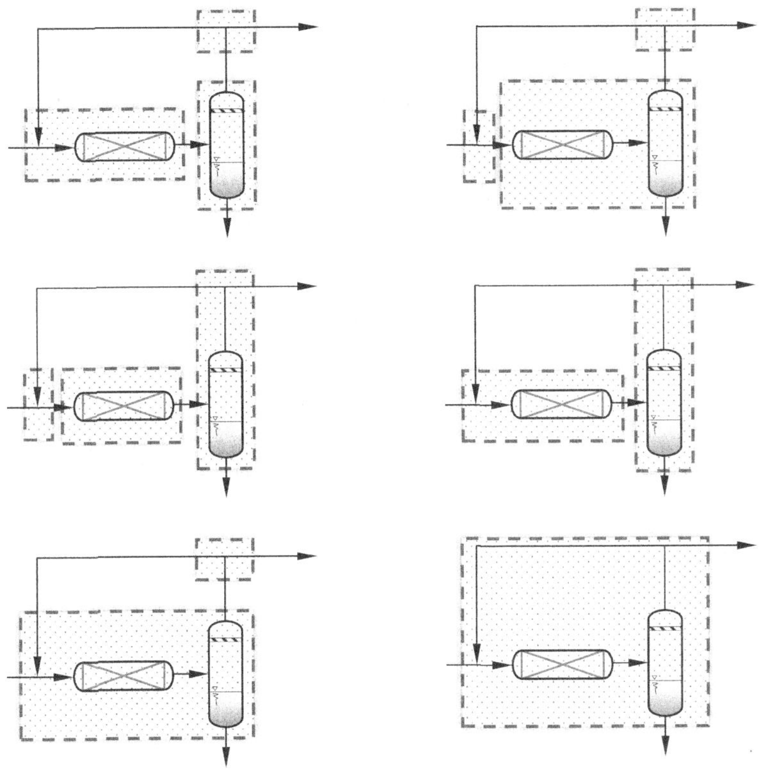

Figure 6.28 A few of the alternative definitions of independent control volumes that can be defined for the process of Fig. 6.25.

can define no more than the maximum number of independent control volumes. To demonstrate this, let us assume that, in addition to the four independent control volumes shown in Fig. 6.27, we also define a control volume for the total system, as shown in Fig. 6.29.

Let us now write the overall material balances for the control volumes 1 to 5:

$$\text{CV 1:}\ \dot{N}_1 + \dot{N}_7 = \dot{N}_2 \quad (6.5.16)$$

$$\text{CV 2:}\ \dot{N}_3 = \dot{N}_2 - 2\xi \quad (6.5.17)$$

$$\text{CV 3:}\ \dot{N}_3 = \dot{N}_4 + \dot{N}_5 \quad (6.5.18)$$

$$\text{CV 4:}\ \dot{N}_4 = \dot{N}_6 + \dot{N}_7 \quad (6.5.19)$$

$$\text{CV 5:}\ \dot{N}_1 - 2\xi = \dot{N}_6 + \dot{N}_5 \quad (6.5.20)$$

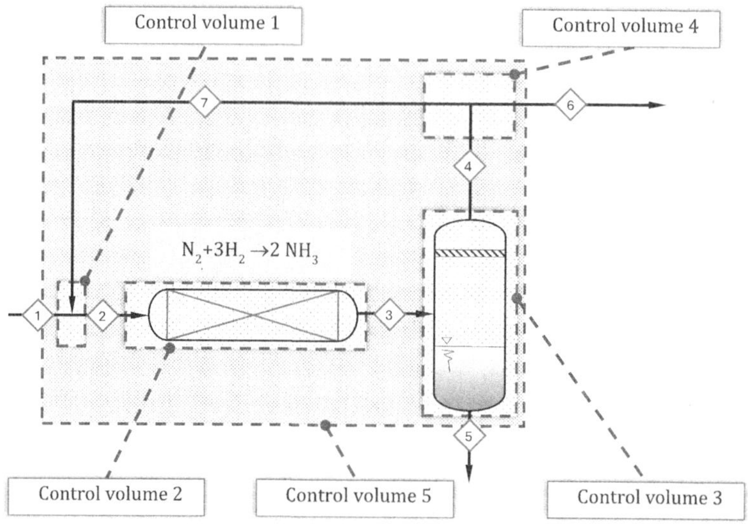

Figure 6.29 Using more than the maximum number of independent control volumes that can be defined for the process of Fig. 6.25 results in dependent material balance equations.

Let us now consider the following combination CV1-CV2+CV3+CV4:

$$\text{CV1} - \text{CV2} + \text{CV3} + \text{CV4}: \dot{N}_1 + \dot{N}_7 - \dot{N}_3 + \dot{N}_3 + \dot{N}_4 = \dot{N}_2 - \dot{N}_2 + 2\xi + \dot{N}_4 + \dot{N}_5 + \dot{N}_6 + \dot{N}_7$$

or

$$\dot{N}_1 = 2\xi + \dot{N}_5 + \dot{N}_6$$

which is identical to Eq. (6.5.20). The same result is obtained if we consider the component mole balances. The meaning is that when we use more control volumes than the maximum number of independent control volumes, the equations are dependent! The conclusion is that you must never use more control volumes than the maximum number of independent control volumes. The independent control volumes that you can define can be less than the maximum number of independent control volumes but can never exceed the maximum number of independent control volumes.

Chapter summary

In this chapter we presented a formal, systematic methodology for analyzing mass balance problems at steady state with the aim of assessing whether the mathematical model is well defined and a unique solution can be obtained. The basic rules that have been presented are the following:

- the number of variables associated with each product stream are equal to the number of components present in this stream
- when chemical reactions occur, we need as many additional variables to fully describe the problem as there are independent chemical reactions (these variables are associated with the reacting unit and not with a process stream)
- the number of independent mass balance equations that can be formulated for a process unit is equal to the number of components present in the streams crossing the system boundaries
- for the mathematical problem to be well defined a necessary condition is that the number of specifications equals the number of degrees of freedom, which is equal to the number of variables minus the number of independent equations
- even if the mathematical problem is well defined, a unique physically realizable solution may not exist as there may be problems with our specifications (they are not attainable) or some of the material balances become dependent due to unfortunate numerical values selected.

Problems

6.1. A single-effect evaporator is to be designed to produce 750 kg/h of potable water from seawater containing 3.5 wt% salt. The process concentrates the seawater to a brine with 14 wt% salt. Determine the required seawater feed rate and the amount of brine produced. The process is illustrated in Fig. P6.1.

6.2. The objective is to design a double-effect evaporator system to produce 750 kg/h of potable water from seawater containing 3.5 wt% salt. The process is intended to concentrate the seawater to a final brine containing 14 wt% salt. Determine the required seawater feed rate and the amount of brine produced. A schematic of the process is provided in Fig. P6.2. Evaluate whether the specifications shown in Fig. P6.2 result in a well-posed problem. If the problem is not well-posed, propose the necessary modifications to make it well-defined. Then, solve the revised problem and

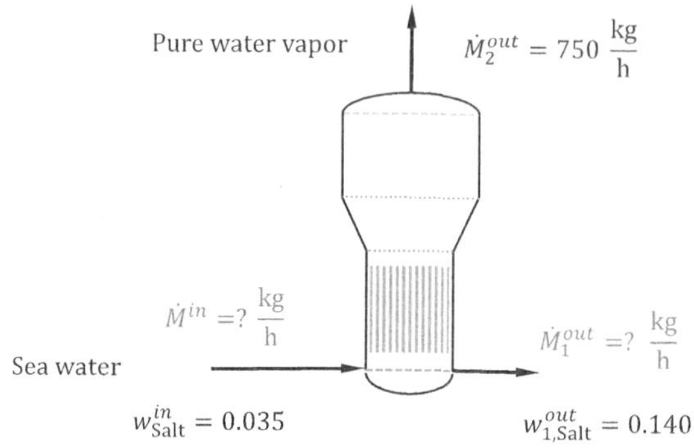

Figure P6.1 A simple evaporator process.

Figure P6.2 A double evaporator process.

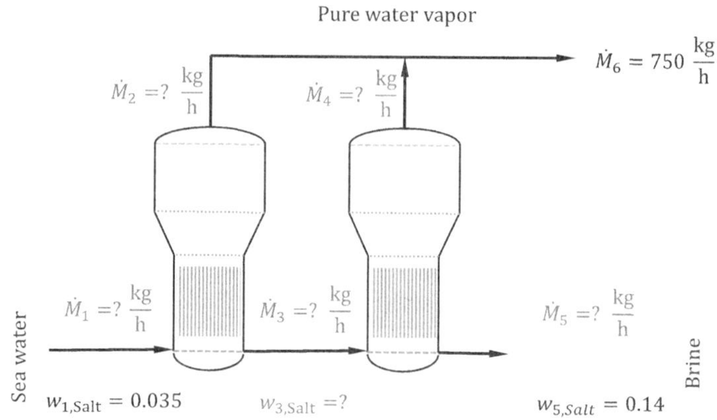

determine the values of all variables involved in the mathematical model.

6.3. The objective is to design a triple-effect evaporator system to produce 750 kg/h of potable water from seawater containing 3.5 wt% salt. The process is intended to concentrate the seawater to an unspecified final brine concentration. A schematic of the process is provided in Fig. P6.3.

Evaluate whether the specifications shown in Fig. P6.3 result in a well-posed problem. If the problem is not well-posed, propose the necessary modifications to make it well-defined. In any case, solve the revised problem and determine the values of all variables involved in the mathematical model.

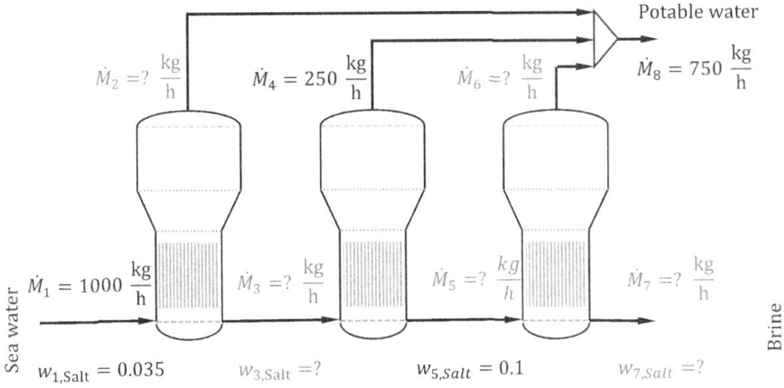

Figure P6.3 A triple evaporator process.

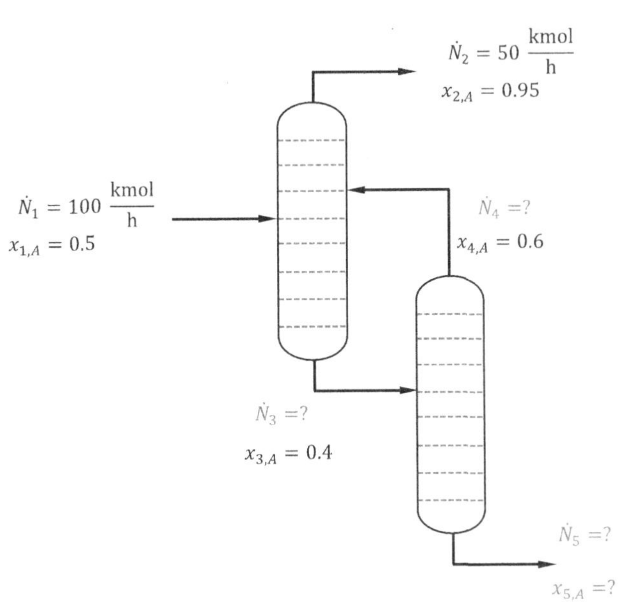

Figure P6.4 A double distillation process.

6.4. Consider the two-column distillation process illustrated in Fig. P6.4. Assess whether the process model is well-posed based on the specifications provided in the figure. If the model is not well-posed, clearly identify the issue(s) and suggest appropriate remedies to rectify them. If the model is well-posed, proceed to solve the model and determine the values of all stream variables.

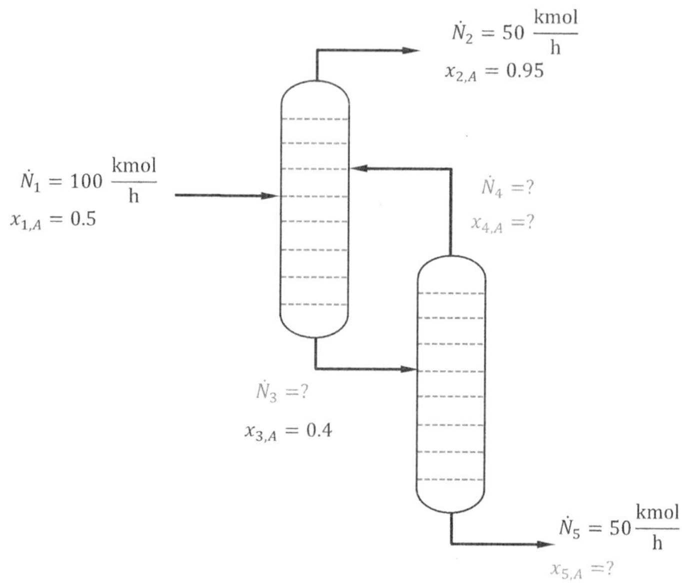

Figure P6.5 A double distillation process.

6.5. Consider the two-column distillation process illustrated in Fig. P6.5. Assess whether the process model is well-posed based on the specifications provided in the figure. If the model is not well-posed, clearly identify the issue(s) and suggest appropriate remedies to rectify them. If the model is well-posed, proceed to solve the model and determine the values of all stream variables.

6.6. Prove that when two streams containing the same three components are mixed, the composition of the resulting mixture lies on the line segment connecting the compositions of the two original streams on a ternary diagram (or in composition space).

6.7. In Fig. P6.7, a solvent extraction process is illustrated. A feed of stream containing 100 kmol/h of a mixture of water (W) and acetic acid (A), with 90 mol% water, is combined with 200 kmol/h of an organic solvent (S). The goal is to extract the acetic acid from the aqueous phase into the organic solvent. The resulting mixture is sent to a decanter, where it separates into two immiscible liquid phases: an organic phase, primarily composed of the solvent and acetic acid (A, S), and an aqueous phase, mainly water with a small amount of organics.

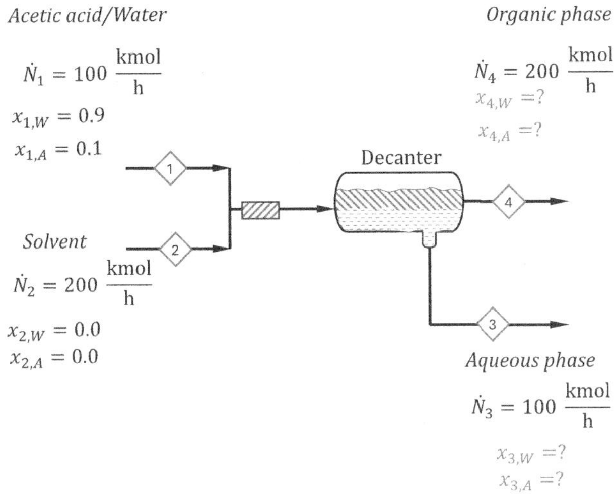

Figure P6.7 A solvent extraction process.

A colleague suggests that by measuring only the flow rates of the two output streams, it is possible to determine their compositions. Can you prove that this is incorrect?

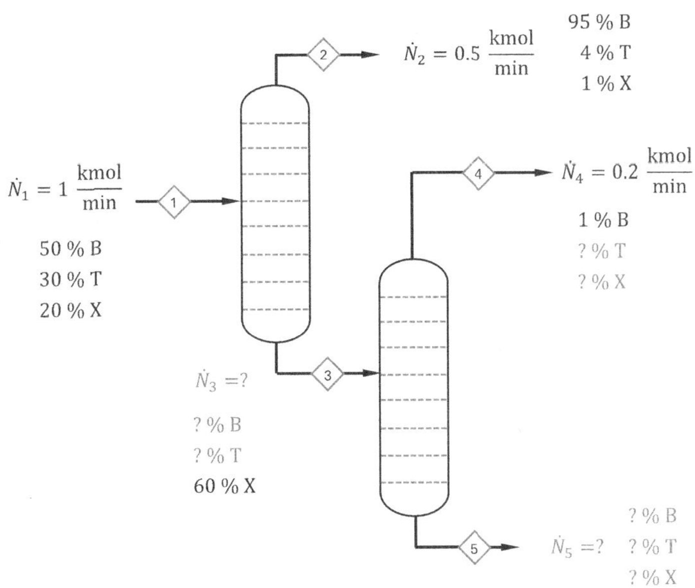

Figure P6.8 A two-distillation process.

How many measurements do you need in order to calculate the composition of the two product streams?

6.8. In Fig. P6.8, a system of two distillation columns in series is presented. Can, for the specifications indicated in Fig. P6.8, you solve the problem and determine all flowrates and compositions?

6.9. In Fig. P6.9, a system of two distillation columns in series is presented. Can you, for the specifications indicated in Fig. P6.8, solve the problem and determine all flowrates and compositions?

6.10. In Fig. P6.10, a system of three distillation columns is presented. Can you, for the specifications indicated in Fig. P6.10, solve the problem and determine all flowrates and compositions?

6.11. An aqueous solution, 20 wt% in sulfate of potash (SOP, K_2SO_3), is fed at a mass flowrate of 1000 kg/min to a crystallizer where part of the water is evaporated and solid crystals of the salt are formed. The concentrated salt solution is fed to a centrifuge where two streams are obtained: a stream that contains crystals and part of the aqueous solution and an aqueous solution stream free of crystals, that is recycled back to the crystallizer.

Are the specifications shown in Fig. P6.11 enough to have a well-posed mathematical model? If your answer is negative, add as many specifications as necessary to have a well-defined problem.

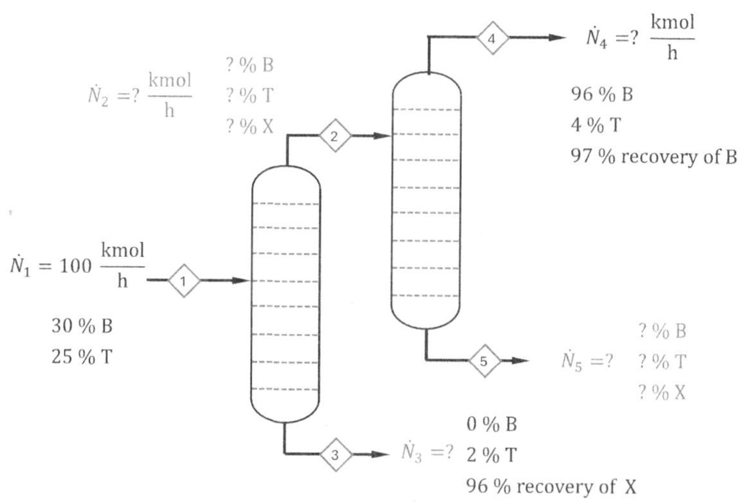

Figure P6.9 A two-distillation process.

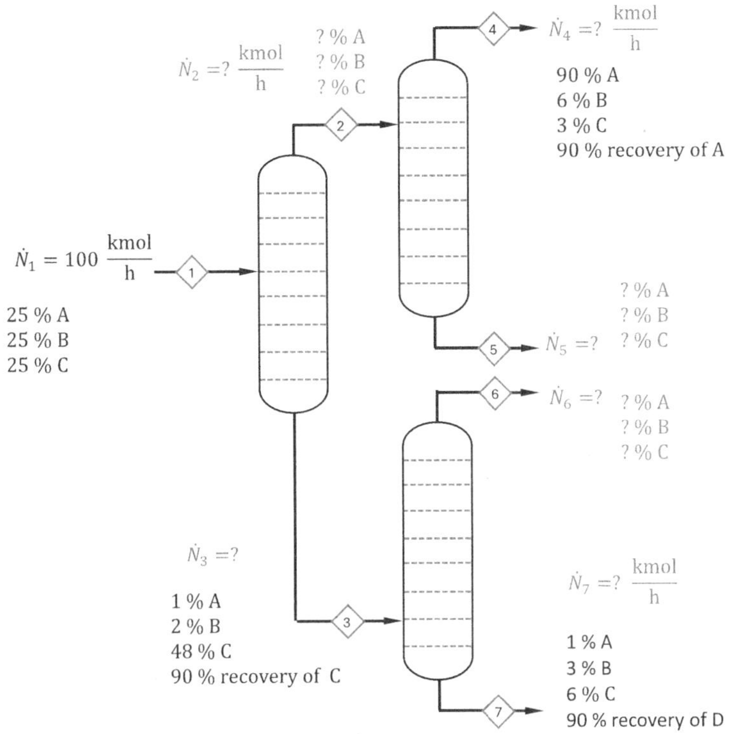

Figure P6.10 A triple distillation process.

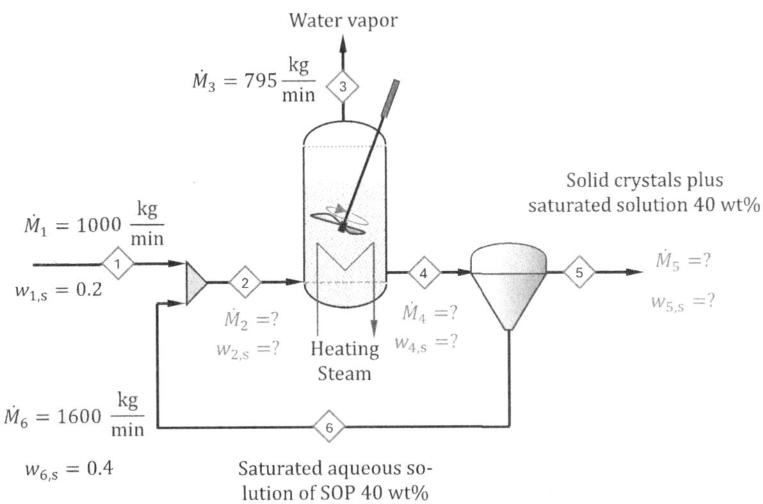

Figure P6.11 A crystallizer with recycling.

Figure P6.12 A simple membrane process.

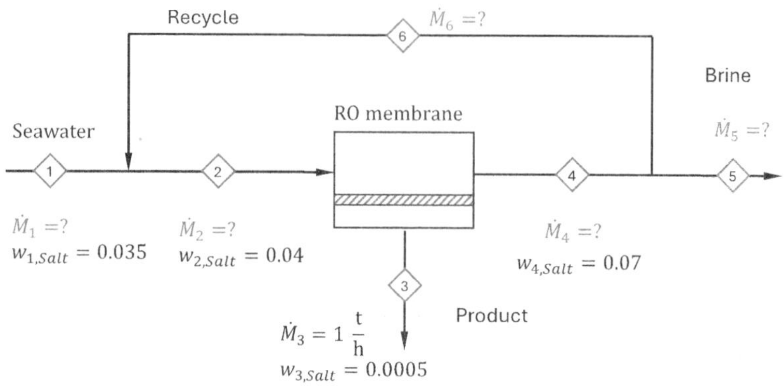

Solve the model to determine all mass flowrates and compositions.

6.12. Reverse osmosis (RO) membranes are commonly used to produce potable water from seawater. In the process illustrated in Fig. P6.12, 1 t/h of potable water is produced, containing a salt concentration of 500 ppm (equivalent to 0.05 wt%), from seawater with an initial salt concentration of 3.5 wt%. The system includes a recycle loop, in which a portion of the by-product stream is returned to the feed of the RO membrane. Using the measurements provided in Figure P6.12, calculate all unknown mass flow rates within the process.

Figure P6.13 A reactor-separator process.

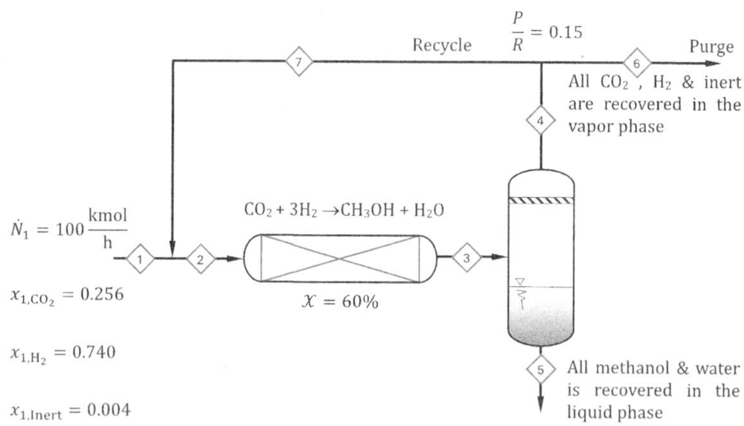

6.13. A process for methanol (M) production from captured CO_2 (C) and hydrogen (H) is pictured in Fig. P6.13. The fresh feed is 25.6 mol% CO_2, 74 mol% H_2, and 0.4 mol% inert. The per-pass conversion achieved in the reactor is 60% while in the separation unit only the methanol and water are completely condensed and recovered as liquids while all other components are either recycled or purged. The ratio of the molar flowrate of the purge to the molar flowrate of the recycle stream is 0.15. Perform a degree of freedom analysis and determine all molar flowrates and compositions.

Further reading

Many definitions have been retrieved from the Wikipedia, the free encyclopedia.

Books on material and energy balances.

Andersen, L.B., Wenzel, L.A., 1961. Introduction to Chemical Engineering. McGraw Hill, NY.

Henley, E.J., Bieber, H., 1959. Chemical Engineering Calculations. McGraw Hill, NY.

Himmelblau, D.M., Bischoff, K.B., 1968. Process Analysis and Simulation, Deterministic Systems. John Wiley & Sons, NY.

Hougen, O.A., Watson, k.m., Ragatz, R.A., 1943. Chemical Process Principles, Part I, Material and Energy Balances. John Wiley and Sons, NY.

Kirkbride, C.G., 1947. Chemical Engineering Fundamentals. McGraw Hill, NY.

Luyben, W.L., Wenzel, L.A., 1988. Chemical Process Analysis: Mass and Energy Balances. Prentice Hall, Englewood Cliffs, NJ.

Myers, A.L., Seider, W.D., 1976. Introduction to Chemical Engineering and Computer Calculations. Prentice Hall, Inc., Englewood Cliffs, NJ.

Reklaitis, G.V., 1983. Introduction to Material and Energy Balances. John Wiley and Sons, NY.

7

Properties of single component systems

7.1 Introduction

Up to this point, we have assumed that all stream properties are readily known and have not considered additional thermodynamic constraints, such as phase equilibrium. In real-world applications, however, calculating stream properties—such as phase coexistence, density, and enthalpy—is essential before formulating accurate mass and energy balances.

Until now, we have also relied solely on the ideal gas model. While useful, this model represents a single-phase, idealized fluid and fails to account for the complex phase behavior observed in real-world systems. In everyday life, multiphase phenomena are commonplace: we boil water while cooking, add ice cubes that melt to cool our drinks, and observe water vapor condensing into rain or fog. These examples highlight the limitations of the ideal gas model and the importance of considering multiphase systems in practical applications.

Before proceeding with the estimation of pure component properties, it is useful to revisit the concept of the ideal gas. The ideal gas model describes a simplified system of noninteracting molecules and serves as a good approximation of real gas behavior under conditions of low pressure and high temperature. This model is governed by the well-known ideal gas law:

$$PV = NRT \tag{7.1.1}$$

where P is the pressure, V is the volume, N is the number of moles of the gas, T is the absolute temperature, and R is the ideal gas constant. In this text we will be using the ideal gas law in the following form where the molar or specific volume $v = V/N$ is introduced

$$Pv = RT \tag{7.1.2}$$

Introduction to Process Systems Engineering. https://doi.org/10.1016/B978-0-443-45444-8.00007-5

The ideal gas law expresses a fundamental physical principle: in a single-phase system, any two of the three variables—pressure, temperature, and specific volume—uniquely determine the third. For example, consider two systems with identical compositions that are brought to the same temperature and pressure, even if they began in different states or followed different paths. According to the ideal gas law, both systems will have the same molar volume. This consistency underscores a key characteristic of ideal gases: their state is completely defined by any two intensive properties. Thus, the ideal gas law can be viewed as a specific case of a more general class of equations of state, which relate pressure, temperature, and specific volume for a given substance

$$f(P, v, T) = 0 \tag{7.1.3}$$

We have also seen that the model of the ideal gas for energy calculations is based on the specific heat capacity of the ideal gas, which is usually expressed as a polynomial of temperature. Then the specific enthalpy, which for the ideal gas is a property of temperature only, can be calculated with the convention that at the reference temperature of $T_{ref} = 298.15$ K the specific enthalpy is equal to the specific enthalpy of formation of the chemical species

$$h(T) = h_f(T_{ref}) + \int_{T_{ref}}^{T} c_p(\theta)d\theta \tag{7.1.4}$$

The internal energy is related to enthalpy through the definition of enthalpy

$$u(T) = h(T) - RT \tag{7.1.5}$$

In order to be able to calculate the properties of a single component stream for given conditions, we must be able to find what is the physical state of the stream. You know that there are three phases: solid, liquid, and vapor or gas, and a pure component can exist in any of them when we vary the conditions. So, the questions that arise naturally are:

- what are the characteristics of a **phase diagram** that shows the physical states (phases) of a substance under different conditions of temperature and pressure and maps out the boundaries between solid, liquid, and vapor (gas) phases (and indicates the conditions under which these phases coexist)?
- what are the thermodynamic models that can be used to describe in mathematical terms these diagrams and how accurate are these (what is the approximation error)?

In this introductory text we will only restrict attention to vapor-liquid systems. More elaborated discussions and mathematical models that cover all possible phases including solids can be found in thermodynamics books.

7.2 Phase diagram of CO_2

To understand the problem we aim to address, let us consider Fig. 7.1, which shows the phase diagram of carbon dioxide, constructed using experimental data from multiple sources. In this diagram, the x-axis represents the density of CO_2, and the y-axis represents pressure. Each curve corresponds to a constant temperature (isothermal) process.

Let's examine the isotherm at 260 K. Imagine a "thought" experiment starting at 10 MPa—this scenario is illustrated in Fig. 7.2. At 260 K and 10 MPa, CO_2 has a density of 1030 kg/m^3 and exists as a subcooled liquid. As we gradually reduce the pressure, the density decreases, as expected. For instance, at 3 MPa, the density has dropped to 1002 kg/m^3, yet CO_2 remains in the liquid phase. However, something critical occurs at 2.44 MPa: the first bubble of vapor forms, marking the onset of boiling. CO_2 enters a two-phase region, where saturated vapor and

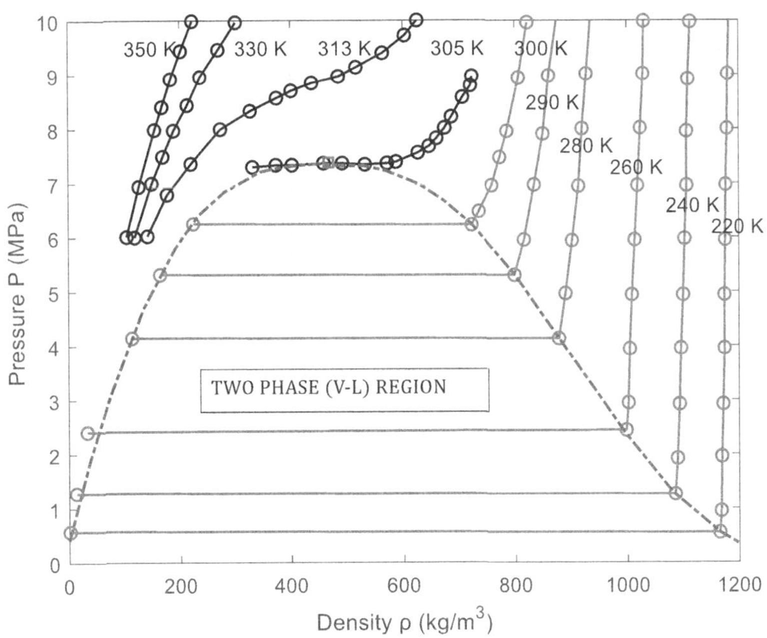

Figure 7.1 Phase diagram of CO_2: illustrates the conditions under which a pure substance exists as either a liquid, a vapor, or a mixture of both (in equilibrium).

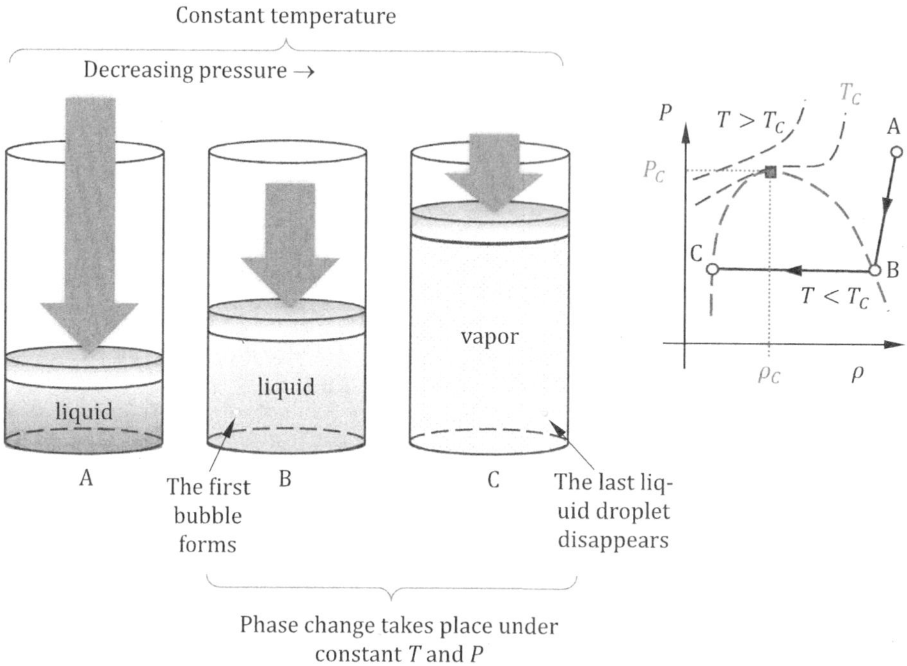

Figure 7.2 Hypothetical experimental determination of an isotherm of a pure fluid.

saturated liquid coexist in equilibrium. As we try to further reduce the pressure, more vapor is generated, but the temperature and pressure remain constant, despite our efforts to decrease the pressure, during this phase change. The system stays in equilibrium until the last drop of liquid evaporates, at which point the substance is entirely in the vapor phase. We are now able to reduce the pressure further. This vapor–liquid phase behavior is observed for all isotherms with temperatures below the critical temperature, $T_C = 304.25$ K.

The **critical temperature** of a substance is the highest temperature at which it can exist as a liquid, regardless of the pressure applied. Above this temperature, the substance cannot be liquefied—no matter how much pressure is applied, it will remain in the gaseous (**supercritical**) state. The key point to remember is that at the critical temperature, the distinction between liquid and gas begins to disappear, and, along with the critical pressure, it defines the critical point—the endpoint of the liquid–vapor phase boundary. Above the critical point, the substance exists as a supercritical fluid, which exhibits properties of both liquids and gases. The critical point plays a pivotal role in all thermodynamic models of pure fluids and mixtures.

Now consider the isotherm at 313 K, which is above the critical temperature. Again, we start at 10 MPa. At these conditions (313 K, 10 MPa), the density of CO_2 is 625 kg/m^3, and the substance exists as a supercritical fluid. As we decrease (or increase) the pressure, the density correspondingly decreases (or increases), but no phase change occurs—no liquid is ever formed.

When plotting isotherms over a range of temperatures, a distinct vapor–liquid coexistence region becomes apparent, delineated by the saturation envelope (indicated with a dashed line in Fig. 7.1). The apex of this envelope corresponds to the critical point, defined by the critical density (ρ_C) and critical pressure (P_C), beyond which the distinction between liquid and vapor phases ceases to exist. A calculation of particular thermodynamic importance involves determining the **saturation pressure** at a given temperature and phase. This allows identification of the conditions at which phase transitions occur—specifically, the bubble point (liquid-to-vapor transition) and the dew point (vapor-to-liquid transition).

In Fig. 7.3, the CO_2 phase diagram is shown again, but with a different x-axis: molar volume ($v = 1/\rho$) instead of mass density.

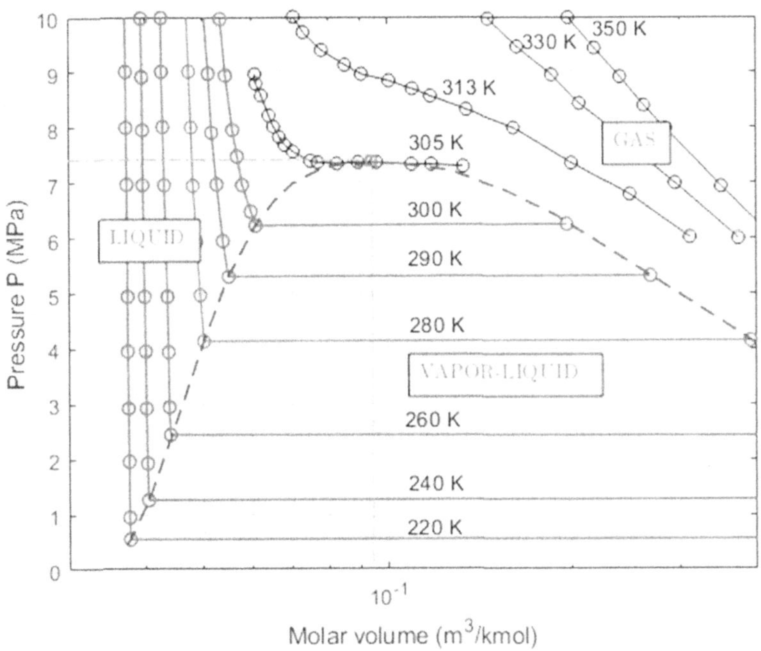

Figure 7.3 Phase diagram of CO_2: illustrates the conditions under which a pure substance exists as either a liquid, a vapor, or a mixture of both (in equilibrium).

The thermodynamic behavior remains the same, but since molar volume is the inverse of molar density, the subcooled liquid region now appears on the left side of the diagram. Using molar volume is more common in phase diagrams for pure components, and that will be the convention used throughout this text. Nevertheless, it's important to be comfortable interpreting diagrams using both density and molar volume, as they are frequently used in thermodynamics and process systems engineering.

Can the ideal gas law be used to generate phase diagrams like the one shown in Fig. 7.3 for CO_2? The answer is clearly no. The ideal gas law is far too simplistic to capture the complex phase behavior exhibited by real substances. The ideal gas law assumes that gas molecules: do not interact with one another (i.e., no intermolecular forces), occupy no volume themselves, and exist in a single phase only—the gaseous phase. These assumptions completely neglect the possibility of phase transitions such as condensation and boiling. As a result, the ideal gas law cannot predict the coexistence of vapor and liquid, the appearance of a saturation region where both liquid and vapor phases exist in equilibrium, or the existence of a critical point. In contrast, real phase diagrams like Fig. 7.3 show that CO_2 can exist in multiple phases, including **subcooled liquid** (at a pressure greater than the saturation pressure), **saturated liquid** and **vapor** (at a pressure equal to the saturation pressure), **supercritical fluid**, and pure (superheated) **vapor**. These behaviors are fundamentally driven by molecular interactions and finite molecular volume, which the ideal gas model ignores.

To accurately model phase behavior, more advanced equations of state (EOS) are required—such as the **van der Waals equation**, **Soave-Redlich-Kwong**, or **Peng-Robinson** equations. These models incorporate nonidealities by accounting for attractive forces between molecules and the finite size of molecules, allowing them to reproduce phase transitions and critical phenomena observed in real substances. We begin with the first equation of state that incorporated molecular interactions, proposed by van der Waals in the late 19th century. Although it is a relatively simple model, it laid the foundation for more advanced equations of state and remains important in thermodynamics for illustrating the conceptual transition from ideal to real fluid behavior. It also qualitatively predicts phenomena such as phase coexistence, critical phenomena, and compressibility effects, making it a cornerstone in the historical and pedagogical development of fluid phase behavior.

7.3 The van der Waals equation of state

7.3.1 van der Waals modification to the ideal gas law

The van der Waals equation is the first equation of state to capture phase behavior of real fluids. It qualitatively predicts liquid-vapor coexistence, the existence of a critical point, and non-ideal $P - v - T$ behavior. The van der Waals equation can be derived from the ideal gas equation, Eq. (7.1.2) by using the following two corrections:

- *Correction for Finite Molecular Volume* — b: gas molecules are not point particles—they occupy finite volume b per mol of the fluid. The actual "free" volume available for motion is less than the total volume, and this is corrected by subtracting b from the molar volume of the fluid

$$P(v - b) = RT \tag{7.3.1}$$

- *Correction for Molecular Attractions* — a/v^2: real gases exhibit attractive forces between molecules (especially at moderate pressures), and these attractions reduce the pressure exerted on the container walls. Van der Waals corrects for this by adding a pressure term

$$P = \frac{RT}{(v - b)} - \frac{a}{v^2} \tag{7.3.2}$$

Eq. (7.3.2) is the van der Waals equation of state. It involves two constants a and b that can be determined for each chemical species using a variety of methods, but the two most often used will be presented below.

7.3.2 Fitting the constants of the van der Waals equation to the critical point

At the critical point, the $P - v$ isotherm has an inflection point, meaning:

$$\left.\frac{dP}{dv}\right|_C = 0 \tag{7.3.3}$$

$$\left.\frac{d^2P}{dv^2}\right|_C = 0 \tag{7.3.4}$$

These mathematical conditions impose two algebraic constraints on any equation of state and can be used to find the two parameters that appear in the van der Waals equation. This can be achieved as follows: we calculate the first and second derivatives of Eq. (7.3.2) with respect to molar volume

$$\left.\frac{dP}{dv}\right|_C = -\frac{RT_C}{(v_C - b)^2} + 2\frac{a}{v_C^3} = 0 \tag{7.3.5}$$

$$\left.\frac{d^2P}{dv^2}\right|_C = 2\frac{RT_C}{(v_C - b)^3} - 6\frac{a}{v_C^4} = 0 \tag{7.3.6}$$

We multiply Eq. (7.3.6) by $\frac{v_C}{3}$ and add the result to Eq. (7.3.5) to obtain

$$b = \frac{v_C}{3} \tag{7.3.7}$$

$$a = \frac{9}{8}RT_C v_C \tag{7.3.8}$$

These equations calculate the constants a and b of the van der Waals equation using measurements of the critical temperature and critical molar volume. However, there is an alternative procedure that relates a and b to the critical temperature and pressure. In this method, we substitute Eq. (7.3.7) into Eq. (7.3.5) to obtain

$$RT_C = \left(\frac{2}{3}\right)^3 \frac{a}{b} \tag{7.3.9}$$

We also substitute Eqs. (7.3.7) and (7.3.9) into Eq. (7.3.2) to obtain

$$P_C = \left(\frac{1}{3}\right)^3 \frac{a}{b^2} \tag{7.3.10}$$

We then solve Eq. (7.3.9) for b and substitute into Eq. (7.3.10) to eliminate b and obtain

$$a = \frac{27}{64}\left(\frac{R^2 T_C^2}{P_C}\right) \tag{7.3.11}$$

Finally, we solve Eq. (7.3.9) for a

$$a = \left(\frac{3}{2}\right)^3 RT_C b$$

and substitute into Eq. (7.3.10) to eliminate a

$$P_C = \left(\frac{1}{3}\right)^3 \frac{1}{b^2}\left(\frac{3}{2}\right)^3 RT_C b = \frac{1}{8}\frac{RT_C}{b}$$

or

$$b = \frac{1}{8}\left(\frac{RT_C}{P_C}\right) \tag{7.3.12}$$

7.3.3 The van der Waals equation in cubic polynomial form

A very important result follows if we multiply Eq. (7.3.2) with $v^2(v-b)$ to obtain

$$v^2(v-b)P = RTv^2 - a(v-b)$$

or

$$v^3 - \left(b+\frac{RT}{P}\right)v^2 + \frac{a}{P}v - \frac{ab}{P} = 0 \tag{7.3.13}$$

We conclude that the van der Waals equation can be rearranged into a cubic polynomial in molar volume, which, like all cubic equations, yields either one real root and two complex conjugates or three real roots. These roots correspond to physically meaningful (and sometimes nonphysical) solutions depending on the thermodynamic conditions. More commonly, cubic equations of state (cEOS), including the van der Waals equation, are expressed in terms of the **compressibility factor**, Z, a dimensionless quantity defined as the ratio of the molar volume of a real gas to that of an ideal gas under the same temperature and pressure:

$$Z = \frac{v}{v^{ig}} = \frac{v}{\left(\frac{RT}{P}\right)} = \frac{Pv}{RT} \tag{7.3.14}$$

This form facilitates the analysis of real fluid behavior and simplifies comparisons across different substances and conditions. The cubic form in Z is particularly useful for identifying multiple phases and predicting phase equilibrium. To change the variable appearing in Eq. (7.3.13) from the molar volume v to the compressibility factor Z we multiply Eq. (7.3.13) by $(P/RT)^3$

$$\left(\frac{P}{RT}\right)^3 v^3 - \left(\frac{P}{RT}\right)^3\left(b+\frac{RT}{P}\right)v^2 + \left(\frac{P}{RT}\right)^3\frac{a}{P}v - \left(\frac{P}{RT}\right)^3\frac{ab}{P} = 0$$

or

$$\left(\frac{Pv}{RT}\right)^3 - \left(\frac{Pb}{RT}+1\right)\left(\frac{Pv}{RT}\right)^2 + \frac{aP}{(RT)^2}\left(\frac{Pv}{RT}\right) - \frac{aP}{(RT)^2}\left(\frac{Pb}{RT}\right) = 0$$

and finally

$$Z^3 - \left(\frac{Pb}{RT}+1\right)Z^2 + \frac{aP}{(RT)^2}Z - \frac{aP}{(RT)^2}\left(\frac{Pb}{RT}\right) = 0 \qquad (7.3.15)$$

We now define two new dimensionless constants

$$\mathscr{A} = a\frac{P}{(RT)^2} \qquad (7.3.16)$$

$$\mathscr{B} = b\frac{P}{RT} \qquad (7.3.17)$$

and the cubic polynomial becomes

$$Z^3 - (\mathscr{B}+1)Z^2 + \mathscr{A}Z - \mathscr{A}\mathscr{B} = 0 \qquad (7.3.18)$$

To solve the cubic polynomial and determine the roots, we may use the analytic expression for the cubic polynomial equation. For the general cubic equation

$$Z^3 + a_1Z^2 + a_2Z + a_3 = 0 \qquad (7.3.19)$$

we first calculate (see Perry's Handbook)

$$\mathscr{Q} = \frac{3a_2 - a_1^2}{3}, \qquad \mathscr{R} = \frac{27a_3 - 9a_1a_2 + 2a_1^3}{27} \qquad (7.3.20)$$

and

$$\mathscr{U} = \sqrt[3]{-\frac{\mathscr{R}}{2} + \sqrt{D}}, \qquad \mathscr{V} = \sqrt[3]{-\frac{\mathscr{R}}{2} - \sqrt{D}} \qquad (7.3.21)$$

where $\mathscr{D} = \left(\frac{\mathscr{Q}}{3}\right)^3 + \left(\frac{\mathscr{R}}{2}\right)^2$, which is the discriminant, and

- if $\mathscr{D} > 0$ there is one real root (the other two are complex conjugates and irrelevant),
- if $\mathscr{D} = 0$ all roots are real and at least two are equal,
- if $\mathscr{D} < 0$ there are three real and unequal roots.

The solutions are

$$Z_1 = \mathscr{U} + \mathscr{V} - \frac{a_1}{3} \qquad (7.3.22)$$

$$Z_{2,3} = -\frac{1}{2}(\mathscr{U}+\mathscr{V}) - \frac{a_1}{3} \pm i\frac{1}{2}\sqrt{3}(\mathscr{U}-\mathscr{V}), \qquad i = \sqrt{-1} \tag{7.3.23}$$

Alternatively, we can use a numerical scheme where the root is calculated iteratively through the following equations

Newton's method (quadratic convergence)

$$Z_{n+1} = Z_n - \frac{f(Z_n)}{f'(Z_n)} \tag{7.3.24}$$

Halley's method (cubic order convergence)

$$Z_{n+1} = Z_n - \frac{2f(Z_n)f'(Z_n)}{2(f'(Z_n))^2 - f(Z_n)f''(Z_n)} \tag{7.3.25}$$

where

$$f(Z) = Z^3 - (\mathscr{B}+1)Z^2 + \mathscr{A}Z - \mathscr{A}\mathscr{B} \tag{7.3.26}$$

$$f'(Z) = 3Z^2 - 2(\mathscr{B}+1)Z + \mathscr{A} \tag{7.3.27}$$

$$f''(Z) = 6Z - 2(\mathscr{B}+1) \tag{7.3.28}$$

Example 7.1. van der Waals constants for CO_2

The critical temperature and pressure for CO_2 are $T_C = 304.14$ K and $P_C = 7.38$ MPa. Find the constants a, b of the van der Waals equation.

Solution

We use $R = 8.314\,463\ \frac{\text{Pa}\cdot\text{m}^3}{\text{mol}\cdot\text{K}}$ and Eqs. (7.3.11) and (7.3.12)

$$a = \frac{27}{64}\left(\frac{R^2T_C^2}{P_C}\right) = \frac{27}{64}\left(\frac{8.314463^2 \times 304.14^2}{7.38\times10^6}\right)\left(\frac{\text{Pa}\cdot\text{m}^3}{\text{mol}\cdot\text{K}}\right)^2\frac{\text{K}^2}{\text{Pa}} = 0.36555\ \frac{\text{Pa}\cdot\text{m}^6}{\text{mol}^2}$$

$$b = \frac{1}{8}\left(\frac{RT_C}{P_C}\right) = \frac{1}{8}\left(\frac{8.314463\times304.14}{7.38\times10^6}\right)\left(\frac{\text{Pa}\cdot\text{m}^3}{\text{mol}\cdot\text{K}}\right)\frac{\text{K}}{\text{Pa}} = 4.2831\times10^{-5}\ \frac{\text{m}^3}{\text{mol}}$$

Example 7.2. Application of the van der Waals equation of state

The critical temperature and pressure for CO_2 are $T_C = 304.14$ K and $P_C = 7.38$ MPa. Experimental data for CO_2 are the following at T = 320 K (Gilgen et al., J. Chem. Thermodynamics 1992, 24, pp. 1243-1250):

Pressure (Pa)	Molar volume (m^3/kmol)
59.70×10^5	0.3188
69.94×10^5	0.2465

Calculate the pressure for the given molar density and temperature.

Solution

We use $R = 8.314\,463\ \frac{\text{Pa}\cdot\text{m}^3}{\text{mol}\cdot\text{K}}$ and the result of Example 7.1:

$$a = 0.365\,547\ \frac{\text{Pa}\cdot\text{m}^6}{\text{mol}^2}$$

$$b = 4.283\,131 \times 10^{-5}\ \frac{\text{m}^3}{\text{mol}}$$

First point

$$P = \frac{RT}{(v-b)} - \frac{a}{v^2} = \frac{8.314\,463\ \frac{\text{Pa}\cdot\text{m}^3}{\text{mol}\cdot\text{K}}\ 320\ \text{K}}{\left(0.3188\frac{\text{m}^3}{\text{kmol}} \times \frac{1\ \text{kmol}}{10^3\ \text{mol}} - 4.283\,131 \times 10^{-5}\ \frac{\text{m}^3}{\text{mol}}\right)} - \frac{0.365\,547\ \frac{\text{Pa}\cdot\text{m}^6}{\text{mol}^2}}{\left(\frac{0.3188}{1000}\right)^2 \frac{\text{m}^6}{\text{mol}^2}} = 60.44 \times 10^5\ \text{Pa}$$

The error in the calculation of the pressure is $\frac{100(60.44 - 59.7)}{59.7} = 1.24\%$

Second point

$$P = \frac{RT}{(v-b)} - \frac{a}{v^2} = \frac{8.314\,463\ \frac{\text{Pa}\cdot\text{m}^3}{\text{mol}\cdot\text{K}}\ 320\ \text{K}}{\left(0.2465\frac{\text{m}^3}{\text{kmol}} \times \frac{1\ \text{kmol}}{10^3\ \text{mol}} - 4.283\,131 \times 10^{-5}\ \frac{\text{m}^3}{\text{mol}}\right)} - \frac{0.365\,547\ \frac{\text{Pa}\cdot\text{m}^6}{\text{mol}^2}}{\left(\frac{0.2465}{1000}\right)^2 \frac{\text{m}^6}{\text{mol}^2}} = 70.47 \times 10^5\ \text{Pa}$$

The error in the calculation of the pressure is $\frac{100(70.47 - 69.94)}{69.94} = 0.75\%$

Example 7.3. van der Waals equation for estimating the molar volume of gaseous CO_2

The critical temperature and pressure for CO_2 are $T_C = 304.14$ K and $P_C = 7.38$ MPa. Experimental data for CO_2 are given in Example 7.2. Calculate the molar volume for the given pressure and temperature for the two experimental points. What is the % error in the estimation of molar volume?

Solution

First point:

We use $R = 8.314\,463 \frac{\text{Pa·m}^3}{\text{mol·K}}$ and the result of Example 7.1:

$$a = 0.365\,547 \frac{\text{Pa·m}^6}{\text{mol}^2}$$

$$b = 4.283\,131 \times 10^{-5} \frac{\text{m}^3}{\text{mol}}$$

We then use Eqs. (7.3.16) and (7.3.17)

$$\mathscr{A} = a\frac{P}{(RT)^2} = 0.365\,547 \frac{\text{Pa·m}^6}{\text{mol}^2} \frac{59.7 \times 10^5 \text{ Pa}}{\left(8.314\,463 \frac{\text{Pa·m}^3}{\text{mol·K}} 320 \text{ K}\right)^2} = 0.30828$$

$$\mathscr{B} = b\frac{P}{RT} = 4.283\,131 \times 10^{-5} \frac{\text{m}^3}{\text{mol}} \times \frac{59.7 \times 10^5 \text{ Pa}}{8.314\,463 \frac{\text{Pa·m}^3}{\text{mol·K}} 320 \text{ K}}$$

$$= 0.096106$$

Our next step is to solve the following equation (Eq. (7.3.26))

$$f(Z) = Z^3 + \underbrace{(-1.096106)}_{a_1}Z^2 + \underbrace{0.30828}_{a_2}Z + \underbrace{(-0.029628)}_{a_3}$$

Analytic solution:

$$\mathscr{Q} = \frac{3a_2 - a_1^2}{3} = \frac{3 \times 0.30828 - (-1.096106)^2}{3} = -9.22 \times 10^{-2}$$

$$\mathscr{R} = \frac{27a_3 - 9a_1a_2 + 2a_1^3}{27} = \frac{27(-0.029628) - 9(-1.096106) \times 0.30828 + 2(-1.096106)^3}{27}$$

$$= -1.454 \times 10^{-2}$$

$$\mathscr{D} = \left(\frac{\mathscr{Q}}{3}\right)^3 + \left(\frac{\mathscr{R}}{2}\right)^2 = \left(\frac{-9.22 \times 10^{-2}}{3}\right)^3 + \left(\frac{-1.454 \times 10^{-2}}{2}\right)^2$$
$$= 2.3826 \times 10^{-5} > 0$$

As $\mathscr{D} > 0$ there is only one real root

$$\mathscr{U} = \sqrt[3]{-\frac{\mathscr{R}}{2} + \sqrt{D}} = 2.2990 \times 10^{-1}$$

$$\mathscr{V} = \sqrt[3]{-\frac{\mathscr{R}}{2} - \sqrt{D}} = 1.3368 \times 10^{-1}$$

$$Z_1 = \mathscr{U} + \mathscr{V} - \frac{a_1}{3} = 0.72895$$

From the definition of the compressibility factor we have that

$$Z = \frac{v}{v^{ig}} \Rightarrow v = Zv^{ig} = Z\left(\frac{RT}{P}\right)$$

$$= 0.72895\left(\frac{8.314\,463\ \frac{\text{Pa}\cdot\text{m}^3}{\text{mol}\cdot\text{K}}\,320\ \text{K}}{59.7 \times 10^5\ \text{Pa}}\right)1000\frac{\text{mol}}{\text{kmol}} = 0.32487\ \frac{\text{m}^3}{\text{kmol}}$$

The error in the calculation of the molar volume is $\frac{100(0.32487 - 0.3188)}{0.3188} = 1.9\%$

Second point:

We then use Eq. (7.3.16) and (7.3.17)

$$\mathscr{A} = a\frac{P}{(RT)^2} = 0.365\,547\ \frac{\text{Pa}\cdot\text{m}^6}{\text{mol}^2}\ \frac{69.94 \times 10^5\ \text{Pa}}{\left(8.314\,463\ \frac{\text{Pa}\cdot\text{m}^3}{\text{mol}\cdot\text{K}}\,320\ \text{K}\right)^2} = 0.36116$$

$$\mathscr{B} = b\frac{P}{RT} = 4.283\,131 \times 10^{-5}\ \frac{\text{m}^3}{\text{mol}} \times \frac{69.94 \times 10^5\ \text{Pa}}{8.314\,463\ \frac{\text{Pa}\cdot\text{m}^3}{\text{mol}\cdot\text{K}}\,320\ \text{K}}$$
$$= 0.11259$$

Our next step is to solve the following equation (Eq. (7.3.26))

$$f(Z) = Z^3 + \underbrace{(-1.11259)}_{a_1}Z^2 + \underbrace{0.36116}_{a_2}Z + \underbrace{(-0.040663)}_{a_3}$$

Note also that

$$f'(Z) = 3Z^2 + 2 \times \underbrace{(-1.11259)}_{a_1}Z + \underbrace{0.36116}_{a_2}$$

$$f''(Z) = 6Z + 2 \times \underbrace{(-1.11259)}_{a_1}$$

Let us assume an initial estimate of $Z^1 = 0.7$ (the superscript denotes iteration number) and then

$$f(Z^1) = 0.7^3 + \underbrace{(-1.11259)}_{a_1} \times 0.7^2 + \underbrace{0.36116}_{a_2} \times 0.7 + \underbrace{(-0.040663)}_{a_3} = 0.0100$$

Note also that

$$f'(Z) = 3 \times 0.7^2 + 2 \times \underbrace{(-1.11259)}_{a_1} \times 0.7 + \underbrace{0.36116}_{a_2} = 0.2735$$

$$f''(Z) = 6 \times 0.7 + 2 \times \underbrace{(-1.11259)}_{a_1} = 1.9748$$

We revise our estimation using Newton's method

$$Z^2 = Z^1 - \frac{f(Z^1)}{f'(Z^1)} = 0.7 - 0.0365 = 0.6635$$

and $f(Z^2) = 0.0013, f'(Z) = 0.2055$

$$Z^3 = Z^2 - \frac{f(Z^2)}{f'(Z^2)} = 0.6635 - 0.0062 = 0.6574$$

and $f(Z^3) = 0.00003308$, which is close enough to zero and the root has been found.

Had we used the Halley's method then

$$Z^2 = Z^1 - \frac{2f(Z^1)f'(Z^1)}{2(f'(Z^1))^2 - f(Z^1)f''(Z^1)} = 0.7 - 0.0420 = 0.6580$$

and $f(Z^2) = 0.0001554$, and a second iteration gives

$$Z^3 = Z^2 - \frac{2f(Z^2)f'(Z^2)}{2\left(f'(Z^2)\right)^2 - f(Z^2)f''(Z^2)} = 0.6580 - 0.00079624 = 0.6572$$

and $f(Z^3) = 1.4 \times 10^{-9}$, which is close enough to zero and the root has been found.

Finally

$$v = Zv^{ig} = Z\left(\frac{RT}{P}\right) = 0.6572\left(\frac{8.314\,463\,\dfrac{\text{Pa}\cdot\text{m}^3}{\text{mol}\cdot\text{K}}320\ \text{K}}{69.94 \times 10^5\ \text{Pa}}\right)1000\frac{\text{mol}}{\text{kmol}}$$

$$= 0.2500\,\frac{\text{m}^3}{\text{kmol}}$$

The error in the calculation of the molar volume is $\frac{100(0.25 - 0.2465)}{0.2465} = 1.4\%$

You should keep in mind that the van der Waals equation is usually not that accurate, especially for the liquid phase, and errors close to 30% are to be expected in the estimation of molar volume and density of liquids, as it is shown in the example that follows.

Example 7.4. Application of the van der Waals equation of state for the liquid state reveals its limited accuracy

The critical temperature and pressure for CO_2 are $T_C = 304.14$ K and $P_C = 7.38$ MPa. Experimental data for CO_2 at the liquid state are the following (Duschek et al., J. Chem. Thermodynamics 1990, 22, pp. 827-840):

	Temperature (K)	Pressure (Pa)	Molar volume (m^3/kmol)
1	220	5000×10^3	0.03743
2	303	7500×10^3	0.06600

a. Calculate the pressure for the given molar volume and temperature.

b. Calculate the molar volume for the given temperature and pressure (use MATLAB to perform the calculations).

Solution

a. We use $R = 8.314\ 463\ \dfrac{\text{Pa·m}^3}{\text{mol·K}}$ and the result of Example 7.1:

$$a = 0.365\ 547\ \frac{\text{Pa·m}^6}{\text{mol}^2}$$

$$b = 4.283\ 131 \times 10^{-5}\ \frac{\text{m}^3}{\text{mol}}$$

whence, for the first point

$$P = \frac{RT}{(v-b)} - \frac{a}{v^2} = \underbrace{\frac{8.314\ 463\ \dfrac{\text{Pa·m}^3}{\text{mol·K}} 220\ \text{K}}{(3.743 - 4.283\ 131\ ) \times 10^{-5} \dfrac{\text{m}^3}{\text{mol}}}}_{<0}$$

$$- \frac{0.365\ 547\ \dfrac{\text{Pa·m}^6}{\text{mol}^2}}{\left(3.743 \times 10^{-5}\right)^2 \dfrac{\text{m}^6}{\text{mol}^2}} < 0!!!$$

The pressure calculated from the van der Waals equation is negative as the experimental value of the molar density is smaller than the constant b, making the first term negative. This is a clear indication of the problems expected when we use the van der Waals equation to predict properties of the liquid phase.

For the second point

$$P = \frac{RT}{(v-b)} - \frac{a}{v^2} = \frac{8.314\ 463\ \dfrac{\text{Pa·m}^3}{\text{mol·K}} 260\ \text{K}}{(6.6 - 4.283\ 131\ ) \times 10^{-5} \dfrac{\text{m}^3}{\text{mol}}} - \frac{0.365\ 547\ \dfrac{\text{Pa·m}^6}{\text{mol}^2}}{\left(6.6 \times 10^{-5}\right)^2 \dfrac{\text{m}^6}{\text{mol}^2}}$$
$$= 248 \times 10^5\ \text{Pa}$$

The % error is $\frac{100(248-75)}{75} = 230\ \%$.

b. We will not study further the first experimental point. For the second point we perform the calculations in MATLAB. We first give the critical constants for CO_2, the ideal gas constant, and the experimental conditions

```
clear all
PC = 73.8E5;     % critical pressure in Pa
TC = 304.14;     % critical temperature in K
Rg = 8.314463;  % Pa m^3 / mol K

T  = 303;        % temperature in K at the experiment
P  = 75E5;       % pressure in Pa at the experiment
```

We then calculate a, b, A, B and finally the coefficients of the cubic polynomial:

```
>> a = (27/64)*(Rg*TC)^2/PC  % Pa m^6 mol^(-2)
a =
    0.3655

>> b = (1/8)*Rg*TC/PC         % m^3 mol^(-1)
b =
   4.2831e-05

>> A = a*P/(Rg*T)^2           % dimensionless
A =
    0.4320

>> B = b*P/(Rg*T)             % dimensionless
B =
    0.1275

>> c(1)  = -(B+1)             % vdW polynomial Z^3+a1*Z^2+a2*Z+a3
c =
   -1.1275

>> c(2)  = A
c =
   -1.1275    0.4320

>> c(3)  = -A*B
c =
   -1.1275    0.4320   -0.0551
```

MATLAB has a powerful command for calculating all roots of polynomials

```
>> r=roots([1 c(1) c(2) c(3)])

ans =
   0.4144 + 0.1126i
   0.4144 - 0.1126i
   0.2987 + 0.0000i
```

Note that there are two complex conjugate roots and one real root. We keep only the real root:

```
>> Z=r(real(r)==r) % keep only the real roots
Z =
    0.2987
```

We finally calculate the molar volume

```
>> v   = Z*(Rg*T/P)*1000 % m^3/kmol
v =
    0.1003
```

The % error in the calculation of the molar volume is $\frac{100(0.1003-0.066)}{0.066} = 52\%$.

7.3.4 The dimensionless form of the van der Waals equation

We will derive a dimensionless form of the van der Waals equation that will help us to clarify the general behavior of the equation. To this end, we first define the reduced variables

$$v_r = \frac{v}{v_C} \tag{7.3.29}$$

$$P_r = \frac{P}{P_C} \tag{7.3.30}$$

$$T_r = \frac{T}{T_C} \tag{7.3.31}$$

We also remind ourselves that $b = \frac{v_C}{3}$ and $a = \left(\frac{9}{8}\right)RT_C v_C$. We start with Eq. (7.3.2) in which we replace a and b and divide by P_C to obtain

$$\frac{P}{P_C} = \frac{RT}{\left(v - \frac{v_C}{3}\right)P_C} - \left(\frac{9}{8}\right)\frac{RT_C v_C}{P_C}\frac{1}{v^2}$$

or

$$P_r = \frac{RT_C T_r}{\frac{v_C}{3}(3v_r - 1)P_C} - \left(\frac{9}{8}\right)\frac{RT_C}{P_C v_C}\frac{1}{v_r^2}$$

This is written as

$$P_r = \frac{1}{8}\left(\frac{3RT_C}{P_C v_C}\right)\left[\frac{8T_r}{(3v_r - 1)} - \frac{3}{v_r^2}\right] \tag{7.3.32}$$

Let us apply the van der Waals equation at the critical point

$$P_C = \frac{RT_C}{\left(v_C - \frac{v_C}{3}\right)} - \left(\frac{9}{8}\right) RT_C v_C \frac{1}{v_C^2} = \left(\frac{3}{2}\right)\frac{RT_C}{v_C} - \frac{3}{4}\left(\frac{3}{2}\right)\frac{RT_C}{v_C} = \left(\frac{3}{8}\right)\frac{RT_C}{v_C}$$

From the last one, it follows that

$$\frac{1}{8}\left(\frac{3RT_C}{P_C v_C}\right) = 1$$

or

$$Z_C = \frac{v_C P_C}{RT_C} = \frac{3}{8} \tag{7.3.33}$$

and Eq. (7.3.32) becomes

$$P_r = \frac{8T_r}{(3v_r - 1)} - \frac{3}{v_r^2} \tag{7.3.34}$$

We can plot this function very efficiently in MATLAB

```
>> vr=linspace(1/3+0.01,3,1001);
>> Tr=1.1;
>> plot(vr,8*Tr./(3*vr-1)-3./vr.^2,'linewidth',1.2)
>> axis([0 3 -0.5 2])
```

The isotherms predicted by the van der Waals equation, when expressed in reduced variables, are shown in Fig. 7.4. For the critical isotherm ($T_r = 1$) and for supercritical temperatures ($T_r > 1$), the predicted behavior aligns reasonably well with the experimental isotherms presented in Fig. 7.3. However, for subcritical temperatures ($T_r < 1$), significant deviations from experimental data become evident. In this regime, the van der Waals isotherms display nonphysical behavior, characterized by a local minimum, followed by an inflection point, and then a local maximum in the $P - v$ curve. In the region between these extrema—specifically, for molar volumes greater than the local minimum or smaller than the local maximum—the equation predicts that increasing the pressure causes the molar volume to increase, violating

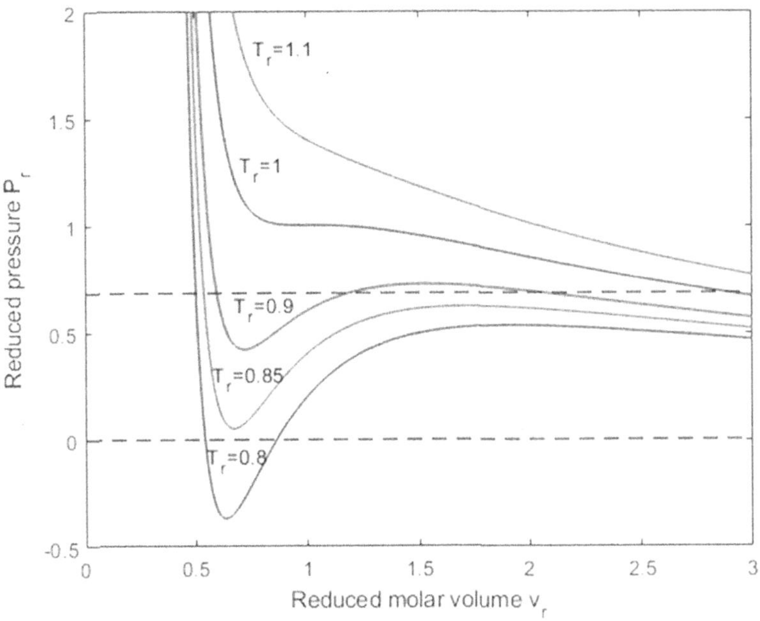

Figure 7.4 Isotherms from the van der Waals equation in reduced variables.

thermodynamic stability criteria. Moreover, for $T_r < 0.84375$, the van der Waals equation even predicts negative pressures, which are physically unrealistic. These limitations underscore the need for corrections or alternative models, particularly in the subcritical region where phase separation occurs.

Let us now examine the isotherm for $T_r = 0.9$. This curve exhibits a local minimum near $v_r \approx 0.7$, associated with a reduced pressure $P_{r,\min}$ and a local maximum near $v_r \approx 1.5$, corresponding to $P_{r,\max}$. For any pressure within the range $[P_{r,\min}; P_{r,\max}]$, the van der Waals equation yields three real roots for the molar volume. This outcome is consistent with the nature of cubic equations, which can produce up to three distinct real solutions for given temperature and pressure conditions. However, this is inconsistent with the experimental behavior pictured in Fig. 7.3.

As illustrated in Fig. 7.4, the intermediate root lies in a region where the slope of the isotherm is positive ($dP/dv > 0$), which violates the criterion for mechanical stability and is therefore not physically meaningful. The smallest of the three roots corresponds to the liquid phase, while the largest represents the vapor phase. These two phases have the same temperature and pressure and correspond to phases in coexistence. However, for any temperature there are infinite pressures in the range $[P_{r,\min};$

$P_{r,\max}]$ that satisfy the same conditions of thermal and mechanical equilibrium. Which is the one with physical meaning that corresponds to two stable phases that coexist in equilibrium? This is the **saturation pressure** denoted by P^{sat} and is the unique pressure at which the two phases coexist in equilibrium at a given subcritical temperature.

7.3.5 Using the van der Waals equation to calculate the saturation pressure

The problem of calculating the saturation pressure has been addressed by Maxwell, and a neat solution has been proposed. For all pressures in the range $[P_{r,\min}; P_{r,\max}]$, with each one corresponding to two roots with physical meaning for the molar volume satisfying $v_L < v_G$ (L denotes liquid and G denotes gas), the saturation pressure is the only pressure that satisfies the equation

$$\int_{v_L(P^{\text{sat}})}^{v_L(P^{\text{sat}})} \left(P(v) - P^{\text{sat}}\right) dv = 0 \tag{7.3.35}$$

The physical interpretation of this condition is pictured in Fig. 7.5.

The Maxwell's condition for calculating the saturation pressure as expressed through the integral in Eq. (7.3.35) is not very practical. We can, however, substitute the pressure from the van der Waals equation and calculate the integral analytically

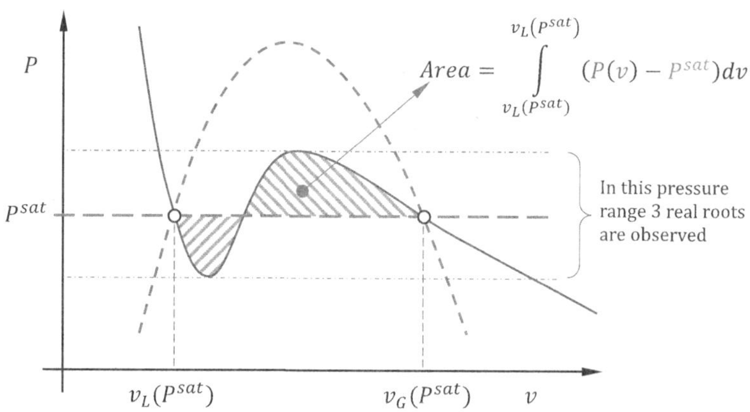

Figure 7.5 Explanation of the Maxwell condition for the saturation pressure.

$$\int_{v_L(P^{sat})}^{v_G(P^{sat})} \left(\frac{RT}{v-b} - \frac{a}{v^2} - P^{sat}\right) dv = RT \int_{v_L(P^{sat})}^{v_G(P^{sat})} d\ln(v-b) + a \int_{v_L(P^{sat})}^{v_G(P^{sat})} d\left(\frac{1}{v}\right) - P^{sat} \int_{v_L(P^{sat})}^{v_G(P^{sat})}$$

$$dv = \left(RT\ln(v_G - b) + \frac{a}{v_G} - P^{sat}v_G\right) - \left(RT\ln(v_L - b) + \frac{a}{v_L} - P^{sat}v_L\right)$$

$$= RT\left[\left(\ln(v_G - b) + \frac{a}{v_G RT} - \frac{P^{sat}v_G}{RT}\right) - \left(\ln(v_L - b) + \frac{a}{v_L RT} - \frac{P^{sat}v_L}{RT}\right)\right]$$

$$= RT\left[\left(\ln\left(Z_G\frac{RT}{P^{sat}} - \mathscr{B}\frac{RT}{P^{sat}}\right) + \frac{\mathscr{A}}{Z_G} - Z_G\right) - \left(\ln\left(Z_L\frac{RT}{P^{sat}} - \mathscr{B}\frac{RT}{P^{sat}}\right) + \frac{\mathscr{A}}{Z_L} - Z_L\right)\right]$$

$$= -RT\left[\left(Z_G - 1 - \ln(Z_G - \mathscr{B}) - \frac{\mathscr{A}}{Z_G}\right) - \left(Z_L - 1 - \ln(Z_L - \mathscr{B}) - \frac{\mathscr{A}}{Z_L}\right)\right] = 0$$

or

$$\psi(Z_G(P^{sat})) - \psi(Z_L(P^{sat})) = 0 \qquad (7.3.36)$$

where we have defined the function

$$\psi(Z) = (Z-1) - \ln(Z - \mathscr{B}) - \frac{\mathscr{A}}{Z} \qquad (7.3.37)$$

This equation is valid for the van der Waals equation only. $\psi(Z)$ turns out to be a more generally useful function as can be shown to be equal to the residual Gibbs function. Its exponential is defined as the **fugacity coefficient** denoted by φ

$$\ln\varphi(Z) = \psi(Z) = (Z-1) - \ln(Z - \mathscr{B}) - \frac{\mathscr{A}}{Z} \qquad (7.3.38)$$

We now summarize the **practical criterion for determining the saturation pressure of a pure component using a cubic equation of state** (EOS). When solving the cubic form of the EOS:

a. If only one real root exists, it is the only physically meaningful solution. If $T > T_C$, then the fluid is a supercritical fluid. If $T < T_C$, then the phase of the fluid can be inferred by comparing the compressibility factor Z to the critical compressibility factor Z_C:

- If $Z < Z_C = \frac{3}{8}$, the fluid is in the liquid phase
- If $Z > Z_C = \frac{3}{8}$, the fluid is in the vapor phase

b. If three real roots are obtained, the intermediate root is thermodynamically unstable and must be discarded. The remaining two roots represent candidate liquid and vapor phases. For each, the fugacity coefficient φ is computed. The thermodynamically stable phase at a given pressure and temperature is the one with the lower fugacity. The phase is determined by the fact that $Z_G > Z_L$. If the fugacity coefficients are equal, the two phases are in equilibrium.

The saturation pressure is the pressure at which the fugacity coefficients of the liquid and vapor roots are equal. This criterion ensures phase equilibrium and reconciles the phase identity based on the critical compressibility factor.

7.3.6 Using the van der Waals equation to calculate the residual enthalpy

To calculate the specific enthalpy, we need to apply the fundamental equation that has been presented in Chapter 5, i. e., Eq. (5.3.7)

$$h(T,P) = h_f^{ig}(T_{\text{ref}}) + \int_{T_{\text{ref}}}^{T} c_p^{ig}(\theta)d\theta + \Delta h^R(T,P) \tag{7.3.39}$$

where $h_f^{ig}\left(T_{\text{ref}}\right)$ is the enthalpy of formation of the component at the reference absolute temperature (usually $T_{\text{ref}} = 298.15$ K), $c_p^{ig}(T)$ is the heat capacity of the component at the ideal gas state and $\Delta h^R(T,P)$ is the **residual enthalpy** of the real fluid at absolute temperature T and pressure P. There is a fundamental equation that can be used to determine the residual enthalpy of a pure fluid using any equation of state. This equation is proved in most chemical thermodynamics books, but the proof is rather involved and lengthy. However, the result and its application are rather simple. The fundamental equation is the following

$$\Delta h^R(T,P) = RT(Z-1) + \int_{T,P\to 0}^{T,P} \left[T\left(\frac{\partial P}{\partial T}\right)_v - P\right] dv \qquad (7.3.40)$$

where the notation $()_v$ denotes that the differentiation is performed by keeping v constant. While this equation may be looking intimidating, the actual application is much simpler than it looks. Let us apply Eq. (7.3.40) for the case of the van der Waals EOS. We calculate the derivative of pressure (Eq. (7.3.2)) with respect to the temperature by keeping the molar volume constant

$$\left(\frac{\partial P}{\partial T}\right)_v = \frac{\partial}{\partial T}\left(\frac{RT}{(v-b)} - \frac{a}{v^2}\right)\bigg|_v = \frac{R}{(v-b)} \qquad (7.3.41)$$

We then note that

$$T\left(\frac{\partial P}{\partial T}\right)_v - P = \frac{RT}{(v-b)} - \left(\frac{RT}{(v-b)} - \frac{a}{v^2}\right) = \frac{a}{v^2} \qquad (7.3.42)$$

and

$$\Delta h^R_{vdW}(T,P) = RT(Z-1) + a\int_{T,P\to 0}^{T,P} \frac{dv}{v^2} = RT(Z-1) - a\int_{T,v\to+\infty}^{T,v} d\left(\frac{1}{v}\right)$$

$$= RT(Z-1) - a\cdot\left[\frac{1}{v}\right]_{v\to+\infty}^{v} = RT(Z-1) - \frac{a}{v} = RT(Z-1)$$

$$-RT\frac{a}{RT}\frac{P}{ZRT}$$

where the equation

$$Z = \frac{vP}{RT}$$

has been used. Finally

$$\frac{\Delta h^R_{vdW}(T,P)}{RT} = (Z-1) - \frac{\mathscr{A}}{Z} \qquad (7.3.43)$$

Example 7.5. Residual enthalpy calculation of isobutane using van der Waals EOS

The critical temperature and pressure for isobutane are T_C = 408.5 K and P_C = 3.648 MPa. Experimental data for the residual enthalpy of isobutane at 0.1 and 1 MPa and 400 K are

$h^R = -125$ and $h^R = -1170$ kJ/kmol, correspondingly. Use the van der Waals equation to estimate the residual enthalpy and calculate the error in the estimation.

Solution

The calculation is very lengthy and very error-prone, and we therefore use MATLAB

```
clear all
%% i-C4
PC = 36.48E5;   % critical pressure in Pa
TC = 408.05;    % critical temperature in K
Rg = 8.314463; % Pa m^3 / mol K = kJ/kmol K
%%
T  = 400;       % temperature in K at the experiment
P  = 10E5;      % pressure in Pa at the experiment
%%
a = (27/64)*(Rg*TC)^2/PC   % Pa m^6 mol^(-2)
b = (1/8)*Rg*TC/PC         % m^3 mol^(-1)
A = a*P/(Rg*T)^2           % dimensionless
B = b*P/(Rg*T)             % dimensionless
%                          % vdW polynomial Z^3+a1*Z^2+a2*Z+a3
r=roots([1 -(B+1) A -A*B])
Z=r(real(r)==r)
v  = Z*(Rg*T/P)*1000 % m^3/kmol
%%
HR = ((Z-1)-A/Z)*8.314*T  % residual enthalpy in kJ/kmol
e  = 100*((-1170)-HR)/(-1170)
```

The MATLAB results for the case of 1 MPa (10×10^5 Pa $= 1000$ kPa $= 10$ bar) are

```
>> residual_enthalpy_vdW

a =
    1.3311
b =
  1.1625e-04
A =
    0.1203
B =
    0.0350
r =
  0.9074 + 0.0000i
  0.0638 + 0.0239i
  0.0638 - 0.0239i
Z =
    0.9074
v =
    3.0180
HR =
 -748.8634
e
  -35.9946
```

The error is 36% and is significant, even though the isobutane is in the gaseous state.

Repeat the calculations manually to validate the results obtained by MATLAB.

7.3.7 MATLAB and Python implementation of the van der Waals EOS

A MATLAB implementation of the van der Waals equation of state is provided in Table 7.1. This function is designed to compute key thermodynamic properties of a fluid based on the following user-defined input parameters: the temperature and pressure at which the properties are to be evaluated, the critical temperature (in K), and the critical pressure (in kPa) of the fluid. Outputs from the function are: the compressibility factor, Z (dimensionless), the fugacity coefficient, (dimensionless), the residual enthalpy (in kJ/kmol), and a flag indicating the identified phase (vapor or liquid).

The implementation begins by calculating the constants a and b, which are derived from the fluid's critical properties. From these, the dimensionless parameters $\mathscr{A}$ and $\mathscr{B}$ are computed. These constants are then used to formulate the cubic equation in the compressibility factor Z, as described by Eq. (7.3.18). To solve the cubic equation, MATLAB's built-in `roots` function is employed. This command locates all roots of the polynomial, after which only the real roots are retained. The number of real roots determines the phase selection procedure: if a single real root exists, the phase is identified based on the critical compressibility factor; if multiple real roots are found, the one corresponding to the minimum fugacity coefficient is selected as the stable phase.

Once the appropriate root is selected, the function computes the fugacity coefficient and residual enthalpy corresponding to that phase.

This implementation is computationally efficient and can be applied to generate the phase diagram of any pure substance with known critical temperature and pressure. However, as discussed in the following sections and illustrated through numerical examples, the van der Waals EOS provides limited accuracy for predicting the thermodynamic properties of the liquid phase. In practical applications, more sophisticated models are typically preferred for accurate liquid-phase property estimation.

Despite its limitations, the van der Waals equation represents a significant milestone in thermodynamic modeling. It was among the first equations of state to account for non-ideal

Table 7.1 MATLAB implementation of the van der Waals EOS.

```
function [Z,IsLiquid,phi,hR]=vdW_EoS(T,P,Tc,Pc)
% implementation of the vdW EOS (IKK, June 2025)
%% INPUT ARGUMENTS
% T,Tc temperature and critical pressure [=] K
% P,Pc pressure and critical pressure    [=] kPa
%% VALUES RETURNED
% Z    the compressibility factor [=] dimensionless
% IsLiquid flag, 1:Liquid, 0: Gas
% phi fugacity coefficient
% hR  residual specific enthalpy [=] kJ/kmol

%%
Zc = 3/8;         % critical compressibility for the vdW EOS
R  = 8.31446262; % i.g. constant m^3 kPa/(kmol K) or kJ/(kmol K)
a = (27/64)*(R*Tc)^2/Pc ; A=a*P/(R*T)^2;
b = (1/8)*R*Tc/Pc ;       B=b*P/(R*T);
%% calculate all roots
r=roots([1 -(1+B) A -A*B]);
%% isolate real roots
Z=r(imag(r)==0);

%% count how many real roots
Nroots = length(Z);
if Nroots==1 % one real root
    IsLiquid=0;
    if Z<Zc
        IsLiquid=1;
    end
else % 3 real roots
    ZL=min(Z); ZG=max(Z);
    if exp(ZL-1-A/ZL-log(ZL-B))<exp(ZG-1-A/ZG-log(ZG-B))
       Z=ZL; IsLiquid =1;
    else
       Z=ZG; IsLiquid =0;
    end
end
phi=exp((Z-1)-A/Z-log(Z-B));
hR = R*T*((Z-1)-A/Z);
end % function
```

behavior in real gases and laid the groundwork for the development of more accurate models.

The MATLAB commands that follow are used to obtain the phase diagram of iso-butane shown in Fig. 7.6:

```
% i-C4
R  = 8.31446262; % i.g. constant m^3 kPa/(kmol K)
TC = 408.05;% K
PC = 3648;   % kPa
T  = linspace(100,500,21);
P=logspace(-1,4,1001);
for j=1:length(T)
   for i=1:length(P)
      [Z,IsLiquid,phi,hR]=vdW_EoS(T(j),P(i),TC,PC);
      v(i) = Z*((R*T(j))/P(i)); % m^3/kmol
   end
   loglog(v,P,'linewidth',1.2)
   hold on
end
axis([0.1 50 10 10000])
grid on
xlabel('Molar volume (m^{3}/kmol)')
ylabel('Pressure (kPa)')
title('Phase diagram of i-C_{4}')
```

In Table 7.2 an implementation of the van der Waals EOS in Python is presented. An example problem is also solved.

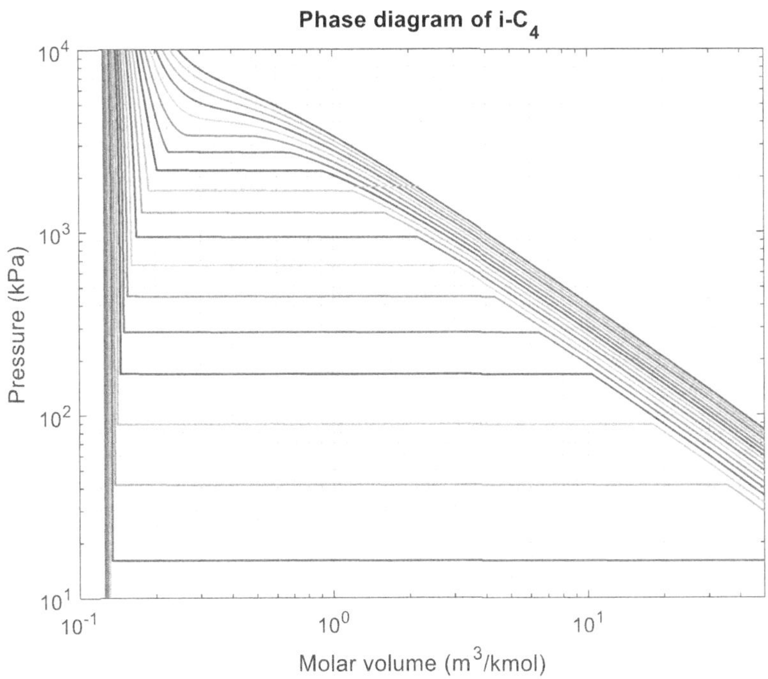

Figure 7.6 The phase diagram of iso-butane generated with the MATLAB file of Table 7.1.

Table 7.2 Python implementation of the van der Waals EOS.

```
"""
van der Waals EOS example problem @author: IKK, June 2025
"""
import math
import numpy as np

def vdW(T,P,Tc,Pc):
    Zc = 3.0/8.0     # critical compresibility for the vdW EOS
    R  = 8.31446262 # ideal gas constant in m^3 kPa / kmol K
    a  = (27.0/64.0)*(R*Tc)**2/Pc ; A  = a*P/(R*T)**2
    b  = (1.0/8.0)*(R*Tc)/Pc ;       B  = b*P/(R*T)
    r  = np.roots([1, -(1+B), A, -A*B])
    Z  = r[r.imag==0].real
    N  = len(Z)
    if N>=2:
        ZL  = min(Z)  ; phiL=math.exp(ZL-1-A/ZL-math.log(ZL-B))
        ZG  = max(Z)  ; phiG=math.exp(ZG-1-A/ZG-math.log(ZG-B))
        if (phiG<phiL):
            Z=ZG
        else:
            Z=ZL
    if (Z<Zc):
        isLiquid = 1
    else:
        isLiquid = 0
    hR = R*T*((Z-1)-A/Z)
    phi=math.exp(Z-1-A/Z-math.log(Z-B))
    return (Z,phi,hR,isLiquid)
'''
calculate the compressibility factor and the residual enthalpy of CO2 @
373.15 K & 40.53 bar exper.values Z=0.9012, h^R = -1150 kJ/kmol
'''
Tc = 304.2  # critial temperature [=] K
Pc = 7380.0 # critical pressure [=] kPa
T  = 373.15 # temperature in K
P  = 4053.0 # pressure [=] kPa
(Z,phi,hR,isLiquid)=vdW(T,P,Tc,Pc)
print('Compressibility factor=',Z,' [-]')
print('Residual enthalpy      =',hR,' kJ/kmol')
print('Is Liquid              =',isLiquid)
```

7.4 The Peng-Robinson equation of state

7.4.1 The fundamental equations of the Peng-Robinson EOS

The Peng-Robinson EOS is a recent variant of the van der Waals equation and has a different pressure correction term. Most equations that are derived from the van der Waals equation offer improvements that are related to the pressure correction term, which in the van der Waals equation is given by the constant a divided by the molar volume squared. In the Peng-Robinson equation the pressure correction term is a function of the temperature and the **acentric factor** ω.

The acentric factor, ω, is a dimensionless empirical parameter introduced by Kenneth Pitzer in 1955 to characterize the non-sphericity (or acentricity) of a molecule and its deviation from idealized behavior. It serves as a correction factor in generalized correlations for thermodynamic properties and equations of state. The acentric factor is defined based on the vapor pressure behavior of a substance:

$$\omega = -\log_{10}\left(\frac{P^{\text{sat}}(T_r = 0.7)}{P_C}\right) - 1 \tag{7.4.1}$$

For simple, nearly spherical molecules like argon or methane, ω is close to zero, but as molecular shape becomes more complex or polar (e.g., larger alkanes, polar compounds), ω increases. The acentric factor captures deviations from the behavior of "simple" fluids and is particularly useful for substances that do not conform well to the law of corresponding states. In cubic EOS models, ω helps tune the attractive term to match experimental vapor pressure data more accurately. For instance, in the Peng-Robinson equation, the temperature-dependent attraction parameter $a(T)$ is a function of ω.

The Peng-Robinson equation of state is the following

$$P = \frac{RT}{(v-b)} - \frac{a\tilde{a}(\omega, T_r)}{v^2 + 2bv - b^2}, \quad a = a(T_C, P_C), \quad b = b(T_C, P_C) \tag{7.4.2}$$

where

$$\tilde{a}(\omega, T_r) = \left[1 + \kappa(\omega)\left(1 - \sqrt{T_r}\right)\right]^2, \quad T_r = \frac{T}{T_C} \tag{7.4.3}$$

$$\kappa(\omega) = 0.37464 + 1.54226\omega - 0.26992\omega^2 \tag{7.4.4}$$

$a(T_C, P_C)$ and $b(T_C, P_C)$ are constants that vary with each substance and are derived from its critical temperature and critical pressure. Consistent with the procedure used for the van der Waals equation, these parameters are determined by requiring the equation of state to reproduce the critical point behavior of the fluid:

$$a(T_C, P_C) = 0.4572355289\left(\frac{R^2 T_C^2}{P_C}\right) \tag{7.4.5}$$

$$b(T_C, P_C) = 0.0777960739\left(\frac{R T_C}{P_C}\right) \tag{7.4.6}$$

The critical compressibility factor for the Peng-Robinson equation is

$$Z_C = 0.3074013087 \tag{7.4.7}$$

$\mathscr{A}$ and $\mathscr{B}$ are again defined as

$$\mathscr{A} = a(T_C, P_C)\,\tilde{a}(\omega, T_r)\frac{P}{(RT)^2} \tag{7.4.8}$$

$$\mathscr{B} = b(T_C, P_C)\frac{P}{RT} \tag{7.4.9}$$

and the cubic polynomial becomes

$$Z^3 - (1-\mathscr{B})Z^2 + \left(\mathscr{A} - 2\mathscr{B} - 3\mathscr{B}^2\right)Z - \left(\mathscr{A}\mathscr{B} - \mathscr{B}^2 - \mathscr{B}^3\right) = 0 \tag{7.4.10}$$

The equations for calculating residual enthalpy and the fugacity coefficient are

$$\frac{h_{PR}^R(T,P)}{RT} = (Z-1) - Q\left(1 + \kappa\sqrt{\frac{T_r}{\tilde{a}(\omega, T_r)}}\right) \tag{7.4.11}$$

$$\ln(\varphi) = (Z-1) - \ln(Z-\mathscr{B}) - Q \tag{7.4.12}$$

$$Q = \frac{\mathscr{A}}{\sqrt{8}\,\mathscr{B}}\ln\left[\frac{Z + (1+\sqrt{2})\mathscr{B}}{Z + (1-\sqrt{2})\mathscr{B}}\right] \tag{7.4.13}$$

7.4.2 MATLAB implementation of the Peng-Robinson EOS

A MATLAB/Python implementation of the Peng Robinson EOS is given in Table 7.3a and b. The implementation details follow closely the van der Waals implementation of Tables 7.1 and 7.2.

Table 7.3a MATLAB implementation of the Peng-Robinson EOS.

```
function [Z,IsLiquid,phi,hR]=PR_EoS(T,P,Tc,Pc,omega)
% implementation of the PR EOS (IKK, June 2025)
%% INPUT ARGUMENTS
% T,Tc temperature and critical pressure [=] K
% P,Pc pressure and critical pressure    [=] kPa
% omega acentric factor (as defined by Pitzer)
%% OUTPUT ARGUMENTS
% Z    the compressibility factor [=] dimensionless
% IsLiquid flag for liquid phase, 1:Liquid, 0:Gas, -1:Supercritical
% phi  fugacity coefficient [=] [-]
% hR   residual specific enthalpy [=] kJ/kmol
Zc = 0.307401308698704;    % critical compressibility for PREoS
R  = 8.31446262; % ideal gas constant m^3 kPa/(kmol K)
kappa = 0.37464+1.54226*omega-0.26992*omega^2;
aPR = (1+kappa*(1-sqrt(T/Tc)))^2; alpha = 0.4572355289*(R*Tc)^2/Pc;
a = alpha*aPR;             A = a*P/(R*T)^2;
b = 0.0777960739*R*Tc/Pc; B = b*P/(R*T);
C = 1+kappa*sqrt((T/Tc)/aPR);
% calculate all roots and isolate real roots
r=roots([1 -(1-B) (A-3*B^2-2*B) -(A*B-B^2-B^3)]);
Z=r( (imag(r)==0) & (real(r)>B));
% count how many real roots
Nroots = length(Z);
if Nroots==1 % one real root
    IsLiquid=0;
    if Z<Zc % check phase (works if subcritical state)
        IsLiquid=1;
    end
    if ((T>Tc)&(P>Pc)) % supercritical state
       IsLiquid = -1 ;
    end
else % 3 real roots
    ZL  = min(Z); ZG  = max(Z);
    QL  = log((ZL+(1+sqrt(2))*B)/(ZL+(1-sqrt(2))*B))*A/(B*sqrt(8));
    QG  = log((ZG+(1+sqrt(2))*B)/(ZG+(1-sqrt(2))*B))*A/(B*sqrt(8));
    phiL= exp(ZL-1-log(ZL-B)-QL); phiG= exp(ZG-1-log(ZG-B)-QG);
    % determine more stable phase
    Z=ZG;IsLiquid =0;
    if ((T>Tc)&(P>Pc)) % supercritical state
       Z = max(Z); IsLiquid = -1 ;
    elseif phiL<phiG
       Z=ZL; IsLiquid = 1;
    end
end
Q   =log((Z+(1+sqrt(2))*B)/(Z+(1-sqrt(2))*B))*A/(B*sqrt(8));
phi = exp(Z-1-log(Z-B)-Q);
hR  = R*T*((Z-1)-Q*C);
end % function
```

Table 7.3b Python implementation of the Peng-Robinson EOS.

```
def PR_EoS(T, P, Tc, Pc, omega):
    """
Peng-Robinson EOS
INPUTS :T, Tc  : temperature and critical temperature [K]
        P, Pc  : pressure and critical pressure [kPa]
        omega  : acentric factor
OUTPUTS:Z        : compressibility factor [-]
        IsLiquid : 1 = liquid, 0 = gas, -1 = supercritical
        phi      : fugacity coefficient [-]
        hR       : residual enthalpy [kJ/kmol]
    """
    Zc = 0.307401308698704  # critical compressibility for PR EOS
    R  = 8.31446262         # (m^3·kPa)/(kmol·K)
    r2 = np.sqrt(2.0)
    kappa = 0.37464 + 1.54226*omega - 0.26992*omega**2
    aPR = (1.0 + kappa*(1.0 - np.sqrt(T/Tc)))**2
    alpha = 0.4572355289 * (R*Tc)**2 / Pc   # a_c
    a = alpha * aPR ;               A = a * P / (R*T)**2
    b = 0.0777960739 * R * Tc / Pc; B = b * P / (R*T)
    C = 1.0 + kappa * np.sqrt( (T/Tc) / aPR )
    # Solve cubic equation for Z
    r = np.roots([1.0,-(1.0-B),A-3.*B**2-2.*B,-(A*B-B**2-B**3)])
    # Keep only real roots greater than B
    Z_real = np.real(r[np.isreal(r) & (np.real(r) > B)])
    Z_real = np.sort(Z_real)
    Nroots = len(Z_real)
    if Nroots == 1:   # 1 physical root
        Z = Z_real[0];  IsLiquid = 0  # default = vapor
        if (T < Tc) and (Z < Zc): # Subcritical liquid check: Z < Zc
            IsLiquid =  1
        if (T > Tc) and (P > Pc): # Supercritical region
            IsLiquid = -1
    else:             # 3 real roots: choose phase by fugacity
        ZL = Z_real[0]; ZG = Z_real[-1]
        QL = np.log((ZL+(1.0+r2)*B)/(ZL+(1.0-r2)*B))*A/(B*2.*r2)
        QG = np.log((ZG+(1.0+r2)*B)/(ZG+(1.0-r2)*B))*A/(B*2.*r2)
        phiL = np.exp(ZL - 1.0 - np.log(ZL - B) - QL)
        phiG = np.exp(ZG - 1.0 - np.log(ZG - B) - QG)
        Z = ZG ; IsLiquid = 0      # Default: vapor
        if (T > Tc) and (P > Pc): # Supercritical
            Z = ZG; IsLiquid = -1
        elif phiL < phiG:          # Liquid more stable
            Z = ZL; IsLiquid = 1
    Q = np.log((Z+(1.0+r2)*B)/(Z+(1.0-r2)*B))*A/(B*2.*r2)
    phi = np.exp(Z-1.0-np.log(Z-B)-Q)
    hR = R*T * ((Z - 1) - Q*C)  #  kJ/kmol
    return Z, IsLiquid, phi, hR
```

We now present several case studies to illustrate the improvements offered by the Peng-Robinson equation of state (EOS) over the classical van der Waals equation. Since its introduction in 1976, numerous equations of state have been proposed in the literature. While some offer enhanced accuracy for specific systems or conditions, relatively few achieve a consistent and systematic improvement across a broad range of applications while maintaining the simplicity and computational efficiency that characterize the Peng-Robinson EOS.

Example 7.6. Application of the Peng-Robinson equation

The critical temperature and pressure for CO_2 (molar mass 44.01) are $T_C = 304.1$ K and $P_C = 7370$ kPa ($\omega = 0.2389$) and for isobutane (molar mass 58.12) are $T_C = 408.05$ K and $P_C = 3648$ kPa ($\omega = 0.1848$). Selected experimental data for the molar volume, the residual enthalpy and the fugacity coefficient are:

		T (K)	P (kPa)	v (m³/kmol)	h^R (kJ/kmol)	φ
a	CO_2	220	5000	0.03743		
b	CO_2	303	7500	0.06600		
c	i-C_4	400	1000		−1170	
d	i-C_4	410	5000			0.519

Calculate the properties given using the Peng-Robinson equation of state and estimate the approximation error.

Solution

a. For this application we will present all intermediate steps.

$$a = 0.4572355289\left(\frac{R^2 T_C^2}{P_C}\right) = 0.4572355289\left(\frac{8.31446262^2 \times 304.1^2}{7370}\right)\left(\frac{\text{m}^3\text{kPa}}{\text{kmol K}}\right)^2 \cdot \frac{\text{K}^2}{\text{kPa}}$$

$$= 396.619161\,\frac{\text{m}^6\,\text{kPa}}{\text{kmol}^2}$$

$$\kappa(\omega) = 0.37464 + 1.54226\omega - 0.26992\omega^2$$

$$= 0.37464 + 1.54226 \times 0.2389 - 0.26992 \times 0.2389^2 = 0.727681$$

$$\tilde{a}(\omega, T_r) = \left[1 + \kappa(\omega)\left(1 - \sqrt{T_r}\right)\right]^2 = \left[1 + 0.727681\left(1 - \sqrt{\frac{220}{304.1}}\right)\right]^2 = 1.229320$$

$$\mathscr{A} = a\tilde{a}(\omega, T_r)\frac{P}{(RT)^2} = 396.619161\ \frac{\text{m}^6\ \text{kPa}}{\text{kmol}^2} \times 1.229320$$
$$\times \frac{5000\ \text{kPa}}{(8.31446262 \times 220)^2}\ \frac{\text{kmol}^2}{\text{m}^6\ \text{kPa}^2} = 0.728610$$

$$b = 0.0777960739\left(\frac{RT_C}{P_C}\right)$$
$$= 0.0777960739\left(\frac{8.31446262 \times 304.1}{7370}\right)\frac{\text{m}^3}{\text{kmol}} = 0.026690\ \frac{\text{m}^3}{\text{kmol}}$$

$$\mathscr{B} = b\frac{P}{RT} = 0.026690 \times \frac{5000}{8.31446262 \times 220} = 0.072955$$

The cubic polynomial

$$Z^3 - (1 - \mathscr{B})Z^2 + \left(\mathscr{A} - 2\mathscr{B} - 3\mathscr{B}^2\right)Z - \left(\mathscr{A}\mathscr{B} - \mathscr{B}^2 - \mathscr{B}^3\right) = 0$$

becomes

$$f(Z) = Z^3 - 0.927Z^2 + 0.5667Z - 0.0474 = 0$$

and

$$f'(Z) = 3Z^2 - 1.854Z + 0.5667$$

We start from $Z^1 = 0.1$ and we calculate

$$f(Z^1) = 0.1^3 - 0.927 \times 0.1^2 + 0.5667 \times 0.1 - 0.0474 = 0.001$$

and

$$f'(Z^1) = 3 \times 0.1^2 - 1.854 \times 0.1 + 0.5667 = 0.4113$$

$$Z^2 = Z^1 - \frac{f(Z^1)}{f'(Z^1)} = 0.09757$$

$$f(Z^2) = 0.1^3 - 0.927 \times 0.1^2 + 0.5667 \times 0.1 - 0.0474$$
$$= -3.1757 \times 10^{-6}$$

We finally calculate the molar volume

$$v = Z\left(\frac{RT}{P}\right) = 0.09757 \times \left(\frac{8.31446262 \times 220}{5000\ \text{kPa}}\right) = 0.0357\ \frac{\text{m}^3}{\text{kmol}}$$

The percent error is $\frac{100(0.03743-0.0357)}{0.03743} = 4.6\%$. The error is relatively significant as CO_2 in the specific conditions is in the liquid state.

b. We will now use MATLAB:

```
>> clear all
>> TC=304.1;
>> PC=7370;
>> omega=0.2389;
>> T=303;
>> P=7500;
>> [Z,IsLiquid,phi,hR]=PR_EoS(T,P,TC,PC,omega)

Z =        0.2257
IsLiquid =      1
phi =     0.6265
hR =     -8.9756e+03

>> R=8.31446262;
>> v=Z*R*T/P
v =        0.0758
>> e=100*(0.066-v)/0.066
e =     -14.8454
```

c. (all intermediate calculations are given so that you can perform the calculations manually):

```
>> TC=408.05;
>> PC=3648;
>> omega=0.1848;
>> T=400;
>> P=1000;
>> [Z,IsLiquid,phi,hR]=PR_EoS(T,P,TC,PC,omega)

kappa = 0.6504
aPR   = 1.0129
alpha = 1.4427e+03
a     = 1.4614e+03
A =      0.1321
b =      0.0724
B =      0.0218
ans =    1.0000    -0.9782     0.0872    -0.0024
Z =      0.8825
IsLiquid =       0
phi =    0.8925
hR =    -1.1879e+03
>> e=100*(hR-(-1170))/ (-1170)
e =      1.5299
```

d.

```
>> T=410;
>> P=5000;
>> [Z,IsLiquid,phi,hR]=PR_EoS(T,P,TC,PC,omega)

kappa = 0.6504
aPR =   0.9969
alpha = 1.4427e+03
a =     1.4382e+03
A =     0.6188
b =     0.0724
B =     0.1061
ans =   1.0000    -0.8939    0.3728    -0.0532
Z =     0.2523
IsLiquid =     -1
phi =   0.5146
hR =   -1.2916e+04
>> e=100*(0.519-0.5146)/0.5146
e =     0.8550
```

Note that the errors for the cases c (superheated vapor) and d (supercritical fluid) are very small but for case b (subcooled liquid) are significant.

The MATLAB/Python implementation of the Peng-Robinson equations can be very useful for determining the phase diagram of pure chemical components. The script that follows has been used to construct the phase diagram of CO_2 pictured in Fig. 7.7, which should be compared with Fig. 7.3:

```
clear all
% CO2
R  = 8.31446262; % i.g. constant m^3 kPa/(kmol K)
TC = 304.1;% K
PC = 7370;  % kPa
omega = 0.2389;
T  = linspace(220,350,27);
P=logspace(2,log10(10000),1001);
for j=1:length(T)
   for i=1:length(P)
      [Z,IsLiquid,phi,hR]=PR_EoS(T(j),P(i),TC,PC,omega);
      v(i) = Z*((R*T(j))/P(i)); % m^3/kmol
   end
   semilogx(v,P/1000,'linewidth',1.2)
   hold on
end
axis([0.03 0.4 0 10])
grid on
xlabel('Molar volume (m^{3}/kmol)')
ylabel('Pressure (MPa)')
title('Phase diagram of CO_{2}')
```

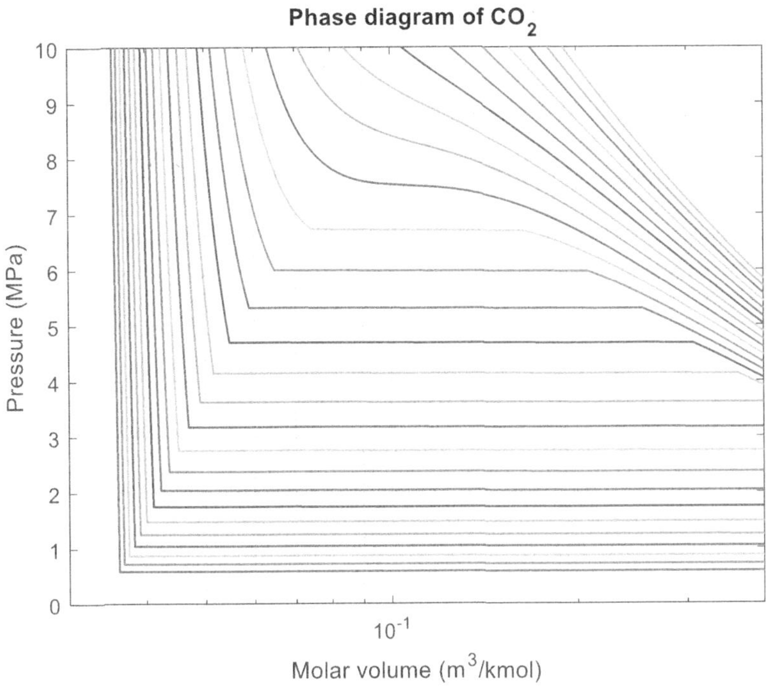

Figure 7.7 Phase diagram of carbon dioxide using the Peng-Robinson EOS.

7.5 Predicting properties of pure components using group contribution methods

The availability of highly accurate equations of state (EOS) is a powerful asset in chemical engineering, enabling the determination of key thermodynamic properties of process streams, including:

- the physical phase (liquid, vapor, or two-phase),
- specific volume and density,
- specific enthalpy and internal energy.

Additional properties, such as the enthalpy of vaporization, can also be estimated indirectly using these models.

A reliable implementation of an accurate EOS—such as the Peng-Robinson equation—opens the door to solving complex and realistic process simulation problems. One such case study will be explored at the end of this chapter.

But what if the critical constants required to use an EOS for a specific chemical component are not available in common reference sources like the NIST Chemistry WebBook or Perry's

Chemical Engineers' Handbook? Addressing this challenge highlights a classic and ingenious example of practical chemical engineering in action.

Imagine you've been asked to determine the critical properties and boiling point of n-octane (CH_3–$(CH_2)_6$–CH_3). However, after an extensive search, it appears that this specific compound has not been previously studied in the literature, and the required data are nowhere to be found. While attempting to locate the properties of n-octane, you manage to find reliable data for all the preceding members of its homologous series, beginning with ethane. These values have been compiled and presented in Table 7.4. Can you estimate the boiling point temperature and critical properties based on Table 7.4?

Note that the boiling point temperature is consistently increasing as -CH_2- structural fragments are added starting from propane to heptane. Let us assume that the addition of each methylene increases the boiling point by a fixed number. Then we may find this contribution of methylene to the boiling point by noticing that when we move from zero methylene fragments in ethane to 5 methylene fragments in heptane the boiling point increases from 184.6 to 371.5 K. We may therefore assume that each methylene increases the boiling point by

$$\delta T_b(-\mathrm{CH_2}-) = \frac{371.5 - 184.6}{5 - 0} = 37.4\ \mathrm{K}$$

Then we can estimate the boiling point of octane

$$T_b(\mathrm{C_8H_{18}}) = T_b(\mathrm{C_2H_6}) + 6 \times 37.4\ \mathrm{K} = 184.6 + 6 \times 37.4\ \mathrm{K} = 409\ \mathrm{K}$$

Table 7.4 Normal boiling point temperature and critical temperature and pressure of some straight-chain alkanes.

Name	Condensed structural formula	T_b (K)	T_C (K)	P_C (bar)
Ethane	CH_3–CH_3	184.6	305.3	49.0
n-Propane	CH_3–(CH_2)–CH_3	231.1	369.9	42.5
n-Butane	CH_3–(CH_2)–(CH_2)–CH_3	273.0	425.0	38.0
n-Pentane	CH_3–(CH_2)–(CH_2)–(CH_2)–CH_3	309.0	469.8	33.6
n-hexane	CH_3–(CH_2)–(CH_2)–(CH_2)–(CH_2)–CH_3	341.9	507.6	30.2
n-heptane	CH_3–(CH_2)–(CH_2)–(CH_2)–(CH_2)–(CH_2)–CH_3	371.5	540.0	27.4
n-octane	CH_3–(CH_2)–(CH_2)–(CH_2)–(CH_2)–(CH_2)–(CH_2)–CH_3	?	?	?

Using similar arguments:

$$\delta T_C(-CH_2-) = \frac{540 - 305.3}{5 - 0} = 47.94 \text{ K}$$

$$\delta P_C(-CH_2-) = \frac{27.4 - 49}{5 - 0} = -4.32 \text{ K}$$

and

$$T_C(C_8H_{18}) = T_C(C_2H_6) + 6 \times 47.94 \text{ K} = 305.3 + 6 \times 47.94 \text{ K} = 592.9 \text{ K}$$

$$P_C(C_8H_{18}) = P_C(C_2H_6) + 6 \times (-4.32) \text{ K} = 49 + 6 \times (-4.32) \text{ K} = 23.1 \text{ bar}$$

Since octane is a very well known organic compound, we can find that

$$T_b(C_8H_{18}) = 398.7 \text{ K}, T_C(C_8H_{18}) = 568.9 \text{ K}, P_C(C_8H_{18}) = 24.9 \text{ bar}$$

The errors are of the order of 2%–7%. Not very satisfactory but at least we have a good estimation of the properties of the component.

This concept can be generalized by assuming that each structural fragment—such as CH_3–, $=C=$, –OH, and others—contributes a fixed amount to certain thermophysical properties of a compound. These contributions may be subject to a few qualifying conditions, such as whether the fragment is part of a ring structure. Despite these simplifications, the idea has proven to be highly valuable and has attracted significant attention within the scientific community and significant results have been published. These techniques are known as **group contribution methodologies**.

One of the earliest and most influential methods based on this principle is the Joback-Reid method, which itself builds upon earlier work by Lydersen. The Joback-Reid method assumes that a given property is a strictly linear function of the number of structural groups present. It treats the total property value as a purely additive sum of individual group contributions, while neglecting any interactions between the groups or between molecules. While this approach offers simplicity and ease of use, it is important to recognize its limitations. In reality, interactions between **structural fragments** can be significant, and the contribution of each group is not always constant, even within members of the same homologous series.

The method of Joback and Reid is summarized in Table 7.5. The best way to demonstrate the method is through an example. Consider determining the boiling point temperature, the critical

Table 7.5 Joback and Reid, group contribution method: contribution of individual groups.

Group	δT_c	δP_c	δV_c	δT_b	δh_f	δa	δb	δc	δd	δh_{vap}
Non-ring groups										
$-CH_3$	0.0141	−0.0012	65	23.58	−76.45	1.95E+1	−8.08E−3	1.53E−4	−9.67E−8	2.373
$-CH_2-$	0.0189	0.0000	56	22.88	−20.64	−9.09E−1	9.50E−2	−5.44E−5	1.19E−8	2.226
$>CH-$	0.0164	0.0020	41	21.74	29.89	−2.30E+1	2.04E−1	−2.65E−4	1.20E−7	1.691
$>C<$	0.0067	0.0043	27	18.25	82.23	−6.62E+1	4.27E−1	−6.41E−4	3.01E−7	0.636
$=CH_2$	0.0113	−0.0028	56	18.18	−9.630	2.36E+1	−3.81E−2	1.72E−4	−1.03E−7	1.724
$=CH-$	0.0129	−0.0006	46	24.96	37.97	−8.00	1.05E−1	−9.63E−5	3.56E−8	2.205
$=C<$	0.0117	0.0011	38	24.14	83.99	−2.81E+1	2.08E−1	−3.06E−4	1.46E−7	2.138
$=C=$	0.0026	0.0028	36	26.15	142.14	2.74E+1	−5.57E−2	1.01E−4	−5.02E−8	2.661
$\equiv CH$	0.0027	−0.0008	46	9.20	79.30	2.45E+1	−2.71E−2	1.11E−4	−6.78E−8	1.155
$\equiv C-$	0.0020	0.0016	37	27.38	115.51	7.87	2.01E−2	−8.33E−6	1.39E-9	3.302
Ring groups										
$-CH_2-$	0.0100	0.0025	48	27.15	−26.80	−6.03	8.54E−2	−8.00E−6	−1.80E−8	2.398
$>CH-$	0.0122	0.0004	38	21.78	8.67	−2.05E+1	1.62E−1	−1.60E−4	6.24E−8	1.942
$>C<$	0.0042	0.0061	27	21.32	79.72	−9.09E+1	5.57E−1	−9.00E−4	4.69E−7	0.644
$=CH-$	0.0082	0.0011	41	26.73	2.09	−2.14	5.74E−2	−1.64E−6	−1.59E−8	2.544
$=C<$	0.0143	0.0008	32	31.01	46.43	−8.25	1.01E−1	−1.42E−4	6.78E−8	3.059
Halogen groups										
$-F$	0.0111	−0.0057	27	−0.03	−251.92	2.65E+1	−9.13E−2	1.91E−4	−1.03E−7	−0.670
$-Cl$	0.0105	−0.0049	58	38.13	−71.55	3.33E+1	−9.63E−2	1.87E−4	−9.96E−8	4.532
$-Br$	0.0133	0.0057	71	66.86	−29.48	2.86E+1	−6.49E−2	1.36E−4	−7.45E−8	6.582
$-I$	0.0068	−0.0034	97	93.84	21.06	3.21E+1	−6.41E−2	1.26E−4	−6.87E−8	9.520
Oxygen groups										
$-OH$ (alcohol)	0.0741	0.0112	28	92.88	−208.04	2.57E+1	−6.91E−2	1.77E−4	−9.88E−8	16.826
$-OH$ (phenol)	0.0240	0.0184	−25	76.34	−221.65	−2.81	1.11E−1	−1.16E−4	4.94E−8	12.499
$-O-$ (non-ring)	0.0168	0.0015	18	22.42	−132.22	2.55E+1	−6.32E−2	1.11E−4	−5.48E−8	2.410
$-O-$ (ring)	0.0098	0.0048	13	31.22	−138.16	1.22E+1	−1.26E−2	6.03E−5	−3.86E−8	4.682
$>C{=}O$ (non-ring)	0.0380	0.0031	62	76.75	−133.22	6.45	6.70E−2	−3.57E−5	2.86E−9	8.972
$>C{=}O$ (ring)	0.0284	0.0028	55	94.97	−164.50	3.04E+1	−8.29E−2	2.36E−4	−1.31E−7	6.645
$O{=}CH-$ (aldehyde)	0.0379	0.0030	82	72.24	−162.03	3.09E+1	−3.36E−2	1.60E−4	−9.88E−8	9.093
$-COOH$ (acid)	0.0791	0.0077	89	169.09	−426.72	2.41E+1	4.27E−2	8.04E−5	−6.87E−8	19.537
$-COO-$ (ester)	0.0481	0.0005	82	81.10	−337.92	2.45E+1	4.02E−2	4.02E−5	−4.52E−8	9.633
$=O$ (other)	0.0143	0.0101	36	−10.50	−247.61	6.82	1.96E−2	1.27E−5	−1.78E−8	5.909

Table 7.5 Joback and Reid, group contribution method: contribution of individual groups.—*continued*

Group	δT_c	δP_c	δV_c	δT_b	δh_f	δa	δb	δc	δd	δh_{vap}
Nitrogen groups										
$-NH_2$	0.0243	0.0109	38	73.23	−22.02	2.69E+1	−4.12E−2	1.64E−4	−9.76E−8	10.788
>NH (non-ring)	0.0295	0.0077	35	50.17	53.47	−1.21	7.62E−2	−4.86E−5	1.05E−8	6.436
>NH (ring)	0.0130	0.0114	29	52.82	31.65	1.18E+1	−2.30E−2	1.07E−4	−6.28E−8	6.930
>N− (non-ring)	0.0169	0.0074	9	11.74	123.34	−3.11E+1	2.27E−1	−3.20E−4	1.46E−7	1.896
−N= (non-ring)	0.0255	−0.0099	n. a.	74.60	23.61	n. a.	n. a.	n. a.	n. a.	3.335
−N= (ring)	0.0085	0.0076	34	57.55	55.52	8.83	−3.84E-3	4.35E−5	−2.60E−8	6.528
=NH	n. a.	n. a.	n. a.	83.08	93.70	5.69	−4.12E−3	1.28E−4	−8.88E−8	12.169
−CN	0.0496	−0.0101	91	125.66	88.43	3.65E+1	−7.33E−2	1.84E−4	−1.03E−7	12.851
$-NO_2$	0.0437	0.0064	91	152.54	−66.57	2.59E+1	−3.74E−3	1.29E−4	−8.88E−8	16.738
Sulfur groups										
−SH	0.0031	0.0084	63	63.56	−17.33	3.53E+1	−7.58E−2	1.85E−4	−1.03E−7	6.884
−S− (non-ring)	0.0119	0.0049	54	68.78	41.87	1.96E+1	−5.61E−3	4.02E−5	−2.76E−8	6.817
−S− (ring)	0.0019	0.0051	38	52.10	39.10	1.67E+1	4.81E−3	2.77E−5	−2.11E−8	5.984

$$T_c[\mathrm{K}] = T_b\left[0.584 + 0.965\sum \delta T_{c,i} - \left(\sum \delta T_{c,i}\right)^2\right]^{-1}$$

$$P_c[\mathrm{bar}] = \left[0.113 + 0.0032N_a - \sum \delta P_{c,i}\right]^{-2},\ N_a = \text{number of atoms in the molecule (including H)}$$

$$v_c\left[\frac{\mathrm{cm}^3}{\mathrm{mol}}\right] = 17.5 + \sum \delta v_{c,i}$$

$$T_b[\mathrm{K}] = 198.2 + \sum \delta T_{b,i}$$

$$h_f^{ig}(298.15\ \mathrm{K})\left[\frac{\mathrm{kJ}}{\mathrm{mol}}\right] = 68.29 + \sum \delta h_{f,i}$$

$$c_p\left[\frac{\mathrm{J}}{\mathrm{mol\ K}}\right] = \left(\sum \delta a_i - 37.93\right) + \left(\sum \delta b_i + 0.21\right)T + \left(\sum \delta c_i - \frac{3.91}{10^4}\right)T^2 + \left(\sum \delta d_i + \frac{2.06}{10^7}\right)T^3$$

$$\Delta h_{vap}(@T_{nb})\left[\frac{\mathrm{kJ}}{\mathrm{mol}}\right] = 15.3 + \sum \delta h_{vap,i}$$

temperature, the critical pressure, the standard heat of formation of ideal gas and the heat of evaporation at the normal boiling point for Acetic acid, ethyl ester

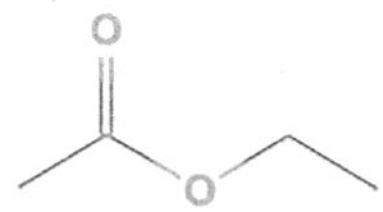

The compound has the following structural fragments, among the ones that appear in Table 7.5:

- 2 fragments of CH_3-
- 1 fragment of –CH2–
- 1 fragment of >C=O
- 1 fragment of –O–

Using these fragments and the corresponding contribution numbers from Table 7.5 we calculate:

Fragment	#	δT_b	δT_C	δP_C	δh_f	δh_{vap}
CH_3-	2	$2 \times 23.58 = 47.16$	$2 \times 0.0141 = 0.0282$	$2 \times (-0.0012) = -0.0024$	$2 \times (-76.45) = -152.9$	$2 \times (2.373) = 4.746$
$-CH_2-$	1	22.88	0.0189	0.0000	−20.64	2.226
>C=O	1	76.75	0.0380	0.0031	−133.22	8.972
–O–	1	22.42	0.0168	0.0015	−132.22	2.410
TOTAL		169.21	0.1019	0.0022	−438.98	18.354

$$T_b[\mathrm{K}] = 198.2 + \sum \delta T_{b,i} = 198.2 + 169.21 = 367.41\ \mathrm{K}$$

$$T_c[\mathrm{K}] = T_b\left[0.584 + 0.965 \sum \delta T_{c,i} - \left(\sum \delta T_{c,i}\right)^2\right]^{-1} = 546.8\ \mathrm{K}$$

$$P_c[\mathrm{bar}] = \left[0.113 + 0.0032\,\underbrace{N_a}_{=14} - \sum \delta P_{c,i}\right]^{-2} = 41.3\ \mathrm{bar}$$

$$h_f^{ig}(298.15\ \mathrm{K})\left[\frac{\mathrm{kJ}}{\mathrm{mol}}\right] = 68.29 + \sum \delta h_{f,i} = -370.7\ \frac{\mathrm{kJ}}{\mathrm{mol}}$$

$$h_{\mathrm{vap}}(@T_{nb})\left[\frac{\mathrm{kJ}}{\mathrm{mol}}\right] = 15.3 + \sum \delta h_{\mathrm{vap},i} = 33.654\ \frac{\mathrm{kJ}}{\mathrm{mol}}$$

From the NIST Chemistry WebBook we find that experimental data are:

$$T_b[\mathrm{K}] = 350.2\ \mathrm{K}$$

$$T_c[\mathrm{K}] = 530\ \mathrm{K}$$

$$P_c[\mathrm{bar}] = 38.82\ \mathrm{bar}$$

$$h_f^{ig}(298.15\ \mathrm{K}) = -445.4\ \frac{\mathrm{kJ}}{\mathrm{mol}}$$

$$h_{\mathrm{vap}}(@T_{nb})\left[\frac{\mathrm{kJ}}{\mathrm{mol}}\right] = 35.2\ \frac{\mathrm{kJ}}{\mathrm{mol}}$$

7.6 An advanced application: Filling an hydrogen Tank[1]

In recent years, there has been significant discussion around technological solutions to address climate change. One promising approach is the use of hydrogen as an energy carrier, particularly in mobile applications such as transportation and hydrogen-powered vehicles. In these applications, hydrogen tanks will replace conventional petrol and diesel storage tanks in vehicles. However, several significant challenges must be overcome before hydrogen can be widely adopted as a transportation fuel. One key issue is the rapid temperature rise in the hydrogen tank during refueling, caused by the compression of hydrogen.

Type III hydrogen tanks are high-pressure gas storage vessels commonly used in hydrogen-powered vehicles and other mobile applications. They consist of a metal liner, typically made of aluminum, which provides a gas-tight barrier, wrapped with a full composite overwrap of carbon fiber-reinforced polymer (CFRP) for strength and durability. These tanks are designed to store hydrogen at pressures of up to 700 bar, making them suitable for automotive applications where lightweight and high-pressure storage is essential. The metal liner helps manage hydrogen permeability and ensures a robust internal structure, while the carbon fiber overwrap significantly reduces the overall weight compared to all-metal tanks.

The filling process of Type III hydrogen tanks has been the subject of extensive research and numerous experimental

[1]As this section addresses a more advanced topic, it may be omitted during an initial reading of the book.

Table 7.6 Tank III specifications, thermal properties, and filling test specifications.

Variable	Symbol	Numerical value	Units
Tank			
Tank Volume	V	0.034	m^3
Inner diameter	D	0.240	m
Inner length	L	0.750	m
Inner heat transfer area	A_i	0.656	m^2
Initial temperature	$T(0)$	20	°C
Initial pressure	$P(0)$	4750 (47.5)	kPa (bar)
Liner			
Thickness	δ_l	0.003	m
Thermal conductivity	k_l	180	$W \cdot m^{-1}K^{-1}$
Sp. heat capacity	$c_{p,l}$	896	$J \cdot kg^{-1}K^{-1}$
Density	ρ_l	2700	$kg \cdot m^{-3}$
Mass	m_l	is calculated	kg
Inner heat transfer area	a_i	0.656	m^2
CFRP			
Thickness	δ_{CF}	0.017	m
Thermal conductivity	k_{CF}	1.14	$W \cdot m^{-1}K^{-1}$
Sp. heat capacity	$c_{p,CF}$	1075	$J \cdot kg^{-1}K^{-1}$
Density	ρ_{CF}	1374	$kg \cdot m^{-3}$
External heat transfer area	A_o	0.8181	m^2
Mass	m_{CF}	is calculated	kg
External heat transfer coefficient	a_o	4.5	$W \cdot m^{-2}K^{-1}$
Filling process			
Hydrogen molar flowrate	$\dot{N}_{in}$	2.91	$mol \cdot s^{-1}$
Hydrogen temperature	T_{in}	22	°C
Environmental temperature	T_{air}	5	°C

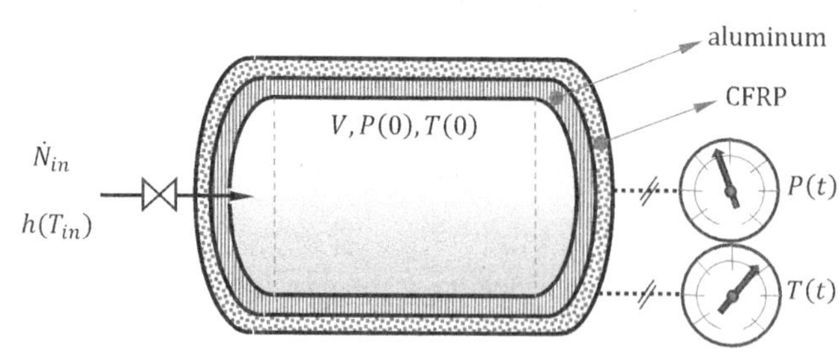

Figure 7.8 Tank III specifications, thermal properties, and filling test specifications.

data are available. On a particular case the experimental data have become available to the open literature (Monde, et al., 2012, *International J. of Hydrogen Energy*, 37, pp 5723-5734). Technical data for the filling process are summarized in Table 7.6. The process is pictured in Fig. 7.8. A stream of constant molar flowrate of hydrogen is fed to the Type III tank resulting in a simultaneous increase in temperature and pressure inside the tank. The aim is to develop a total mole balance for hydrogen, an energy balance for hydrogen, and an energy balance for the composite wall of the tank (the notation used is introduced in Table 7.6).

We will first develop the simplest possible model that does have an analytic solution, but it fails to represent reality. In this model we assume that the process is fast enough to be considered as an adiabatic process. In this case the model is described by the total mole balance:

$$\frac{dN(t)}{dt} = \dot{N}_{in} - \dot{N}_{\text{out}} \tag{7.6.1}$$

As there is no stream exiting the tank is simplifies to

$$\frac{dN(t)}{dt} = \dot{N}_{in} \tag{7.6.2}$$

and as $\dot{N}_{in}$ is constant it follows that

$$N(t) = N(t) + \dot{N}_{in}t \tag{7.6.3}$$

For the energy balance, we start with the general energy balance equation

$$\frac{dU(t)}{dt} = \dot{N}_{in}h(T_{in}) - \dot{N}_{\text{out}}h(t) + \dot{Q} + \dot{W}_s \tag{7.6.4}$$

there is no exiting stream ($\dot{N}_{\text{out}} = 0$), we have assumed that the process is adiabatic ($\dot{Q} = 0$) and there is no work interaction ($\dot{W}_s = 0$)

$$\frac{dU(t)}{dt} = \dot{N}_{in}h(T_{in}) \tag{7.6.5}$$

The right-hand side is constant and we obtain the solution

$$N(t)u(t) = N(0)u(0) + \dot{N}_{in}h(T_{in})t \tag{7.6.6}$$

To get an initial solution, we may assume ideal gas behavior with

$$h(t) = h_{\text{ref}} + c_p(T - T_{\text{ref}}) \tag{7.6.7}$$

If we consider $h_{\text{ref}} = 0$ at $T_{\text{ref}} = 298.15$, then

$$u(t) = u_{\text{ref}} + c_v(T - T_{\text{ref}}) \tag{7.6.8}$$

Since from the definition of enthalpy

$$h_{\text{ref}} = u_{\text{ref}} + RT_{\text{ref}}$$

it follows that

$$u(t) = -RT_{\text{ref}} + c_v(T - T_{\text{ref}}) \tag{7.6.9}$$

We replace Eqs. (7.6.7) and (7.6.9) into Eq. (7.6.5) to obtain

$$-N(t)RT_{\text{ref}} + N(t)c_v(T(t) - T_{\text{ref}}) = -N(0)RT_{\text{ref}} + N(0)c_v(T(0) - T_{ref}) + \dot{N}_{in}c_p(T_{in} - T_{\text{ref}})t$$

After some algebraic manipulations

$$T(t) = T_{\text{ref}} + \left(\frac{\dot{N}_{in}}{N(0) + \dot{N}_{in}t}\right)\left[\frac{c_p}{c_v}T_{in} - T_{\text{ref}}\right]t + \left(\frac{N(0)}{N(0) + \dot{N}_{in}t}\right)(T(0) - T_{\text{ref}}) \tag{7.6.10}$$

```
clear all
% general constants
Rg    = 8.31446261815324;    % J / mol K
Rgv   = Rg/1e5 ;             % bar m^3 / K mol
Cp    = (7/2)*Rg; Cv    = (5/2)*Rg;
T_ref = 25+273.15;           % K
mw    = 2.01568;             % g / mol
% gas in the tank at t=0
T_0   = 20+273.15;           % K
P_0   = 21.0;                % bar
V     = 0.034;               % m^3
N_0   = P_0*V/(Rgv*T_0);     % mol
% feed
T_in  = 22+273.15;           % K
P_in  = 350;                 % bar
N_in  = 2.91;                % mol/s
tf    = 110;                 % s

t=linspace(0,tf,31);
T=T_ref+(N_in./(N_0+N_in*t)).*((Cp*T_in/Cv-T_ref)*t+(T_0-T_ref));
P=(N_0+N_in*t)*Rgv.*T/V;

figure(1)
plot(t,T-273.15,'-o','linewidth',1.2,'Color',[0,0.45,0.74])
grid on
xlabel('Time (s)')
ylabel('Temperature (degre celsius)')
text(10,120,'Adiabatic, ideal gas','Color',[0,0.45,0.74])
text(20,55,'Experimental data','Color',[0.64,0.08,0.18])
figure(2)
plot(t,P,'-o','linewidth',1.2)
grid on
xlabel('Time (s)')
ylabel('Pressure (bar)')
axis([0 110 0 400])
```

To determine the pressure, we may use the ideal gas law

$$P(t)V = \left(N(0) + \dot{N}_{in}t\right)RT(t) \tag{7.6.11}$$

These two equations can be used to predict the evolution of temperature and pressure in the hydrogen tank:

The response of the temperature of the hydrogen in the tank is shown in Fig. 7.9, while the evolution of the pressure is shown in Fig. 7.10. While pressure is close to the reported value of 350 bar at the end of the filling process, the temperature deviates significantly from the experimental results shown also in Fig. 7.9. There are two reasons for this. The first one is related to the ideal gas assumption, which is not valid for the pressures observed in this process, and the second one is the adiabatic process assumption.

To relax the assumption of adiabatic operation, we model the composite tank walls as having a uniform temperature and allow for heat exchange between the gas and the tank walls, as well as between the walls and the surrounding environment. Heat transfer from the gas to the walls occurs over an internal surface area A_i with a heat transfer coefficient a_i, while heat loss from the walls to the ambient air takes place across an external area A_o with coefficient a_o (see Fig. 7.11).

In place of the ideal gas assumption, we employ the Peng-Robinson equation of state to more accurately calculate the

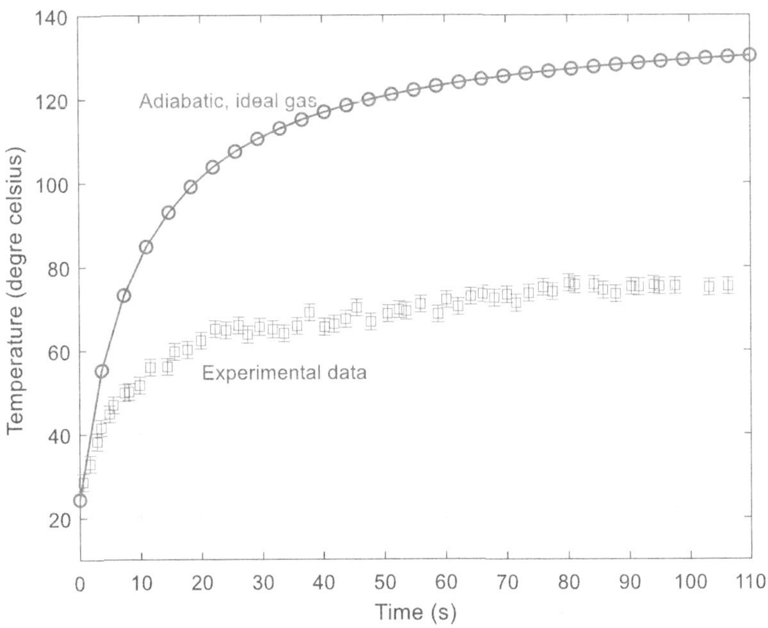

Figure 7.9 Evolution of the temperature in the hydrogen storage tank under the assumptions of adiabatic operation and ideal gas behavior.

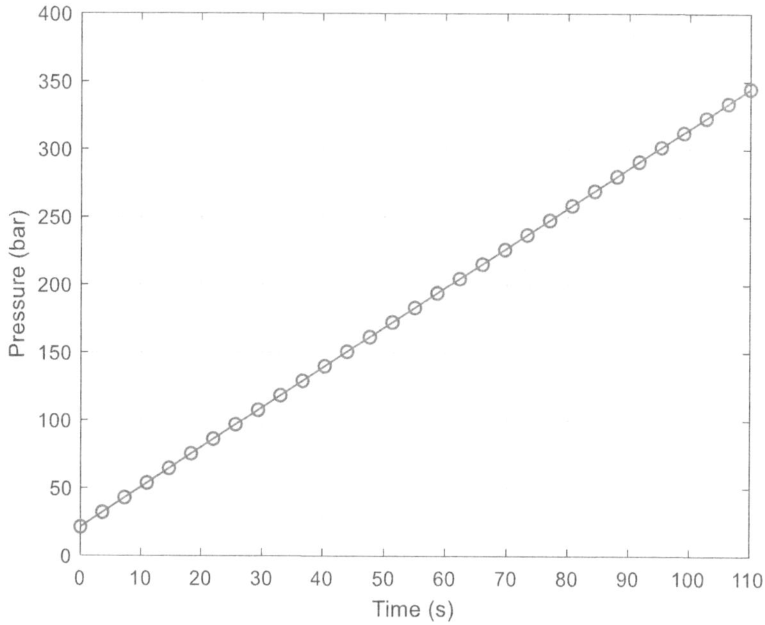

Figure 7.10 Evolution of the pressure in the hydrogen storage tank under the assumptions of adiabatic operation and ideal gas behavior.

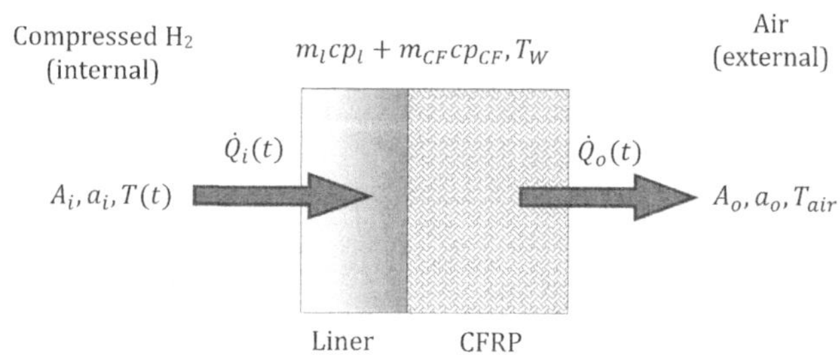

Figure 7.11 Evolution of the pressure in the hydrogen storage tank under the assumptions of adiabatic operation and ideal gas behavior.

thermodynamic properties of hydrogen, including specific enthalpy, specific internal energy, and density.

We begin by formulating the total mass balance for hydrogen within the tank

$$\frac{dN(t)}{dt} = \dot{N}_{in} \tag{7.6.12}$$

where $N(t)$ is the molar holdup of hydrogen in the tank. The energy balance of the contents of the hydrogen tank can be written as

$$\frac{dU(t)}{dt} = \dot{N}_{in}h(T_{in}) - A_i a_i(T(t) - T_W(t)) \tag{7.6.13}$$

There is also an energy balance for the tank composite walls

$$\begin{aligned}(m_l c_{p,l} + m_{CF} c_{p,CF})\frac{dT_W(t)}{dt} &= A_i a_i(T(t) - T_W(t)) \\ &\quad - A_o a_o(T_w(t) - T_{\text{air}})\end{aligned} \tag{7.6.14}$$

where we assume that the composite walls have uniform temperature $T_W(t)$. The temperature is measured with a temperature sensor with a time constant of $\tau_m = 1$ s

$$\tau_m \frac{dT_m(t)}{dt} + T_m(t) = T(t) \tag{7.6.15}$$

where T_m is the temperature measurement. The model is completed by two algebraic equations that relate the total internal energy to the temperature and pressure

$$0 = U(t) - N(t)\, u_{PR_{\text{EOS}}}(T, P) \tag{7.6.16}$$

and the pressure with the temperature, total volume, total molar holdup, and the compressibility factor

$$0 = P(t)V - Z_{PR_{\text{EOS}}}(T, P)N(t)RT(t) \tag{7.6.17}$$

The subscript PR_{EOS} denotes that the specific property is calculated by the Peng-Robinson EOS.

This is a mathematical model of four differential equations and two algebraic equations in four differential variables $(N(t), U(t), T_W(t), T_m(t))$ and two algebraic variables $(P(t), T(t))$. All variables are known (or can be calculated) at the beginning of the filling process. At any time, if the differential variables are known from the solution of the differential equations, then the two algebraic variables are calculated by solving the two algebraic equations.

MATLAB can solve this type of problem, known as **a differential-algebraic equations** (DAEs) problem, in practically the same way that it can solve the ordinary differential equations (ODEs) problem. The only difference is that we need to indicate which equations are differential equations and which equations are algebraic equations.

We first create the function that implements the DAE model, which is given in Table 7.7. In this function, we initially give values to all constants appearing in the model and then calculate several constants that are needed in the DAE system. Then the right-hand side of the DAEs is calculated. Note that we make use of a parser function to the Peng-Robinson function that has been presented in Table 7.3. We then use the function of Table 7.7 to simulate the filling process

Table 7.7 MATLAB implementation of the DAE model for hydrogen tank dynamics.

```
function dxdt = FillingTank_PR_dae(t,x,N_in,h_in,a_i)
%% state to local variables
N  = x(1);              % moles of H2 in the tank      in mol
U  = x(2);              % internal energy of the H2    in the tank in J
T_w= x(3);              % temperature of the wall      in K
Tm = x(4);              % temperature measurement      in K
T  = x(5);              % temperature of the gas       in K
P  = x(6);              % pressure of the gas          in bar
[Z_PREos,h_PREos,u_PREos]=PREoS(T,P);
%% model constants
V      = 0.034;         % internal volume of the tank in m^3
L      = 0.75;          % length of the tank          in m
D      = 0.24;          % diameter of the tank        in m
t_al   = 0.003;         % thickness of the liner      in m
t_cf   = 0.017;         % thickness of carbon fiber   in m
Rg     = 8.31446261815324; % ideal gas constant       in J / mol K
Rgv    = Rg/1e5 ;       % --//--                      in bar m^3/K mol
cp_al  =  896;          % heat capacity of aluminium  in J / kg K
rho_al= 2700;           % density of aluminium alloy  in kg / m^3
cp_cf = 1075;           % heat capacity of CFRP       in J / kg K
rho_cf= 1374;           % density of CFRP             in kg / m^3
T_ref = 25+273.15;      % reference temperature       in K
T_inf = 5 + 273.15;     % temperature of air          in K
a_o    = 4.5;           % outside heat transfer coef. in J/(s m^2 K)
tau_m = 1;              % time constant of the sensor in s
%% preliminary calculations
Di = D;                      Li=L;
Dm = Di + 2* t_al;           Lm = Li+2* t_al;
Do = Di + 2*(t_al+t_cf); Lo = Li+2*(t_al+t_cf);
% internal and external areas available for heat transfer [=] m^2
Area_i = 2*pi*Di^2/4 + pi*Di*Li;
Area_o = 2*pi*Do^2/4 + pi*Do*Lo;
% volume al aluminum [=] m^3
Vol_al= (Lm*pi*Dm^2/4 - Li*pi*Di^2/4) + 2*t_al*pi*Dm^2/4;
Vol_cf= (Lo*pi*Do^2/4 - Lm*pi*Dm^2/4) + 2*t_cf*pi*Do^2/4;
mCp    = Vol_al*cp_al*rho_al + Vol_cf*cp_cf*rho_cf;
Qi     = Area_i *a_i * (T  -T_w);
Qo     = Area_o *a_o * (T_w-T_inf);
%% differential equations
dxdt(1,1) = N_in;              % material balance [=] mol/s
dxdt(2,1) = N_in*h_in - Qin; % energy balance    [=] J/s
dxdt(3,1) = (Qi - Qo)/mCp;% energy balance composite wall [=] K/s
dxdt(4,1) = (T-Tm)/tau_m;      % temperature sensor [=] K/s
%% algebraic equations
dxdt(5,1) = (U/N)-u_PREos;     % definition of spec. internal energy
dxdt(6,1) = P*V-Z_PREos*N*Rgv*T; % definition of compr. factor
end % function FillingTank(t,x,a_in)
```

```
clear all
% general constants
Rg    = 8.31446261815324;    % J / mol K
Rgv   = Rg/1e5 ;             % bar m^3 / K mol
T_ref = 25+273.15;           % K
mw    = 2.01568;             % g / mol

% gas in the tank at t=0
T_0   = 20+273.15;           % K
P_0   = 21.0;                % bar
V     = 0.034;               % m^3
[Z_0,h_0,u_0]=PREoS(T_0,P_0);
N_0   = P_0*V/(Z_0*Rgv*T_0);% mol

% feed
T_in  = 22+273.15;           % K
P_in  = 350;                 % bar
N_in  = 2.91;                % mol/s
[Z_in,h_in,u_in]=PREoS(T_in,P_in);

% solve system of DAEs
M=diag([ones(1,4) zeros(1,2)])
opt = odeset('AbsTol',1E-6,'RelTol',1E-6,'Mass',M)
a_in = 205;
x0 = [N_0; N_0*u_0; T_0; T_0; T_0; P_0];
[time,y]=ode15s(@FillingTank_PR_dae,[0 110],x0,opt,N_in,h_in,a_in);

% plot results
figure(1)
plot(time,y(:,4)-273.15,'b-','LineWidth',1.2)
hold on
plot(time,y(:,3)-273.15,'k-.o')
grid on
xlabel('Time (s)')
ylabel('Temperature (deg C)')
axis([0 110 10 90])

figure(2)
plot(time,y(:,6),'r--o','linewidth',1.2)
grid on
xlabel('Time (s)')
ylabel('Pressure (bar)')
axis([0 110 0 400])

function [Z,h,u]=PREoS(T,Pbar)
%
Rg=8.31446261815324;% kJ/(kmol K)
% hydrogen
TC = 33.44; PC = 1316; omega  = -0.1201;
CP = 4.1868*[6.952 -0.04576E-2 0.09563E-5 -0.2079E-9];
[Z,IsLiquid,phi,hR]=PR_EoS(T,100*Pbar,TC,PC,omega);
Tref=273.15+25;
hig=0;          % @ Tref
for i=1:4
    hig=hig+CP(i)*(T^i-Tref^i)/i;
end
h=hig+hR;       % kJ/kmol
u=h-Rg*T*Z;     % kJ/kmol
end
```

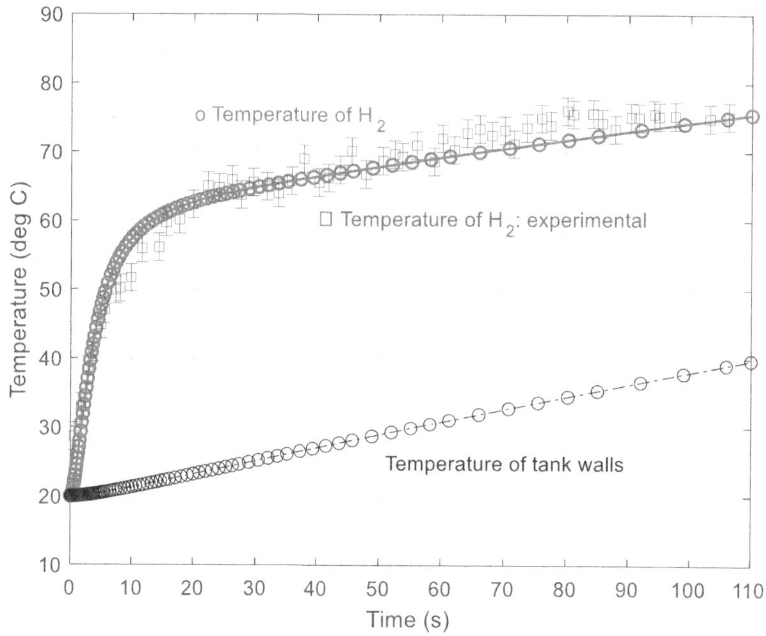

Figure 7.12 Hydrogen tank temperature dynamics and comparison with the experimental data.

The simulation results are presented in Fig. 7.12. The internal heat transfer coefficient was determined through a trial-and-error approach to achieve the best fit with experimental data. The close agreement between the simulated and experimental results is striking and highlights the progress made in this chapter. Specifically, it demonstrates the power of integrating advanced thermodynamic models with fundamental mass and energy balance equations to effectively simulate complex processes and systems.

Chapter summary

In this chapter, we introduced the Peng-Robinson equation of state (EOS), one of the most widely used cubic equations of state in chemical and process engineering. To lay a solid foundation, we began by deriving the basic thermodynamic relations using the van der Waals equation—the first cubic EOS, developed by Dutch physicist Johannes Diderik van der Waals in 1873 as part of his doctoral thesis. Due to its relative simplicity, the van der Waals equation is especially useful for pedagogical purposes, providing a clear and accessible framework for deriving fundamental thermodynamic principles.

We then extended these concepts to more advanced cubic equations of state, focusing on the behavior and predictive capabilities of models such as the Peng-Robinson EOS. The predictive performance of these equations was analyzed, particularly in terms of phase behavior and thermodynamic consistency.

In addition, we presented analytical formulations for determining the most stable phase of a pure component, along with expressions for key thermodynamic properties such as specific enthalpy, specific internal energy, and fugacity. These formulations are essential for applying cubic EOS models in the analysis and simulation of real fluids in practical engineering systems.

Problems

7.1 Prove Eqs. (7.4.5) and (7.4.6) for the Peng-Robinson equation of state. Assume knowledge of the critical compressibility factor (see Eq. (7.4.7)).

7.2 The following data for water are obtained from NIST Chemistry WebBook, SRD 69

Isobaric Data for $P = 50$ bar

Temperature (K)	Density (mol/m3)	Volume (m3/mol)	Internal energy u (kJ/mol)	Enthalpy h(kJ/mol)	Phase
400.00	52173.	1.9167e-05	9.5643	9.6601	liquid
450.00	49568.	2.0174e-05	13.434	13.535	liquid
500.00	46267.	2.1614e-05	17.474	17.582	liquid
537.09	43151.	2.3175e-05	20.685	20.801	**liquid**
537.09	1407.2	0.00071063	46.785	50.338	**vapor**
550.00	1329.5	0.00075217	47.519	51.280	vapor
600.00	1132.0	0.00088340	49.734	54.151	vapor
650.00	1005.9	0.00099413	51.578	56.548	vapor
700.00	912.69	0.0010957	53.286	58.765	vapor
750.00	838.88	0.0011921	54.941	60.901	vapor
800.00	778.05	0.0012853	56.576	63.002	vapor
850.00	726.60	0.0013763	58.210	65.091	vapor
900.00	682.24	0.0014658	59.855	67.183	vapor
950.00	643.46	0.0015541	61.516	69.286	vapor
1000.0	609.18	0.0016416	63.197	71.405	vapor

Auxiliary Data
Reference States, default for fluid

Enthalpy	H = 45.957191 kJ/mol at 300.0 K and 0.010 bar.

Additional fluid properties

Critical temperature (T_c)	647.096 K
Critical pressure (P_c)	220.640 bar
Critical density (D_c)	17,873.727996 mol/m3
Acentric factor	0.3443
Normal boiling point	373.1243 K

a. Use the ideal gas law to estimate the density of steam and the error involved. What is your conclusion?
b. Use the ideal gas model to estimate the enthalpy and the internal energy of steam. What is your conclusion?

7.3 For the data of Problem 7.2 calculate the compressibility factor.

7.4 From the data of Problem 7.2 calculate the heat of evaporation of water at 50 bar. From the NIST Chemistry WebBook we also collect the following data

Pressure (bar)	Saturation (boiling) Temperature (K)	Enthalpy of evaporation (kJ/kmol)
1	372.76	40668
5	424.98	37977
10	454.03	36293
100	584.15	23733

a. Derive an equation for calculating the saturation pressure of water as a function of the absolute temperature. An equation that is commonly used is the Antoine equation

$$\ln P^{sat} = A - \frac{B}{C + T}$$

where A, B, and C are constants specific to each chemical species.

b. Can you find what is the enthalpy of evaporation of water at 50 bar?

c. Derive an equation for calculating the enthalpy of evaporation of water as a function of the absolute temperature. An equation that is commonly used is the Watson's equation

$$\Delta h_{\text{evap}}(T) = \Delta h_{\text{evap}}(T_0)\left(\frac{T - T_C}{T_0 - T_C}\right)^n$$

Where usually $n = 0.38$.

7.5 Data for the fugacity coefficient of i-butane at constant temperature of 410 K, as a function of pressure are given in the following table

P(kPa)	101.3	1000	2600	5000	10000	12000	14000	16000	20000
φ	0.983	0.892	0.755	0.519	0.322	0.290	0.269	0.254	0.235

Estimate the error in calculating the fugacity coefficient using the van der Waals EOS.

7.6 Repeat Problem 7.5 using the Peng-Robinson EOS.

7.7 A very fast and accurate method to generate isotherms in a $P - v$ diagram can be based on solving a DAE system of equations. For the case of the reduced form of the van der Waals equation of state the DAE system is the following:

$$\frac{dv_r}{dt} = 1, v_r(0) = v_{r,0}$$

$$P_r - \frac{8T_r}{3v_r - 1} + \frac{3}{v_r^2} = 0, \qquad P(0) = P_0(T_r, v_{r,0})$$

Select an appropriate value for $v_{r,0} > \frac{1}{3}$ and integrate the system for $0.8 < T_r < 1.1$ to generate Fig. 7.4.

7.8 The ability of a relatively simple equation to capture complex physical behavior is a testament to the brilliance of van der Waals. His approach to distinguishing subcritical and supercritical fluids was particularly insightful. He observed that at the critical point, both the first and second derivatives of pressure with respect to molar volume are zero. This led to two equations that relate the equation of state parameters a and b in terms of the critical constants

T_C and P_C. When this method is extended to more complex equations of state, the algebra can become extremely cumbersome—as you likely experienced in Problem 7.1. Fortunately, a powerful simplification arises from another defining feature of the critical point: at this condition, the liquid and vapor roots of the cubic equation of state are equal, and both coincide with the so-called spurious middle root. This insight allows us to specify the condition $(Z - Z_C)^3$ = lhs of Equation (7.4.10) at the critical point. Use this method to evaluate the numerical constants appearing in Eqs. (7.4.5)–(7.4.7).

7.9 Water is among the substances whose thermodynamic properties are particularly challenging to predict accurately with equations of state. In Tables P7.9a and b, data for two subcritical isotherms are given, taken from NIST Chemistry WebBook, SRD 69. Use the Peng-Robinson EOS to estimate the properties of water at these two temperatures and compare the results obtained with the data obtained from NIST (estimated using highly accurate EOS specific to water).

Table P7.9.a Data for n-octane at T = 300 K.

Isothermal data for T = 300 K					
Temperature (K)	**Pressure (bar)**	**Density (mol/L)**	**Volume (L/mol)**	**Enthalpy (kJ/mol)**	**Phase**
300.00	0.010000	0.00040111	2493.1	45.957	vapor
300.00	0.035368	0.0014204	704.01	45.936	vapor
300.00	0.035368	55.315	0.018078	2.0279	liquid
300.00	1.0100	55.317	0.018078	2.0295	liquid
300.00	10.010	55.340	0.018070	2.0444	liquid
300.00	20.010	55.365	0.018062	2.0610	liquid
300.00	30.010	55.389	0.018054	2.0776	liquid
300.00	40.010	55.414	0.018046	2.0941	liquid
300.00	50.010	55.439	0.018038	2.1106	liquid
300.00	60.010	55.463	0.018030	2.1271	liquid
300.00	70.010	55.488	0.018022	2.1437	liquid
300.00	80.010	55.513	0.018014	2.1601	liquid
300.00	90.010	55.537	0.018006	2.1766	liquid
300.00	100.01	55.562	0.017998	2.1931	liquid

Table P7.9.b Data for n-octane at T = 500 K.

Isothermal data for T = 500 K					
Temperature (K)	**Pressure (bar)**	**Density (mol/L)**	**Volume (L/mol)**	**Enthalpy (kJ/mol)**	**Phase**
500.00	0.010000	0.00024056	4157.0	52.826	vapor
500.00	10.010	0.25184	3.9707	52.085	vapor
500.00	20.010	0.53196	1.8798	51.187	vapor
500.00	26.392	0.73265	1.3649	50.487	**vapor**
500.00	26.392	46.145	0.021671	17.573	**liquid**
500.00	30.010	46.164	0.021662	17.574	liquid
500.00	40.010	46.216	0.021638	17.578	liquid
500.00	50.010	46.267	0.021614	17.582	liquid
500.00	60.010	46.318	0.021590	17.586	liquid
500.00	70.010	46.368	0.021566	17.591	liquid
500.00	80.010	46.419	0.021543	17.595	liquid
500.00	90.010	46.468	0.021520	17.600	liquid
500.00	100.01	46.518	0.021497	17.604	liquid

7.10 The vapor pressure of n-octane for different temperatures is given in the following table (Fluid Phase Equilibria, 220(2), p. 277, 2003):

T (K)	323.2	340.9	352.9	377.4	401.6	453.9	501.0	552.0
P^{sat}(kPa)	6.73	14.51	23.2	53.7	109.5	377.6	902.4	1965.1

Modify the Peng-Robinson implementation presented in Table 7.3 to return the fugacity coefficient of both phases and build an algorithm to determine the saturation pressure as the pressure for which

$$\varphi^L = \varphi^G$$

Use $\omega = 0.398$ and the critical constants reported by NIST. Use also the Antoine equation for an initial estimation of the saturation (or vapor) pressure.

7.11 Use the definition of the acentric factor and the experimental data for n-octane to determine the acentric factor of n-octane. Use also the data in NIST for n-octane to estimate ω. Compare with the value $\omega = 0.398$ given in the excellent book by Reid et al., 1988, *The Properties of Gases and Liquids*, 4th ed., McGraw-Hill.

7.12 A method commonly used to estimate the saturation pressure of chemical species (for which the critical pressure P_C, the critical temperature T_C, and the acentric factor ω are known) is the method of Lee and Kesler. In this method we initially determine the reduced temperature at which we want to calculate the vapor pressure

$$T_r = \frac{T}{T_C}$$

Then we calculate two factors

$$f^{(0)} = 5.92714 - \frac{6.09648}{T_r} - 1.28862 \ln T_r + 0.169347\, T_r^6$$

$$f^{(1)} = 15.2518 - \frac{15.6875}{T_r} - 13.4721 \ln T_r + 0.43577\, T_r^6$$

and finally

$$P^{\text{sat}} = P_C \exp\left(f^{(0)} + \omega f^{(1)}\right)$$

Apply the Lee-Kesler equation for the experimental data for the vapor pressure of n-octane from Problem. 7.11. Give your conclusions.

7.13 Methane is expanded adiabatically from upstream conditions of 40°C and 20 bar to a downstream pressure of 1 bar. What is the downstream temperature? All necessary data can be obtained from NIST Chemistry WebBook.

7.14 For the process of natural gas (assumed pure methane) liquefaction that produces liquefied natural gas (LNG) of Fig. P7.14 and Table P7.14, use the Peng-Robinson equation of state to determine:

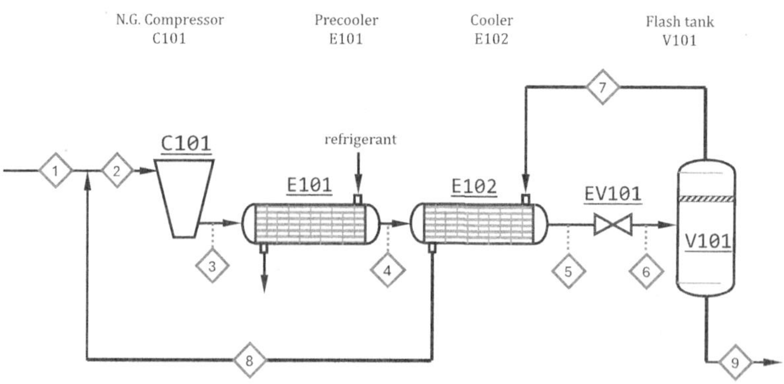

Figure P7.14 LNG liquefication process.

Table P7.14 Stream data for LNG liquefication process.

Stream	1	2	3	4	5
Temperature (°C)	25			−85	
Pressure (kPa)	100	100	6000	6000	6000
Molar flowrate (mol/s)	1	1.9	1.9	1.9	1.9
State	S.H.G.	S.H.G.	S.H.G.	S.H.G.	S.H.G
Spec. enthalpy (J/mol)					

Stream	6	7	8	9
Temperature (°C)			−72.4	
Pressure (kPa)	100	100	100	100
Molar flowrate (mol/s)	1.9	0.9	0.9	1
State	M.G.L.	Sat.G	S.H.G	Sat.L
Spec. enthalpy (J/mol)				

M.G.L: gas-liquid mixture (in equilibrium), *Sat.G*: saturated gas, *Sat.L*: saturated liquid; *S.H.G*: superheated gas.

a. the temperature of Stream 2,
b. the temperature of streams 6, 7, and 9,
c. the fraction of stream 6 in the vapor state,
d. the rate of heat transfer at heat exchanger E102,
e. the temperature of stream 5.

7.15 A method for estimating the compressibility factor for the component in the gaseous state with $v > 2v_C$ or $T_r > 0.686 + 0.439P_r$ has been presented by Abbott and van Ness and is based on the second virial coefficient:

$$Z = 1 + B\left(\frac{P_r}{T_r}\right)$$

$$B = \left(B^{(0)} + \omega B^{(1)}\right)$$

$$B^{(0)} = 0.083 - \frac{0.422}{T_r^{1.6}}$$

$$B^{(1)} = 0.139 - \frac{0.172}{T_r^{4.2}}$$

An alternative method for estimating the second virial coefficient for nonpolar components is the following:

$$B^{(0)} = 0.1445 - \frac{0.330}{T_r} - \frac{0.1385}{T_r^2} - \frac{0.0121}{T_r^3} - \frac{0.000607}{T_r^8}$$

$$B^{(1)} = 0.0637 + \frac{0.331}{T_r^2} - \frac{0.423}{T_r^3} - \frac{0.008}{T_r^8}$$

Use this method to calculate the compressibility factor of CO_2 at 344.3 K and several pressures given in Table P7.15 (Sage, B. H., Lacey, W. N., 1955, API Research Project 37, American Petroleum Institute, NY).
It has been suggested that both approximations can be valid for $P_r < T_r/2$. Comment on whether this suggestion agrees with your results.

Table P7.15 Compressibility data of CO_2 at 433.3 K.

CO_2T = 344.4 K			
P (bar)	Z	P (bar)	Z
13.98	0.9464	138.68	0.4714
27.59	0.9069	172.73	0.4481
41.57	0.8626	206.77	0.4737
55.19	0.8013	275.94	0.5514
69.16	0.7555	345.09	0.6492
103.57	0.6135		

7.16 The two methods for estimating the second virial coefficient for nonpolar components can also be used to calculate the fugacity coefficient for the gaseous state under the same applicability constraints using the following equation:

$$\ln \varphi = B\left(\frac{P_r}{T_r}\right)$$

Check the accuracy of the method using the data from Problem 7.5.

7.17 A methodology to calculate the residual enthalpy of a gas can be based on the second virial coefficient. Use the equations presented in Problem 7.15:

$$\frac{h^R}{RT} = P_r\left(\frac{B}{T_r} - \frac{dB}{dT_r}\right)$$

where

$$\frac{dB}{dT_r} = \frac{dB^{(0)}}{dT_r} + \omega\frac{dB^1}{dT_r}$$

$$\frac{dB^{(0)}}{dT_r} = \frac{0.6752}{T_r^{2.6}}, \qquad \frac{dB^{(1)}}{dT_r} = \frac{0.7224}{T_r^{5.2}}$$

to calculate the residual enthalpy of isobutane ($T_C = 408.05$ K, $P_C = 3648$ kPa, $\omega = 0.1848$) at 400 K and 1000 kPa. The experimental value is $h^R = -1170$ kJ/kmol.

7.18 The Soave-Redlich-Kwong (SRK) equation of state is a commonly used EOS that belongs to the van der Waals family of cubic EOS. The basic equation of the SRK EOS is the following:

$$P = \frac{RT}{v - b(T_C, P_C)} - \frac{a(P_C, T_C)\,\tilde{a}(\omega, T, T_C)}{v(v + b(T_C, P_C))}$$

where

$$\tilde{a}(\omega, T, T_C) = \left[1 + \kappa(\omega)\left(1 - \sqrt{\frac{T}{T_C}}\right)\right]^2$$

$$\kappa(\omega) = 0.48508 + 1.55171\omega - 0.15613\omega^2$$

a. Find the expressions for $a(T_C, P_C)$, $b(T_C, P_C)$ using the method that fits the equation to the critical point.
b. What are the cubic polynomials in v and Z?
c. Derive the equation for calculating the residual enthalpy.

7.19 A 10 m^3 tank initially contains steam at 500°C and 20 bar. The steam escapes the tank at a molar flowrate of 4 mol/s until the pressure in the tank is reduced to 1 bar. Calculate the evolution of temperature and pressure in the tank.

References

Many definitions have been retrieved from the Wikipedia, the free encyclopedia.

Books on thermodynamics

Abbott, M.M., Van Ness, H.C., 1989. Theory and problems of thermodynamics. In: Schaums Outline Series, second ed. McGraw-Hill Book Company, NY.

Balzhiser, R.E., Samuels, M.R., Eliassen, J.D., 1972. Chemical Engineering Thermodynamics. Prentice Hall, Englewood Cliffs, NJ.

Daubert, T.E., 1985. Chemical Engineering Thermodynamics. McGraw-Hill Book Company, NY.

Elliot, J.R., Lira, C.T., 2012. Introductory Chemical Engineering Thermodynamics, second ed. Prentice Hall, Upper Saddle River, NJ.
Gmehling, J., Kolbe, B., Kleiber, M., Rarey, J., 2012. Chemical Thermodynamics for Process Simulation. Wiley-VCH, Weinheim.
Hall, K.R., Iglesias-Silva, G.A., 2022. Thermodynamics for Chemical Engineers. Wiley-VCH, Weinheim.
Koretsky, M.D., 2013. Engineering and Chemical Thermodynamics, second ed. John Wiley & Sons, NJ.
Kyle, B.G., 1999. Chemical and Process Thermodynamics. Prentice-Hall, NJ.
Matsoukas, T., 2012. Fundamentals of Chemical Engineering Thermodynamics. Prentice Hall, Upper Saddle River, NJ.
Moran, M.J., Shapiro, H.N., 1988. Fundamentals of Engineering Thermodynamics. John Wiley and Sons, NY.
Reid, C.E., 1990. Chemical Thermodynamics. Mc Graw Hill, NY.
Sandler, S.I., 1977. Chemical and Engineering Thermodynamics. John Wiley and Sons, NY.
Tassios, D.P., 2003. Applied Chemical Engineering Thermodynamics. Springer-Verlag Berlin Heidelberg.

8

Vapor-liquid equilibrium calculations

8.1 Introduction

In this chapter, we extend the methodologies introduced in the previous chapter—originally applied to pure components—to binary or multicomponent mixtures and solutions. Once again, we limit our focus to systems involving only two coexisting phases: vapor and liquid.

We begin with ideal mixtures, which are defined as mixtures where all thermodynamic properties (such as molar volume and specific enthalpy) can be expressed as weighted averages of the corresponding pure-component properties. In an ideal solution, for example, mixing does not result in a volume change —the total volume is simply the sum of the individual volumes of the pure components. The same principle applies to enthalpy: the enthalpy of the ideal mixture is equal to the sum of the enthalpies of the components in their pure states. This concept is illustrated in Fig. 8.1.

The assumption of ideality applies to all additive (extensive) properties, including total mass, total number of moles, total volume, total enthalpy, internal energy, and others. Ideal mixtures are also characterized by their compliance with Raoult's law, which will be discussed later in this chapter. According to Raoult's law, the vapor pressure of a solvent above an ideal solution is equal to the vapor pressure of the pure solvent (at the same temperature), scaled by its mole fraction in the liquid phase.

Before getting into the detailed presentation of the properties of ideal mixtures, we will review an important result in thermodynamics that will help us to understand further the phase diagrams of pure components and binary mixtures: the Gibbs phase rule.

Introduction to Process Systems Engineering. https://doi.org/10.1016/B978-0-443-45444-8.00002-6

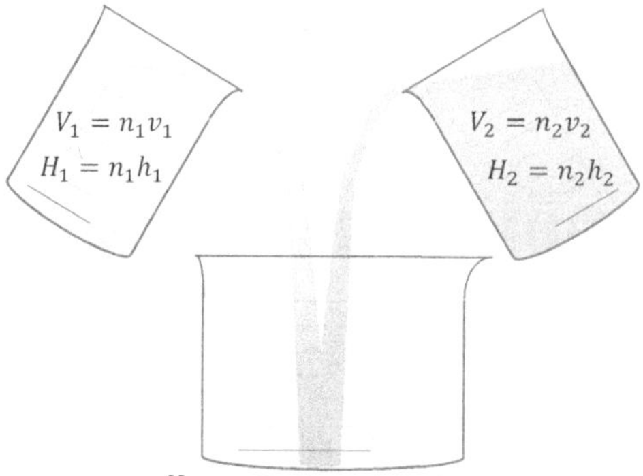

$$V_{mix} = n_m v_m = n_1 v_1 + n_2 v_2$$

$$H_{mix} = n_m h_m = n_1 h_1 + n_2 h_2$$

Figure 8.1 Properties of ideal solutions are the sum of the properties of the constituting components in their pure state.

8.2 The Gibbs phase rule

The Gibbs Phase Rule is a fundamental principle in thermodynamics that provides a relationship between the number of phases ($\mathcal{NP}$), the number of components ($\mathcal{NC}$), and the number of independent intensive variables ($\mathcal{F}$) in a system at equilibrium without chemical reactions. It tells us how many intensive variables (like temperature, pressure, or composition) we can independently change without changing the number of phases in a system. The number $\mathcal{F}$ is the number of independent intensive properties, that is, the largest number of thermodynamic parameters such as temperature or pressure that can be varied simultaneously and independently of each other, and is calculated as follows:

$$\mathcal{F} = \mathcal{NC} - \mathcal{NP} + 2 \tag{8.2.1}$$

For a pure substance, the Gibbs phase rule simplifies to $\mathcal{F} = 3 - \mathcal{NP}$, meaning that for pure substances the number of independent variables depends only on the number of phases present. In a single-phase region, such as liquid or vapor alone, two intensive variables—typically temperature and pressure—can be independently varied (see Fig. 7.3). However, when the system enters a two-phase region, such as liquid-vapor equilibrium, only one degree of freedom remains. Temperature and pressure are no longer independent and are related by a specific relationship defined by the phase boundary curve, such as the one described by the Antoine equation. Along this curve, increasing the temperature requires an increase in pressure to maintain equilibrium, and the system follows this constraint until the critical point is reached. At the critical point, the liquid and vapor phases become indistinguishable, and above this point, the substance exists as a single supercritical fluid phase where temperature and pressure can again vary independently. At a specific temperature and pressure—the triple point—three phases (solid, liquid, and vapor) coexist in equilibrium ($\mathcal{NC} = 1, \mathcal{NP} = 3$), and $\mathcal{F} = 0$. At this point, no variables can be independently changed.

For binary mixtures consisting of two chemically independent components, the number of components is $\mathcal{NC} = 2$, so the Gibbs phase rule becomes $\mathcal{F} = 4 - \mathcal{NP}$. This means that up to three independent intensive variables—typically temperature, pressure, and composition—can be varied when only one phase is present ($\mathcal{NP} = 1$). When two phases coexist ($\mathcal{NP} = 2$), such as a liquid and its vapor in equilibrium, the degrees of freedom reduce to two ($\mathcal{F} = 2$). In this two-phase region, any two thermodynamic variables—commonly temperature and overall composition—can be selected independently, and the other properties will be constrained by equilibrium conditions.

An illustrative example involves a binary liquid mixture of cyclopentane and benzene, two completely miscible substances. At constant pressure, such a system can be represented by a boiling point diagram, which plots the mole fraction of a component (e.g., cyclopentane) on the horizontal axis and temperature on the vertical axis (see Fig. 8.2 and the experimental data of Table 8.1). The diagram typically includes two curves: the bubble point curve (the temperature at which liquid becomes saturated and starts to vaporize) and the dew point curve (where vapor becomes saturated and begins to condense). The region between these curves corresponds to the two-phase equilibrium, where both liquid and vapor coexist.

8.3 Txy diagrams for ideal binary mixtures

Fig. 8.3 shows the Txy diagram for the cyclopentane–benzene mixture, with dashed lines illustrating the steps of a hypothetical constant-pressure experiment. Consider starting with an equimolar mixture of cyclopentane and benzene in a vessel fitted with a movable piston to maintain constant pressure. Throughout the process, the pressure is held constant at 1 bar. The initial state of the system, at 45 °C, is indicated as point A on the diagram. As the mixture is gradually heated—while maintaining constant pressure and overall composition—the system

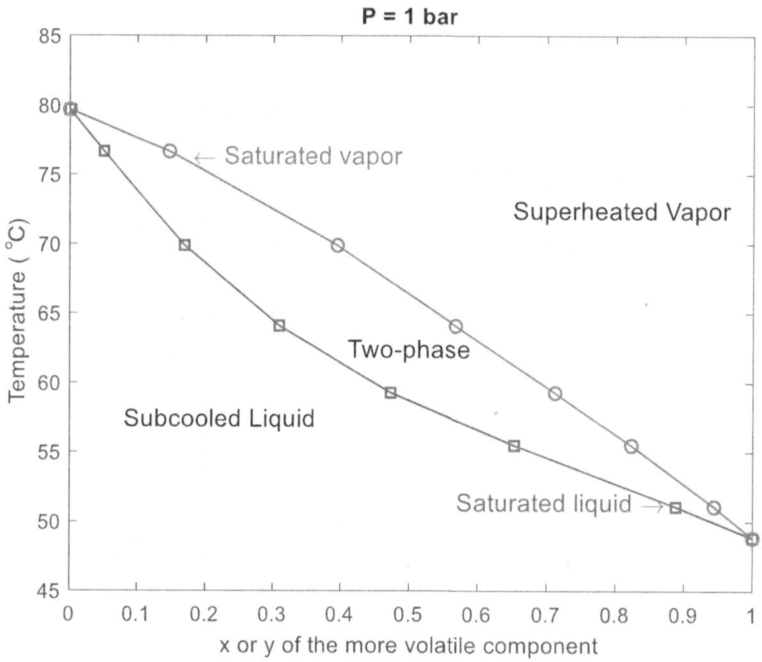

Figure 8.2 Txy diagram of cyclopentane and benzene at 1 bar.

Table 8.1 Equilibrium data for the binary mixture of cyclopentane (1) and benzene (2) at 1 bar.

T (°C)	x_1	y_1
79.7	0.000	0.000
76.7	0.050	0.147
69.9	0.169	0.395
64.2	0.309	0.568
59.3	0.473	0.712
55.5	0.653	0.823
51.1	0.888	0.944
48.9	1.000	1.000

Data from Van Winkle, Distillation, McGraw Hill (1962).

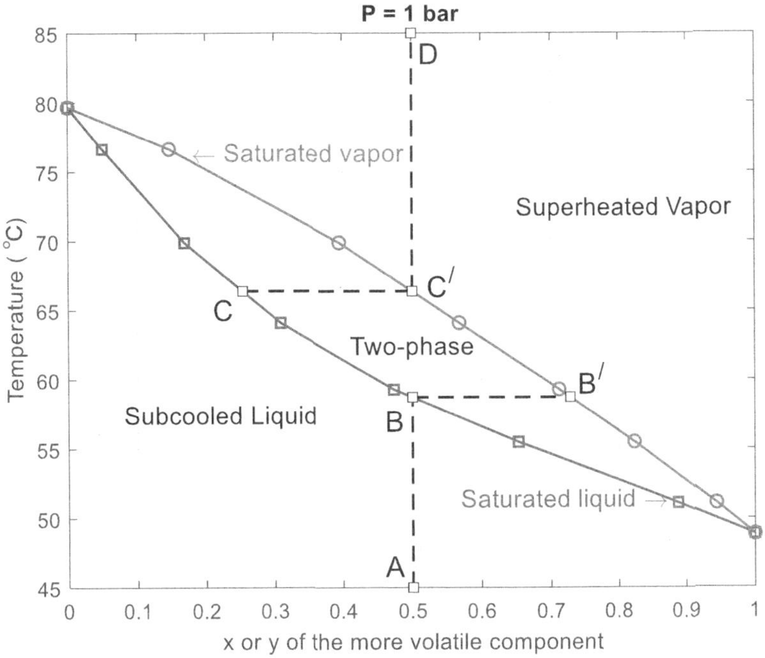

Figure 8.3 A hypothetical experiment pictured (*dashed* lines) on a Txy diagram of cyclopentane and benzene at 1 bar.

follows the trajectory from point A to point B. Initially, the mixture exists as a subcooled liquid. Upon reaching point B, located on the saturated liquid (bubble point) curve, the liquid becomes saturated. At this stage, any additional heat input initiates vaporization, forming the first bubble of vapor. This is why point B is referred to as the bubble point for the given composition—it marks the onset of phase transition from liquid to vapor at equilibrium.

At the **bubble point** (point B), the first bubble of vapor forms. The composition of this vapor is represented by point B′. Although points B and B′ correspond to different phases—liquid and vapor, respectively—they are in equilibrium, sharing the same temperature and pressure. Their compositions differ, as dictated by phase equilibrium. As heating continues, the system remains in the two-phase region, where the liquid stays saturated and its composition follows the liquid saturation curve from B to C. Simultaneously, the vapor composition evolves along the vapor saturation curve from B′ to C′. At each stage along this path, the compositions of the coexisting liquid and vapor phases are given by corresponding pairs of points on the liquid and vapor curves, remaining in equilibrium throughout the process.

The final droplet of saturated liquid has the composition and temperature corresponding to point C on the diagram. Beyond this point, the mixture becomes fully vaporized, with its composition and temperature now defined by point C′. At this stage, the equimolar cyclohexane–benzene mixture exists entirely as a saturated vapor. Any further heat added to the system increases the temperature of the vapor, causing it to move from point C′ to point D, along the superheated vapor-phase region.

This hypothetical experiment can now be reversed by cooling the mixture. Starting from point D in the superheated vapor region, the system follows a cooling path back toward point C′ on the saturated vapor curve. Point C′ now represents the **dew point** of the vapor mixture—the temperature and composition at which the first droplet of liquid forms. This initial droplet has a composition corresponding to point C on the saturated liquid line. As cooling continues, the vapor phase composition moves along the vapor saturation curve from C′ toward B′, while the liquid phase composition follows the liquid saturation curve from C to B. Throughout this two-phase region, the coexisting liquid and vapor phases remain in equilibrium, with their compositions determined by the tie lines connecting corresponding points on the two curves. Eventually, the last bubble of vapor condenses, and the system reaches point B, where it exists entirely as a saturated liquid. Further cooling shifts the system from point B to point A, entering the subcooled liquid region where no vapor remains.

At constant pressure, each composition in a binary mixture has two key temperatures represented on the Txy diagram:

- the first is the bubble-point temperature (T_{BP}), which marks the onset of vaporization—when the first bubble of vapor forms in a saturated liquid mixture,
- the second is the dew-point temperature (T_{DP}), at which the first droplet of liquid appears in a saturated vapor mixture, indicating the start of condensation.

If the temperature of the mixture is below the bubble point, the system exists as a subcooled liquid. Conversely, if the temperature is above the dew point, the mixture is in the superheated vapor state. Between these two temperatures, the system exists in a two-phase equilibrium, with both liquid and vapor present.

It is therefore essential to efficiently calculate the bubble-point and dew-point temperatures in order to accurately locate a given mixture on the Txy diagram at known pressure and composition. Determining the thermodynamic state of the mixture is critical, as it allows us to compute all relevant intensive properties—such as density, specific enthalpy, and internal energy—which are necessary for performing reliable mass and energy balances.

This approach is conceptually similar to property calculations for a pure component, where the saturation pressure must be determined at a given temperature. In that case, if the system pressure exceeds the saturation pressure, the substance exists in the liquid state; if the pressure is below the saturation pressure, the substance is in the vapor state. For mixtures, the analogous goal is to determine whether the system lies in the subcooled liquid region, the two-phase region, or the superheated vapor region, based on a comparison between the mixture's temperature and its bubble- and dew-point temperatures at the given pressure.

An important difference between the phase-change behavior of pure components and mixtures is that a pure component changes phase at constant temperature for a given pressure, while a mixture changes phase under variable temperature (increasing temperature during vaporization and decreasing temperature during condensation).

In the following sections, we will outline the methodology for calculating the bubble-point and dew-point temperatures, along with the corresponding compositions of the first bubble of vapor and the first droplet of liquid. To perform these calculations, a vapor–liquid equilibrium (VLE) model is required—one that describes the relationship between the compositions of the coexisting phases at equilibrium. The simplest such model is Raoult's law, which we will examine first as a foundation. Following that, we will briefly introduce more advanced models that account for nonideal behavior in both the liquid and vapor phases, providing more accurate predictions for systems that deviate from ideality.

8.4 Raoult's law for VLE of ideal solutions

Raoult's Law is one of the most fundamental models used to describe vapor–liquid equilibrium (VLE) in ideal liquid mixtures. It relates the partial pressure of a component in the vapor phase of a mixture to its mole fraction in the liquid phase and its pure-component vapor pressure. It is particularly useful in modeling binary or multicomponent systems when the liquid phase behaves ideally. Raoult's Law is based on several simplifying assumptions:

- ideal liquid phase behavior: the liquid behaves ideally, meaning there are no significant intermolecular interactions beyond those present in the pure substances,
- ideal gas behavior in the vapor phase: the vapor phase is assumed to follow the ideal gas law,
- equilibrium conditions: the system is at equilibrium, so temperature and pressure are uniform, and the chemical potential of each component is equal in both phases.

Raoult's law refers to the situation pictured in Fig. 8.4. A vapor-liquid mixture of NC components is in equilibrium state at given pressure P and temperature T. Raoult's law states that **the partial pressure of each component of an ideal mixture of liquids is equal to the vapor pressure of the pure component (a function of the temperature only) multiplied by its mole fraction in the liquid phase**. Mathematically, Raoult's law for a component in an ideal solution of NC components is stated as

$$P_i = x_i P_i^{\text{sat}}(T) \tag{8.4.1}$$

where P_i is the partial pressure of component i in the gaseous mixture above the solution, P_i^{sat} is the equilibrium vapor pressure (or saturation pressure) of pure component i, and x_i is the mole fraction of component i in the liquid solution.

When NC components are mixed to form a solution, the vapor phase consists of all components of the solution, and in equilibrium, the total vapor pressure of the solution can be determined by combining Raoult's law with Dalton's law of partial pressures to give:

$$P_i = y_i P = x_i P_i^{\text{sat}}(T), \quad i = 1, 2, \ldots, NC \tag{8.4.2}$$

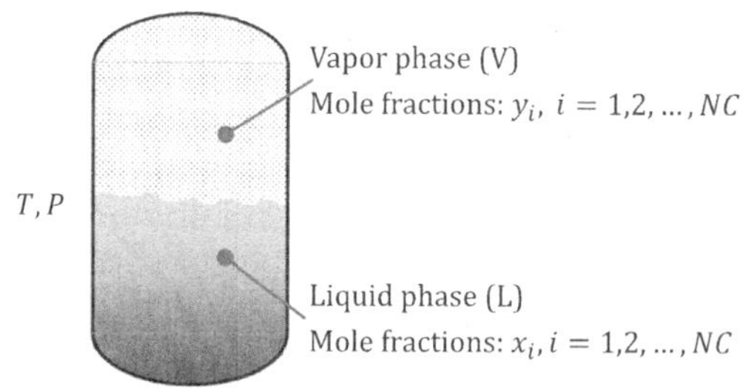

Figure 8.4 A vapor-liquid mixture of NC components in equilibrium.

According to Eq. (8.4.2), the compositions of component i in the liquid phase (x_i) and in the vapor phase (y_i) are related through an equation of the form:

$$y_i = K_i(T, P)x_i, \quad i = 1, 2, \ldots, NC \tag{8.4.3}$$

where K_i is the equilibrium ratio that depends (when Raoult's law is applicable) only on temperature and pressure

$$K_i(T, P) = \frac{y_i}{x_i} = \frac{P_i^{sat}(T)}{P}, \quad i = 1, 2, \ldots, NC \tag{8.4.4}$$

8.5 BUBBLE- and DEW-point calculations using Raoult's law

Most applications in process systems engineering deal with VLE problems that center on dew-point and bubble-point calculations (see Fig. 8.4 for notation).

- BUBBLE T: Given $\{x_i, P\}$ calculate $\{y_i, T\}$, i.e., for a liquid solution of known composition x_i at a given pressure P calculate the temperature T_{BP} at which the first bubble will be formed and its composition y_i,
- DEW T: Given $\{y_i, P\}$ calculate $\{x_i, T\}$, i.e., for a vapor mixture of known composition y_i at a given pressure P calculate the temperature T_{DP} at which the first liquid droplet will be formed and its composition x_i,

8.5.1 BUBBLE-point calculations

We refer to Fig. 8.4 and our aim is to solve the bubble-point problem using Raoult's law as summarized by Eq. (8.4.4). We are given $\{x_i, P\}$ and we seek to calculate $\{y_i, T\}$. There are $NC + 1$ unknowns, and we, therefore, need an equal number of equations. These equations are:

NC equilibrium equations

$$y_i = \frac{P_i^{sat}(T_{BP})}{P} x_i, \quad i = 1, 2, \ldots, NC \tag{8.5.1}$$

Summation equation

$$\sum_{i=1}^{NC} y_i = 1 \tag{8.5.2}$$

We replace Eq. (8.5.1) into Eq. (8.5.2) to obtain one equation in one unknown, which is the T_{BP}.

$$f(T_{BP}) = \sum_{i=1}^{NC} \frac{P_i^{sat}(T_{BP})}{P} x_i - 1 = 0 \tag{8.5.3}$$

The saturation pressure is determined using the Antoine's equation, which is of the form:

$$\ln P_i^{sat}(T) = A_i - \frac{B_i}{C_i + T} \tag{8.5.4}$$

Values for the constants A_i, B_i, and C_i can be obtained from several sources, including the NIST Chemistry WebBook.

One method to solve Eq. (8.5.3) is by using Newton's method, which is based on the derivative of the function $f(T_{BP})$. To determine the derivative we first note that

$$\frac{dP_i^{sat}}{dT_{BP}} = \frac{d}{dT_{BP}}\left[\exp\left(A_i - \frac{B_i}{C_i + T_{BP}}\right)\right]$$
$$= \exp\left(A_i - \frac{B_i}{C_i + T_{BP}}\right)\frac{d}{dT_{BP}}\left(A_i - \frac{B_i}{C_i + T_{BP}}\right)$$

or

$$\frac{dP_i^{sat}}{dT_{BP}} = P_i^{sat}(T_{BP})\frac{B_i}{(C_i + T_{BP})^2} \tag{8.5.5}$$

Then, by using Eq. (8.5.3), we have that

$$\frac{df(T_{BP})}{dT_{BP}} = \sum_{i=1}^{NC} \frac{x_i}{P}\frac{dP_i^{sat}}{dT_{BP}} = \sum_{i=1}^{NC} \frac{x_i P_i^{sat}(T_{BP})}{P}\frac{B_i}{(C_i + T_{BP})^2}$$
$$= \sum_{i=1}^{NC} y_i \frac{B_i}{(C_i + T_{BP})^2} > 0 \tag{8.5.6}$$

The derivative is always positive, and as a result the function is a strictly increasing function of temperature. This ensures that a unique solution exists. This solution is always between the extreme boiling temperatures of the pure components in the ideal solution, which can be calculated using the Antoine's equation

$$T_{b,i} = \frac{B_i}{A_i - \ln(P)} - C_i \tag{8.5.7}$$

It then holds true for ideal mixtures that

$$\min_i\{T_{b,i}\} < T_{BP} < \max_i\{T_{b,i}\} \tag{8.5.8}$$

An initial estimation of the temperature can be obtained as a weighted average of the boiling temperatures of the constituting components

$$T_{BP}^0 = \sum_{i=1}^{NC} x_i T_{b,i} \tag{8.5.9}$$

Example 8.1. Calculation of the bubble-point temperature for a binary mixture

For an equimolar mixture of cyclopentane (C) and benzene (B), calculate the bubble-point temperature at 1 bar. The Antoine's equations for the two components are the following

$$\ln P_C^{sat} = 9.2170 - \frac{2577.1}{T - 42.412}$$

$$\ln P_B^{sat} = 9.2521 - \frac{2771.9}{T - 53.226}$$

where P^{sat} is given in bar and the temperature T in K.

Solution

We use Eq. (8.5.7) to determine the boiling temperature of the two components at 1 bar:

$$T_{b,C} = \frac{B_C}{A_C - \ln(P)} - C_C = \frac{2577.1}{9.2170 - \ln(1)} - (-42.412)$$
$$= 322.013 \text{ K}$$

$$T_{b,B} = \frac{B_B}{A_B - \ln(P)} - C_B = \frac{2771.9}{9.2521 - \ln(1)} - (-53.226)$$
$$= 352.826 \text{ K}$$

An initial estimation of the bubble-point temperature is obtained using the Eq. (8.5.9)

$$T_{BP}^0 = \sum_{i=1}^{NC} x_i T_{b,i} = 0.5 \times 322.013 + 0.5 \times 352.826 = 337.42 \text{ K}$$

First Iteration.

We then calculate the saturation pressure of the two components:

$$P_C^{sat} = \exp\left(9.2170 - \frac{2577.1}{T_{BP}^0 - 42.412}\right) = 1.6183 \text{ bar}$$

$$P_B^{sat} = \exp\left(9.2521 - \frac{2771.9}{T_{BP}^0 - 53.226}\right) = 0.6056 \text{ bar}$$

The equilibrium ratios are

$$K_C = \frac{P_C^{sat}}{P} = \frac{1.6183}{1} = 1.6183$$

$$K_B = \frac{P_B^{sat}}{P} = \frac{0.6056}{1} = 0.6056$$

and the mole fractions in the vapor phase:

$$y_C = K_C x_C = 1.6183 \times 0.5 = 0.8091$$

$$y_B = K_B x_B = 0.6056 \times 0.5 = 0.3028$$

The value of the function is

$$f(T_{BP}^0) = \sum_{i=1}^{NC} \frac{P_i^{sat}(T_{BP})}{P} x_i - 1 = y_C + y_B - 1$$
$$= 0.8091 + 0.3028 - 1 = 0.1119$$

The value of the derivative is (see Eq. (8.5.6))

$$\frac{df(T_{BP})}{dT_{BP}} = \sum_{i=1}^{NC} y_i \frac{B_i}{(C_i + T_{BP})^2} = y_C \frac{B_C}{(T_{BP}^0 + C_C)^2} + y_B \frac{B_B}{(T_{BP}^0 + C_B)^2}$$
$$= 0.8091 \times \frac{2577.1}{(T_{BP}^0 - 42.412)^2} + 0.3028 \times \frac{2771.9}{(T_{BP}^0 - 53.226)^2}$$
$$= 0.0344$$

We calculate the new estimation of the bubble-point temperature

$$T_{BP}^1 = T_{BP}^0 - \frac{f(T_{BP}^0)}{\frac{df(T_{BP}^0)}{dT_{BP}}} = 337.42 - \frac{0.1119}{0.0344} = 334.1613$$

Second Iteration.
We then calculate the saturation pressure of the two components:

$$P_C^{sat} = \exp\left(9.2170 - \frac{2577.1}{T_{BP}^1 - 42.412}\right) = 1.4679 \text{ bar}$$

$$P_B^{sat} = \exp\left(9.2521 - \frac{2771.9}{T_{BP}^1 - 53.226}\right) = 0.5408 \text{ bar}$$

The equilibrium ratios are

$$K_C = \frac{P_C^{sat}}{P} = \frac{1.6183}{1} = 1.4679$$

$$K_B = \frac{P_B^{sat}}{P} = \frac{0.6056}{1} = 0.5408$$

and the mole fractions in the vapor phase:

$$y_C = K_C x_C = 1.4679 \times 0.5 = 0.7339$$

$$y_B = K_B x_B = 0.5408 \times 0.5 = 0.2704$$

The value of the function is

$$f(T_{BP}^1) = y_C + y_B - 1 = 0.7339 + 0.2704 - 1 = 0.0043$$

The value of the derivative is (see Eq. (8.5.6))

$$\frac{df(T_{BP})}{dT_{BP}} = 0.7339 \times \frac{2577.1}{(T_{BP}^1 - 42.412)^2} + 0.2704 \times \frac{2771.9}{(T_{BP}^1 - 53.226)^2} = 0.0317$$

We calculate the new estimation of the bubble-point temperature

$$T_{BP}^2 = T_{BP}^1 - \frac{f(T_{BP}^1)}{\frac{df(T_{BP}^1)}{dT_{BP}}} = 334.1613 - \frac{0.0043}{0.0317} = 334.0248$$

Third Iteration.

We then calculate the saturation pressure of the two components:

$$P_C^{sat} = \exp\left(9.2170 - \frac{2577.1}{T_{BP}^2 - 42.412}\right) = 1.4618 \text{ bar}$$

$$P_B^{\text{sat}} = \exp\left(9.2521 - \frac{2771.9}{T_{BP}^2 - 53.226}\right) = 0.5382 \text{ bar}$$

The equilibrium ratios are

$$K_C = \frac{P_C^{\text{sat}}}{P} = \frac{1.6183}{1} = 1.4618$$

$$K_B = \frac{P_B^{\text{sat}}}{P} = \frac{0.6056}{1} = 0.5382$$

and the mole fractions in the vapor phase:

$$y_C = K_C x_C = 1.4618 \times 0.5 = 0.7309$$

$$y_B = K_B x_B = 0.5382 \times 0.5 = 0.2691$$

The value of the function is

$$f(T_{BP}^2) = y_C + y_B - 1 = 0.7309 + 0.2691 - 1 = 0.000007321 \approx 0$$

Convergence has been achieved as $f(T_{BP}^2) \approx 0$. The bubble-point temperature is 334.025 K or 60.87 °C.

Example 8.2. Calculation of the bubble-point temperature for a multicomponent mixture

For the following mixture.

Component	Formula	mol%	A	B	C
1: Methane	CH_4	5	9.186163	1020.1097	−0.490
2: Ethane	C_2H_6	10	9.068386	1519.1052	−16.719
3: Propane	C_3H_8	15	9.171012	1886.4988	−24.417
4: n-butane	C_4H_{10}	20	8.865000	2094.5465	−36.146
5: Benzene	C_6H_6	25	9.252109	2771.9325	−53.226
6: Toluene	C_7H_8	25	9.536317	3171.9906	−50.507

where A, B, and C are the constants of the Antoine equation (T in K and P^{sat} in bar). Calculate the bubble-point temperature at 5 bar.

Solution.

This is a multicomponent mixture, and performing the calculations manually is not an option. We will use MATLAB to solve the problem. We will perform the calculations line-by-line initially, and we will then organize them in a dedicated function.

We assign values to the constants of the Antoine equation, the liquid mole fractions, and the pressure.

```
clear all
% 1: METHANE  C1H4 (111.0 K)
% 2: ETHANE   C2H6 (184.6 K)
% 3: PROPANE  C3H8 (231.1 K)
% 4: n-BUTANE C4H10(273.1 K)
% 5: BENZENE  C6H6 (353.3 K)
% 6: TOLUENE  C7H8 (383.8 K)
% Antoine's constants from NIST Chemistry Webbook
%       CH4     C2H6     C3H8    n-C4H10 C6H6     C7H8
A  = [3.9895  3.93835 3.98292 3.85002 4.01814  4.14157 ]*log(10);
B  = [443.028 659.739 819.296 909.65  1203.835 1377.578]*log(10);
C  = [ -0.490 -16.719 -24.417 -36.146 -53.226  -50.507 ];
x  = [  0.050   0.100   0.150   0.200   0.250    0.250 ];
P  = 5;
```

It is important to note that we obtain the constants of the Antoine equation from the NIST Chemistry WebBook. But the form of the Antoine equation for which the values of the constants are given on the webpage is

$$\log_{10} P^{sat} = A' - \frac{B'}{C' + T}$$

In this book, we use the following form of the Antoine equation:

$$\ln P^{sat} = A - \frac{B}{C + T}$$

Using the fact that $\frac{\ln x}{\ln 10} = \log_{10} x$ we see that

$$A = A' \log 10$$

$$B = B' \log 10$$

$$C = C'$$

We then calculate the boiling point of the components at the given pressure of 5 bar.

```
>> format short g
>> Tb  = -C+B./(A-log(P))

Tb =
135.13        220.38        273.9        324.83        415.92        450.66
```

We then calculate the first estimation of the bubble-point temperature.

```
>> TBP = Tb*x(:)

TBP =
        351.49
```

We then calculate the vapor pressures of the components at this initial estimation of the temperature.

```
>> Ps = exp(A-B./(C+TBP))

Ps =
533.72      92.819      30.06      9.2341      0.95942      0.36698
```

Then we calculate the mole fractions of the components in the vapor phase.

```
>> y  = (Ps/P).*x

y =
5.3372      1.8564      0.9018      0.36937      0.047971      0.018349
```

We finally determine the value of the function and its derivative.

```
>> e  = sum(y)-1

e =
        7.5311

>> de = y * (B(:)./(C(:)+TBP).^2)

de =
      0.095175
```

The new estimation of the bubblepoint temperature is as follows.

```
>> TBP  = TBP - e/de

TBP =
        272.36
```

These commands are organized in a function shown in Table 8.2.

Table 8.2 MATLAB function for solving the bubble-point problem for multicomponent ideal mixtures using Raoult's law.

```
function [TBP,y] = BubblePoint_Raoult(x,P,A,B,C,tol)

% Calculates the bubble point temperature of
% an ideal mixture using Raoult's law

% x     : mole fraction in the liquid phase
% P     : pressure in bar
% A,B,C: vectors with Antoine's constants:
%        ln(Ps)=A-B/(C+T), T in K Ps in bar
% tol   : tolerance for convergence

% TBP   : bubblepoint temperature in K
% y     : mole fractions in the first bubble

%% initial estimate of bubblepoint temperature
Tb  = -C+B./(A-log(P)); Tb=Tb(:);
TBP = Tb'*x(:);

%% apply Newton's method
e   = 1E6; iter=0;
while (abs(e)>tol)
% calculate vapor pressures
    Ps = exp(A-B./(C+TBP)); Ps=Ps(:);
% calculate mole fraction in vapor phase
    y  = (Ps/P).*x(:);
% error function and its derivative
    e  = sum(y)-1;
    de = y' * (B(:)./(C(:)+TBP).^2);
% Newton's iteration
    TBP  = TBP - e/de;
    iter = iter + 1;
    fprintf('Iter = %4d T=%12.6f e=%16.8f \n',iter,TBP,e)
end
```

We can now use the function of Table 8.2 to solve the problem.

```
>> [TBP,y]=BubblePoint_Raoult(x,P,A,B,C,1e-8)
Iter =    1 T=  272.361779 e=       7.53106677
Iter =    2 T=  232.306984 e=       1.93144562
Iter =    3 T=  219.127525 e=       0.38857404
Iter =    4 T=  217.676774 e=       0.03514961
Iter =    5 T=  217.659922 e=       0.00039904
Iter =    6 T=  217.659920 e=       0.00000005
Iter =    7 T=  217.659920 e=       0.00000000

TBP =

       217.66

y =

      0.89021
     0.090392
     0.016608
    0.0027582
    2.489e-05
    3.973e-06
```

8.5.2 DEW-point calculations

We refer to Fig. 8.4 and our aim is to solve the dew-point problem using Raoult's law as summarized by Eq. (8.4.4). We are given $\{y_i, P\}$ and we seek to calculate $\{x_i, T\}$. There are $NC+1$ unknowns, and we, therefore, need an equal number of equations. These equations are:

NC equilibrium equations

$$x_i = \frac{P}{P_i^{\text{sat}}(T_{DP})} y_i, \quad i = 1, 2, \ldots, NC \tag{8.5.10}$$

Summation equation

$$\sum_{i=1}^{NC} x_i = 1 \tag{8.5.11}$$

We replace Eq. (8.5.10) into Eq. (8.5.11) to obtain one equation in one unknown, which is the T_{DP}.

$$f(T_{DP}) = \sum_{i=1}^{NC} \frac{P}{P_i^{\text{sat}}(T_{BP})} y_i - 1 = 0 \tag{8.5.12}$$

The saturation pressure is determined using the Antoine's equation (see Eq. (8.5.4)).

One method to solve Eq. (8.5.12) is by using Newton's method, which is based on the derivative of the function $f(T_{BP})$. To determine the derivative we first note that

$$\frac{d}{dT_{DP}}\left(\frac{1}{P_i^{\text{sat}}}\right) = \frac{d}{dT_{DP}}\left[\exp\left(-A_i + \frac{B_i}{C_i + T_{DP}}\right)\right]$$
$$= \exp\left(-A_i + \frac{B_i}{C_i + T_{DP}}\right) \frac{d}{dT_{BP}}\left(-A_i + \frac{B_i}{C_i + T_{DP}}\right)$$

or

$$\frac{dP_i^{\text{sat}}}{dT_{DP}} = -\frac{1}{P_i^{\text{sat}}(T_{DP})} \frac{B_i}{(C_i + T_{DP})^2} \tag{8.5.13}$$

We then, using Eq. (8.5.12), have that

$$\frac{df(T_{DP})}{dT_{DP}} = \sum_{i=1}^{NC} y_i P \frac{d}{dT_{DP}}\left(\frac{1}{P_i^{\text{sat}}}\right) = -\sum_{i=1}^{NC} x_i \frac{B_i}{(C_i + T_{DP})^2} < 0 \tag{8.5.14}$$

The derivative is always negative and as a result the function is a strictly decreasing function of temperature. This ensures that a unique solution exists. Eqs. (8.5.7) and (8.5.8) are valid also in the case of dew-point calculations by replacing T_{BP} by T_{DP}, while an initial estimation for the dew-point temperature can be obtained by

$$T_{DP}^{0} = \sum_{i=1}^{NC} x_i T_{b,i} \tag{8.5.15}$$

Example 8.3. Calculation of the dew-point temperature for a binary mixture

For an equimolar mixture of cyclopentane (C) and benzene (B), calculate the dew-point temperature at 1 bar. The Antoine's equations for the two components are the following

$$\ln P_C^{sat} = 9.2170 - \frac{2577.1}{T - 42.412}$$

$$\ln P_B^{sat} = 9.2521 - \frac{2771.9}{T - 53.226}$$

where P^{sat} is given in bar and the temperature T in K.

Solution.

We use Eq. (8.5.7) to determine the boiling temperature of the two components at 1 bar:

$$T_{b,C} = \frac{B_C}{A_C - \ln(P)} - C_C = \frac{2577.1}{9.2170 - \ln(1)} - (-42.412)$$
$$= 322.013 \text{ K}$$

$$T_{b,B} = \frac{B_B}{A_B - \ln(P)} - C_B = \frac{2771.9}{9.2521 - \ln(1)} - (-53.226)$$
$$= 352.826 \text{ K}$$

An initial estimation of the bubble-point temperature is obtained using Eq. (8.5.15)

$$T_{DP}^{0} = \sum_{i=1}^{NC} y_i T_{b,i} = 0.5 \times 322.015 + 0.5 \times 352.826 = 337.42 \text{ K}$$

First Iteration.

We then calculate the saturation pressure of the two components:

$$P_C^{sat} = \exp\left(9.2170 - \frac{2577.1}{T_{BP}^0 - 42.412}\right) = 1.6183 \text{ bar}$$

$$P_B^{sat} = \exp\left(9.2521 - \frac{2771.9}{T_{BP}^0 - 53.226}\right) = 0.6056 \text{ bar}$$

The equilibrium ratios are

$$K_C = \frac{P_C^{sat}}{P} = \frac{1.6183}{1} = 1.6183$$

$$K_B = \frac{P_B^{sat}}{P} = \frac{0.6056}{1} = 0.6056$$

and the mole fractions in the liquid phase:

$$x_C = \frac{y_C}{K_C} = \frac{0.5}{1.6183} = 0.30897$$

$$x_B = \frac{y_B}{K_B} = \frac{0.5}{0.6056} = 0.82566$$

The value of the function is

$$f\left(T_{DP}^0\right) = x_C + x_B - 1 = 0.30897 + 0.82566 - 1 = 0.13463$$

The value of the derivative is (see Eq. (8.5.6))

$$\begin{aligned}
\frac{df(T_{DP})}{dT_{DP}} &= -\sum_{i=1}^{NC} x_i \frac{B_i}{(C_i + T_{BP})^2} \\
&= -x_C \frac{B_C}{\left(T_{DP}^0 + C_C\right)^2} - x_B \frac{B_B}{\left(T_{DP}^0 + C_B\right)^2} \\
&= -0.30897 \times \frac{2577.1}{\left(T_{DP}^0 - 42.412\right)^2} - 0.82566 \\
&\quad \times \frac{2771.9}{\left(T_{DP}^0 - 53.226\right)^2} \\
&= -0.037486
\end{aligned}$$

We calculate the new estimation of the bubble-point temperature

$$T_{DP}^{1} = T_{DP}^{0} - \frac{f(T_{DP}^{0})}{\dfrac{df(T_{DP}^{0})}{dT_{DP}}} = 337.42 - \frac{0.134593}{(-0.037485)} = 341.01$$

Second Iteration.

We then calculate the saturation pressure of the two components:

$$P_C^{sat} = \exp\left(9.2170 - \frac{2577.1}{T_{DP}^{1} - 42.412}\right) = 1.7976 \text{ bar}$$

$$P_B^{sat} = \exp\left(9.2521 - \frac{2771.9}{T_{DP}^{1} - 53.226}\right) = 0.68396 \text{ bar}$$

The equilibrium ratios are

$$K_C = \frac{P_C^{sat}}{P} = \frac{1.7976}{1} = 1.7976$$

$$K_B = \frac{P_B^{sat}}{P} = \frac{0.68396}{1} = 0.68396$$

and the mole fractions in the vapor phase:

$$x_C = \frac{y_C}{K_C} = \frac{0.5}{1.7976} = 0.27816$$

$$x_B = \frac{y_B}{K_B} = \frac{0.5}{0.68396} = 0.73103$$

The value of the function is

$$f(T_{DP}^{1}) = x_C + x_B - 1 = 0.27816 + 0.73103 - 1 = 0.0091888$$

The value of the derivative is (see Eq. (8.5.6))

$$\frac{df(T_{BP})}{dT_{BP}} = -0.27816 \times \frac{2577.1}{(T_{DP}^{1} - 42.412)^2} - 0.73103 \times \frac{2771.9}{(T_{DP}^{1} - 53.226)^2} = -0.032507$$

We calculate the new estimation of the bubble-point temperature

$$T_{DP}^{2} = T_{DP}^{1} - \frac{f(T_{BP}^{1})}{\dfrac{df(T_{BP}^{1})}{dT_{BP}}} = 341.01 - \frac{0.0091888}{(-0.032507)} = 341.29$$

Third Iteration.

We then calculate the saturation pressure of the two components:

$$P_C^{sat} = \exp\left(9.2170 - \frac{2577.1}{T_{DP}^2 - 42.412}\right) = 1.8123 \text{ bar}$$

$$P_B^{sat} = \exp\left(9.2521 - \frac{2771.9}{T_{DP}^2 - 53.226}\right) = 0.69046 \text{ bar}$$

The equilibrium ratios are

$$K_C = \frac{P_C^{sat}}{P} = \frac{1.8123}{1} = 1.8123$$

$$K_B = \frac{P_B^{sat}}{P} = \frac{0.69046}{1} = 0.69046$$

and the mole fractions in the vapor phase:

$$x_C = \frac{y_C}{K_C} = \frac{0.5}{1.8123} = 0.27589$$

$$x_B = \frac{y_B}{K_B} = \frac{0.5}{0.69046} = 0.72416$$

The value of the function is

$$f\left(T_{DP}^2\right) = x_C + x_B - 1 = 0.27589 + 0.72416 - 1 = 0.000050727$$

Convergence has been achieved as $f\left(T_{DP}^2\right) \approx 0$. The dew-point temperature is 341.29 K or 68.14 °C.

8.6 Flash calculations using Raoult's law

One of the most important practical applications of vapor–liquid equilibrium (VLE) is the flash calculation. The name comes from a physical phenomenon: when a liquid is at a high pressure—equal to or above its bubble-point pressure—and the pressure is suddenly reduced, part of the liquid quickly "flashes" into vapor. This results in a two-phase system where liquid and vapor coexist in equilibrium. In this section, we focus specifically on what is called a $P - T$ flash calculation. This refers to determining how much vapor (V) and how much liquid (L) are present in a system, as well as the composition of each phase, given three known quantities:

- the pressure (P),
- the temperature (T),
- and the overall composition of the mixture (z_i; $i = 1, 2, \ldots, NC$), i.e., how much of each chemical is present in total.

To explain the process more clearly, imagine we have F mole of a nonreactive mixture containing several chemical components. The overall composition is described by the mole fractions $\{z_i\}$, which indicate how much of each component is present in the whole mixture. We assume, with reference to the Txy diagram of a binary mixture shown in Fig. 8.3, that the conditions are such that the mixture corresponds to the two-phase region, i.e., part of the mixture is in the vapor phase and the remainder in the liquid phase. Although the two coexisting phases have the same temperature T and pressure P, they have different compositions indicated by the VLE conditions. We assume that in a flash unit the mixture splits into two phases:

- the liquid phase contains L mole, with its own composition described by mole fractions $\{x_i\}$,
- the vapor phase contains V mole, with mole fractions $\{y_i\}$.

The goal of the flash calculation is to determine

- the amounts of vapor and liquid formed,
- the composition of each phase,

while ensuring that the total material is conserved and that each component is in phase equilibrium. As we will see shortly, this is arguably the most important calculation in process systems engineering.

The problem involves $2NC + 2$ unknowns $\{x_i, y_i, L, V\}$ and we need an equal number of equations, which are the following (see Fig. 8.5):

Component mole balances

$$z_i F = x_i L + y_i V, \quad i = 1, 2, \ldots, NC \tag{8.6.1}$$

Equilibrium relationships

$$y_i = K_i(T, P)\, x_i, \quad i = 1, 2, \ldots, NC \tag{8.6.2}$$

Summation relationships

$$\sum_{i=1}^{NC} x_i - 1 = 0 \tag{8.6.3}$$

$$\sum_{i=1}^{NC} y_i - 1 = 0 \tag{8.6.4}$$

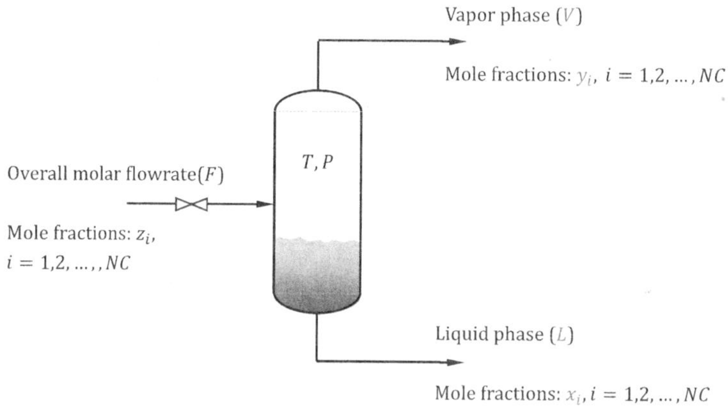

Figure 8.5 Summary of the "flash" calculation problem.

These are $2NC + 2$ independent equations whose solution gives the solution of the flash calculation problem. It is important to note at this point the overall mole balance

Total mole balance

$$F = L + V \tag{8.6.5}$$

can be obtained by adding all component mole balances and then using the summation constraints.

A particularly efficient method for solving Eqs. (8.6.1)–(8.6.4) was proposed by Rachford and Rice (RR) in 1952. To develop the RR method we first replace Eq. (8.6.2) into Eq. (8.6.1) and then solve for x_i:

$$z_i F = x_i L + x_i V K_i$$

or

$$x_i = \frac{z_i F}{F + V(K_i - 1)} \tag{8.6.6}$$

and

$$y_i = \frac{z_i F K_i}{F + V(K_i - 1)} \tag{8.6.7}$$

At this point we define the vapor fraction ψ

$$\psi = \frac{V}{F} \tag{8.6.8}$$

We subtract Eq. (8.6.3) from (8.6.4) and replace Eqs. (8.6.6), (8.6.7), and (8.6.8) to obtain

$$0 = \left(\sum_{i=1}^{NC} y_i - 1\right) - \left(\sum_{i=1}^{NC} x_i - 1\right) = \sum_{i=1}^{NC} \frac{z_i K_i}{1 + \psi(K_i - 1)} - \sum_{i=1}^{NC} \frac{z_i}{1 + \psi(K_i - 1)}$$

or

$$f_{RR}(\psi) = \sum_{i=1}^{NC} \frac{z_i(K_i - 1)}{1 + \psi(K_i - 1)} = 0 \qquad (8.6.9)$$

This function is known as the Rachford and Rice (RR) equation and is a single nonlinear equation in one unknown: the vapor fraction ψ. The Rachford-Rice equation is widely used due to its robust and desired mathematical characteristics, which make it especially suitable for phase equilibrium calculations.

The first derivative of the RR equation is the following

$$\frac{df_{RR}(\psi)}{d\psi} = -\sum_{i=1}^{NC} \frac{z_i(K_i - 1)^2}{[1 + \psi(K_i - 1)]^2} < 0 \qquad (8.6.10)$$

The Rachford-Rice function is monotonically decreasing function of the vapor fraction, denoted as ψ. This behavior arises from the fact that the derivative of the RR function with respect to ψ is always negative, assuming physically meaningful values of equilibrium ratios K_i. Consequently, the function decreases continuously across the domain.

Furthermore, the vapor fraction ψ is physically constrained to lie within the interval [0, 1], representing a system ranging from entirely liquid ($\psi = 0$) to entirely vapor ($\psi = 1$). Due to the monotonic nature of the RR function and the bounded domain, if a solution exists within the interval [0,1], it is guaranteed to be unique. This makes the equation particularly well-suited for numerical root-finding methods.

For a two-phase equilibrium solution to exist—that is, one in which both vapor and liquid are present—the RR function must exhibit a change of sign across the interval. Mathematically, this means that:

$$f_{RR}(0) f_{RR}(1) < 0 \qquad (8.6.11)$$

It is these properties of the Rachford-Rice equation that allow us to build a powerful algorithm for solving VLE problems of multicomponent mixtures when the equilibrium ratios do not depend on composition but on temperature and pressure only, i.e., $K_i = K_i(T, P)$. The steps of the algorithm are as follows (remember that the known variables are $\{P, T, z_i, F\}$ and the unknown $\{x_i, L, y_i, V\}$):

STEP 0: For the given temperature and pressure calculate the equilibrium ratios K_i.

STEP 1: Calculate $f_{RR}(0)$ and $f_{RR}(1)$,

STEP 2: If $f_{RR}(0)f_{RR}(1) > 0$ then the mixture is a single phase mixture. Terminate the calculation as no two-phase coexistence solution exists. Otherwise set $k = 1$, select an initial vapor fraction $\psi^k = \frac{f_{RR}(0)}{f_{RR}(0) - f_{RR}(1)}$ and continue to STEP 3.

STEP 3: Calculate $f_{RR}(\psi^k)$ and $\frac{df_{RR}(\psi^k)}{d\psi}$.

STEP 4: If $|f_{RR}(\psi^k)| <$ `tolerance`, the solution has been found, go to STEP 6.

STEP 5: Update the estimation of the vapor fraction $\psi^{k+1} = \psi^k - \frac{f_{RR}(\psi^k)}{\frac{df_{RR}}{d\psi}(\psi^k)}$, continue in STEP 3.

STEP 6: Calculate $V = \psi F$, and x_i, y_i from Eqs. (8.6.6) and (8.6.7).

Example 8.4. Flash calculations for a binary mixture

For an equimolar mixture of cyclopentane (C) and benzene (B), calculate the vapor fraction and the composition of the vapor and liquid phases at a temperature of $T = 65$ °C and pressure of $P = 1$ bar. The Antoine's equations for the two components are the following

$$\ln P_C^{sat} = 9.2170 - \frac{2577.1}{T - 42.412}$$

$$\ln P_B^{sat} = 9.2521 - \frac{2771.9}{T - 53.226}$$

where P^{sat} is given in bar and the temperature T in K.

Solution.

We first calculate the saturation pressure of the two components:

$$P_C^{sat} = \exp\left(9.2170 - \frac{2577.1}{(273.15 + 65) - 42.412}\right) = 1.6536 \text{ bar}$$

$$P_B^{\text{sat}} = \exp\left(9.2521 - \frac{2771.9}{(273.15+65) - 53.226}\right) = 0.6209 \text{ bar}$$

The equilibrium ratios are

$$K_C = \frac{P_C^{\text{sat}}}{P} = \frac{1.6536}{1} = 1.6536$$

$$K_B = \frac{P_B^{\text{sat}}}{P} = \frac{0.6209}{1} = 0.6209$$

The Rachford-Rice function is the following

$$f_{RR}(\psi) = \sum_{i=1}^{NC} \frac{z_i(K_i-1)}{1+\psi(K_i-1)} = \frac{z_C(K_C-1)}{1+\psi(K_C-1)} + \frac{z_B(K_B-1)}{1+\psi(K_B-1)} = \frac{0.3268}{1+0.6536\psi} - \frac{0.1895}{1-0.3791\psi}$$

We then calculate $f_{RR}(0)$ and $f_{RR}(1)$:

$$f_{RR}(0) = \frac{0.3268}{1+0.6536\times 0} - \frac{0.1895}{1-0.3791\times 0} = 0.1372$$

$$f_{RR}(1) = \frac{0.3268}{1+0.6536\times 1} - \frac{0.1895}{1-0.3791\times 1} = -0.1076$$

We note that $f_{RR}(0)\cdot f_{RR}(1) < 0$ and we calculate ψ^1.

$$\psi^1 = \frac{f_{RR}(0)}{f_{RR}(0) - f_{RR}(1)} = \frac{0.1372}{0.1372 - (-0.1076)} = 0.5604$$

We calculate the value and the derivative of the Rachford-Rice equation at ψ^1

$$f_{RR}(\psi^1) = \frac{0.3268}{1+0.6536\psi^1} - \frac{0.1895}{1-0.3791\psi^1} = -0.0015$$

$$\frac{df_{RR}}{d\psi}(\psi^1) = -\frac{0.5(1.6536-1)^2}{[1+0.5604\times(1.6536-1)]^2} - \frac{0.5(0.6209-1)^2}{[1+0.5604\times(0.6209-1)]^2} = -0.2303$$

We calculate a new estimate of the vapor fraction

$$\psi^2 = \psi^1 - \frac{f_{RR}(\psi^k)}{\frac{df_{RR}}{d\psi}(\psi^k)} = 0.5604 - \frac{(-0.0015)}{(-0.2303)} = 0.5539$$

We note that

$$f_{RR}(\psi^2) = \frac{0.3268}{1+0.6536\psi^2} - \frac{0.1895}{1-0.3791\psi^2} = -2.9207\ 10^{-8}$$

which is practically zero, and we terminate the calculation of the vapor fraction. We assume that $F = 100$ kmol and then $V = \psi F = 55.39$ kmol and using Eqs. (8.6.6) and (8.6.7)

$$x_C = \frac{z_C F}{F + V(K_C - 1)} = \frac{z_C}{1+\psi(K_C-1)}$$
$$= \frac{0.5}{1+0.5539 \times (1.6536-1)} = 0.3671$$

$$y_C = K_C x_C = 1.6536 \times 0.3671 = 0.6070$$

$$x_B = \frac{z_B}{1+\psi(K_B-1)} = \frac{0.5}{1+0.5539 \times (0.6209-1)} = 0.6329$$

$$y_B = K_B x_B = 0.6209 \times 0.6329 = 0.3930$$

Example 8.5. Flash calculations for a multicomponent mixture

100 kmol/h of the mixture studied in Example 8.2 are fed to a flash unit (see Fig. 8.5) at temperature $T = 25$ °C and pressure $P = 5$ bar. Determine the flowrates and the compositions of the two product streams.

Use the data for the Antoine's equation provided in Example 8.2.

Solution.

As this is a multicomponent mixture, performing the calculations manually is not an option. We therefore will be using MATLAB. We first create a function that evaluates the Rachford-Rice equation.

```
function [f,df]=RR(psi,z,K)
% calculate the Rachford Rice
% equation and its derivative
f = 0; df=0;
for i=1:length(z);
    term = (K(i) -1)/( 1 + psi*(K(i)-1) );
    f  = f  + z(i)*term;
    df = df - z(i)*term^2;
end
```

Then we generate a function that applies the steps of the algorithm presented in this section.

```
function [psi,x,y]=flash_Raoult(z,K,tol)
% flash calculations for constant K values
f0 = RR(0,z,K); f1 = RR(1,z,K);
if (f0*f1>0)
    psi=[]; x=[], y=[];
    return
end
psi  = f0/(f0-f1);
f = max(abs(f0),abs(f1));
iter = 1;
while (abs(f)>tol)
    [f,df] = RR(psi,z,K);
    psi = psi - f/df;
    fprintf('Iter = %4d ψ = %12.8f f_RR =%14.8f \n',iter,psi,f)
    iter = iter + 1;
end
x = z(:)./(1+psi*(K(:)-1));
y = K(:).*x(:);
```

Finally, we generate the script file that solves the case study.

```
clear all
% 1: METHANE  C1H4 (111.0 K)
% 2: ETHANE   C2H6 (184.6 K)
% 3: PROPANE  C3H8 (231.1 K)
% 4: n-BUTANE C4H10(273.1 K)
% 5: BENZENE  C6H6 (353.3 K)
% 6: TOLUENE  C7H8 (383.8 K)
% Antoine's constants from NIST Chemistry Webbook
%       CH4     C2H6     C3H8     n-C4H10 C6H6     C7H8
A  = [  3.9895 3.93835  3.98292      3.85002 4.01814   4.14157]*log(10);
B  = [443.028  659.739  819.296 909.65   1203.835 1377.578]*log(10);
C  = [ -0.490  -16.719  -24.417 -36.146 -53.226  -50.507];
z  = [  0.050   0.100    0.150   0.200    0.250    0.250];
P  = 5;
T  = 273.15+25;

K = exp(A-B./(C+T))/P

[psi,x,y]=flash_Raoult(z,K,1E-8)

>> EXAMPLE_8_5

K =   63.4104     7.8551     1.9539     0.4777     0.0254     0.0076

Iter =     1 ψ =    0.12962267 f_RR =     0.51041271
Iter =     2 ψ =    0.16487028 f_RR =     0.15794092
Iter =     3 ψ =    0.17089245 f_RR =     0.02059544
Iter =     4 ψ =    0.17101001 f_RR =     0.00038716
Iter =     5 ψ =    0.17101005 f_RR =     0.00000014
Iter =     6 ψ =    0.17101005 f_RR =     0.00000000

psi =    0.1710

x =
    0.0043
    0.0460
    0.1290
    0.2196
    0.3000
    0.3011

y =
    0.2716
    0.3616
    0.2520
    0.1049
    0.0076
    0.0023
```

8.7 The modified Raoult's law for nonideal liquid phase

8.7.1 Nonideal solutions

Ideal mixtures or solutions are those in which the components obey Raoult's law across the entire composition range. This implies that intermolecular interactions between unlike molecules (A–B) are essentially equal to those between like molecules (A–A and B–B). Typical characteristics of components in ideal mixtures are the following:

- Similar molecular size and shape
- Similar intermolecular forces (e.g., both are nonpolar or both are polar)
- Chemically and structurally similar
- No significant hydrogen bonding or specific interactions between different components

Common examples of chemical components forming ideal mixtures are nonpolar components such as benzene - toluene, n-hexane - n-heptane, cyclohexane – methylcyclohexane, ethane + propane (in the gaseous phase). Most gases approximate ideal behavior under low pressure and moderate temperatures, such as oxygen-nitrogen, methane and carbon dioxide, etc. Mixtures of polar components exhibit usually nonideal behavior; classical examples are mixtures of alcohols with water.

Fig. 8.6 illustrates a binary mixture that exhibits ideal behavior. The T–x–y diagram presents experimental data alongside theoretical predictions based on Raoult's law, showing close agreement between the experimental data and the thermodynamic model. The corresponding x–y diagram is also included, clearly demonstrating the system's ideality (the xy diagram appears to be symmetric).

However, such behavior is uncommon—most real mixtures deviate from ideality due to disparities in intermolecular interactions, leading to either positive or negative deviations from Raoult's law. A representative nonideal system is shown in Fig. 8.7, which depicts the T–x–y and x–y diagrams for the methanol–water binary mixture. In this case, the deviations between experimental data and Raoult's law predictions are substantial, underscoring the significant impact of hydrogen bonding and molecular polarity on phase behavior. It is interesting to note that the xy diagram lacks the symmetry observed in Fig. 8.6b.

The nonideality demonstrated in Fig. 8.7 is due to nonidealities appearing in the liquid phase as molecular interactions of dissimilar components are significant, mainly in the liquid phase.

We will not get into a detailed presentation of the underlying theory on the thermodynamics of nonideal solutions as this tends to be a rather demanding concept. However, for the needs of this introductory text, we can simply say that when nonideality is mainly due to the liquid phase, a simple modification of Raoult's law is sufficient for engineering applications. More specifically, for nonideal solutions the use of the **modified Raoult's law** is satisfactory (compare with Eq. (8.4.2)):

$$y_i P = \gamma_i(x_1, x_2, \ldots, x_{NC}) x_i P_i^{\mathrm{sat}}(T), \quad i = 1, 2, \ldots, NC \tag{8.7.1}$$

where $\gamma_i(x_1, x_2, \ldots, x_{NC})$ is the **activity coefficient** of component i in a mixture of NC components, which accounts for nonidealities in the liquid phase and depends on the composition of the liquid phase (and possibly on temperature).

The data that have been used to create Fig. 8.7 for methanol and water are presented in Table 8.3, together with the activity coefficients of the two components calculated by Eq. (8.7.1):

$$\gamma_i(x_1, x_2, \ldots, x_{NC}) = \frac{y_i P}{x_i P_i^{\mathrm{sat}}(T)}, \quad i = 1, 2, \ldots, NC \tag{8.7.2}$$

The activity coefficients are also presented in Fig. 8.8.

One of the earlier models for approximating the activity coefficients is the van Laar model, which for a binary mixture has the form

$$\ln \gamma_1 = a\left(\frac{bx_2}{ax_1 + bx_2}\right)^2 \tag{8.7.3.a}$$

$$\ln \gamma_2 = b\left(\frac{ax_1}{ax_1 + bx_2}\right)^2 \tag{8.7.3.b}$$

where a, b are constants different for each binary mixture. These constants are derived through analysis of experimental data, such as that presented in Table 8.3 for the methanol–water mixture. The process of determining these values is known as parameter estimation, a fundamental technique in process systems engineering. Parameter estimation involves using mathematical models and optimization techniques to adjust model parameters so that the predicted behavior aligns closely with experimental observations.

In the remainder of this section, we will introduce and apply this methodology in detail. While this material provides valuable insights into model calibration and is crucial for the development of predictive models in thermodynamics and transport

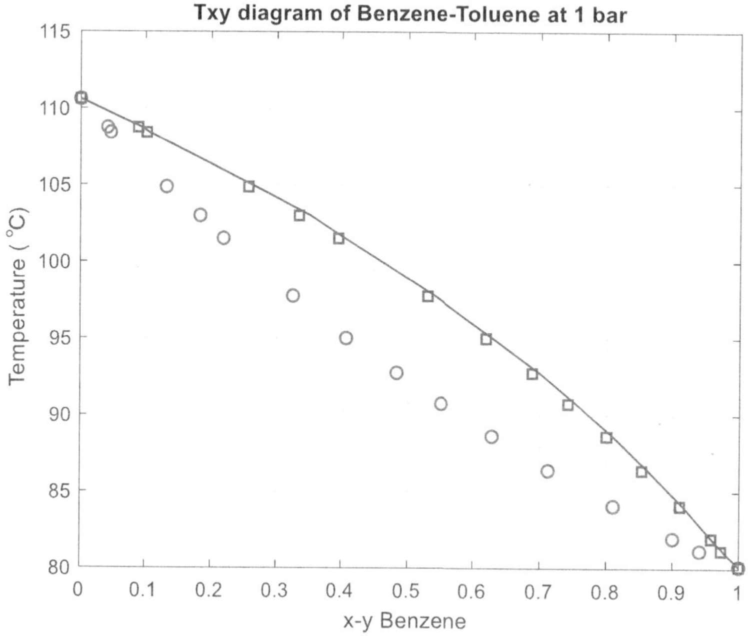

(a)

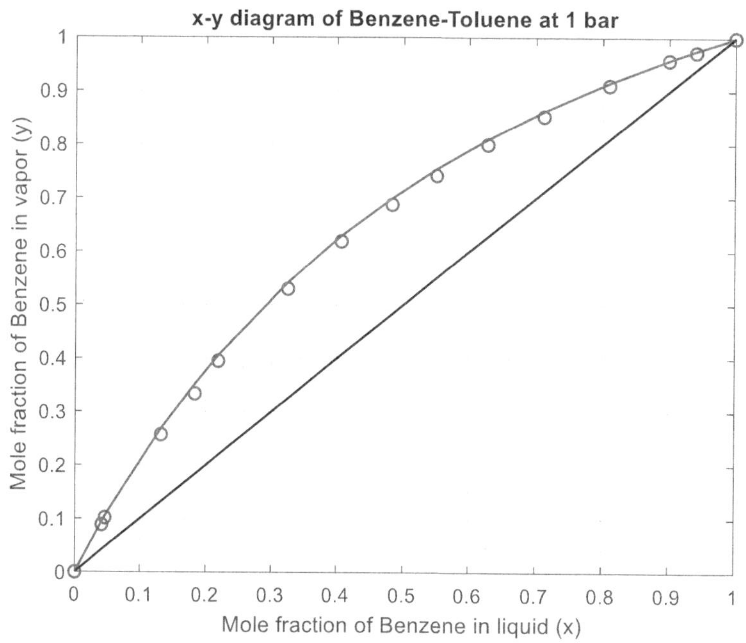

(b)

Figure 8.6 T–x–y and x–y phase diagrams for the binary system benzene–toluene, exhibiting ideal solution behavior. Experimental data are represented by symbols, while the continuous curves correspond to theoretical predictions based on Raoult's law.

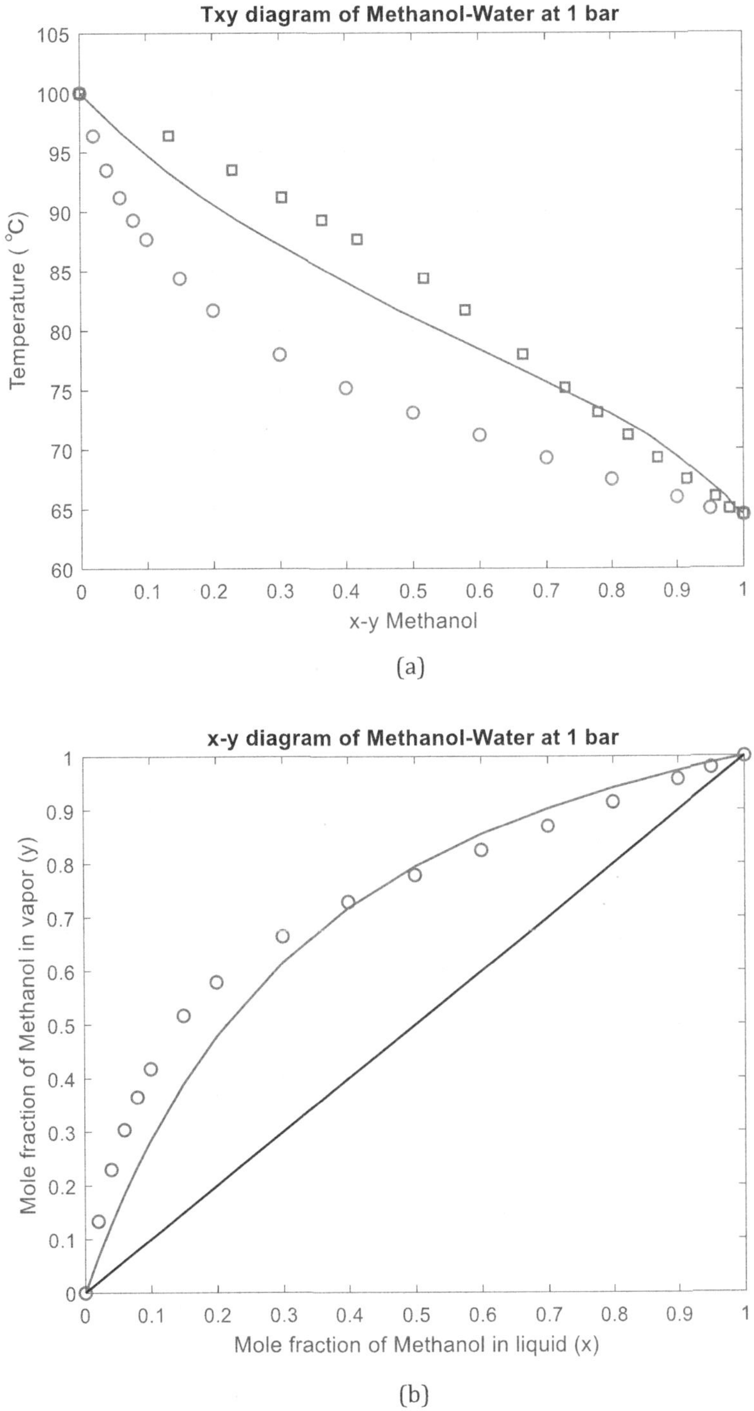

Figure 8.7 T–x–y and x–y phase diagrams for the binary system methanol–water, exhibiting nonideal solution behavior. Experimental data are represented by symbols, while the continuous curves correspond to theoretical predictions based on Raoult's law.

phenomena, it is recognized as a more advanced topic. Readers who are primarily interested in the final model results may choose to skip the methodological discussion and proceed directly to the summary of findings at the end of Section 8.7.2 that follows.

8.7.2 Nonlinear parameter estimation

In this section, we address the following fundamental question:

- given the experimental values of the activity coefficients provided in Table 8.3, as a function of the liquid-phase composition, and
- a thermodynamic model (see Eqs. (8.7.3.a) and (8.7.3.b)) for estimating activity coefficients that depends on two parameters, a and b, how can we determine the optimal values of a and b such that the model accurately captures the behavior of the real system?

Table 8.3 VLE data for the nonideal binary mixture of methanol and water at 1 bar.

T (°C)	x_{MeOH}	y_{MeOH}	γ_{MeOH}	γ_{Water}
100.0	0.00	0.000		1.0000
96.4	0.02	0.134	2.1541	1.0060
93.5	0.04	0.230	2.0315	1.0160
91.2	0.06	0.304	1.9314	1.0223
89.3	0.08	0.365	1.8534	1.0241
87.7	0.10	0.418	1.7924	1.0202
84.4	0.15	0.517	1.6552	1.0194
81.7	0.20	0.579	1.5279	1.0509
78.0	0.30	0.665	1.3350	1.1104
75.2	0.40	0.729	1.2154	1.1767
73.1	0.50	0.779	1.1229	1.2579
71.2	0.60	0.825	1.0641	1.3502
69.3	0.70	0.870	1.0337	1.4517
67.5	0.80	0.915	1.0193	1.5404
66.0	0.90	0.958	1.0054	1.6267
65.0	0.95	0.979	1.0122	1.7009
64.5	1.00	1.000	1.0000	

Data from Perry's Handbook.

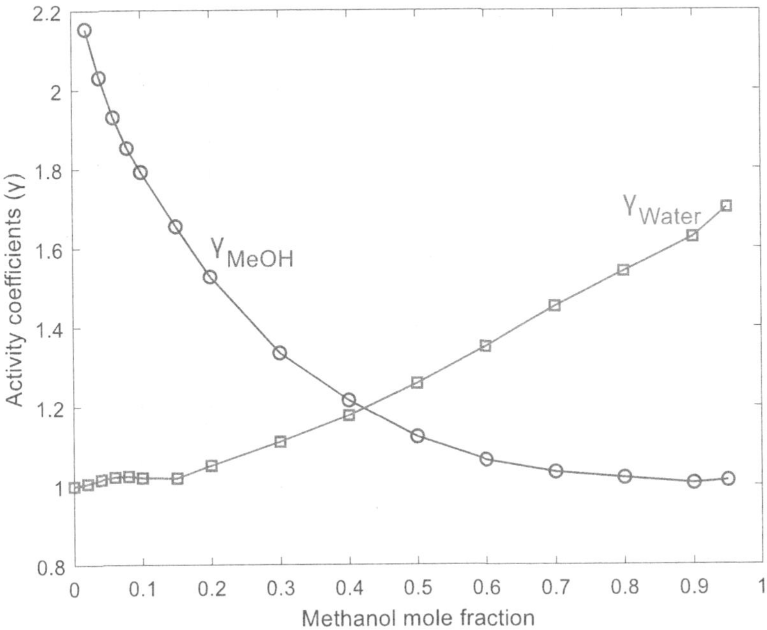

Figure 8.8 Activity coefficients in the nonideal mixture of methanol and water.

This process, known as parameter estimation, is central to developing reliable predictive models in thermodynamics and process systems engineering. The goal is to adjust the model parameters so that the computed values (the activity coefficients predicted by the model) closely match the experimental data.

To quantify how well the model fits the data, we use objective functions to assign a score to a particular choice of parameters. The objective function is typically based on the differences or deviations between experimental and predicted values (called the residuals). A common choice is the sum of squared errors (SSE), defined as:

$$\mathrm{SSE}(a,b)=\sum_{i=1}^{N}\left[\left(\gamma_1^{\exp}(x_i)-\gamma_1^{\text{model}}(x_i;a,b)\right)^2+\left(\gamma_2^{\exp}(x_i)-\gamma_2^{\text{model}}(x_i;a,b)\right)^2\right] \tag{8.7.4}$$

where $\gamma_1^{\exp}(x_i)$ is the experimentally measured value of the activity coefficient of component 1 (methanol in our case) at the i-th experimental point at which the composition of the liquid phase is x_i

$\gamma_2^{exp}(x_i)$ is the experimentally measured value of the activity coefficient of component 2 (water in our case) at the i-th experimental point at which the composition of the liquid phase is x_i

$\gamma_1^{model}(x_i; a, b)$ is the calculated value of the activity coefficient of component 1 (methanol in our case) at the i-th experimental point at which the composition of the liquid phase is x_i , using the mathematical model and its parameters a, b,

$\gamma_2^{model}(x_i; a, b)$ is the calculated value of the activity coefficient of component 2 (water in our case) at the i-th experimental point at which the composition of the liquid phase is x_i, using the mathematical model and its parameters a, b.

The squared values of the residuals, i.e., the differences between experimental and model-predicted values, are used to ensure that all deviations, regardless of sign, contribute positively to the overall error metric. This approach prevents large positive and negative errors from offsetting one another, which could otherwise result in a deceptively low total deviation and mask significant discrepancies in the model's predictive performance.

The idea on how to determine the constants a and b is relatively simple:

a. assume arbitrary values for the two constants a, b,

b. use these values to determine the activity coefficients at all experimental points,

c. determine the residuals at each experimental point and evaluate the SSE,

d. change the values of the two model parameters a, b and repeat the process.

After completing the procedure, we select the values of the model parameters that result in the minimum value of the objective function SSE as the optimal values.

To demonstrate the steps, let us assume that we were given that $a = 1$ and $b = 0.5$, then for the first experimental point of Table 8.3, where $x = 0.02$, we may use Eq. (8.73.a) and (8.7.3.b) to calculate the activity coefficients:

$$\gamma_1^{model}(x_1; a, b) = \exp\left[a\left(\frac{b(1-x_1)}{ax_1 + b(1-x_1)}\right)^2\right]$$

$$= \exp\left[1\left(\frac{0.5 \times 0.98}{1 \times 0.02 + 0.5 \times 0.98}\right)^2\right] = 2.5171$$

$$\gamma_2^{\text{model}}(x_1; a, b) = \exp\left[b\left(\frac{ax_1}{ax_1 + b(1 - x_1)}\right)^2\right]$$

$$= \exp\left[0.5\left(\frac{1 \times 0.02}{1 \times 0.02 + 0.5 \times 0.98}\right)^2\right] = 1.0008$$

We then calculate the $SSE\ (x_1; a, b)$:

$$SSE(x_1; a, b) = \left(\gamma_1^{\text{exp}}(x_1) - \gamma_1^{\text{model}}(x_1; a, b)\right)^2 + \left(\gamma_2^{\text{exp}}(x_1) - \gamma_2^{\text{model}}(x_1; a, b)\right)^2$$

$$= (2.1541 - 2.5171)^2 + (1.0060 - 1.0008)^2 = 0.1318$$

We need to repeat the calculations for all experimental points and finally add all $SSE(x_i)$ to determine the objective function. This cannot be performed manually, and we need to rely on the computer.

We first create the following function that calculates the activity coefficients.

```
function gama = vanLaar(x, a, b)
term = a*x+b*(1-x);
gama(1) = exp(a*( b*(1-x) / term)^2 );
gama(2) = exp(b*( a*x     / term)^2 );
end
```

We then create a function that calculates the SSE given the parameters a, b and the experimental data available

```
function f = SSE(x,a,b,gamaExp)
sse = 0;
for i=1:length(x)
    gamaModel = vanLaar(x(i),a,b);
    se = ( gamaModel(1) - gamaExp(i,1) )^2 + ...
         ( gamaModel(2) - gamaExp(i,2) )^2;
    sse = sse + se;
end
f = sse;
end
```

We then read the experimental data, generate several combinations of the two parameters, evaluate the objective function, and plot the results in a 3D diagram.

```
clear all
d =[96.4     0.02  0.134 2.1541      1.0060 % experimental data
    93.5     0.04  0.230 2.0315      1.0160
    91.2     0.06  0.304 1.9314      1.0223
    89.3     0.08  0.365 1.8534      1.0241
    87.7     0.10  0.418 1.7924      1.0202
    84.4     0.15  0.517 1.6552      1.0194
    81.7     0.20  0.579 1.5279      1.0509
    78.0     0.30  0.665 1.3350      1.1104
    75.2     0.40  0.729 1.2154      1.1767
    73.1     0.50  0.779 1.1229      1.2579
    71.2     0.60  0.825 1.0641      1.3502
    69.3     0.70  0.870 1.0337      1.4517
    67.5     0.80  0.915 1.0193      1.5404
    66.0     0.90  0.958 1.0054      1.6267
    65.0     0.95  0.979 1.0122      1.7009];
t=d(:,1); x=d(:,2); y=d(:,3); gamaExp=d(:,4:5)
a = linspace(0.70,0.90,61)
b = linspace(0.45,0.70,61)
for i=1:length(a)
    for j=1:length(b)
        f(j,i) = SSE(x,a(i),b(j),gamaExp);
    end
end

figure(1)
[aa,bb]=meshgrid(a,b);
meshc(aa,bb,f)
grid on
xlabel('a')
ylabel('b')
zlabel('SSE')
axis([min(a) max(a) min(b) max(b) 0.01 0.07])
```

The results are presented in the 3D plot of Fig. 8.9. Note that $SSE(a, b)$ has a clear minimum. The best values of the parameters are $a = 0.7982$ and $b = 0.5637$.

8.7.3 Flash calculations using the modified Raoult's law

When the modified Raoult's law applies, then the equilibrium ratios are defined as

$$K_i(x_1, x_2, \ldots, x_{NC}, P, T) = \frac{y_i}{x_i} = \gamma_i(x_i)\frac{P_i^{sat}(T)}{P}, \quad i = 1, 2, \ldots, NC \tag{8.7.5}$$

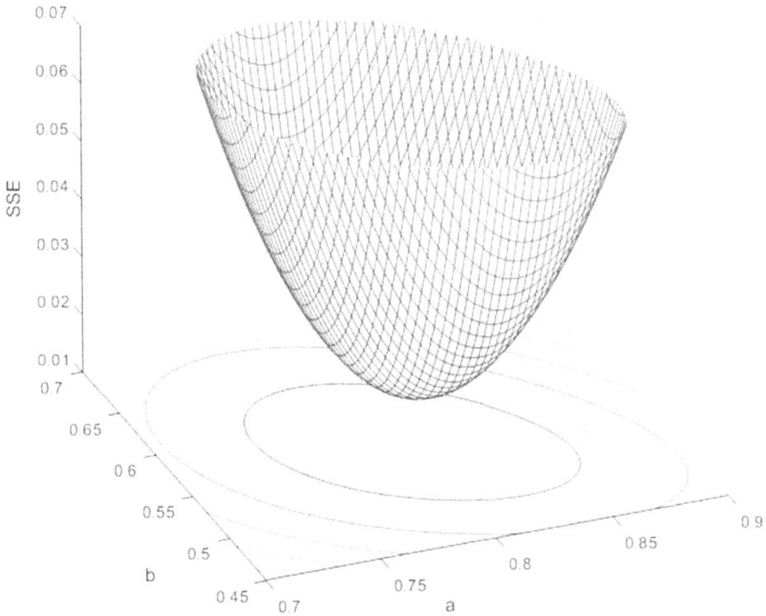

Figure 8.9 3D diagram of the SSE as a function of the two model parameters.

In the case of Raoult's law, the equilibrium ratios depend only on temperature and pressure. This is advantageous because the equilibrium ratios can be easily calculated and remain constant during the solution of the Rachford-Rice equation. However, under the modified Raoult's law, the equilibrium ratios also depend on the liquid-phase compositions, which are unknown. To handle this added complexity, an iterative approach can be used.

1. Assume initial values for the liquid-phase compositions $\{x_i\}$.
2. Using the given pressure and temperature, calculate the corresponding equilibrium ratios.
3. Solve the Rachford-Rice equation to obtain updated estimates of the liquid-phase compositions.
4. Repeat the process until the solution converges.

This algorithm has proven to be particularly efficient in practice. The initial estimates for the liquid-phase composition can be obtained using Raoult's law, assuming ideal behavior by setting the activity coefficients to unity. However, one potential issue should be noted: the algorithm may fail if, at any intermediate step, the Rachford-Rice equation yields a single-phase solution. In such cases, the algorithm becomes stuck. Importantly, this single-phase result may not represent the correct physical solution, as it could stem from an inappropriate initial guess for the liquid-phase composition. When this occurs, it is necessary to fall back on the general mathematical formulation of the equilibrium flash calculation to ensure accuracy. In these cases, we need to rely on the general mathematical formulation of the equilibrium flash calculation:

Component mole balances

$$z_i = (1-\psi)x_i + \psi y_i, \quad i = 1, 2, \ldots, NC \tag{8.7.6}$$

Equilibrium relationships

$$y_i P - \gamma_i(x_1, x_2, \ldots, x_{NC}, T, P)\, x_i P_i^{\text{sat}}(T), \quad i = 1, 2, \ldots, NC \tag{8.7.7}$$

Summation relationships

$$\sum_{i=1}^{NC} x_i - \sum_{i=1}^{NC} y_i = 0 \tag{8.7.8}$$

This is a set of $2NC + 1$ equations in $2NC + 1$ unknowns x_i, y_i, ψ all of which obtain values between 0 and 1.

Example 8.6. Flash calculations for a binary mixture of methanol and water

For an equimolar mixture of methanol (Me) and water (W), calculate the vapor fraction and the composition of the vapor and liquid phases at a temperature of $T = 78\ ^\circ\text{C}$ and pressure of $P = 1$ bar. The Antoine's equations for the two components are the following

$$\ln P_{\text{Me}}^{\text{sat}} = 11.9829 - \frac{3641.2}{T - 33.5}$$

$$\ln P_{\text{W}}^{\text{sat}} = 11.7053 - \frac{3829.5}{T - 45.622}$$

where P^{sat} is given in bar and the temperature T in K. Use the van Laar model with $a = 0.7982$ and $b = 0.5637$.

Solution.

We first calculate the saturation pressure of the two components:

$$P_{\text{Me}}^{\text{sat}} = \exp\left(11.9829 - \frac{3641.2}{(273.15 + 78) - 33.5}\right) = 1.6821 \text{ bar}$$

$$P_{\text{W}}^{\text{sat}} = \exp\left(11.7053 - \frac{3829.5}{(273.15 + 78) - 45.622}\right) = 0.4366 \text{ bar}$$

The equilibrium ratios are

$$K_C = \frac{P_{\text{Me}}^{\text{sat}}}{P} = \frac{1.653456}{1} = 1.6821$$

$$K_B = \frac{P_{\text{W}}^{\text{sat}}}{P} = \frac{0.620978}{1} = 0.4366$$

The Rachford-Rice function is the following

$$f_{RR}(\psi) = \sum_{i=1}^{NC} \frac{z_i(K_i - 1)}{1 + \psi(K_i - 1)} = \frac{0.34105}{1 + 0.6821\psi} - \frac{0.2817}{1 - 0.5634\psi}$$

We then calculate $f_{RR}(0)$ and $f_{RR}(1)$:

$$f_{RR}(0) = \frac{0.34105}{1 + 0.6821 \times 0} - \frac{0.2817}{1 - 0.5634 \times 0} = 0.05935$$

$$f_{RR}(1) = \frac{0.34105}{1 + 0.6821 \times 1} - \frac{0.2817}{1 - 0.5634 \times 1} = -0.4424$$

We note that $f_{RR}(0) \cdot f_{RR}(1) < 0$ and we calculate ψ^1, $\psi^1 = \frac{f_{RR}(0)}{f_{RR}(0) - f_{RR}(1)} = \frac{0.05935}{0.05935 - (-0.44246)} = 0.1182$.

We solve the Rachford-Rice equation and find that $\psi = 0.1544$ and

$$x_{Me} = \frac{z_{Me}}{1 + \psi(K_{Me} - 1)} = \frac{0.5}{1 + 0.1544 \times (1.6821 - 1)} = 0.4524$$

$$x_W = 1 - x_{Me} = 0.5476$$

Since we have a first estimation of the liquid-phase compositions, we may use Eq. (8.7.3a) and (8.7.3b) to calculate the activity coefficients:

$$\gamma_{Me} = \exp\left[\alpha\left(\frac{bx_2}{\alpha x_1 + bx_2}\right)^2\right] = \exp\left[0.7982\left(\frac{0.5637 \times 0.5476}{0.7982 \times 0.4524 + 0.5637 \times 0.5476}\right)^2\right] = 1.1847$$

$$\gamma_W = \exp\left[b\left(\frac{\alpha x_1}{\alpha x_1 + bx_2}\right)^2\right] = \exp\left[0.5637\left(\frac{0.7982 \times 0.4524}{0.7982 \times 0.4524 + 0.5637 \times 0.5476}\right)^2\right] = 1.1780$$

We correct the values of the equilibrium ratios to account for the activity coefficients:

$$K_C = \gamma_{Me}\frac{P_{Me}^{sat}}{P} = 1.1847\frac{1.6821}{1} = 1.9929$$

$$K_B = \gamma_W\frac{P_W^{sat}}{P} = 1.1780\frac{0.4366}{1} = 0.5134$$

We solve the Rachford-Rice equation again to obtain that $\psi = 0.5259$ and

$$x_{Me} = \frac{z_{Me}}{1 + \psi(K_{Me} - 1)} = \frac{0.5}{1 + 0.5259 \times (1.9929 - 1)} = 0.3285$$

$$x_W = 1 - x_{Me} = 0.6715$$

The significant change in the liquid-phase composition indicates that further iteration is required. However, performing these iterations manually is inefficient and susceptible to numerical inaccuracies. Therefore, computational implementation is employed to automate the iterative procedure and ensure both accuracy and convergence efficiency. To this end, use is made of the following script file in MATLAB.

```
clear all

t = 78;
T = t+273.15;
P = 1;
z = [0.5 0.5];
A = [5.20409  5.08354 ]*log(10);
B = [1581.341 1663.125]*log(10);
C = [-33.50   -45.622 ] ;
a = 0.7982;
b = 0.5637;
Ps = exp(A-B./(C+T))
K  = Ps/P

iter = 0;
while (iter==0)|(max(abs(K-Kold))>1e-6)
   iter = iter + 1;
   Kold = K;
   [psi,x,y]=flash_Raoult(z,K,1e-8);
   gama=vanLaar(x(1),a,b);
   K=(Ps/P).*gama;
   fprintf('Iter = %4d x_{Me} =%12.6f ψ=%12.8f \n',iter,x(1),psi)
end
```

```
>> EXAMPLE_8_6

Iter =     1 x_{Me} =    0.452360 ψ=  0.15440804
Iter =     2 x_{Me} =    0.328477 ψ=  0.52592219
Iter =     3 x_{Me} =    0.298495 ψ=  0.55221707
Iter =     4 x_{Me} =    0.289273 ψ=  0.56242321
Iter =     5 x_{Me} =    0.286299 ψ=  0.56578936
Iter =     6 x_{Me} =    0.285327 ψ=  0.56689603
Iter =     7 x_{Me} =    0.285008 ψ=  0.56726006
Iter =     8 x_{Me} =    0.284903 ψ=  0.56737983
Iter =     9 x_{Me} =    0.284868 ψ=  0.56741925
Iter =    10 x_{Me} =    0.284857 ψ=  0.56743222
Iter =    11 x_{Me} =    0.284853 ψ=  0.56743649
Iter =    12 x_{Me} =    0.284852 ψ=  0.56743789
Iter =    13 x_{Me} =    0.284851 ψ=  0.56743836
Iter =    14 x_{Me} =    0.284851 ψ=  0.56743851
```

Note that after an initial abrupt change in the estimated value of the liquid-phase composition and of the vapor fraction, a slow convergence of the algorithm to the solution is observed.

The alternative is to solve Eqs. (8.7.6) to (8.7.8) as a system of equations. We first create a function that returns the right-hand side of the equations given the estimation of x_i, y_i, ψ:

```
function f=FlashBinary_vanLaar(X,z,P,T,A,B,C,a,b)

x = X(1:2); y = X(3:4); psi = X(5);
Ps       = exp(A-B./(C+T)) ;
gama(1)  = exp(a*(b*x(2)/(a*x(1)+b*x(2)))^2);
gama(2)  = exp(b*(a*x(1)/(a*x(1)+b*x(2)))^2);

for i=1:2
    f(i)    = z(i) - (1-psi)*x(i) - psi*y(i);
    f(i+2)  = y(i)*P - gama(i)*x(i)*Ps(i);
end
f(5) = sum(x) - sum(y);
```

and then use the `fsolve` command to obtain the solution.

```
>> fsolve(@FlashBinary_vanLaar,[z z 0.5],optimset('Display','it-
er'),z,P,T,A,B,C,a,b)

                                         Norm of      First-order   Trust-region
 Iteration  Func-count     f(x)          step         optimality    radius
     0          6         0.270797                          0.732          1
     1         12         0.100147       0.190039            0.4           1
     2         18        0.0060782       0.337113         0.0509           1
     3         24      6.76355e-06       0.167062        0.00139           1
     4         30      4.86013e-13     0.00459444       3.73e-07           1

Equation solved.

fsolve completed because the vector of function values is near zero
as measured by the value of the function tolerance, and
the problem appears regular as measured by the gradient.

ans =     0.2849     0.7151     0.6640     0.3360     0.5674
```

The convergence is fast, and the solution agrees well with the solution obtained with the iterative methodology.

8.7.4 Azeotropes

An azeotrope is a mixture of liquids that boils at a constant temperature and retains the same composition in both the liquid and vapor phases at that boiling point. This characteristic makes azeotropes behave like pure substances during phase transitions, despite being mixtures. The defining feature of an azeotrope is its constant boiling point, which can be either a maximum or a minimum compared to the boiling points of the pure components. These are known respectively as maximum-boiling and minimum-boiling azeotropes.

The existence of azeotropes is attributed to nonideal intermolecular interactions in the liquid phase. Unlike ideal mixtures that obey Raoult's law across the entire composition range, azeotropic mixtures deviate significantly due to strong or weak interactions between dissimilar molecules. These deviations can be either positive (leading to minimum-boiling azeotropes, where vapor pressure is higher than predicted) or negative (leading to maximum-boiling azeotropes, where vapor pressure is lower than predicted).

Azeotropes are of critical importance in both theoretical and applied chemistry. They play a significant role in industrial processes such as petroleum refining, pharmaceutical production, and alcohol purification, where they can either hinder or be exploited for efficient separation. For example, the ethanol-water azeotrope, which boils at 78.2 °C and consists of approximately 90 mol% (96% ethanol by volume), presents a well-known challenge in the production of absolute ethanol. Overcoming azeotropic behavior in such systems often requires advanced separation techniques such as azeotropic distillation, extractive distillation, pressure-swing distillation, or the use of entrainers and membrane-based methods.

Understanding the thermodynamic principles that govern azeotropy is essential for designing and optimizing separation processes. Moreover, modern computational tools and phase diagram analyses are increasingly employed to predict azeotropic behavior and guide the selection of appropriate separation strategies.

In Fig. 8.10 the Txy diagram for the ethanol-water mixture is presented. This is a typical minimum-boiling azeotrope that represents an extreme case of nonideal VLE behavior.

To determine the azeotropic point computationally, we use the fact that at the azeotropic point $x_i = y_i$, which gives the following VLE equations

Equilibrium relationships

$$P - \gamma_i\left(x_1^{\text{azeo}}, x_2^{\text{azeo}}, \ldots, x_{NC}^{\text{azeo}}, T^{\text{azeo}}, P\right) P_i^{\text{sat}}\left(T^{\text{azeo}}\right) = 0, \quad i = 1, 2, \ldots, NC \tag{8.7.9}$$

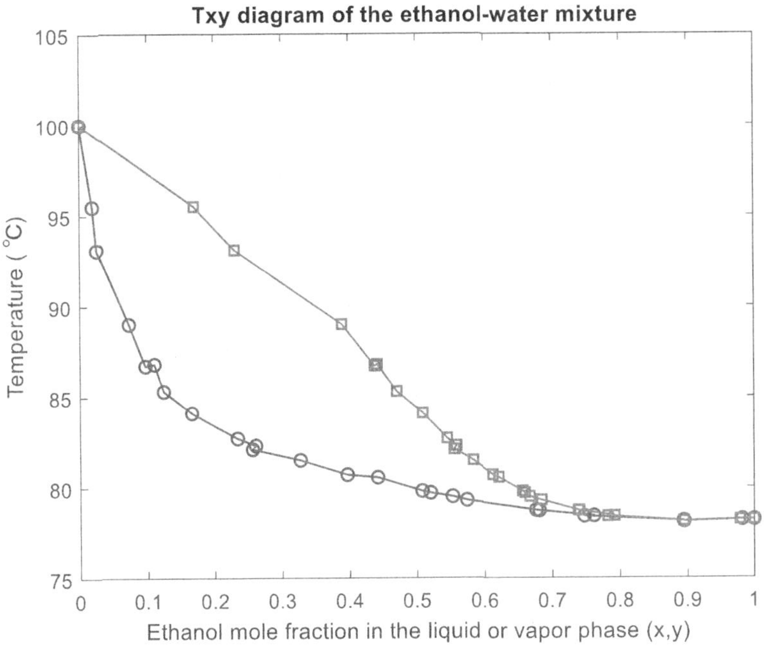

Figure 8.10 Txy diagram for the ethanol-water binary mixture. An azeotrope appears at 78.2 °C and 89 mol% ethanol at 1 bar.

Summation relationships

$$\sum_{i=1}^{NC} x_i^{\text{azeo}} - 1 = 0 \tag{8.7.10}$$

These are $NC+1$ equations in $NC+1$ unknowns $x_i^{\text{azeo}}, T^{\text{azeo}}$. Since all equations used to estimate the activity coefficients are expressed in terms of the natural logarithm of the activity coefficient, and the vapor pressure is also expressed in terms of its natural logarithm, Eq. (8.7.9) can be rewritten as follows:

$$\begin{aligned}&\ln P - \ln \gamma_i\left(x_1^{\text{azeo}}, x_2^{\text{azeo}}, \ldots, x_{NC}^{\text{azeo}}, T^{\text{azeo}}, P\right) - \ln P_i^{\text{sat}}\left(T^{\text{azeo}}\right)\\&= 0, i = 1, 2, \ldots, NC\end{aligned} \tag{8.7.11}$$

If the van Laar equation is used for the activity coefficients and the Antoine equation for the vapor pressure, then Eq. (8.7.11) can be written as:

$$\ln P - a\left(\frac{b\left(1 - x_1^{\text{azeo}}\right)}{ax_1^{\text{azeo}} + b\left(1 - x_1^{\text{azeo}}\right)}\right)^2 - A_1 + \frac{B_1}{C_1 + T^{\text{azeo}}} = 0 \tag{8.7.12}$$

$$\ln P - b\left(\frac{ax_1^{\text{azeo}}}{ax_1^{\text{azeo}} + b(1 - x_1^{\text{azeo}})}\right)^2 - A_2 + \frac{B_2}{C_2 + T^{\text{azeo}}} = 0 \quad (8.7.13)$$

For the case of a binary ethanol (EtOH)-water (W) mixture we are given the following

$$\ln P_{\text{EtOH}}^{\text{sat}} = 12.0590 - \frac{3667.7}{T - 46.966}$$

$$\ln P_{\text{W}}^{\text{sat}} = 11.9648 - \frac{3984.9}{T - 39.724}$$

and $a = 1.6798$, $b = 0.9227$. The azeotropic point is found by solving Eqs. (8.7.12) and (8.7.13) in MATLAB:

```
clear all
x0 = [0.9; 273.15+50]
x = fsolve(@ Azeo_EtOH_Water,x0,optimset('Display','iter'))

function f = Azeo_EtOH_Water(x)
A=[12.0590 11.9648];
B=[3667.7 3984.9];
C=[-46.966 -39.724];
a = 1.6798   ;
b = 0.9227   ;
P = 1.013;
f(1) = log(P) - a *  (b*(1-x(1))/ (a*x(1) + b*(1-x(1))) )^2 + ...
      - A(1) + B(1)/(C(1)+x(2)) ;
f(2) = log(P) - b *  (a*x(1)     / (a*x(1) + b*(1-x(1))) )^2 + ...
      - A(2) + B(2)/(C(2)+x(2)) ;
f=f(:);
end
```

```
>> azeotrope

                                          Norm of      First-order   Trust-reg
 Iteration  Func-count     f(x)            step        optimality    radius
     0           3         3.10362                         1.19
     1           6         1.18618              1         0.615
     2           9        0.427799            2.5         0.095              2
     3          12        0.295541           6.25       0.00871             6.
     4          15        0.157253         15.625         0.383             15
     5          18      0.00104159        4.33869         0.011             15
     6          21     4.83963e-07       0.687302      0.000303             15
     7          24     1.59019e-13      0.0146364      1.73e-07             15
Equation solved.
x =

    0.9161
  351.0151
```

Chapter summary

In this chapter, we introduced the basis of VLE calculations of bubble-point and dew-point calculations and the most important one, which is the flash calculation. We restricted attention to vapor-liquid equilibrium and to ideal mixtures or mixtures that exhibit nonideality in the liquid phase. More complex cases and thermodynamic models are usually covered in chemical thermodynamics courses. These calculations, especially the flash calculation, are among the most fundamental calculations in chemical and process engineering.

Problems

8.1 For a particular flash distillation unit the feed composition of the four components and the corresponding equilibrium ratios are.

Component	mol% in feed	Equilibrium ratio
1	30	7.0
2	10	2.4
3	15	0.8
4	45	0.3

a. Write down the Rachford-Rice function.
b. Plot the Rachford-Rice function for $\psi \in [0, 1]$.
c. Apply the bisection method to find the solution.
d. Apply Newton's method to find the solution.
e. Apply Halley's method to find the solution.
f. State your general conclusions.

8.2 A flash distillation unit operates at 50 °C and 3 bar and is fed with 100 kmol/h of a feed that is 30 mol% propane, 10 mol% n-butane, 15 mol% n-pentane, and 45 mol% n-hexane. These components are considered to form an ideal mixture. Find the vapor and liquid product compositions and flowrates.

8.3 A flash distillation unit operates at 100 °C and 5 bar and is fed with 100 kmol/h of a feed that is 10 mol% ethane, 10 mol% propane, 10 mol% n-butane, 20 mol% n-pentane, 20 mol% n-hexane, 20 mol% n-heptane, and 10 mol% n-octane. These components are considered to form an ideal mixture. Find the vapor and liquid product compositions and flowrates.

8.4 An equimolar mixture of n-pentane and n-hexane is fed to a flash distillation unit which operates at 60 °C and one half of the feed is obtained as vapor product. What is the operating pressure of the flash distillation unit?

8.5 An equimolar mixture of methane and n-hexane is fed to a train of two flash distillation units which operate at a constant pressure of 10 bar and both have $\psi = \frac{3}{4}$. Calculate the temperature of both units and the composition of all streams.

8.6 In Table P8.6 experimental VLE data for the nonideal mixture of benzene (1) and cyclohexane (2) at 283.15 K (10 °C) are given. Use these experimental data to determine the constants of the van Laar model for the estimation of the activity coefficients (Fig. 8.11).

8.7 In Table P8.7 experimental VLE data for the nonideal mixture of n-hexane (1) and ethanol (2) at 1.013 bar are given. Use these experimental data to determine the constants of the van Laar model for the estimation of the activity coefficients.

8.8 A mixture of n-hexane (1) and ethanol (2), 30 mol% in hexane, is fed to a flash distillation unit at $T = 65$ °C and $P = 1.013$ bar. Find the vapor fraction and the composition of the vapor-phase and liquid-phase product streams. Use the van Laar model for calculating the activity coefficients with $a = 1.908$ and $b = 2.78$ and the following Antoine equations:

$$\ln P_1^{sat} = 9.2921 - \frac{2739.2}{T - 46.87}$$

Table P8.6 VLE data for benzene and cyclohexane.

$x_{benzene}$	$y_{benzene}$	P (bar)
0.0610	0.0953	0.0659
0.2149	0.2710	0.0698
0.3187	0.3600	0.0714
0.4320	0.4453	0.0717
0.5246	0.5106	0.0722
0.6117	0.5735	0.0714
0.7265	0.6626	0.0697
0.8040	0.7312	0.0684
0.8830	0.8200	0.0662
0.8999	0.8382	0.0656

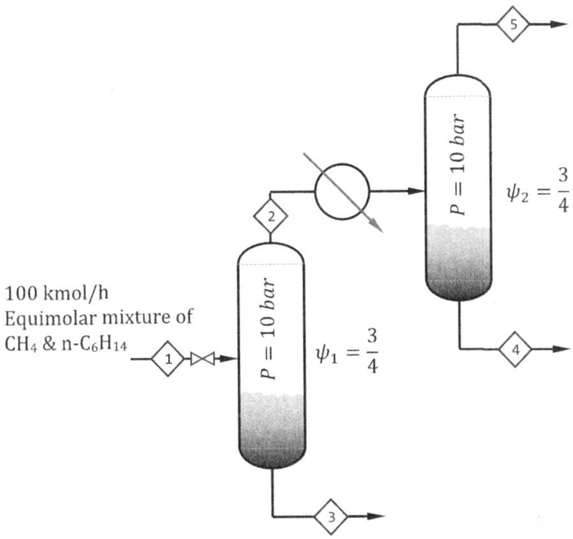

Figure P8.5 Two flash distillation units in series.

Table P8.7 VLE data for n-hexane and ethanol.

x	y	T (°C)
0.000	0.000	78.30
0.010	0.095	76.00
0.020	0.193	73.20
0.060	0.365	67.40
0.080	0.420	65.90
0.152	0.532	61.80
0.245	0.605	59.40
0.333	0.630	58.70
0.452	0.640	58.35
0.588	0.650	58.10
0.670	0.660	58.00
0.725	0.670	58.25
0.765	0.675	58.45
0.898	0.710	59.15
0.955	0.745	60.20
0.990	0.840	63.50
0.994	0.935	66.70
1.000	1.000	68.70

$P = 1.013$ bar

$$\ln P_2^{sat} = 12.059 - \frac{3667.7}{T - 46.97}$$

8.9 Find the azeotropic composition of a mixture of n-hexane (1) and ethanol (2) at $P = 1.013$ bar. Use the van Laar model for calculating the activity coefficients with $a = 1.908$ and $b = 2.78$ and the following Antoine equations:

$$\ln P_1^{sat} = 9.2921 - \frac{2739.2}{T - 46.87}$$

$$\ln P_2^{sat} = 12.059 - \frac{3667.7}{T - 46.97}$$

8.10 Show that for the flash separation problem of binary mixtures at given pressure P and temperature T with feed composition z there is an analytic solution to the problem given by

$$x_1 = \frac{1 - K_1}{K_1 - K_2}$$

$$y_1 = K_1 x_1$$

$$\psi = \frac{z_1}{1 - K_2} - \frac{z_2}{K_1 - 1}$$

8.11 An alternative equation to the Rachford-Rice equation for calculating the vapor fraction ψ in a flash distillation problem is the following:

$$f(\psi) = \sum_{i=1}^{NC} \frac{z_i K_i}{1 + \psi(K_i - 1)} - 1$$

a. How is this equation derived?

b. Use the data given in Table P8.11 to plot the Rachford-Rice equation and the alternative one. What are your observations?

c. Explain the reasons why the Rachford-Rice equation is preferred over the alternative.

8.12 In Table P8.12 experimental data for the VLE of acetone (1) – chloroform (2) nonideal binary mixture at 1.013 bar are presented.

a. Create the Txy plot and the xy plot for this binary mixture.

Table P8.11 Data for Problem 8.11.

Component	z_i	K_i
1	0.04	10
2	0.08	5
3	0.12	2
4	0.16	1
5	0.20	0.5
6	0.30	0.25
7	0.10	0.125

Table P8.12 VLE data for acetone and chloroform.

$x_{acetone}$	$y_{acetone}$	T (°C)
0.0817	0.05	62.50
0.1390	0.10	62.82
0.2338	0.20	63.83
0.3162	0.30	64.30
0.3535	0.35	64.37
0.3888	0.40	64.35
0.4582	0.50	64.02
0.5299	0.60	63.33
0.6106	0.70	62.23
0.7078	0.80	60.72
0.8302	0.90	58.71
0.9075	0.95	57.48

b. Use the following Antoine equations to determine the parameters a and b of the van Laar model for estimating the activity coefficients:

$$\ln P_1^{sat}[\text{bar}] = 9.7677 - \frac{2787.5}{T[\text{K}] - 43.486}$$

$$\ln P_2^{sat}[\text{bar}] = 9.3936 - \frac{2696.2}{T[\text{K}] - 46.918}$$

c. Solve the equations to determine the azeotropic point.

d. Solve the bubble-point problem at each experimental point and compare the results with the experimental data given in Table P8.12. Use the optimal values of a and b of the van Laar model estimated in b.

8.13 A mixture of acetone and chloroform is fed to a flash distillation unit that operates at $P = 1.013$ bar and $T = 64$ °C. What is the molar flowrate and the composition of the vapor phase and of the liquid-phase products if the feed is $F = 100$ kmol/h and the mole fraction of acetone in the feed is $z_1 = 0.479$: (a) by using the van Laar model and the complete set of equations (b) using the experimental points of Table P8.12?

Further reading

Books on thermodynamics

Abbott, M.M., Van Ness, H.C., 1989. Theory and problems of thermodynamics. In: Schaums Outline Series, second ed. McGraw-Hill Book Company, NY.

Balzhiser, R.E., Samuels, M.R., Eliassen, J.D., 1972. Chemical Engineering Thermodynamics. Prentice Hall, Englewood Cliffs, NJ.

Daubert, T.E., 1985. Chemical Engineering Thermodynamics. McGraw-Hill Book Company, NY.

Elliot, J.R., Lira, C.T., 2012. Introductory Chemical Engineering Thermodynamics, second ed. Prentice Hall, Upper Saddle River, NJ.

Gmehling, J., Kolbe, B., Kleiber, M., Rarey, J., 2012. Chemical Thermodynamics for Process Simulation. Wiley-VCH, Weinheim.

Hall, K.R., Iglesias-Silva, G.A., 2022. Thermodynamics for Chemical Engineers. Wiley-VCH, Weinheim.

Koretsky, M.D., 2013. Engineering and Chemical Thermodynamics, second ed. John Wiley & Sons, NJ.

Kyle, B.G., 1999. Chemical and Process Thermodynamics. Prentice-Hall, NJ.

Matsoukas, T., 2012. Fundamentals of Chemical Engineering Thermodynamics. Prentice Hall, Upper Saddle River, NJ.

Sandler, S.I., 1977. Chemical and Engineering Thermodynamics. John Wiley and Sons, NY.

Tassios, D.P., 2003. Applied Chemical Engineering Thermodynamics. Springer-Verlag Berlin Heidelberg.

9

Process flowsheeting

9.1 Introduction

Flowsheeting refers to the use of computer tools to carry out mass and energy balances, followed by equipment sizing and cost estimation for process systems. Process and plant design typically involve three main activities: synthesis, analysis, and optimization.

- **Synthesis** focuses on selecting the best flowsheet from a range of alternatives through a structured and systematic approach.
- For each alternative generated, **analysis** is performed to solve the mass and energy balances, size and cost the equipment, and evaluate the overall process economics.
- Finally, **optimization**—both structural and parametric—is conducted to refine the design and identify the most promising options for further evaluation.

The goal of this book is to help you develop the skills required to perform mass and energy balances efficiently, as this foundational task underpins all stages of process and plant design.

In the previous chapters, we developed the fundamental tools needed to formulate mass and energy balances for relatively simple—and in some cases, idealized—process flowsheets. This approach helped build a foundational understanding of the problem and its inherent complexity.

In this chapter, we turn to a more challenging task: how to systematically organize and perform mass and energy balances within an integrated framework—one that **is general, reusable, scalable, and applicable to complete process flowsheets across a wide range of systems**.

Process flowsheeting dates back to the 1950s and has evolved significantly with advances in computer hardware and software. It was soon recognized that any flowsheeting program typically relies on three core components: numerical methods, thermophysical property prediction, and a user interface for building and interacting with the flowsheet. Through the user interface—which has advanced dramatically over the years—the user defines the **process topology** and specifies all necessary

Introduction to Process Systems Engineering. https://doi.org/10.1016/B978-0-443-45444-8.00012-9

parameters to fully describe the process model in mathematical terms. The mathematical models of the individual process units are then assembled, and numerical solvers are employed to find a solution to the complete set of mass and energy balances. At this stage, the role of thermophysical property libraries is crucial. Though often unnoticed by the user, these libraries provide essential data that underpin accurate calculations. Once a solution is found, the user interface reports the results to the user, completing a flowsheeting cycle.

The focus of this chapter is on the general structure of a flowsheeting system that is responsible for achieving a solution to the mass and energy balances of complete chemical plants. More specifically, we will present the two most fundamental approaches to process flowsheeting known as the **sequential modular approach** and the **equation-based approach** (or equation-solving approach).

Before exploring these two alternative approaches in detail through a simple yet comprehensive example, it's worth noting that the method we've used so far for solving mass and energy balances is an example of the equation-based approach. In this approach, the complete set of equations—representing all process units along with user-defined specifications—is assembled into a (typically large) system of nonlinear equations. These equations are then solved either simultaneously or using intelligent decomposition strategies. This approach is a natural and rigorous way to address mass and energy balance problems. However, it was initially less favored in industrial settings due to several practical disadvantages. That said, it also offers significant benefits, many of which have become increasingly evident in recent years as computational tools and techniques have advanced.

In the sequential modular approach, the simulation of a process flowsheet is structured around individual modules, each representing a specific process unit. These modules are built on a straightforward and intuitive logic: **if all required information about the incoming streams and relevant unit specifications is available, the module can be solved independently to determine all properties of the outgoing streams**.

To simulate an entire process, **the user introduces process units one at a time, in the sequence dictated by the process flow**—typically starting from feed streams and progressing toward final products. For each unit, the user specifies the necessary design and operating parameters, and the software solves the mass and energy balances locally for that module. **Once a unit is solved, its output streams become the input to the**

next downstream unit, and the process continues until all units in the flowsheet have been resolved. This modular, step-by-step approach offers simplicity, flexibility, and ease of debugging, making it especially attractive for process engineers during early design stages or for troubleshooting. However, it does rely on a solvable path through the flowsheet—meaning that the sequence of unit operations must allow for a clear progression of known variables. As a result, certain complex flowsheets with strong recycles or tightly coupled units may require tear streams and iterative convergence methods to ensure accurate and consistent results across the entire system.

After this extended—yet necessary—introduction, we now turn our attention to a representative process flowsheet that will serve as a practical case study. We will begin by clearly describing the process, outlining its key components, unit operations, and stream connections. Then, we will demonstrate how to solve the associated mass and energy balances using both of the alternative approaches discussed earlier: the sequential modular and the equation-based methods. This comparative analysis will not only illustrate the differences between the two approaches but also highlight their respective strengths, limitations, and practical considerations in real-world process simulation.

9.2 A representative process

Fig. 9.1 presents a representative process flow diagram. The feed to the system is an equimolar gas mixture of methane (CH_4) and n-hexane (n-C_6H_{14}), available at 130 °C and 10 bar, with a total molar flow rate of 100 kmol/h. The proposed separation process consists of two flash distillation units arranged in series, with a recycle stream from the second unit returning to the feed stream, as illustrated in Fig. 9.1.

The primary objective is to assess the feasibility of achieving both 95% purity and 95% recovery of methane in the product stream (stream 5). A 95% recovery criterion indicates that 95% of the methane introduced in the feed stream (stream 1) must be recovered in the final product. To evaluate whether these performance targets can be met—and to explore potential optimization opportunities—we will develop a computer simulation of the proposed process.

Our analysis will proceed in two stages. First, we will formulate an equation-based approach, where all governing equations are solved simultaneously. Then, we will implement a sequential

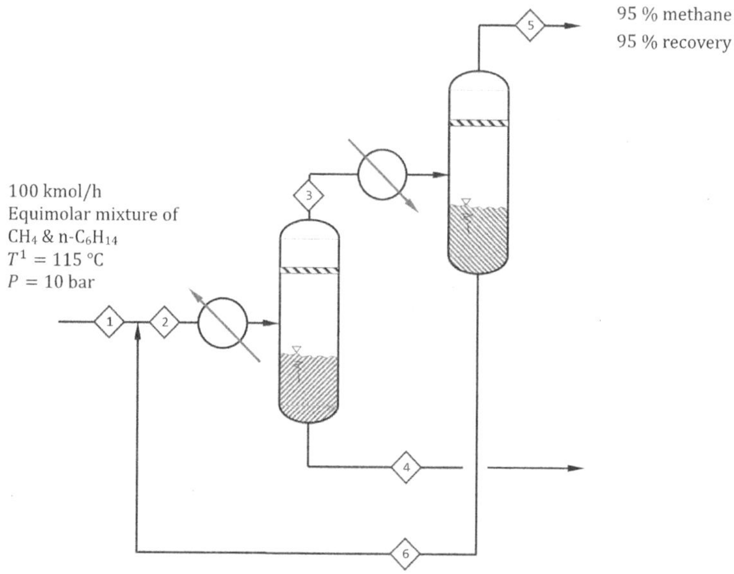

Figure 9.1 Presents a representative process flow diagram.

modular approach, where individual process units are solved in sequence using stream-wise information. To focus on the fundamental differences between these two flowsheeting methodologies and to keep the model tractable, we will neglect energy balances and restrict our analysis to mass balances only. This simplification allows us to concentrate on the structural behavior and solution strategies of the flowsheet without introducing the added complexity of energy considerations.

9.3 Application of the equation-solving approach

9.3.1 Application to the representative case study

This approach is the one that we have already followed in our introduction to mass and energy balances. In this approach we formulate all mass balances that can be written for the process, and count the number of variables and the number of equations in order to calculate the number of degrees of freedom. For the system of Fig. 9.1 we can formulate a model for (a) the mixing point (streams 1 and 6 are mixed to form stream 2), (b) the first flash distillation unit, and (c) the second flash distillation unit.

We present all relevant equations (note that at this chapter we drop the overdot notation for simplification and the symbol N is used for the total molar flowrate and the symbol x for the mole fraction of methanol, irrespective of whether there is a liquid or vapor phase):

Mixing point:
Total mole balance

$$f_1 = N_1 + N_6 - N_2 = 0 \tag{9.3.1}$$

Methane mole balance

$$f_2 = N_1 x_1 + N_6 x_6 - N_2 x_2 = 0 \tag{9.3.2}$$

First flash distillation unit.
Total mole balance

$$f_3 = N_2 - N_3 - N_4 = 0 \tag{9.3.3}$$

Methane mole balance

$$f_4 = N_2 x_2 - N_3 x_3 - N_4 x_4 = 0 \tag{9.3.4}$$

Equilibrium relationships

$$f_5 = x_3 - K_1(T_1, P)\, x_4 = 0 \tag{9.3.5}$$

$$f_6 = (1 - x_3) - K_2(T_1, P)\,(1 - x_4) = 0 \tag{9.3.6}$$

Second flash distillation unit.
Total mole balance

$$f_7 = N_3 - N_5 - N_6 = 0 \tag{9.3.7}$$

Methane mole balance

$$f_8 = N_3 x_3 - N_5 x_5 - N_6 x_6 = 0 \tag{9.3.8}$$

Equilibrium relationships

$$f_9 = x_5 - K_1(T_2, P)\, x_6 = 0 \tag{9.3.9}$$

$$f_{10} = (1 - x_5) - K_2(T_2, P)\,(1 - x_6) = 0 \tag{9.3.10}$$

Note that there are 14 variables appearing in the model

$$N_1, N_2, N_3, N_4, N_5, N_6, x_1, x_2, x_3, x_4, x_5, x_6, T_1, T_2$$

and that we have 10 equations (Eqs. (9.3.1)(9.3.10)). The feed stream is fully specified with two independent parameters: the total molar flow rate, $N_1 = 100$ kmol/h, and the composition, $x_1 = 0.5$ (mole fraction of methane). This leaves two additional degrees of freedom that can be independently specified within the system. While several combinations of two independent variables can be selected from the total set of 12 process variables, it is common practice to choose the temperatures of the two flash units (T_1 and T_2) as the independent variables.

However, an alternative and equally valid specification is to use the vapor fractions ($\psi_1 = N_3/N_2$ and $\psi_2 = N_5/N_3$) of the two flash units as the independent variables. This alternative offers a significant practical advantage: by specifying the vapor fractions directly, the existence of two-phase equilibrium in each flash unit is inherently ensured. In contrast, choosing temperatures as independent variables carries the risk that one or both flash units may fall outside the two-phase region—resulting in either a single liquid or single vapor phase—under which the flash model would no longer be valid.

Therefore, while both approaches are theoretically acceptable, specifying vapor fractions provides greater robustness and physical consistency in the context of this separation process. The disadvantage is that Eqs. (9.3.5), (9.3.6), (9.3.9), and (9.3.10) remain nonlinear in this case, while when the temperature is fixed, they are linear.

A MATLAB implementation of the mathematical model of the process shown in Fig. 9.1 is presented in Table 9.1. Using several combinations of values of the two vapor fractions achieved in the two flash distillation units, Figs. 9.2 and 9.3 are obtained. Note the 0.95 line for the methane mole fraction in the product stream. All combinations of ψ_1 and ψ_2 that result in mole fractions of methane larger than 0.95 are potentially acceptable solutions. Note also the 95% line in Fig. 9.3. All combinations of ψ_1 and ψ_2 that result in recoveries greater than 95% are potentially acceptable solutions. The selection $\psi_1 = 0.6$, $\psi_2 = 0.7$ satisfies both criteria and can be chosen. The MATLAB implementation of Table 9.1 is used to find all molar flowrates and compositions:

```
clear all
N0  = [150; 100;  50;  50; 50]
x0  = [0.45;0.8; 0.1; 0.9; 0.4]
T0  = [120; 50] + 273.15;
X0  = [N0;x0;T0]
X=fsolve(@Model,X0,optimset('display','iter'),[0.6;0.7])
fprintf('Methane in product stream    = %12.8f mol per cent \n',100*X(9))
fprintf('Methane recovery in product = %12.8f     per cent
\n',100*X(9)*X(4)/50)
```

Table 9.1 MATLAB implementation of Eqs. (9.3.1) to (9.3.10) and the two specifications on the vapor fractions.

```
function f=Model(X,psi)
N(1) = 100;          % molar flowrate of the feed in kmol/h
x(1) = 0.5;          % mole fraction of methane in feed
P = 10;              % system pressure
N(2:6) = X(1:5,1);   % molar flowrates of streams 2 to 6
x(2:6) = X(6:10);    % mole fractions of methane in streams 2 to 6
T = X(11:12);        % temperatures in the flash units in K
%% calculate equilibrium rations in the flash units
K1 = EqRatios(T(1),P);
K2 = EqRatios(T(2),P);
%% equations (9.3.1) to (9.3.10)
f(1) = N(1)      + N(6)      - N(2);
f(2) = N(1)*x(1) + N(6)*x(6) - N(2)*x(2);
f(3) = N(2)      - N(3)      - N(4);
f(4) = N(2)*x(2) - N(3)*x(3) - N(4)*x(4);
f(5) = x(3)     - K1(1)*x(4);
f(6) = (1-x(3)) - K1(2)*(1-x(4));
f(7) = N(3)      - N(5)      - N(6);
f(8) = N(3)*x(3) - N(5)*x(5) - N(6)*x(6);
f(9) = x(5)     - K2(1)*x(6);
f(10)= (1-x(5)) - K2(2)*(1-x(6));
%% specs
f(11)= N(3) - psi(1)*N(2);
f(12)= N(5) - psi(2)*N(3);
end

function K=EqRatios(T,P)
% calculate equilibrium ratios of the
% methane / n-hexane mixture using Raoult's law
A = [4.22061 4.00266]'*log(10);
B = [516.689 1171.53]'*log(10);
C = [11.223  -48.784]';
K = exp(A-B./(C+T))/P;
K = K(:);
end
```

Iteration	Func-count	f(x)	Norm of step	First-order optimality	Trust-region radius
0	13	1426.51		3.24e+03	1
1	26	916.109	1	1.08e+03	1
2	39	727.982	2.5	691	2.5
3	52	516.884	6.25	477	6.25

Figure 9.2 Plots of methane mole fraction in the product stream as a function of the two vapor fractions in the flash distillation units.

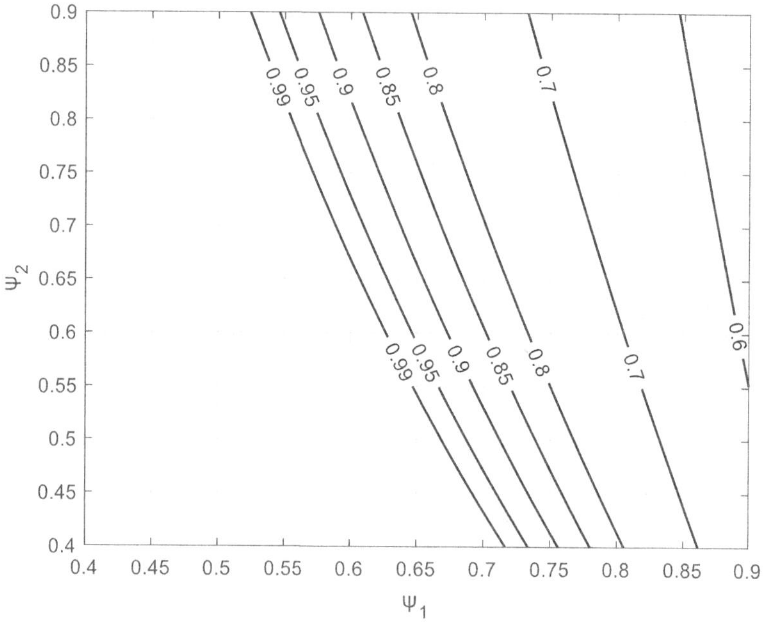

Figure 9.3 Plots of methane percent recovery in the product stream as a function of the two vapor fractions in the flash distillation units.

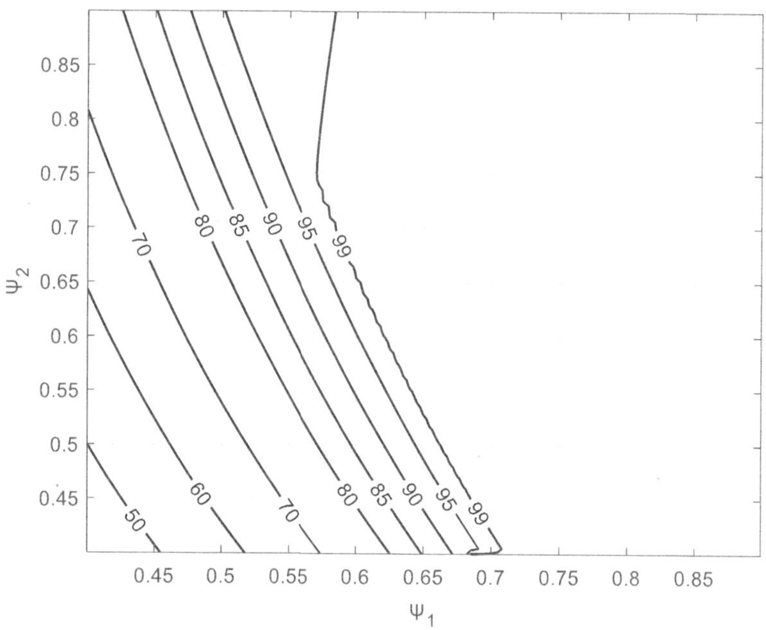

```
4        65        187.769        15.625        366        15.6
5        78        10.3073        25.0584       231        39.1
6        91        1.53609e-05    3.65422       0.154      62.6
7        104       2.11191e-12    0.122963      8.18e-05   62.6
8        117       4.44543e-24    0.000136842   1.33e-10   62.6

Equation solved.

X =
  121.9512
   73.1707
   48.7805
   51.2195
   21.9512
    0.4143
    0.6849
    0.0084
    0.9682
    0.0237
  383.4604
  309.7203

Methane in product stream    =  96.81935816 mol per cent
Methane recovery in product  =  99.18080592     per cent
```

This concludes our case study on how to apply the equation-solving approach for simulating the process shown in Fig. 9.1. The presence of the recycle stream caused no complications or additional computational problems. One point that needs further development is related to the numerical solution of systems of n nonlinear equations with an equal number of unknowns. This is presented next.

9.3.2 Solution of systems of nonlinear equations

In Chapter 5, Newton's method for solving nonlinear equations has been presented. For a real function of a real variable

$$f(x): R \rightarrow R \tag{9.3.11}$$

the Taylor series of f centered at a point x^k is given by:

$$f(x) = f(x^k) + \left.\frac{df}{dx}\right|_{x^k} (x - x^k) + \left.\frac{1}{2}\frac{d^2f}{dx^2}\right|_{x^k} (x - x^k)^2$$

$$+\ldots+\frac{1}{n!}\frac{df^n}{dx^n}\bigg|_{x^k}\left(x-x^k\right)^n+\ldots \tag{9.3.12}$$

To develop Newton's method we keep only the first derivative term and set the result equal to zero:

$$f(x)\approx f\left(x^k\right)+\frac{df}{dx}\bigg|_{x^k}\left(x-x^k\right)=0 \tag{9.3.13}$$

We solve for the Newton step:

$$s^k=x-x^k=-\frac{f\left(x^k\right)}{\left.\frac{df}{dx}\right|_{x^k}} \tag{9.3.14}$$

In this way we develop an iterative method (see Chapter 5) that is based on the following equation for generating improved estimations of the root of $f(x)=0$:

$$x^{k+1}=x^k-\frac{f\left(x^k\right)}{\left.\frac{df}{dx}\right|_{x^k}} \tag{9.3.15}$$

We now consider the case of two equations f_1 and f_2 in two variables x_1 and x_2 with the aim of solving the problem:

$$f_1(x_1,x_2)=0 \tag{9.3.16}$$

$$f_2(x_1,x_2)=0 \tag{9.3.17}$$

The Taylor series of the first equation is now approximated by

$$f_1(x_1,x_2)\approx f_1\left(x_1^k,x_2^k\right)+\frac{\partial f_1}{\partial x_1}\bigg|_{\left(x_1^k,x_2^k\right)}\left(x_1-x_1^k\right)+\frac{\partial f_1}{\partial x_2}\bigg|_{\left(x_1^k,x_2^k\right)}\left(x_2-x_2^k\right) \tag{9.3.18}$$

For the second function we have that

$$f_2(x_1,x_2)\approx f_2\left(x_1^k,x_2^k\right)+\frac{\partial f_2}{\partial x_1}\bigg|_{\left(x_1^k,x_2^k\right)}\left(x_1-x_1^k\right)+\frac{\partial f_2}{\partial x_2}\bigg|_{\left(x_1^k,x_2^k\right)}\left(x_2-x_2^k\right) \tag{9.3.19}$$

From Eqs. (9.3.16) to (9.3.19) we obtain the following

$$f_1\left(x_1^k,x_2^k\right)+\frac{\partial f_1}{\partial x_1}\bigg|_{\left(x_1^k,x_2^k\right)}\left(x_1-x_1^k\right)+\frac{\partial f_1}{\partial x_2}\bigg|_{\left(x_1^k,x_2^k\right)}\left(x_2-x_2^k\right)=0 \tag{9.3.20}$$

$$f_2\left(x_1^k, x_2^k\right) + \left.\frac{\partial f_2}{\partial x_1}\right|_{\left(x_1^k, x_2^k\right)} \left(x_1 - x_1^k\right) + \left.\frac{\partial f_2}{\partial x_2}\right|_{\left(x_1^k, x_2^k\right)} \left(x_2 - x_2^k\right) = 0 \quad (9.3.21)$$

Using vectors and matrices we can write these equations in a compact form

$$\begin{bmatrix} \frac{\partial f_1}{\partial x_1} & \frac{\partial f_1}{\partial x_2} \\ \frac{\partial f_2}{\partial x_1} & \frac{\partial f_2}{\partial x_2} \end{bmatrix}_{\left(x_1^k, x_2^k\right)} \begin{bmatrix} s_1^k \\ s_2^k \end{bmatrix} = -\begin{bmatrix} f_1 \\ f_2 \end{bmatrix}_{\left(x_1^k, x_2^k\right)} \quad (9.3.22)$$

where

$$\mathbf{s}^k = \begin{bmatrix} s_1^k \\ s_2^k \end{bmatrix} = \begin{bmatrix} x_1 - x_1^k \\ x_2 - x_2^k \end{bmatrix}$$

is the Newton step in two dimensions. The square matrix of partial derivatives that multiplies the Newton step is known as the Jacobian matrix denoted by $\boldsymbol{F}$:

$$\boldsymbol{F}\left(\boldsymbol{x}^k\right) = \begin{bmatrix} \frac{\partial f_1}{\partial x_1} & \frac{\partial f_1}{\partial x_2} \\ \frac{\partial f_2}{\partial x_1} & \frac{\partial f_2}{\partial x_2} \end{bmatrix}_{\left(x_1^k, x_2^k\right)} \quad (9.3.23)$$

In the general case of n equations in n unknowns the Newton step is calculated by solving the set of linear equations:

$$\boldsymbol{F}\left(\boldsymbol{x}^k\right) \mathbf{s}^k = -\boldsymbol{f}\left(\boldsymbol{x}^k\right) \quad (9.3.24)$$

In every iteration of Newton's method the solution is updated as follows:

$$\boldsymbol{x}^{k+1} = \boldsymbol{x}^k + \mathbf{s}^k \quad (9.3.25)$$

Newton's method for systems of nonlinear equations follows the same general procedure outlined in Chapter 5 for single-variable cases. MATLAB provides an efficient implementation of Newton's method through the `fsolve` function, which has already been used in several examples. Similarly, Python offers a comparable implementation via the `fsolve` function in the `scipy.optimize` module.

Consider the following example (see Fig. 9.4)

$$f_1(x_1, x_2) = x_1^2 + x_2^2 - 1 \quad (9.3.26)$$

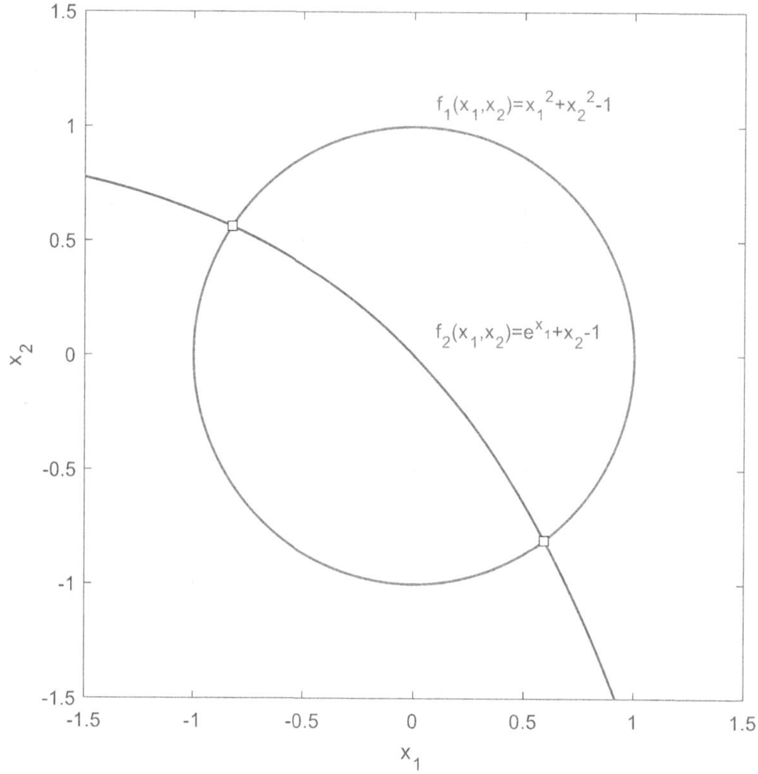

Figure 9.4 A problem of two nonlinear algebraic equations in two variables.

$$f_2(x_1, x_2) = e^{x_1} + x_2 - 1 \tag{9.3.27}$$

The Jacobian matrix is

$$\boldsymbol{F}(\boldsymbol{x}) = \begin{bmatrix} \dfrac{\partial f_1}{\partial x_1} & \dfrac{\partial f_1}{\partial x_2} \\ \dfrac{\partial f_2}{\partial x_1} & \dfrac{\partial f_2}{\partial x_2} \end{bmatrix} = \begin{bmatrix} 2x_1 & 2x_2 \\ e^{x_1} & 1 \end{bmatrix}$$

If we select $\boldsymbol{x}^0 = \begin{bmatrix} 0 & \frac{1}{2} \end{bmatrix}^T$, then

$$f_1\left(x_1^0, x_2^0\right) = 0 + \frac{1}{2^2} - 1 = -\frac{3}{4}$$

$$f_2\left(x_1^0, x_2^0\right) = e^0 + \frac{1}{2} - 1 = \frac{1}{2}$$

$$\boldsymbol{F}(x_1^0, x_2^0) = \begin{bmatrix} 0 & 1 \\ 1 & 1 \end{bmatrix}, \quad \boldsymbol{F}^{-1} = \begin{bmatrix} -1 & 1 \\ 1 & 0 \end{bmatrix}$$

$$s^0 = -\boldsymbol{F}^{-1}\boldsymbol{f} = -\begin{bmatrix} -1 & 1 \\ 1 & 0 \end{bmatrix}\begin{bmatrix} -\frac{3}{4} \\ +\frac{2}{4} \end{bmatrix} = \frac{1}{4}\begin{bmatrix} -5 \\ 3 \end{bmatrix}$$

$$\boldsymbol{x}^1 = \boldsymbol{x}^0 + \boldsymbol{s}^0 = \frac{1}{4}\begin{bmatrix} 0 \\ 2 \end{bmatrix} + \frac{1}{4}\begin{bmatrix} -5 \\ 3 \end{bmatrix} = \frac{1}{4}\begin{bmatrix} -5 \\ 5 \end{bmatrix}$$

The steps of Newton's method for this case study are presented in Fig. 9.5.

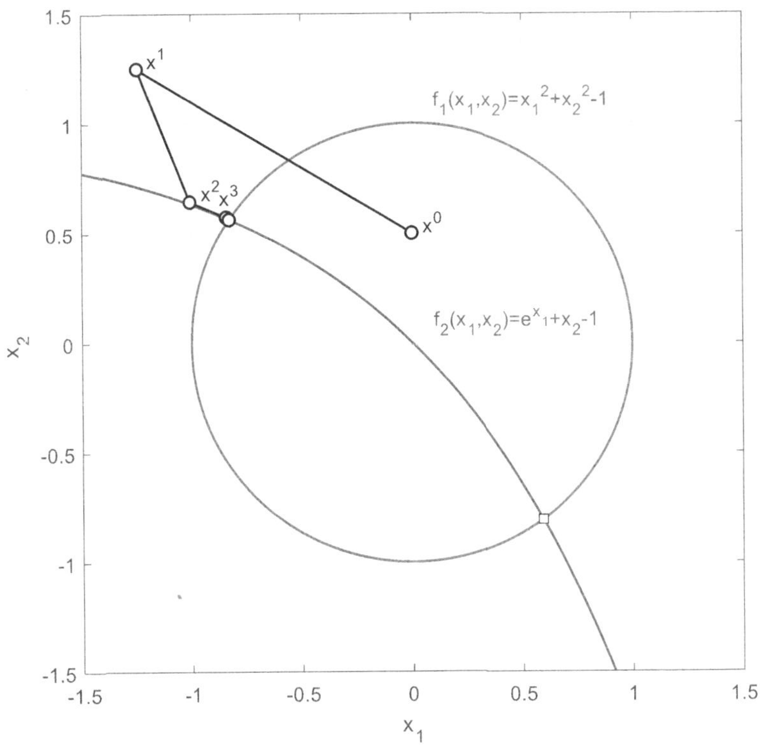

Figure 9.5 A problem of two nonlinear algebraic equations in two variables: steps of Newton's method starting from point $[0\ 0.5]^T$.

9.3.3 Solving sets of sparse nonlinear equations by precedence-ordering methods

In the previous section, Newton's method was introduced as a general iterative approach for solving systems of n nonlinear equations in n unknowns—that is, **square systems**. While powerful and broadly applicable, Newton's method does not inherently exploit the structural characteristics of many practical problems, such as the sparsity that commonly arises in chemical process flowsheeting. A **sparse system** refers to one in which the majority of the coefficients in the equations are zero, often reflecting limited interdependencies between variables.

In such sparse square systems, it is frequently possible to identify a subset of m equations that can be solved independently for m variables. Once these variables are determined, the remaining $n - m$ equations can be solved for the remaining $n - m$ unknowns. This hierarchical approach can often be applied recursively, leading to a **decomposition** of the original large system into a series of smaller, more manageable subsystems. Each subsystem can then be solved in a stepwise manner, significantly improving computational efficiency.

This strategy of breaking down a large system into smaller parts is known as **partitioning**, and the specific sequence in which these partitions are solved is referred to as the **precedence order**. Identifying an effective precedence order is critical, as it ensures that each subsystem can be solved with the information available from previously solved partitions, thereby avoiding the need to simultaneously solve the entire system.

Let us consider the following set of five equations $f_1, f_2, \ldots, f_5$ in five unknowns $x_1, x_2, \ldots, x_5$

$$\begin{aligned}
&f_1(x_1, x_4): && x_1 + x_4 - 2 &&= 0 \\
&f_2(x_2, x_3, x_4, x_5): && x_2 x_3 x_4 - x_5 &&= 0 \\
&f_3(x_1, x_3, x_4): && x_1(x_4 - 2)x_3^2 + 4 &&= 0 \\
&f_4(x_1, x_4): && x_4 - 2x_1 + 1 &&= 0 \\
&f_5(x_1, x_3, x_5): && x_1 x_3 - x_5 + 2 &&= 0
\end{aligned} \tag{a}$$

At first glance, the solution of this system does not appear to be obvious. However, let us change the order of the equations:

$$
\begin{aligned}
&f_1(x_1, x_4): && x_1 + x_4 - 2 && = 0 \\
&f_4(x_1, x_4): && x_4 - 2x_1 + 1 && = 0 \\
&f_3(x_1, x_3, x_4): && x_1(x_4 - 2)x_3^2 + 4 && = 0 \\
&f_5(x_1, x_3, x_5): && x_1x_3 - x_5 + 2 && = 0 \\
&f_2(x_2, x_3, x_4, x_5): && x_2x_3x_4 - x_5 && = 0
\end{aligned}
\tag{b}
$$

To see more clearly what is the structural difference between formulations (a) and (b) we can use the occurrence or **incidence matrix**. An incidence matrix is an $n - by - n$ matrix with each column corresponding to a variable and each row corresponding to an equation. A "1" in this matrix indicates that the corresponding variable appears in the corresponding equation. The incidence matrix for the systems (a) and (b) are the following:

$$
\mathscr{A}_a = \begin{array}{c|ccccc} & x_1 & x_2 & x_3 & x_4 & x_5 \\ \hline f_1 & 1 & & & 1 & \\ f_2 & & 1 & 1 & 1 & 1 \\ f_3 & 1 & & 1 & 1 & \\ f_4 & 1 & & & 1 & \\ f_5 & 1 & & 1 & & 1 \end{array}
\qquad
\mathscr{A}_b = \begin{array}{c|ccccc} & x_1 & x_4 & x_3 & x_5 & x_2 \\ \hline f_1 & 1 & 1 & & & \\ f_4 & 1 & 1 & & & \\ f_3 & 1 & 1 & 1 & & \\ f_5 & 1 & & 1 & 1 & \\ f_2 & & 1 & 1 & 1 & 1 \end{array}
$$

It can be noted from $\mathscr{A}_b$ that in Equations f_1 and f_4 only the two variables x_1 and x_4 appear and a 2-by-2 subsystem can be identified. Solving this subsystem is relatively simple as these are linear equations. Subtract f_1 from f_4 to obtain $x_1 = x_4 = 1$. Then equation f_3 can be solved for x_3

$$x_3^2 = \frac{-4}{x_1(x_4 - 2)} = 4$$

and $x_3 = 2$ (assuming all variables to be positive). Finally, equations f_5 and f_2 are solved to obtain $x_5 = 4$, $x_2 = 2$.

It is therefore essential to address the question of whether a suitable partitioning of the system can be developed in order to arrange the equations in a **precedence order**—an order that enables sequential resolution based on dependency relationships among the variables. However, an even more fundamental question is whether it is possible to assign a unique variable—referred to as the **output variable**—to each equation in the system. This output variable represents the unknown that the equation is intended to solve. If such an **assignment** does not exist, the system is said to be structurally singular, meaning that the equations are either redundant, insufficiently specified, or otherwise entangled in a way that prevents a consistent solution from being found. Structural singularity typically arises due to two causes: either some equations are linearly dependent (i.e., redundant), or some variables are over- or under-specified. In either case, the absence of a one-to-one correspondence between equations and output variables renders the system unsolvable using sequential methods. Fortunately, there exists a straightforward and computationally efficient method for solving this so-called **assignment problem**, based on well-established principles from linear programming. However, as this is generally an advanced topic, we will not get into the details, but the interested reader can consult the references given at the end of this chapter.

While advanced algorithms can handle large systems automatically, our focus here is on a simple manual technique that can be applied effectively to small- to medium-sized systems of equations. We will present a simple technique that involves analyzing the structure of the system through its incidence matrix. The method proceeds as follows:

STEP 1: If an equation involves only a single variable, remove the corresponding row (equation) and column (variable) from the incidence matrix. Assign the variable as the output of that equation. Since it is independent of other variables, this equation can be solved first.

STEP 2: If a variable appears in only one equation, assign it to that equation as its output. Then, remove the corresponding row and column from the incidence matrix. This equation will be solved last, as it depends on the values of all other variables involved in it. In general, the order of solution is the reverse of the deletion order in this step.

Steps 1 and 2 are applied iteratively. If all equations and variables are successfully eliminated, a sequential solution can be constructed. In this case, each equation can be solved for a single variable in a step-by-step manner, leading to an efficient and systematic resolution of the entire system.

To illustrate this method in practice, let us now consider a simple example that demonstrates the process of partitioning the system, assigning output variables, and determining the appropriate precedence order for solving the equations. Let us consider a simple example to demonstrate the method just presented

$$
\begin{aligned}
f_1(x_1): &\quad x_1 - 4 &= 0\\
f_2(x_2, x_3, x_4): &\quad x_2x_3 - x_4 - 6 &= 0\\
f_3(x_1, x_2, x_3): &\quad x_1(x_2 - 5)x_3^2 - 16 &= 0\\
f_4(x_1, x_2): &\quad -3x_1 + x_2 + 6 &= 0\\
f_5(x_1, x_4, x_5): &\quad x_1x_5 - x_4 + 6 &= 0
\end{aligned} \tag{c}
$$

which has the following incidence matrix

$$
\mathscr{A}_c = \begin{array}{c|ccccc} & x_1 & x_2 & x_3 & x_4 & x_5 \\ \hline f_1 & 1 & & & & \\ f_2 & & 1 & 1 & 1 & \\ f_3 & 1 & 1 & 1 & & \\ f_4 & 1 & 1 & & & \\ f_5 & 1 & & & 1 & 1 \end{array}
$$

We note that in equation f_1 only one variable (x_1) appears. We delete the first row and first column:

$$
\mathscr{A}_{c,1} = \begin{array}{c|cccc} & x_2 & x_3 & x_4 & x_5 \\ \hline f_2 & 1 & 1 & 1 & \\ f_3 & 1 & 1 & & \\ f_4 & 1 & & & \\ f_5 & & & 1 & 1 \end{array}
$$

and generate the first line of the solution list:

$$(f_1 \to x_1)$$

We note that there is no other equation in which only one variable appears. We therefore terminate STEP 1.

We note that variable x_5 appears only in equation f_5, and we delete the corresponding row and column of the incidence matrix:

$$
\mathscr{A}_{c,2} = \begin{array}{c|ccc} & x_2 & x_3 & x_4 \\ \hline f_2 & 1 & 1 & 1 \\ f_3 & 1 & 1 & \\ f_4 & 1 & & \end{array}
$$

and update the solution list (note that assignments made at STEP 2 are placed at the end of the list in reverse order):

$$(f_1 \rightarrow x_1)$$

$$(f_5 \rightarrow x_5)$$

We note that variable x_4 appears only in equation f_2, and we delete the corresponding row and column of the incidence matrix:

$$\mathscr{A}_{c,3} = \begin{array}{c|cc} & x_2 & x_3 \\ f_3 & 1 & 1 \\ f_4 & 1 & \end{array}$$

and update the solution list (note that the assignment discovered in this stage is placed before the assignment made in the previous stage of STEP 2 as the order of solution is the inverse of the order of elimination):

$$(f_1 \rightarrow x_1)$$

$$(f_2 \rightarrow x_4)$$

$$(f_5 \rightarrow x_5)$$

We note that variable x_3 appears only in equation f_3, and we delete the corresponding row and column of the incidence matrix:

$$\mathscr{A}_{c,4} = \begin{array}{c|c} & x_2 \\ f_4 & 1 \end{array}$$

and update the solution list:

$$(f_1 \rightarrow x_1)$$

$$(f_3 \rightarrow x_3)$$

$$(f_2 \rightarrow x_4)$$

$$(f_5 \rightarrow x_5)$$

There is only one equation and variable, and we finally have the following solution list:

$$(f_1 \rightarrow x_1)$$

$$(f_4 \rightarrow x_2)$$

$$(f_3 \rightarrow x_3)$$

$$(f_2 \rightarrow x_4)$$

$$(f_5 \rightarrow x_5)$$

We rearrange the columns and rows of the incidence matrix as indicated by the solution list:

$$\mathscr{A}_c = \begin{array}{c|ccccc} & x_1 & x_2 & x_3 & x_4 & x_5 \\ f_1 & \mathbf{1} & & & & \\ f_4 & 1 & \mathbf{1} & & & \\ f_3 & 1 & 1 & \mathbf{1} & & \\ f_2 & & 1 & 1 & \mathbf{1} & \\ f_5 & 1 & & & 1 & \mathbf{1} \end{array}$$

Note the lower triangular structure of the incidence matrix—this is a hallmark of systems that can be solved sequentially, one equation at a time. Such a structure is highly desirable, as it allows for straightforward, step-by-step computation. Unfortunately, this ideal configuration is relatively rare in practice.

When the previously described method fails to produce a lower triangular incidence matrix—indicating that the system cannot be directly solved through sequential substitution—an alternative approach is to use **tearing methods**. These methods are particularly useful for solving complex systems of nonlinear equations that are structurally coupled. The tearing process begins by analyzing the incidence matrix to determine the minimum number of equations in which any variable appears. Specifically, let $\rho(x_i)$ denote the number of equations in which variable x_i appears. This metric helps identify candidates for "**tear variables**"—variables to be temporarily removed (or treated as knowns) in order to decouple the system and simplify its structure. Once such variables are identified, the system is partitioned by removing these variables and the equations they appear in, effectively breaking dependency cycles. The reduced system can then often be reordered into a lower triangular form and solved. After the simplified system is solved, the previously "torn" variables are reintroduced and their values computed through substitution. The number of tear variables is $r = \min\{\rho(x_i)\} - 1$. We delete r equations and r variables until we find a selection that results in a set of remaining $n - r$ equations and variables that have an incidence matrix with the desirable lower triangular structure. These remaining equations can be solved by solving one equation in one variable at a time. Values are assumed for the r variables, then the $n - r$ equations are solved for the remaining variables. Finally, the r equations that were deleted are used to correct our estimations of the tear variables. If deleting r variables and equations does not result in a lower triangular structure in the incidence matric of the remaining equations, then we increase r ($r = r+1$) and continue.

A classical example of the application of the tearing method is the solution of the flash distillation problem. The equations of the flash distillation unit for a problem of two components are the following (it can be generalized to an arbitrary number of components)

Total mole balance

$$f_1 = F - V - L = 0 \tag{9.3.28}$$

Component mole balances

$$f_2 = z_1 F - (K_1 V - L)\, x_1 = 0 \tag{9.3.29}$$

$$f_3 = z_2 F - (K_2 V - L)\, x_2 = 0 \tag{9.3.30}$$

Equilibrium relationships

$$f_4 = y_1 - K_1(T, P)\, x_1 = 0 \tag{9.3.31}$$

$$f_5 = y_2 - K_2(T, P) x_2 = 0 \tag{9.3.32}$$

Rachford–Rice equation

$$f_6 = (y_1 + y_2) - (x_1 + x_2) = 0 \tag{9.3.33}$$

The incidence matrix is.

$$\mathscr{A} = \begin{array}{c|cccccc} & V & L & x_1 & x_2 & y_1 & y_2 \\ f_1 & 1 & \mathbf{1} & & & & \\ f_2 & 1 & 1 & \mathbf{1} & & & \\ f_3 & 1 & 1 & & \mathbf{1} & & \\ f_4 & & & 1 & & \mathbf{1} & \\ f_5 & & & & 1 & & \mathbf{1} \\ f_6 & & & 1 & 1 & 1 & 1 \end{array}$$

Application of STEP 1 and STEP 2 fail to achieve any simplification of the problem as there is no equation in which only one variable appears or any variable that appears in only one equation. We note that

$$\begin{aligned} r &= \min\{\rho(V), \rho(L), \rho(x_1), \rho(x_2), \rho(y_1), \rho(y_2)\} - 1 \\ &= \min\{3, 3, 3, 3, 2, 2\} - 1 = 2 - 1 = 1 \end{aligned}$$

We therefore conclude that selecting one tear variable might be enough for solving the system using this methodology. As can be seen from the incidence matrix, selecting V to be the tear variable and deleting the Rachford–Rice equation results in an incidence matrix that has the desired lower triangular structure. A value of V is therefore selected and

- L is calculated from Eq. (9.3.28)
- x_1 is calculated from Eq. (9.3.29)
- x_2 is calculated from Eq. (9.3.30)
- y_1 is calculated from Eq. (9.3.31)
- y_2 is calculated from Eq. (9.3.32)

All known variables are substituted in Eq. (9.3.33) and the value of V is corrected (using, for instance, Newton's method) until Eq. (9.3.33) is satisfied.

9.4 Application of the sequential modular approach

The core idea of the sequential modular approach is to construct the flowsheet using modules with predefined structures, each representing a specific type of process unit. Once all necessary information about the input streams and equipment parameters for a module is provided, the module computes all associated output streams. Modules are typically arranged in a sequential (series) order, allowing for step-by-step calculation through the flowsheet. In cases involving recycle streams, an iterative method is applied to reconcile the input and output values of the recycle loop until convergence is achieved.

From Fig. 9.1 it is noted that the representative process that we have selected needs three modules for its simulation:

- a mixing module
- a flash distillation unit module (to be used twice)
- a recycle module

9.4.1 A mixing process module

We will develop a general-purpose module for simulating the mixing of multiple input streams with known flowrates and compositions. A schematic representation of this mixing module is shown in Fig. 9.6. For the purposes of this introductory discussion, energy balances will be neglected to simplify the analysis. The module assumes that all p input streams are fully specified. Using the overall mole balance and the component-wise mole balances for NC components, the molar flowrate and composition of the output stream can then be readily calculated:

Total mole balance

$$N = N_1 + N_2 + \ldots + N_p \tag{9.4.1}$$

Component mole balances

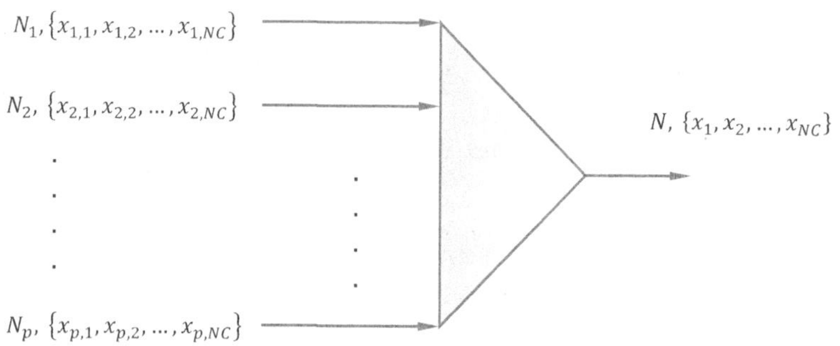

Figure 9.6 Generalized representation of a mixing process module.

$$x_iN = x_{1,i}N_1 + x_{2,i}N_2 + \ldots + x_{p,i}N_p, i = 1, 2, \ldots, NC \qquad (9.4.2)$$

It is important to note that the basic mixing process requires no additional specifications beyond the known properties of the input streams—namely, their flowrates and compositions. This simplicity makes the mixing module particularly straightforward to model and implement. Furthermore, the underlying mathematical formulation of the mixing process requires no specialized numerical techniques or iterative procedures to compute the output stream.

However, before concluding the discussion of the mixing process module, it is worth emphasizing that a complete module—including the energy balance—is significantly more complex than the simplified version described above. This added complexity arises from the need to determine not only the temperature of the output stream but also its phase. For example, when a subcooled liquid is mixed with a superheated vapor, the phase of the resulting stream cannot be known a priori; it depends on the relative flow rates and thermal properties of the incoming streams. In such cases, additional thermodynamic calculations are necessary to determine whether the output will be a liquid, vapor, or two-phase mixture.

9.4.2 An isothermal flash distillation process module

This process involves feeding a stream with total molar flow rate F and component mole fractions $\{z_i, i = 1, 2, \ldots, NC\}$ into a phase equilibrium unit operating at a specified temperature T and pressure P, which uniquely determine the equilibrium ratios K_i. The vapor product stream has molar flowrate V and mole fractions $\{y_i, i = 1, 2, \ldots, NC\}$ and the liquid product stream has molar flowrate L and mole fractions $\{x_i, i = 1, 2, \ldots, NC\}$. This model has been analyzed in detail in Chapter 8, Section 8.6, and is illustrated in Fig. 8.5.

The module does not require any additional parameters apart from the equilibrium ratios and the feed stream molar flowrate and composition. Under these conditions, for known equilibrium ratios K_i, the Rachford-Rice equation is solved iteratively to determine the vapor fraction, denoted by ψ. Once ψ has been determined, the molar flow rates and compositions of the vapor and liquid product streams can be calculated directly using Eqs. (8.6.6) and (8.6.7). The module must calculate the output stream even in cases where the input stream corresponds to a single-phase stream.

9.4.3 Application to the representative case study

This approach is pictured in Fig. 9.7. To implement the sequential modular approach we follow the steps:

1. guess values for the flowrate and composition of stream 6
2. solve the module of mixing streams 1 and 6 to determine stream 2
3. solve the flash distillation module for stream 2 to determine streams 3 and 4
4. solve the flash distillation module for stream 3 to determine streams 5 and 6
5. update the estimation for the flowrate and composition of stream 6. If the new estimation of stream 6 is acceptably close to the assumed values, STOP as the solution has been achieved. Otherwise, return to 2.

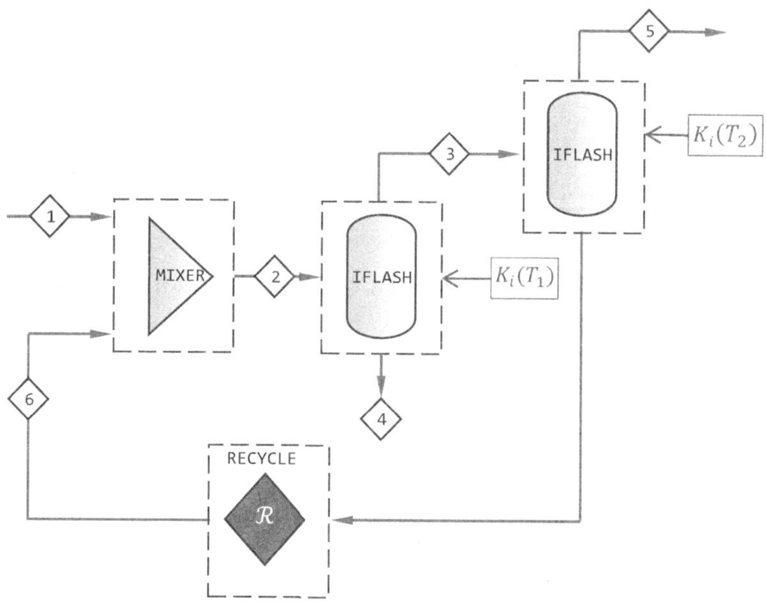

Figure 9.7 Structure of the sequential modular approach to simulate the process of Fig. 9.1.

The MATLAB implementation of the sequential modular approach is as follows:

```
clear all
% feed stream
N(1) = 100;
x(1,:) =[0.5 0.5];
P     = 10;
% specifications
T(1) = 383.4604;
T(2) = 309.7203;
K1    = EqRatios(T(1),P);
K2    = EqRatios(T(2),P);
% initial guess for recylce stream
N(6) = 50;
x(6,:) = [0.5 0.5];
iter = 0;

while (iter==0)|(error>1e-8)
   X = [N(6) x(6,:)];
   % MIXER -> stream 2
   [N(2),x(2,:)]=MIXER([N(1);N(6)],[x(1,:); x(6,:)]);
   % IFLASH 1 -> streams 3 & 4
   [N(4),x(4,:),N(3),x(3,:)]=IFLASH(x(2,:),N(2),K1) ;
   % IFLASH 2 -> streams 5 & 6
   [N(6),x(6,:),N(5),x(5,:)]=IFLASH(x(3,:),N(3),K2) ;
   % calculate criterion for convergence
   error = max(abs(X-[N(6) x(6,:)]));
   iter = iter + 1;
   fprintf('Iteration =%4d error =%16.8f \n',iter,error)
end

N
x

function K=EqRatios(T,P)
% calculate equilibrium ratios o fth e
% methane / n-hexane mixture using Raoult's law
A = [4.22061 4.00266]'*log(10);
B = [516.689 1171.53]'*log(10);
C = [11.223  -48.784]';
K = exp(A-B./(C+T))/P;
K = K(:);
end

function [Nout,xout]=MIXER(Nin,xin)
Nout = sum(Nin);
xout = Nin'*xin/Nout;
end

function [L,x,V,y]=IFLASH(z,F,K)
f0 = sum(z(:).*K(:)) - 1;
f1 = 1 - sum(z(:)./K(:)) ;
if (f0<0)      % subcooled liquid
    L = F; x=z; V = 0; y=z;
    return
elseif (f1>0) % superheated vapor
    V = F; y=z; L = 0; x=z;
    return
end
% two phase mixture
psi = f0/(f0-f1);
```

```
e   = 1e+12;
while (abs(e)>1e-12)
    term = (K(:)-1)./(1+psi*(K(:)-1));
    e  =  sum(z(:).*term);
    de = -sum(z(:).*term.^2);
    psi = psi - e/de;
end
V = psi*F; L=F-V;
x = z(:)./(1+psi*(K(:)-1)); x=x';
y = K(:).*x(:); y=y';
return
end

>> Ch9_sequential_modular
Iteration =    1 error =      17.29702697
Iteration =    2 error =      10.67863746
Iteration =    3 error =       0.07260603
Iteration =    4 error =       0.00049366
Iteration =    5 error =       0.00000336
Iteration =    6 error =       0.00000002
Iteration =    7 error =       0.00000000

N =
  100.0000  121.9512   73.1707   48.7805   51.2195   21.9512

x =
    0.5000    0.5000
    0.4143    0.5857
    0.6849    0.3151
    0.0084    0.9916
    0.9682    0.0318
    0.0237    0.9763
```

In Fig. 9.8 the solution of the process pictured in Fig. 9.1 in a commercial simulator (UniSim from Honeywell) is shown. As the simulator is based on the sequential modular approach, the similarity of Fig. 9.7 with Fig. 9.8 comes as no surprise.

9.4.4 Application to the ammonia production case study

In Section 6.5, we presented a realistic case study involving the production of ammonia from a stoichiometric mixture of hydrogen and nitrogen. The process flow diagram, initially shown in Fig. 6.25, includes both a purge stream—to remove inert components introduced with the feed—and a recycle stream, which ensures maximum conversion of the reactants by returning unreacted hydrogen and nitrogen to the reactor. Only a small

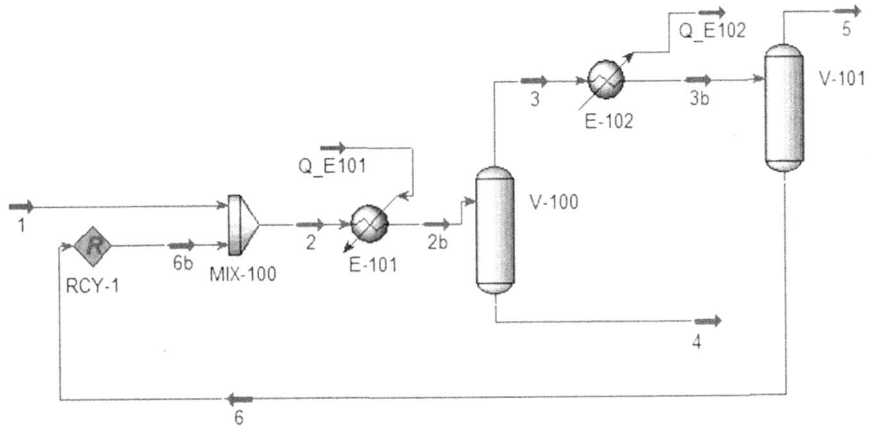

4		
Temperature	110.4	C
Pressure	1000	kPa
Molar Flow	48.78	kgmole/h
Comp Mole Frac (Methane)	0.008306	

5		
Temperature	37.30	C
Pressure	1000	kPa
Molar Flow	51.22	kgmole/h
Comp Mole Frac (Methane)	0.968244	

Figure 9.8 The process of Fig. 9.1 solved in a commercial simulator.

portion of the feed is lost through the purge to prevent the accumulation of inerts in the system.

For ease of reference, the same process flow diagram is reproduced in Fig. 9.9. In Section 6.5, we solved this case study using an equation-based modeling approach, which relies on the

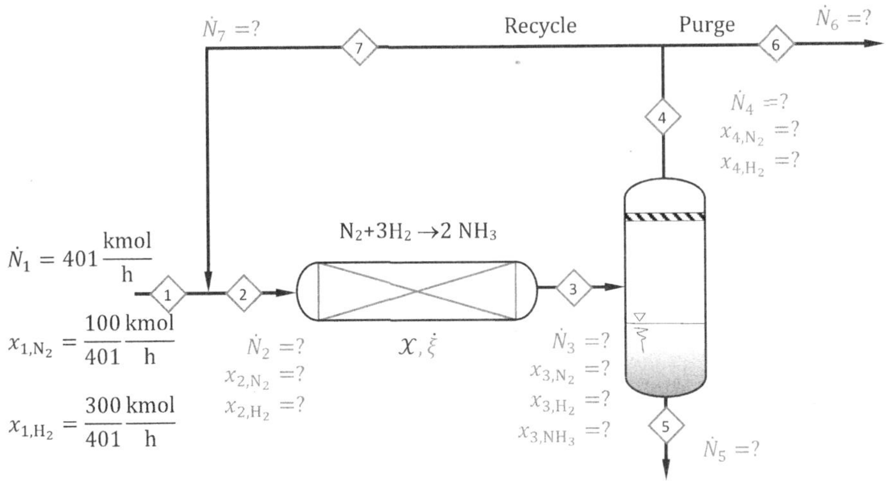

Figure 9.9 A simplified ammonia reaction process with an inert component in the fresh feed, a recycle stream for complete conversion of the reactants, and a purge stream for removing an inert component present in the feed stream.

simultaneous solution of mass and energy balances. In this section, we will re-examine the same process using the sequential modular approach, where unit operations are modeled and solved in a step-by-step sequence.

To simulate the flowsheet for the production of ammonia, we need two additional modules:

- a conversion reactor module
- a splitter module

For the conversion reactor module the mathematical model consists of the *NC* mole balances:

$$x_i^{\text{out}} N^{\text{out}} = x_i^{\text{in}} N^{\text{in}} + \sigma_i \xi, i = 1, 2, \ldots, NC \tag{9.4.3}$$

and the definition of conversion

$$X = \frac{x_{ref}^{\text{in}} N^{\text{in}} - x_{ref}^{\text{out}} N^{\text{out}}}{x_{ref}^{\text{in}} N^{\text{in}}} \tag{9.4.4}$$

where *ref* denotes the reference component (reactant) that is used to define conversion.

The module needs one specification only, which is the fraction conversion of the reference component. We again ignore the energy balance equation to facilitate clarity of presentation.

The splitter module has been analyzed in Chapter 6, Section 6.5 and Fig. 6.26. All output streams inherit the composition of the input stream, and we only need the split ratios of all output streams but the last one.

The MATLAB implementation of the sequential modular approach is as follows:

```
clear all
% 1: nitrogen 2: hydrogen 3: ammonia 4: methane

% feed stream
N(1)    = 401;
x(1,:) = [100 300 0 1]/N(1);
P       = 10;

% specifications
T = 273.15-70;
SR = [1 19]/20;
K=EqRatios(T,P)

% initial guess for recylce stream
N(7) = 400;
x(7,:) = [90 290 5 15 ]/N(7);
iter = 0;

while (iter==0)|(error>1e-5)
   X = [N(7) x(7,:)];
   % MIXER -> stream 2
   [N(2),x(2,:)]=MIXER([N(1);N(7)],[x(1,:); x(7,:)]);

   % CREACTOR -> stream 3
   [N(3),x(3,:)]=CREACTOR(N(2),x(2,:),[-1 -3 2 0],0.5,1);

   % IFLASH 1 -> streams 4 & 5
   [N(5),x(5,:),N(4),x(4,:)]=IFLASH(x(3,:),N(3),K) ;

   % SPLITTER 4 -> streams 6 & 7
   [N(6:7), x(6:7,:)]=SPLITTER(N(4),x(4,:),[0.05; 0.95]) ;

   % calculate criterion for convergence
   error = max(abs(X-[N(7) x(7,:)]));
   iter = iter + 1;
   fprintf('Iteration =%4d error =%16.8f \n',iter,error)
end

N
x

function K=EqRatios(T,P)
% calculate equilibrium ratios o fth e
% nitrogen / hydrogen / ammonia / methane using Raoult's law
A = [ 3.7362  3.54314 3.18757 4.22061 ]'*log(10);
B = [ 264.651 99.395  506.713   516.689 ]'*log(10);
C = [ -6.788  7.726   -80.78    11.223  ]';
K = exp(A-B./(C+T))/P;
K = K(:);
end

function [Nout,xout]=MIXER(Nin,xin)
Nout = sum(Nin);
xout = Nin'*xin/Nout;
end

function [Nout,xout]=CREACTOR(Nin,xin,sigma,FractionConversion,iref)
ksi  = FractionConversion*Nin*xin(iref);
nout = Nin*xin+sigma*ksi;
```

```
Nout = sum(nout);
xout = nout(:)/Nout;
return

function [L,x,V,y]=IFLASH(z,F,K)
f0 = sum(z(:).*K(:)) - 1;
f1 = 1 - sum(z(:)./K(:)) ;
if (f0<0)      % subcooled liquid
    L = F; x=z; V = 0; y=z;
    return
elseif (f1>0) % superheated vapor
    V = F; y=z; L = 0; x=z;
    return
end
% two phase mixture
psi = f0/(f0-f1);
e   = 1e+12;
while (abs(e)>1e-12)
    term = (K(:)-1)./(1+psi*(K(:)-1));
    e  =  sum(z(:).*term);
    de = -sum(z(:).*term.^2);
    psi = psi - e/de;
end
V = psi*F; L=F-V;
x = z(:)./(1+psi*(K(:)-1)); x=x';
y = K(:).*x(:); y=y';
return
end

function [Nout,xout]=SPLITTER(Nin,xin,SRatios)
% Nin  is a NS by 1 vector with the molar flowrates of the input streams
% xin  is a NS by NC matrix with the molar fractions in the input streams
% Nout is the molar flowrate of the product stream
% xout is the 1 by NC vector with the mole fractions in th eproduct stream
NPS  = length(SRatios);
NC   = length(xin);
Nout = Nin*SRatios(:);
xout = kron(ones(NPS,1),xin);
end

>> Ch9_sequential_modular_ammonia

Iteration =   1 error =      4.88493102
Iteration =   2 error =      2.02799348
...
Iteration = 181 error =      0.00001023
Iteration = 182 error =      0.00000961

N =
  401.0000  798.3606  610.3542  418.2744  192.0799   20.9137  397.3607

x =
    0.2494    0.7481         0    0.0025
    0.2355    0.7477    0.0057    0.0112
    0.1540    0.5159    0.3154    0.0146
    0.2215    0.7472    0.0114    0.0199
```

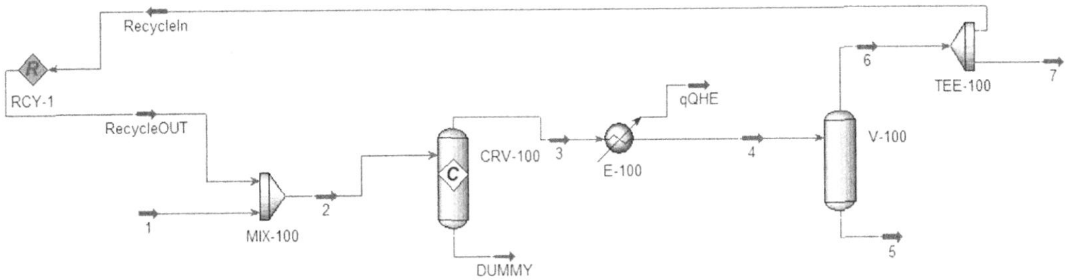

RecycleIn		
Temperature	-70.00	C
Pressure	1000	kPa
Molar Flow	397.3	kgmole/h
Comp Molar Flow (Nitrogen)	88.014	kgmole/h
Comp Molar Flow (Hydrogen)	296.88	kgmole/h
Comp Molar Flow (Ammonia)	4.5228	kgmole/h
Comp Molar Flow (Methane)	7.9192	kgmole/h

5		
Temperature	-70.00	C
Pressure	1000	kPa
Molar Flow	192.1	kgmole/h
Comp Molar Flow (Nitrogen)	1.3652	kgmole/h
Comp Molar Flow (Hydrogen)	2.3638	kgmole/h
Comp Molar Flow (Ammonia)	187.79	kgmole/h
Comp Molar Flow (Methane)	0.58277	kgmole/h

7		
Temperature	-70.00	C
Pressure	1000	kPa
Molar Flow	20.91	kgmole/h
Comp Molar Flow (Nitrogen)	4.6323	kgmole/h
Comp Molar Flow (Hydrogen)	15.625	kgmole/h
Comp Molar Flow (Ammonia)	0.23804	kgmole/h
Comp Molar Flow (Methane)	0.41680	kgmole/h

Figure 9.10 The process of Fig. 9.9 solved in a commercial simulator.

```
0.0071    0.0123    0.9776    0.0030
0.2215    0.7472    0.0114    0.0199
0.2215    0.7472    0.0114    0.0199
```

In Fig. 9.10 the solution of the process pictured in Fig. 9.9 in a commercial simulator (UniSim from Honeywell) is shown. The results agree reasonably well (as expected). Differences stem from the different thermodynamic models.

Chapter summary

In this chapter, we introduced two alternative strategies for simulating complete process flowsheets: the equation-based approach and the sequential modular approach. We also reviewed Newton's method for solving systems of nonlinear algebraic equations and briefly discussed a technique for solving sparse systems of such equations. This material is fundamental to understanding how modern commercial process flowsheeting software operates.

Problems

9.1 For the process of Fig. P9.1 apply the sequential modular approach and solve the mathematical model with the specifications $\psi = \psi_1 = \psi_2 = \psi_3$. Plot the mole fraction of methane and its recovery in the product stream as a function of ψ.

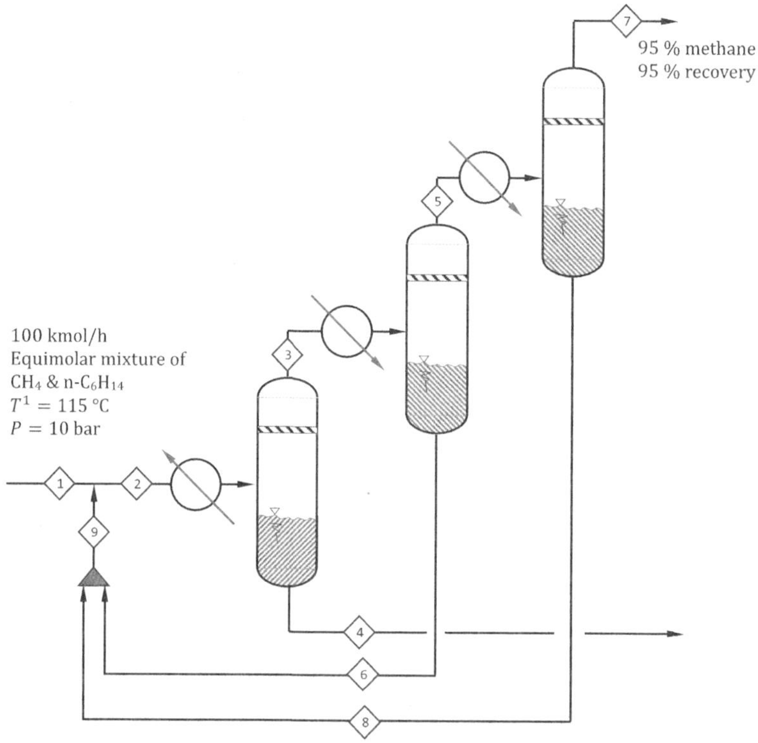

Figure P9.1 A process for methane recovery.

9.2 For the process of Fig. P9.1 apply the equation-based approach and solve the mathematical model with the specifications $T_1 = 135\ ^\circ C$, $T_2 = 115\ ^\circ C$, $T_3 = 55\ ^\circ C$.

9.3 For the process of ammonia production from stoichiometric amounts of hydrogen and nitrogen, shown in Fig. P9.3, develop a simulation based on the equation–solving approach. The reaction achieves equilibrium at a temperature of 350 °C and pressure of 10 bar. The equilibrium constant is given by the following equation.

$$K_{eq} = \exp\left(-13.65 + \frac{6240}{T}\right) = \frac{p_{NH_3}}{p_{H_2}^{3/2} p_{N_2}^{1/2}}, \text{ partial pressures in bar}$$

The flash distillation unit operates at −73.15 °C.

9.4 Repeat Problem 9.3 using the sequential modular approach.

9.5 For the process shown in Fig. P9.5 prove that four specifications are necessary to fully define the mathematical model.

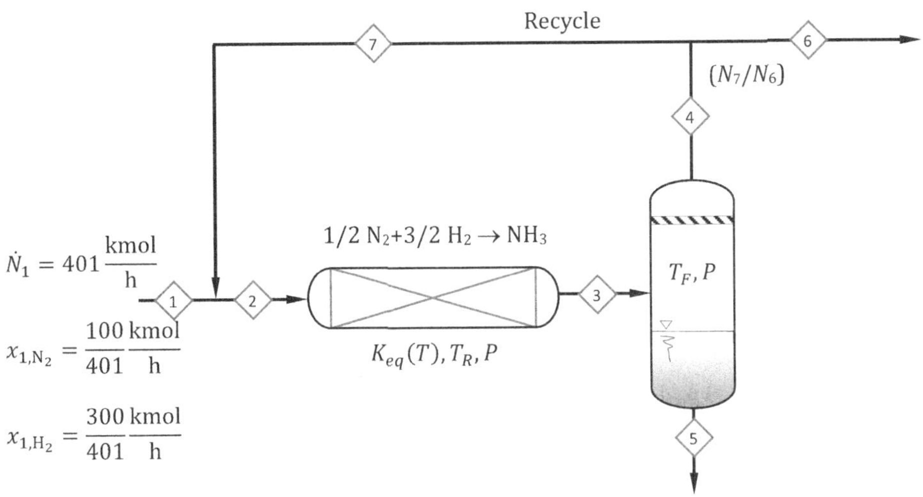

Figure P9.3 A simplified process for ammonia production.

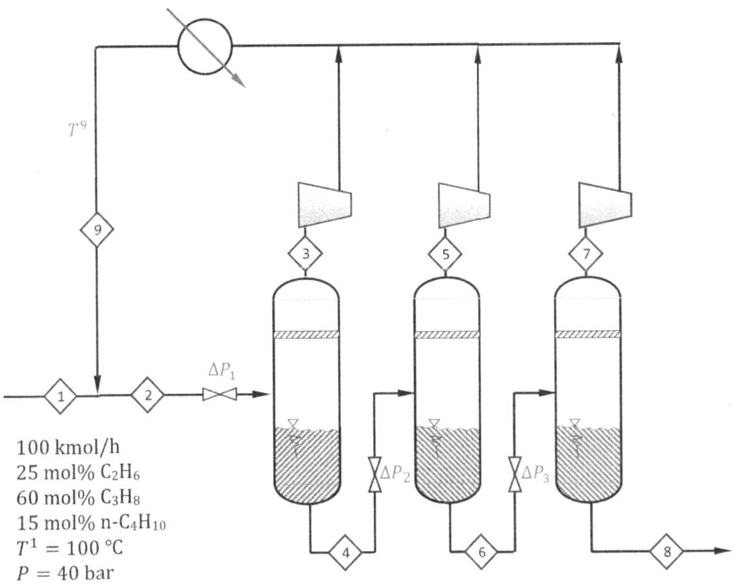

Figure P9.5 Process studied in Problem 9.5.

Choose $\Delta P_1 = 20$ bar, $\Delta P_2 = 10$ bar, $\Delta P_3 = 5$ bar and $T^9 = 10$ °C and use a commercial simulator to validate your findings.

9.6 For the process shown in Fig. P9.5 build your implementation in MATLAB or Python to solve the steady–state simulation problem using the ideal mixture assumption. Choose $\Delta P_1 = 20$ bar, $\Delta P_2 = 10$ bar, $\Delta P_3 = 5$ bar and $T^9 = 10$ °C. Compare your results with the results obtained from a commercial simulator (using the same thermodynamic model) and discuss the potential causes for any discrepancies.

9.7 An LPG (liquefied petroleum gas) vaporizer is shown in Fig. P9.6. A fresh feed of n-butane is fed to the vaporizer with molar flowrate of $\dot{N}_{\text{in}} = 0.1$ kmol/s at a temperature of $T_{\text{in}} = 25$ °C. The volume of the vaporizer is $V = 2\ \text{m}^3$ and is occupied by either liquid butane (V_L) with density ρ_L and specific enthalpy h_L, and gaseous butane (V_G) with density ρ_G and specific enthalpy h_G. The vapor pressure of n-butane can be calculated by the following equation:

$$\ln(P^s[\text{kPa}]) = A - \frac{B}{\text{C} + T[^\circ\text{C}]} = 13.9836 - \frac{2292.44}{T + 245.2877}$$

The density of the liquid butane can be assumed constant $\rho_L = 8.617\ \text{kmol/m}^3$. It is also given that the specific enthalpy of the liquid and gaseous butane at 1000 kPa can be calculated as follows:

$$h_L\left[\frac{\text{kJ}}{\text{kmol}}\right] = -151414 + 148.65T[^\circ\text{C}]$$

$$h_G\left[\frac{\text{kJ}}{\text{kmol}}\right] = -133110 + 133.35T[^\circ\text{C}]$$

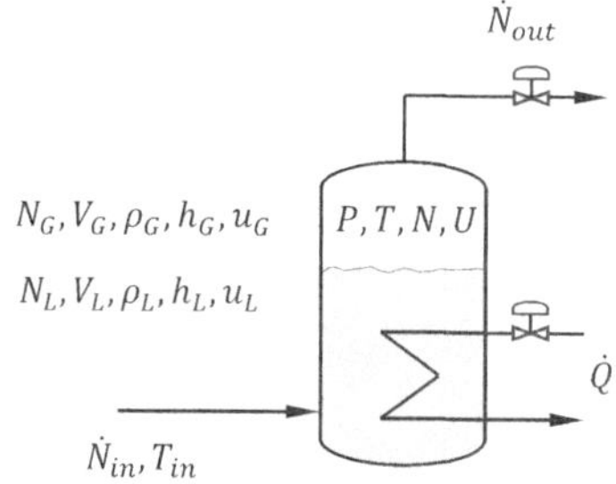

Figure P9.6 An LPG vaporizer.

a. If the pressure is $P = 1000$ kPa and the volume of the liquid $V_L = 1\ \text{m}^3$, calculate all variables appearing in Fig. P9.6 (u denotes specific internal energy, U the total internal energy, N the molar holdup, $\dot{Q}$ the heat transfer rate, the subscript L denotes liquid, and the subscript G denotes gas). Assume that gas and liquid butane are in equilibrium and develop all 14 algebraic equations that describe the steady state of the system.

b. Develop the dynamic model of the vaporizer by assuming that the liquid level is controlled by a proportional controller

$$\dot{Q}(t) = \dot{Q}(0) + k_{c,1}\left(V_L^{SP} - V_L(t)\right), V_L^{SP} = V_L(0), k_{c,1} = -2.8 \times 10^4 \frac{\text{kW}}{\text{m}^3}$$

and that the pressure is controlled by a proportional controller as well:

$$N_{\text{out}}(t) = N_{\text{out}}(0) + k_{c,2}\left(P^{SP} - P(t)\right), P^{SP} = P(0), k_{c,2} = -0.001\ \frac{\text{kmol/s}}{\text{kPa}}$$

c. Simulate the system when the feed flowrate changes from $\dot{N}_{\text{in}} = 0.1$ kmol/s to 0.15 kmol/s (+50% step increase of the molar flowrate of the feed).

Further reading

Books on material and energy balances and process flowsheeting

Andersen, L.B., Wenzel, L.A., 1961. Introduction to Chemical Engineering. McGraw Hill, NY.

Felder, R.M., Rousseau, R.W., 2000. Elementary Principles of Chemical Processes, third ed. John Wiley & Sons, NJ.

Henley, E.J., Bieber, H., 1959. Chemical Engineering Calculations. McGraw Hill, NY.

Henley, E.J., Rosen, E.M., 1969. Material and Energy Balance Calculations. John Wiley & Sons, NY.

Himmelblau, D.M., Bischoff, K.B., 1968. Process analysis and simulation. In: Deterministic Systems. John Wiley & Sons, NY.

Himmelblau, D.M., 1989. Basic Principles and Calculations in Chemical Engineering, fifth ed. Prentice Hall, NJ.

Hougen, O.A., Watson, k.m., Ragatz, R.A., 1943. Chemical Process Principles, Part I, Material and Energy Balances. John Wiley and Sons, NY.

Kirkbride, C.G., 1947. Chemical Engineering Fundamentals. McGraw Hill, NY.

Luyben, W.L., Wenzel, L.A., 1988. Chemical Process Analysis: Mass and Energy Balances. Prentice Hall, Englewood Cliffs, NJ.
Mickley, H.S., Sherwood, T.K., Reed, C.E., 1957. Applied Mathematics in Chemical Engineering. McGraw Hill, NY.
Myers, A.L., Seider, W.D., 1976. Introduction to Chemical Engineering and Computer Calculations. Prentice Hall, Inc., Englewood Cliffs, NJ.
Reklaitis, G.V., 1983. Introduction to Material and Energy Balances. John Wiley and Sons, NY.
Rudd, D.F., Watson, C.C., 1968. Strategy of Process Engineering. John Wiley & Sons, NY.
Westerberg, A.W., Hutchison, H.P., Motard, R.L., Winter, P., 1979. Process Flowsheeting. Cambridge University Press, Cambridge.

10

Graphical design techniques

10.1 Introduction

Graphical design techniques are powerful tools in chemical engineering that allow for the visualization and analysis of complex processes without the need for intensive numerical computation. These techniques are especially valuable during the preliminary design and analysis stages, where they help engineers understand process behavior, estimate key parameters, and make informed decisions quickly.

One of the most widely used graphical methods is the McCabe-Thiele method, which is used for designing and analyzing equilibrium-based separation units. This technique involves plotting the vapor-liquid equilibrium (VLE) data on an x-y diagram, where the x-axis represents the mole fraction of the more volatile component in the liquid phase, and the y-axis represents the same in the vapor phase. Operating lines are drawn to represent the material balances in the column. Theoretical stages are then determined by stepping off stages between the equilibrium curve and the operating lines, providing a clear visual of how separation progresses in the column. Graphical techniques are also used in leaching, absorption, and crystallization design. In absorption, for example, the operating and equilibrium lines are plotted similarly to distillation, enabling stepwise analysis of the number of stages needed for a desired separation.

The Levenspiel plot is a graphical tool used in chemical reactor design to determine the required reactor volume for a given conversion, particularly in flow reactors such as Continuous Stirred-Tank Reactors (CSTRs) and Plug-Flow Reactors (PFRs). Named after the chemical engineer Octave Levenspiel, this method provides a visual and intuitive approach to reactor sizing based on material balances.

Although many modern chemical engineering designs rely on simulation software, graphical methods remain essential for building intuition, checking simulation results, and teaching fundamental principles.

Introduction to Process Systems Engineering. https://doi.org/10.1016/B978-0-443-45444-8.00005-1

In this chapter, two of the most well-known graphical techniques will be presented, namely the use of the Levenspiel plot for reactor design and the use of McCabe-Thiele method for the design of equilibrium separation systems.

10.2 Graphical design of ideal reactors

10.2.1 The continuous stirred-tank reactor

A CSTR is a reactor where (see Fig. 10.1):

- reactants are continuously fed into the tank,
- the contents are well mixed (usually by mechanical stirring),
- products are continuously removed from the reactor.

Because of continuous stirring, the contents of the reactor are assumed to be perfectly mixed, meaning the composition throughout the reactor is uniform and the same as that of the outlet stream. The volumetric flowrate of the feed to the reactor is $\dot{V}_{in}$ and the reactant concentration $C_{in,R}$ while the volumetric flowrate of the product stream is $\dot{V}$ and the reactant concentration C_R. For liquid systems the density is practically constant and from the overall mole balance, it follows that $\dot{V}_{in} = \dot{V}$. The mole balance of the CSTR reactor can be written as:

$$\dot{V}C_R = \dot{V}C_{in,R} + \sigma_R V\dot{R}''' \tag{10.2.1}$$

or

$$V = \dot{V}(C_{in,R} - C_R)\cdot\frac{1}{(-\sigma_R)\dot{R}'''} \tag{10.2.2}$$

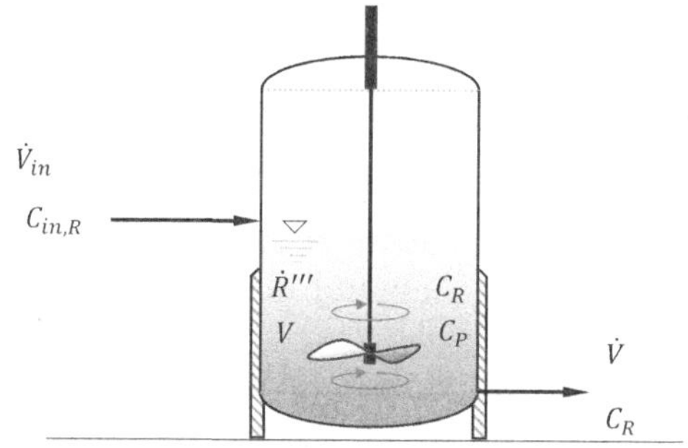

Figure 10.1 A continuous stirred tank reactor.

where σ_R is the stoichiometric number of the reactant and $\dot{R}'''$ is the reaction rate. We remember that the fraction conversion X for constant density systems, has been defined as

$$X = \frac{C_{in,R} - C_R}{C_{in,R}} \tag{10.2.3}$$

When Eq. (10.2.3) is replaced into Eq. (10.2.2) it follows that

$$V = \dot{V} C_{in,R} X \frac{1}{(-\sigma_R)\dot{R}'''}$$

or

$$\frac{V_{\text{CSTR}}}{\dot{n}_{in,R}} = \frac{1}{(-\sigma_R)\dot{R}'''} X \tag{10.2.4}$$

where $\dot{n}_{in,R} = \dot{V} C_{in,R}$ is the molar flowrate of the reactant in the feed. This is the basic equation for calculating the volume of a CSTR. To determine the reactor volume for a given molar flowrate of the reactant in the feed $\dot{n}_{in,R}$ and a given conversion X, the reaction rate, which is a function of conversion, needs to be determined. Let us consider a few indicative examples.

We first consider the general irreversible, first order reaction $R \rightarrow$products

$$\dot{R}''' = k(T) C_R \tag{10.2.5}$$

Using the definition of the fraction conversion given by (10.2.3) we obtain

$$C_R = C_{in,R}(1 - X) \tag{10.2.6}$$

and

$$\dot{R}''' = k(T) C_{in,R}(1 - X) \tag{10.2.7}$$

We consider $\sigma_R = -1$ and we replace Eq. (10.2.7) into (10.2.4) to obtain

$$V_{\text{CSTR}} = \frac{\dot{V}}{k(T)} \frac{X}{(1 - X)} \tag{10.2.8}$$

From Eq. (10.2.7), we observe that the reaction rate begins at $k(T)C_{in,R}$ when the conversion fraction $X = 0$, and decreases continuously as conversion increases. Conversely, the inverse of the reaction rate starts at $1/\left(k(T)C_{in,R}\right)$ for $X = 0$ and increases

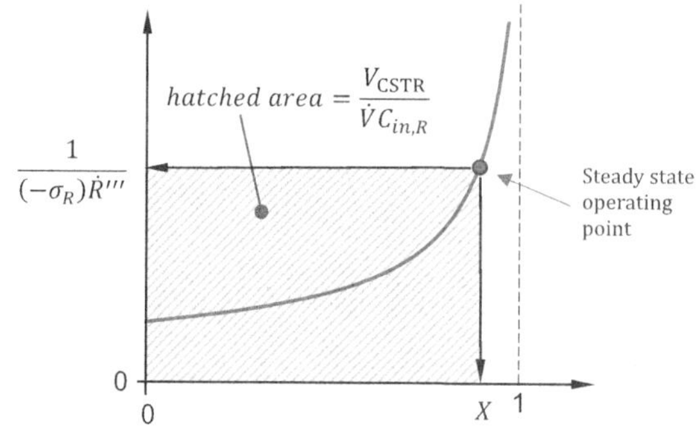

Figure 10.2 Inverse of the rate of consumption of the reactant as a function of fraction conversion and a CSTR operating point.

steadily with conversion. This behavior is illustrated in Fig. 10.2, which plots the inverse of the reactant consumption rate as a function of the conversion fraction.

An important observation is that when the inverse of the reactant consumption rate is known as a function of fractional conversion—either from calculations or experimental data—then, for each value of conversion, the product of the conversion and the inverse of the reaction rate yields, according to Eq. (10.2.4), the reactor volume of a CSTR divided by the molar flow rate of the reactant in the feed ($V/\dot{n}_{in,R}$).

Thus, the required reactor volume can be represented as the hatched area in Fig. 10.2. This graphical approach clearly illustrates how reactor volume evolves with increasing target conversion. A key note is that as the fractional conversion approaches 1, the reaction rate approaches zero and the inverse of the reaction rate tends to infinity, causing the required reactor volume also to approach infinity. This highlights a practical limitation: while increasing conversion initially leads to a manageable increase in reactor volume, the volume grows disproportionately as conversion nears completion, making very high conversions increasingly impractical from a design perspective.

Let us now examine how the required reactor volume can be reduced by using multiple CSTRs in series. The case of two CSTRs is illustrated in Fig. 10.3. For a given overall conversion, there are infinitely many possible combinations of conversion levels in the first reactor, each corresponding to a different total reactor volume. Importantly, the total volume for two CSTRs in series is always less than that of a single CSTR achieving the same final conversion. The shaded area in Fig. 10.3 represents this reduction in total volume.

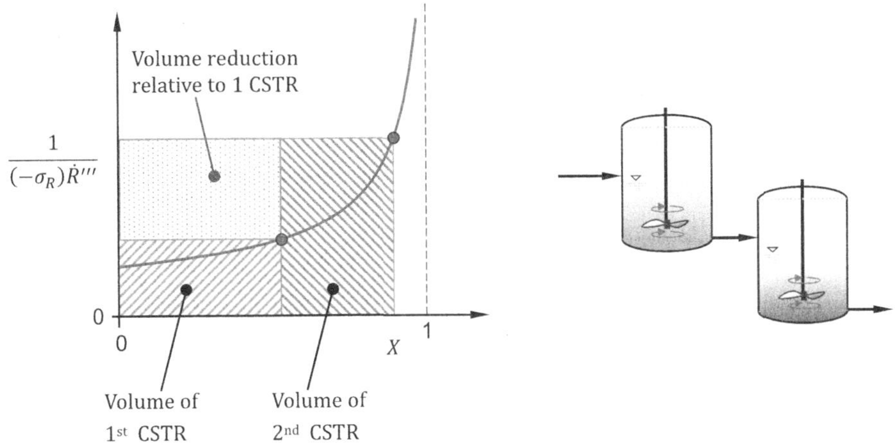

Figure 10.3 The inverse of the rate of consumption of the reactant as a function of fraction conversion for a CSTR can be used to determine the reactor volume: case of two CSTRs in series.

Given the significant volume savings, we extend the approach to three CSTRs in series, as shown in Fig. 10.4. As the number of CSTRs increases, the total reactor volume continues to decrease and asymptotically approaches a minimum. This minimum corresponds to the area under the curve $\sim \frac{1}{(-\sigma_R)\dot{R}'''}$, which, by the definition of the definite integral, can be calculated as

$$\left.\frac{V_{\text{CSTR}}}{\dot{n}_{in,R}}\right|_{\min} = \left[\text{Area under the } \frac{1}{(-\sigma_R)\dot{R}'''(X)} \text{ curve}\right] = \int_0^X \frac{d\chi}{(-\sigma_R)\dot{R}'''(\chi)} \qquad (10.2.9)$$

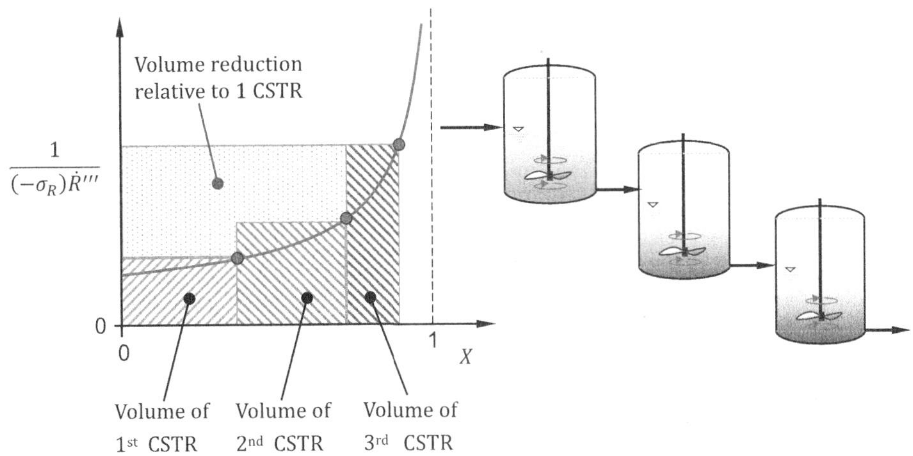

Figure 10.4 The inverse of the rate of consumption of the reactant as a function of fraction conversion for a CSTR can be used to determine the reactor volume: case of three CSTRs in series.

The maximum volume always corresponds to a single CSTR that achieves the desired conversion.

The calculation implied in Eq. (10.2.9) can be performed even when experimental data are available to calculate directly the minimum reactor volume that can achieve the desired concentration (this holds true only when the reaction rate is a strictly decreasing function of conversion). The distance between the minimum and the maximum volume gives an indication of the potential for using multiple CSTR reactors.

Example 10.1. Calculate the CSTR volume from experimental data

The following experimental data are obtained for the reaction $R \rightarrow P$:

Conversion X	0	0.1	0.2	0.3	0.4	0.5	0.6	0.7	0.8	0.9
Reaction rate $\dot{R}'''$ (kmol/(m^3 h))$\times 10^3$	25.0	22.5	20.0	17.5	15.0	12.5	10.0	7.5	5.0	2.5

Calculate the volume of a single CSTR for achieving conversions 0.1, 0.2, ..., 0.9 when the volumetric flowrate to the reactor is $\dot{V} = 0.5 \dfrac{m^3}{h}$ and the reactant concentration is $C_{in,R} = 1 \dfrac{kmol}{m^3}$. What is the minimum and what is the maximum reactor volume for achieving 60% conversion?

Solution

We will perform an indicative calculation for $X = 0.6$. From Fig. 10.5 we note that the inverse of the reaction rate for $X = 0.6$ is 100 h m^3/kmol. If one CSTR is used then Eq. (10.2.4) can be applied

$$\frac{V_{CSTR}}{\dot{n}_{in,R}} = \frac{1}{(-\sigma_R)\dot{R}'''} X = 100 \frac{h \cdot m^3}{kmol} \times 0.6 = 60 \frac{h \cdot m^3}{kmol}$$

or

$$V_{CSTR} = \dot{n}_{in,R} \times 60 \frac{h \cdot m^3}{kmol} = 0.5 \frac{m^3}{h} \times 1 \frac{kmol}{m^3} \times 60 \frac{h \cdot m^3}{kmol} = 30 \ m^3$$

As we have discussed, this is the maximum volume for achieving a 60% conversion.

We repeat the calculations for all fraction conversions to obtain:

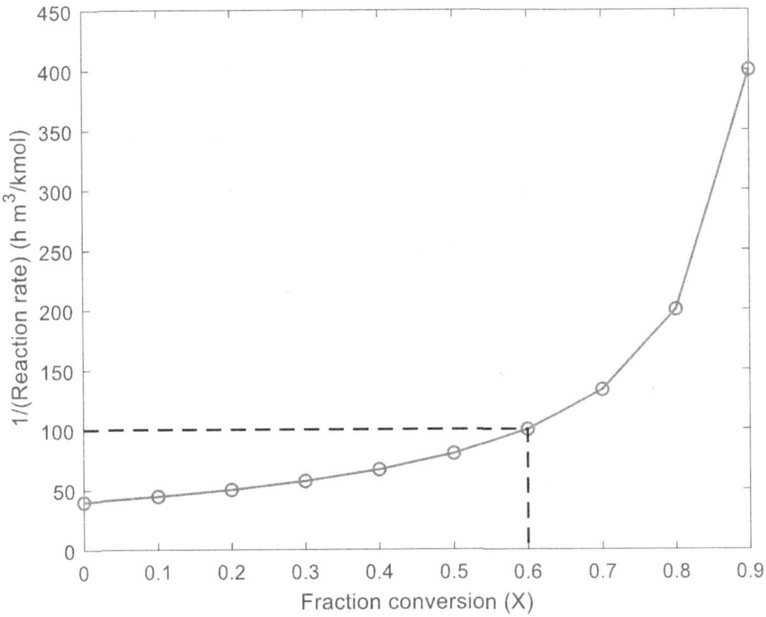

Figure 10.5 Experimental data on the reaction rate as a function of fraction conversion for the reaction R→P.

Conversion X	0	0.1	0.2	0.3	0.4	0.5	0.6	0.7	0.8	0.9
Reaction rate $\dot{R}'''$ (kmol/(m^3 h))× 10^3	25.0	22.5	20.0	17.5	15.0	12.5	10.0	7.5	5.0	2.5
$\frac{1}{\dot{R}'''}X, \frac{\text{m}^3\ \text{h}}{\text{kmol}}$	0	4.44	10	17.14	26.67	40	60	93.33	160	360
$V_{CSTR}(X)$, m^3	0	2.22	5	8.57	13.33	20	30	46.67	80	160

To find what is the minimum reactor volume for achieving 60% conversion using an infinite number of CSTRs in series we need to perform the calculation indicated by Eq. (10.2.9). Then

$$\int_0^{0.6} \frac{dX}{\dot{R}'''(X)} = \int_0^{0.1} \frac{dX}{\dot{R}'''} + \int_{0.1}^{0.2} \frac{dX}{\dot{R}'''} + \ldots + \int_{0.5}^{0.6} \frac{dX}{\dot{R}'''}$$

We note that

$$\int_0^{0.1} \frac{dX}{\dot{R}'''} = \text{area of a trapezoid with basis } \frac{1}{25 \times 10^{-3}} \textit{ and } \frac{1}{22.5 \times 10^{-3}} \text{ and height } 0.1 = \frac{1}{2}\left(\frac{1}{25 \times 10^{-3}} + \frac{1}{22.5 \times 10^{-3}}\right) \times 0.1 = 4.222$$

Using the trapezoid rule to approximate all individual integrals we obtain

$$\begin{aligned}\int_0^{0.6} \frac{dX}{\dot{R}'''} &= \frac{1}{2}\left(\frac{1}{25 \times 10^{-3}} + \frac{1}{22.5 \times 10^{-3}}\right) \times 0.1 + \frac{1}{2}\left(\frac{1}{22.5 \times 10^{-3}}\right.\\ &\quad \left. + \frac{1}{20 \times 10^{-3}}\right) \times 0.1 + \frac{1}{2}\left(\frac{1}{20 \times 10^{-3}} + \frac{1}{17.5 \times 10^{-3}}\right) \times 0.1\\ &\quad + \frac{1}{2}\left(\frac{1}{17.5 \times 10^{-3}} + \frac{1}{15 \times 10^{-3}}\right) \times 0.1 + \frac{1}{2}\left(\frac{1}{15 \times 10^{-3}}\right.\\ &\quad \left. + \frac{1}{12.5 \times 10^{-3}}\right) \times 0.1 + \frac{1}{2}\left(\frac{1}{12.5 \times 10^{-3}} + \frac{1}{10 \times 10^{-3}} + \right)\\ &\quad \times 0.1\\ &= 4.222 + 4.722 + 5.357 + 6.1905 + 7.334 + 9\\ &= 36.826 \ \frac{\text{h}\cdot\text{m}^3}{\text{kmol}}\end{aligned}$$

The total volume is

$$\begin{aligned}V_{\text{min,CSTR}} = \dot{n}_{in,R} \times 36.826 \ \frac{\text{h}\cdot\text{m}^3}{\text{kmol}} &= 0.5 \frac{\text{m}^3}{\text{h}} \times 1 \frac{\text{kmol}}{\text{m}^3}\\ &\quad \times 36.826 \ \frac{\text{h}\cdot\text{m}^3}{\text{kmol}}\\ &= 18.413 \ \text{m}^3\end{aligned}$$

We conclude that, depending on the number of CSTRs, the total volume will be between the minimum volume of 18.413 m^3 and the maximum volume of 30 m^3.

Example 10.2. Calculate the volume of 2 CSTRs in series from experimental data

Calculate the volume of two CSTRs in series that achieve 60%. Select arbitrarily the conversion achieved in the first reactor.

Solution

We select arbitrarily the fraction conversion in the first reactor to be 0.4. The volume of the two reactors are

$$V_{\text{CSTR},1} = \dot{n}_{in,R} \times \frac{1}{\dot{R}'''(X)} X = 0.5\frac{\text{m}^3}{\text{h}} \times 1\frac{\text{kmol}}{\text{m}^3} \times \frac{1000}{15}\frac{\text{h}\cdot\text{m}^3}{\text{kmol}} \times 0.4$$
$$= 13.334\ \text{m}^3$$

$$V_{\text{CSTR},2} = \dot{n}_{in,R} \times \frac{1}{\dot{R}'''(X)} X$$
$$= 0.5\frac{\text{m}^3}{\text{h}} \times 1\frac{\text{kmol}}{\text{m}^3} \times \frac{1000}{10}\frac{\text{h}\cdot\text{m}^3}{\text{kmol}} \times (0.6 - 0.4) = 10\ \text{m}^3$$

The results of Examples 10.1 and 10.2 can be validated in MATLAB:

```
clear all
X     = [ 0.0  0.1  0.2  0.3  0.4  0.5  0.6 0.7 0.8 0.9];
RRate = [25.0 22.5 20.0 17.5 15.0 12.5 10.0 7.5 5.0 2.5]/1000; % m^3 h / kmol
V_dot = 0.5; % m^3 / h
C_in  = 1.0; % kmol/m^3
% one CSTR
i  = 7 ;% 60% conversion
V_CSTR = V_dot*C_in*(1/RRate(i))*X(i) ;
fprintf('Conversion =%6.2f per cent, Maximum CSTR volume (1 CSTR)= %6.2f
m^3\n',100*X(i),V_CSTR)
% two CSTR
i1 = 5;
i2 = 7;
V_CSTR(1)  = V_dot*C_in*(1/RRate(i1))*X(i1);
V_CSTR(2)  = V_dot*C_in*(1/RRate(i2))*(X(i2)-X(i1));
fprintf('Conversion =%6.2f per cent, CSTR volume (2 CSTR) V_{1}= %6.2f V_{2}= %6.2f
m^3\n',100*X(i2),V_CSTR(1),V_CSTR(2))

V_CSTR_min = V_dot*C_in*trapz(1./RRate(1:i2))*0.1
```

10.2.2 The plug-flow reactor

A plug-flow reactor (PFR) is a reactor where (see Fig. 10.6):

- the reacting fluid moves as if it were a series of plugs or slices moving forward through the pipe,
- no back-mixing occurs: each plug moves forward without mixing with the fluid in front or behind it,
- axial mixing is negligible, but complete radial mixing is assumed (so the composition is uniform across any cross-section).

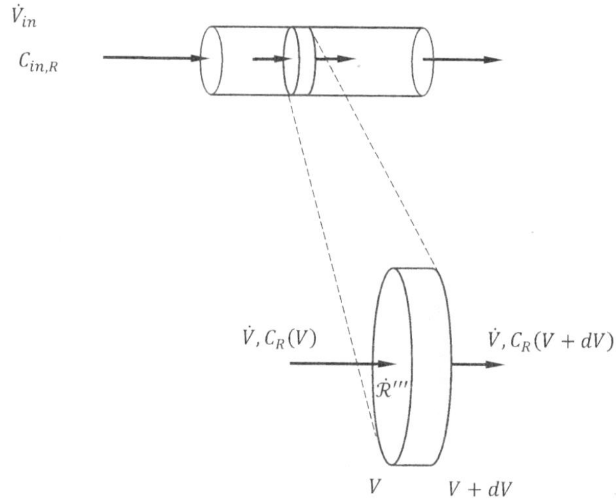

Figure 10.6 A continuous stirred tank reactor.

The volumetric flowrate of the feed to the reactor is $\dot{V}_{in}$ and the reactant concentration $C_{in,R}$. The volumetric flowrate of the product stream, after volume V of the reactor has been traveled by the reacting mixture, is $\dot{V}$ and the reactant concentration C_R. For liquid systems the density is practically constant but for gas streams this assumption is not valid. The mole balance of the slice shown in Fig. 10.6 can be written as:

$$\dot{V}C_R(V + dV) = \dot{V}C_R(V) + \sigma_R dV\dot{R}''' \quad (10.2.10)$$

or

$$\frac{\dot{V}C_R(V + dV) - \dot{V}C_R(V)}{dV} = (-\sigma_R)\dot{R}''' \quad (10.2.11)$$

where σ_R is the stoichiometric number of the reactant and $\dot{R}'''$ is the reaction rate. We use the definition of the fraction conversion X and

$$\frac{\dot{V}C_{in,R}\,X(V + dV) - \dot{V}C_{in,R}\,X(V)}{dV} = (-\sigma_R)\dot{R}''' \quad (10.2.12)$$

We solve for the reactor volume and integrate to obtain:

$$V_{\text{PRF}} = \dot{V}C_{in,R}\int_0^X \frac{d\chi}{(-\sigma_R)\dot{R}'''} \quad (10.2.13)$$

or

$$\frac{V_{\text{PFR}}}{\dot{n}_{in,R}} = \int_0^X \frac{d\chi}{(-\sigma_R)\dot{R}'''} \tag{10.2.14}$$

Eq. (10.2.14) demonstrates that if the reaction rate is known as a function of the fraction conversion, calculating the volume of a plug-flow reactor (PFR) required to achieve a specific conversion reduces to finding the area under the curve of the inverse reaction rate. We have already seen that this is equal to the minimum volume of an infinite number of CSTRs in series.

10.2.3 Combination of CSTR and PFR reactors

In the previous two sections, we examined how to design CSTRs and PFRs using plots of the inverse reaction rate as a function of conversion. These plots can be generated either from a kinetic model or directly from experimental data, without requiring prior identification of a specific rate law. The general shape of such a curve is illustrated in Figs. 10.2 and 10.7, which also demonstrate the graphical methods for determining the reactor volume for both CSTRs and PFRs. For typical reaction systems, the reaction rate decreases monotonically as conversion increases—this is the expected behavior and is reflected in the downward-sloping nature of the curves shown in Figs. 10.2 and 10.7. In such cases, the PFR provides the minimum reactor volume required to achieve a given conversion. Configurations consisting of a small number of CSTRs in series can approximate the performance of a PFR, with the total volume

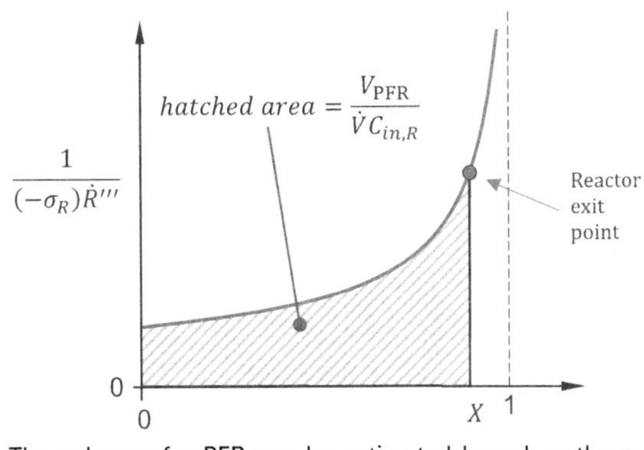

Figure 10.7 The volume of a PFR can be estimated based on the area below the inverse of the reaction rate as a function of the fraction conversion.

approaching that of a true PFR as the number of tanks increases (often impractical from an engineering standpoint).

Combinations of CSTRs and PFRs in series can also be analyzed using the same graphical approach. Fig. 10.8 presents two possible configurations involving one PFR and one CSTR in series, highlighting the different ways these elements can be arranged. The optimal configuration (e.g., CSTR followed by PFR, or vice versa) depends on the specific shape of the inverse rate versus conversion curve derived from the experimental data. The graphical method allows for a straightforward comparison

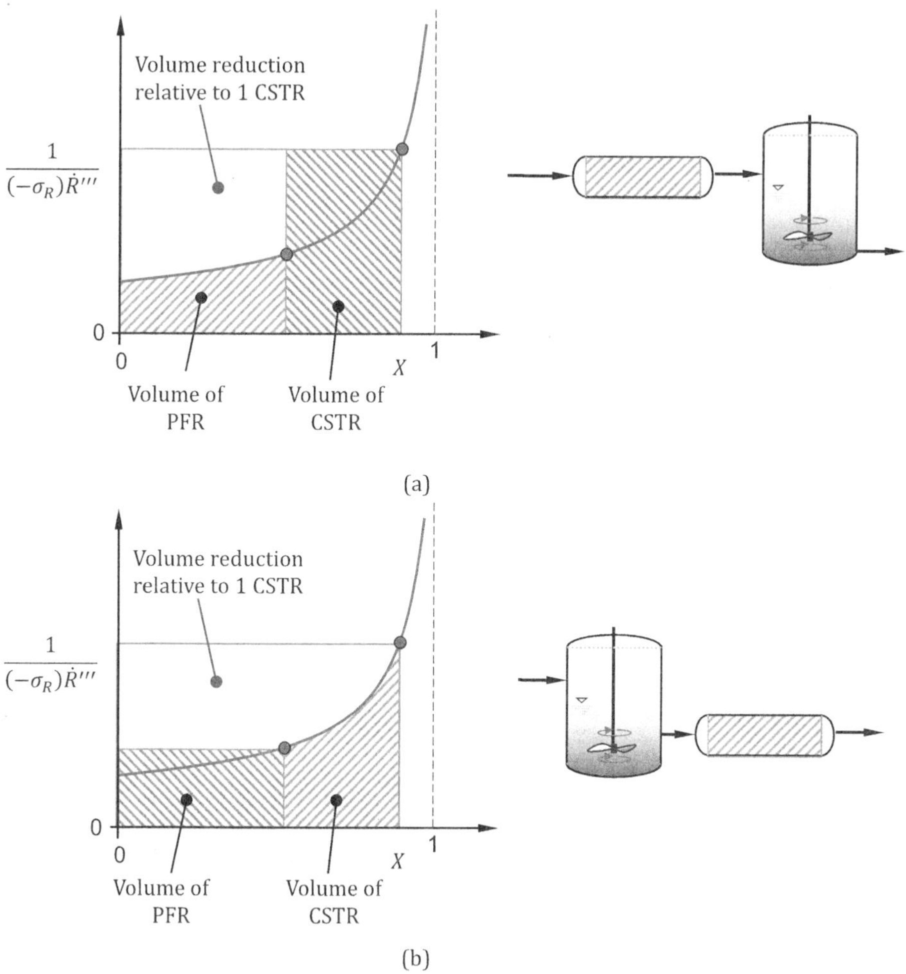

Figure 10.8 Graphical representation of solutions corresponding to using one PFR and one CSTR in series.

of these topologies, facilitating selection of the most efficient design.

A more interesting case study is when the plot of the inverse reaction rate as a function of conversion is not monotonic. A classical example is shown in Fig. 10.9. The optimum solution that corresponds to the minimum overall reactor volume is given in this case by a CSTR followed by a PFR. The CSTR operates at the point that corresponds to the minimum of the inverse reaction rate (or maximum reaction rate) as pictured in Fig. 10.10.

We have seen in this section that classical CSTR and PFR reactors can be designed based solely on experimental data regarding the reaction rate as a function of fractional conversion. These experimental data are combined with the mole balance of

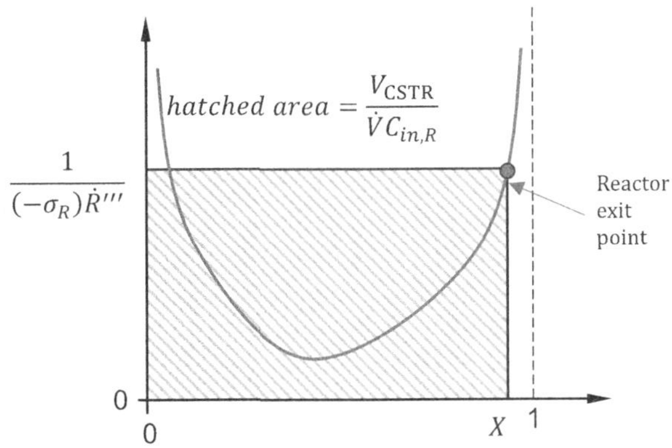

Figure 10.9 Inverse of the rate of consumption of the reactant as a function of fraction conversion and a CSTR operating point: the case of a nonmonotonic behavior.

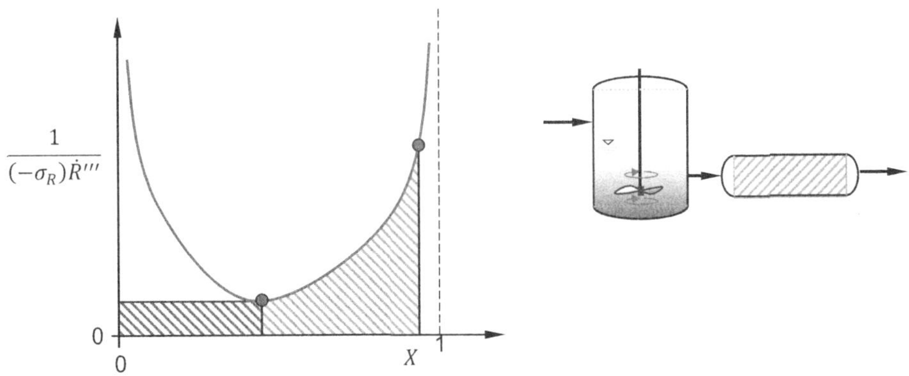

Figure 10.10 A CSTR followed by a PFR is the best (minimum overall volume solution) when the inverse of the rate of consumption of the reactant as a function of fraction conversion exhibits a nonmonotonic behavior.

the reference reactant in order to develop equations for the reactor size. This has a nice and transparent graphical representation. Mastering this graphical representation can be an invaluable tool in structural and parametric optimization of reacting systems that allows us to solve a particularly challenging problem through simple graphical means. The use of computers and process simulation software has antiquated classical graphical design techniques but they are still in common use for easy visualization and preliminary design of complex process systems. In the section that follows we present briefly similar techniques for the preliminary design of equilibrium-stage separation processes.

10.3 Graphical design of equilibrium-stage separation operations

10.3.1 Graphical method for equilibrium flash separation units

The flash separation units have been discussed in Chapter 8. In this section we will consider the graphical design of flash separation units, which consist of a single equilibrium-stage separation unit. A flash separation unit is shown in Fig. 10.11. A binary mixture with molar flowrate F and composition z (mole fraction of more volatile component) is fed to the flash distillation unit. Two product streams are obtained: a vapor stream with molar flowrate V and composition y and a liquid stream with molar

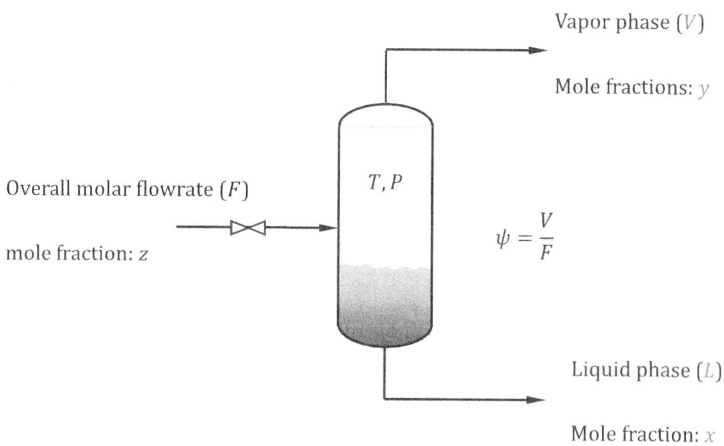

Figure 10.11 Summary of the "flash" calculation problem for a binary mixture.

flowrate L and composition x. The (light) component mole balance is as follows:

$$zF = xL + yV \tag{10.3.1}$$

We divide by F

$$z = x\frac{L}{F} + y\frac{V}{F} \tag{10.3.2}$$

We define the vapor fraction ψ:

$$\psi = \frac{V}{F} \tag{10.3.3}$$

and from the overall mole balance ($F = L + V$) it follows that

$$\frac{L}{F} = \frac{F - V}{F} = 1 - \psi \tag{10.3.4}$$

We substitute Eqs. (10.3.3) and (10.3.4) into (10.3.2) to obtain:

$$z = x(1 - \psi) + y\psi \tag{10.3.5}$$

which is solved for y:

$$y = -\frac{(1 - \psi)}{\psi}x + \frac{1}{\psi}z \tag{10.3.6}$$

As z is fixed, for a given value of the vapor fraction ψ this is a straight line that can be defined by the following two points: (z, z) and $(0, z/\psi)$ in a $x - y$ diagram. Note that when $x = z$

$$y = -\frac{(1 - \psi)}{\psi}z + \frac{1}{\psi}z = -\frac{1}{\psi}z + \frac{\psi}{\psi}z + \frac{1}{\psi}z = z \tag{10.3.7}$$

In Fig. 10.12 a hypothetical $x - y$ diagram for a binary mixture is given ($y_e = f(x_e)$). In the same figure the composition of the feed stream is also shown (point (z, z) on the diagonal). The feed composition is z and the straight line that represents Eq. (10.3.7) is presented in the figure for $0 < \psi < 1$, together with the two extreme values ($\psi = 1$ and $\psi = 0$). The part of the equilibrium line that can be achieved by all possible values of ψ for a given z is denoted by the grayed area. The graphical procedure for solving the mole balance for a single equilibrium-stage separation unit is as follows:

- use experimental equilibrium data to construct the $x - y$ diagram, ($y_e = f(x_e)$)
- use the composition of the feed to locate the point (z, z),
- select a vapor fraction ψ and locate the point $\left(0, \frac{z}{\psi}\right)$,

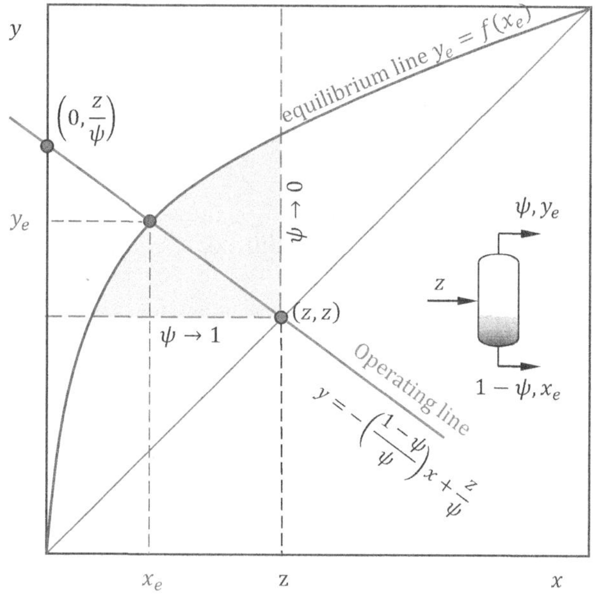

Figure 10.12 Operating line of an equilibrium flash distillation unit for a binary mixture.

- construct the **operating line** given by Eq. (10.3.6) by connecting the points (z, z) and $\left(0, \frac{z}{\psi}\right)$,
- find the point (x_e, y_e) where the operating line crosses the equilibrium line.

If several flash distillation units are connected in series and all operate at the same pressure, we can use the same x-y diagram to solve the mole balances for each unit. But if the units work at different pressures, the equilibrium curves will be different. In that case, we can draw a combined x-y diagram that includes the curves for each pressure. This lets us draw the operating lines for all the units on one diagram and analyze them together.

Example 10.3. Design of two flash distillation units in series from experimental data

In Table 10.1 constant-pressure VLE data for the binary mixture of tetrahydrofuran (THF) and water (W) are presented. A mixture of THF and water, 40% in THF, is fed at a train of two flash distillation units operated in series at a pressure of 1.013 bar with $\psi_1 = \psi_2 = \frac{1}{2}$, as shown in Fig. 10.13. What is the composition and flowrate of streams 2 to 5? What is the recovery of THF in the vapor stream of the second flash distillation

Table 10.1 Constant-pressure VLE data for THF-water.

T (°C)	x_{THF}	y_{THF}
100.00	0.00	0.0000
73.00	0.02	0.6523
66.50	0.04	0.7381
65.58	0.06	0.7516
64.94	0.10	0.7587
64.32	0.20	0.7625
64.27	0.30	0.7635
64.23	0.40	0.7643
64.16	0.50	0.7658
63.94	0.60	0.7720
63.70	0.70	0.7831
63.54	0.80	0.8085
63.53	0.82	0.8180
63.57	0.84	0.8260
63.64	0.86	0.8368
63.87	0.90	0.8660
64.29	0.94	0.9070
65.07	0.98	0.9625
65.39	0.99	0.9805
66.00	1.00	1.0000

Data from Perry's Handbook.

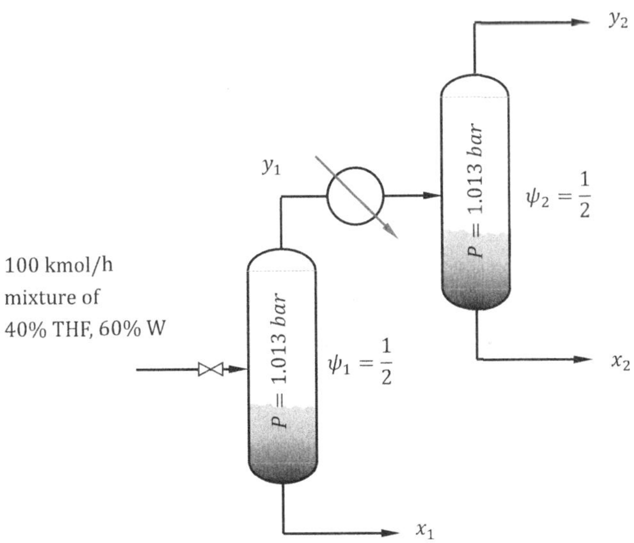

Figure 10.13 Two flash distillation units in series for separating a binary mixture of THF and water.

unit? What are the operating temperatures? Use the data of Table 10.1 to answer all questions.

Solution

We start with the first flash unit and place the points (z, z)= (0.4,0.4) and $(0, z/\psi)$=(0,0.8) on the x-y diagram as shown in Fig. 10.14. We find the point that the line connecting these points crosses the equilibrium line, which is approximately the point $(x_1, y_1) = (0.05, 0.74)$ (from Table 10.1 we see that the temperature in the first unit is 66 °C). The vapor stream from the first line is partially condensed and is fed to the second flash unit. We locate the points (z, z)=(0.74,0.74) and $(0, z/\psi)$=(0,1.48) for the second flash unit on the x-y diagram, as shown in Fig. 10.14. As point $(0, 1.48)$ is outside the scale used we determine the point that corresponds to $y = 1$ which is (by solving the operating line for x) the point (0.48,1). We find the point

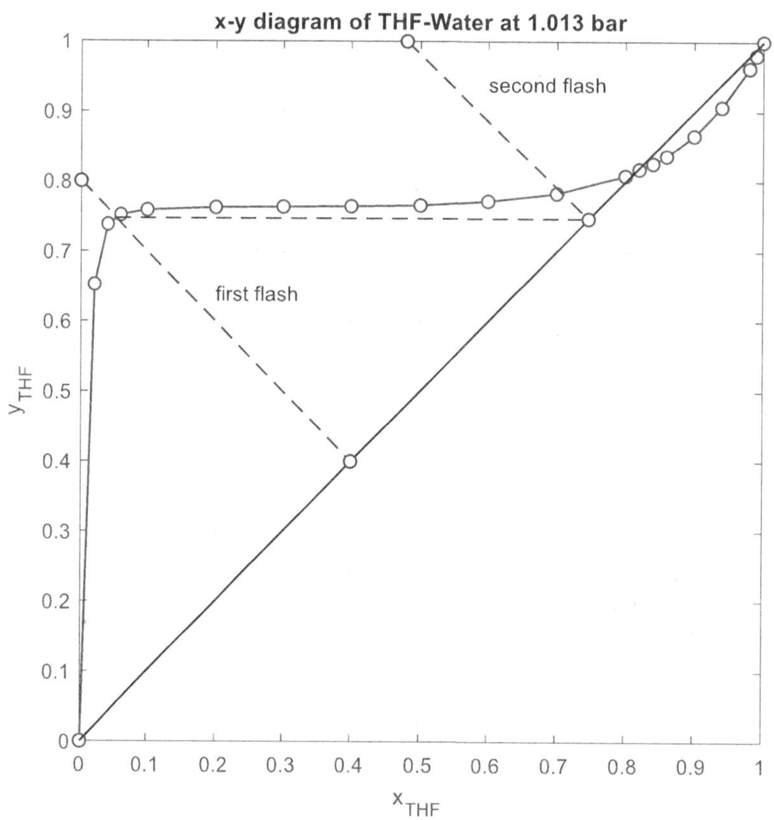

Figure 10.14 Graphical solution of the THF-W separation in a series of two flash distillation units shown in Fig. 10.13.

that the line connecting these points crosses the equilibrium line, which is approximately the point $(x_2, y_2) = (0.70, 0.78)$ (from Table 10.1 we see that the temperature in the first unit is 63.7° C). The recovery is 48.75% and is calculated as follows:

$$\%\ THF\ \text{recovery} = 100\frac{y_2(\psi_1\psi_2 F)}{zF} = \frac{0.78 \times \left(\frac{1}{2}\times\frac{1}{2}\times 100\right)}{0.4\times 100} = 48.75\ \%$$

10.3.2 Graphical method for cascades of equilibrium stages

Cascades of contacting stages are usually employed in process industries for achieving high purity and high recovery of targeted components from mixtures. In each stage of such cascades equilibrium can be achieved between the product streams, which can be of different phases. The design of a cascade such as a distillation column, absorption column, stripping column, or a liquid—liquid extraction cascade involves a consideration of the performance of each contacting stage and of the interactions of the various stages to produce the performance of the cascade as a whole. In this regard, the contacting stage of a physical separation process is usually susceptible to description by a relatively simple mathematical model.

In general, equations are written that describe, for each stage in the cascade, the mole balances and the equilibrium conditions. When experimental data for equilibrium are available, then we need to develop a methodology for combining the experimental data for the equilibrium with the mass balances. A practically useful methodology is obtained when we can depict the material balances on the same diagram with the experimental data and solve their combination by graphical means, as we have already demonstrated for the case of single equilibrium-stage processes. The graphical methods are especially useful as a method of visualizing and understanding the relationships among the various variables involved in the cascade design.

A classical example of an equilibrium-stage cascade is obtained if we extend the two-stage process of Fig. 10.13 to several contacting stages connected as shown in Fig. 10.15. The liquid ("heavy" phase) product stream from each stage is fed to the stage

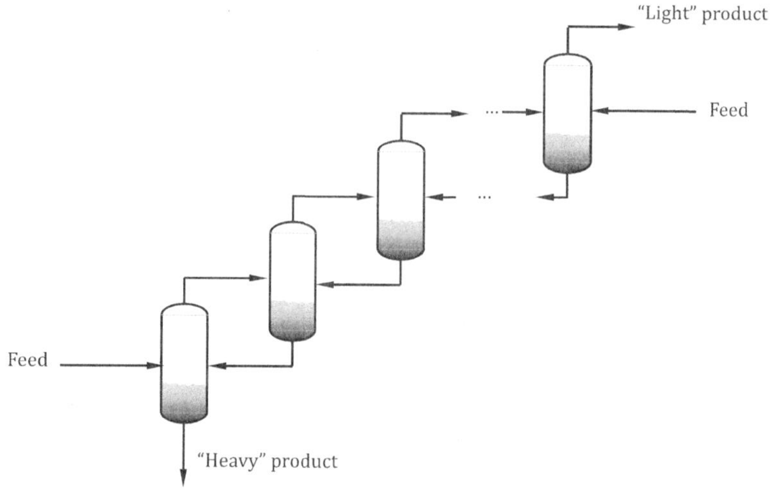

Figure 10.15 A cascade of equilibrium stages.

below while the vapor ("light" phase) stream is fed to the stage above. This creates a cascade of stages that can be represented as shown in Fig. 10.16, which can be used for clarity of presentation. We restrict attention to binary systems, and we will be using the following notation (see also Fig. 10.16):

- L is the molar flowrate of the liquid ("heavy") phase in kmol/h
- V is the molar flowrate of the vapor ("light") phase in kmol/h
- x_i is the mole fraction of the volatile component (the component that tends to concentrate in the vapor phase) in the liquid stream leaving the i-th stage
- y_i is the mole fraction of the volatile component (the component that tends to concentrate in the vapor phase) in the vapor stream leaving the i-th stage

Note that we have assumed that the molar flowrates of the vapor (V) and liquid phase (L) are constant along the cascade. This is a basic assumption that needs to be satisfied for the development of a simple graphical method to be possible. This assumption is known as the constant molar overflow assumption in distillation.

We now consider the control volume shown in Fig. 10.16 between stage 1 and stage i. We write the mole balance of the component of interest

$$Lx_{i+1} + Vy_0 = Lx_1 + Vy_i \tag{10.3.8}$$

We can solve for x_{i+1}:

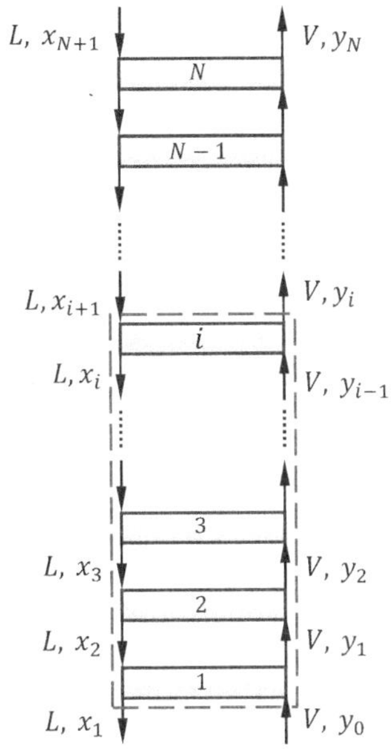

Figure 10.16 A cascade of equilibrium stages: representation for developing general theory.

$$x_{i+1} = ay_i + b \tag{10.3.9}$$

where

$$a = \frac{V}{L}, b = x_1 - \frac{V}{L}y_0$$

Eq. (10.3.9), that relates the composition of passing streams, is a very powerful linear equation. It can be used to calculate the composition of the liquid on tray $i+1$ if the composition of the vapor on tray i is known. This is true as x_1 and y_0 are compositions of external streams that are assumed known and L, V are constants. The straight line described by Eq. (10.3.9) is called the **operating line** of the cascade and can easily plotted on an experimental $x-y$ diagram giving the possibility to find by graphical means the number of equilibrium stages necessary to achieve a specific separation.

One of the most classical applications of stagewise separation processes is the design of absorbers. An **absorber** is a device used to selectively remove one or more components (called solutes) from a gas mixture by dissolving them into a liquid phase known as the solvent. This process, known as gas absorption, typically occurs in a vertical column referred to as an absorber column or simply an absorber.

In a typical absorber, a gas stream—such as an air stream containing a valuable or hazardous component—is introduced at the bottom of the column and flows upward through a series of stages. Simultaneously, a nonvolatile solvent is introduced at the top of the column and flows downward in countercurrent fashion. As the gas and liquid phases come into contact at each stage, the component of interest is transferred from the gas into the liquid solvent.

The key assumptions and characteristics of the absorption process are:

- the solvent does not evaporate into the gas stream,
- the bulk gas (solute-free carrier gas) does not dissolve into the liquid phase,

and as a result, the molar flow rates of the solute-free gas and the solute-free solvent remain constant throughout the absorber. The symbols L' and G' are used to denote the molar flowrate of the solute-free solvent and solute-free gas and the symbols X and Y are used to denote the mole ratio of the solute to the solute-free solvent (liquid) and the mole ratio of the solute to the solute-free carrier gas. The stage cascade representation of an absorption column is shown in Fig. 10.17. The mole balance of the solute written around the bottom of the cascade between the bottom stage and an arbitrary intermediate stage i is the following

$$L'X_{i+1} + G'Y_0 = G'Y_i + L'X_1$$

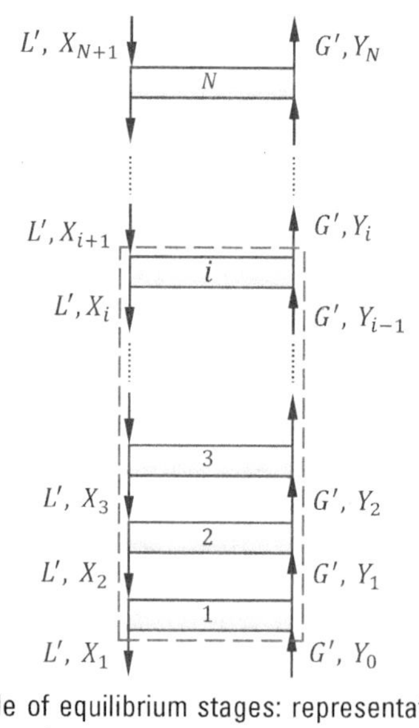

Figure 10.17 A cascade of equilibrium stages: representation of an absorption column as a stage cascade.

or

$$X_{i+1} = aY_i + b \tag{10.3.10}$$

where

$$a = \frac{G'}{L'}, b = X_1 - \frac{G'}{L'}Y_0$$

Let us consider a particular case study where an air stream that contains $Y_0 = 0.06$ mol of a valuable compound per mol of solute-free air with a molar flowrate of $G' = 1$ mol/s is fed to the bottom of an absorption column. Water ($X_{N+1} = 0.01$) is fed at the top of the absorption column with a molar flowrate of $L' = 0.3$ mol/s. The aim is to recover 90% of the valuable component in the water stream. The equilibrium distribution of the valuable component between the gas and the water is described by the following equation (polynomial fit of experimental data):

$$Y = X(0.2881 - 0.1563X) \tag{10.3.11}$$

Our aim is to determine the number of equilibrium stages that are needed in order to achieve the given separation.

The analysis starts with the recovery specification. As 90% of the valuable component is recovered, 10% is lost and the composition at the exiting gas stream is

$$Y_N = 0.10 \times 0.06 = 0.006$$

The next step is to formulate the overall mole balance

$$L'X_1 + G'Y_N = L'X_{N+1} + G'Y_0 \tag{10.3.12}$$

or

$$0.3 \times X_1 + 1.0 \times 0.006 = 0.3 \times 0.01 + 1.0 \times 0.06$$

whence

$$X_1 = 0.19$$

The a and b constants appearing in the operating line are

$$a = \frac{G'}{L'} = \frac{1}{0.3}, b = X_1 - \frac{G'}{L'}Y_0 = 0.19 - \frac{1}{0.3}0.06 = -0.01$$

and the operating line becomes

$$X_{i+1} = \frac{Y_i}{0.3} - 0.01 \tag{10.3.13}$$

Let us perform stage-by-stage calculations. We summarize the state of our calculations for stage 1 in Fig. 10.18. L', G', Y_0, and X_1 are known and we need to determine Y_1 and X_2. We perform the following calculations:

1. *Equilibrium calculation (calculate Y_1 in equilibrium with X_1)*

$$i = 1, Y_i = X_i(0.2881 - 0.1563X_i) = 0.0491 \qquad (10.3.14.a)$$

2. *Operating line (mole balance, calculate X_{i+1} given Y_i)*

$$i = 1, X_{i+1} = \frac{Y_i}{0.3} - 0.01 = 0.1537 \qquad (10.3.14.b)$$

With X_2 known, we increase $i = i + 1 = 2$ and repeat the steps indicated above to determine X_3, and the calculations are repeated until the composition of the gas stream falls below the specification of 0.006 mol/mol air. The value of i is equal to the number of stages in the cascade.

The results of the stage-by-stage calculations are summarized in Table 10.2 and Fig. 10.19. Fig. 10.19 suggests a simple graphical determination of the necessary number of equilibrium stages: we first plot the experimental data on a $x - y$ diagram and then, on the same diagram, we plot the operating line (Eq. (10.3.10)). We start from the bottom of the cascade and

a) move vertically from the operating line to the equilibrium line,

b) move horizontally from the equilibrium line to the operating line.

We continue until the composition of the gas streams falls below the specification. Each pair of vertical and horizontal movement corresponds to an equilibrium stage.

Fractional distillation is the most important equilibrium-stage separation process. The physical configuration of a distillation column is illustrated in Fig. 10.20. The fresh feed is introduced near the middle of the column, dividing it into two main sections: the rectifying section (above the feed) and the stripping section (below the feed). At the bottom of the column, a reboiler provides heat, generating vapor that rises through the column. The portion of the liquid that is not vaporized is withdrawn as the bottom product. At the top of the column, a

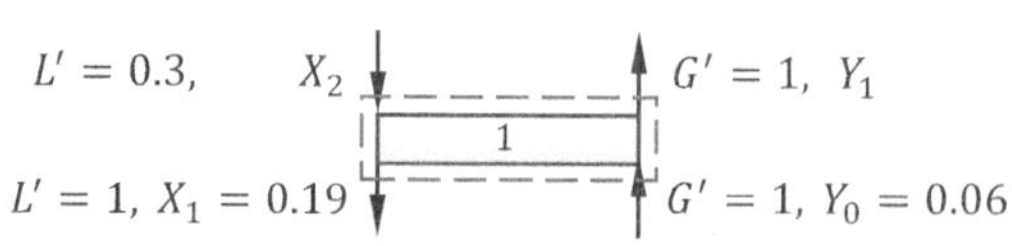

Figure 10.18 Summary of calculations for stage 1.

Table 10.2 Results of the stage-by-stage calculations for the absorption column calculations.

Stage *i*	*Y*	*X*
0	0.0600	–
1	0.0491	0.1900
2	0.0406	0.1537
3	0.0336	0.1253
4	0.0278	0.1021
5	0.0227	0.0826
6	0.0183	0.0658
7	0.0143	0.0509
8	0.0106	0.0376
9	0.0072	0.0253
$N = 10$	0.0040	0.0140
$N + 1$	–	0.0100

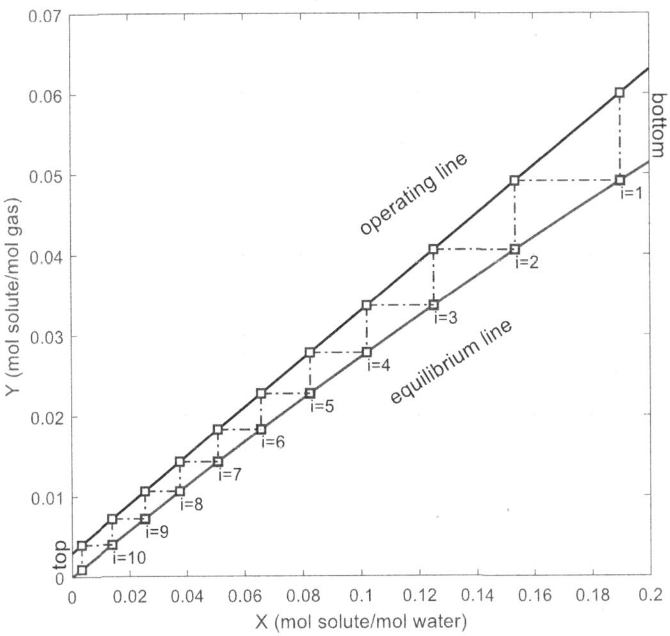

Figure 10.19 The equilibrium line, the operating line and the graphical demonstration of the method for determining the number of equilibrium stages in the cascade.

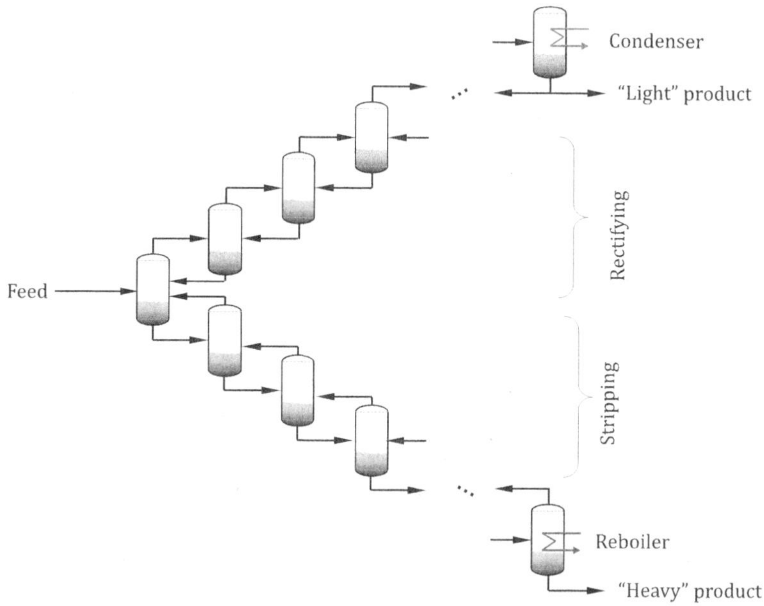

Figure 10.20 Stage cascade representation of a distillation column.

condenser removes heat, partially or fully condensing the vapor. A portion of this condensate is returned to the column as reflux, while the remainder is collected as the top product. The reflux ratio—defined as the ratio of the molar flow rate of the reflux to that of the top product—is a critical parameter that significantly influences the separation efficiency and operation of the distillation column.

A representative case for a binary distillation column with a saturated liquid feed is shown in Fig. 10.21. The reader must be familiar with the notation used:

- L is the molar flowrate of the liquid phase
- V is the molar flowrate of the vapor phase
- x_i is the mole fraction of the volatile component in the liquid stream leaving the i-th stage
- y_i is the mole fraction of the volatile component in the vapor stream leaving the i-th stage
- F is the molar flowrate of the fresh feed
- z is the mole fraction of the volatile component in the feed
- D is the molar flowrate of the top product (distillate)
- B is the molar flowrate of the bottom product

We have assumed that the molar flowrates of the liquid and vapor streams are constant along the cascade and only the molar flowrate of the liquid stream changes at the feed, as the liquid

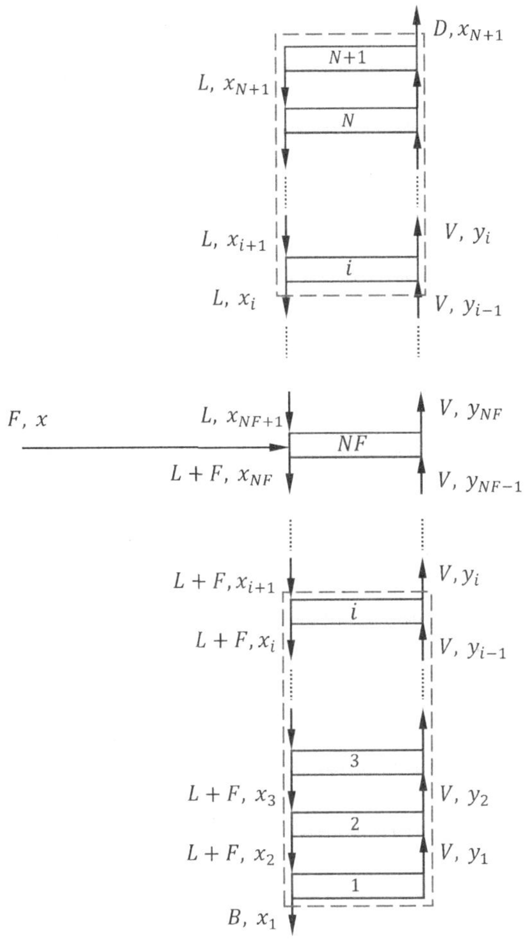

Figure 10.21 A stage cascade representation of a fractional distillation column.

feed (F) is added to the liquid stream in the rectifying section (L) to form the liquid in the stripping section ($F + L$).

There are two operating lines that are associated with a distillation column. One for the rectifying section and one for the stripping section. A mole balance around the control volume between the top stage and the i-th stage of the rectifying section gives:

$$Vy_{i-1} = Dx_{N+1} + Lx_i$$

or

$$y_{i-1} = \frac{L}{V}x_i + \left(\frac{D}{V}x_{N+1}\right) \tag{10.3.15}$$

We define the **reflux ratio** as follows:

$$rr = \frac{L}{D} \tag{10.3.16}$$

and it follows that

$$L = rrD$$

$$V = D + L = (1 + rr)D$$

Eq. (10.3.15) can be written in the compact form

$$y_{i-1} = \left(\frac{rr}{1 + rr}\right)x_i + \left(\frac{x_{N+1}}{1 + rr}\right) \tag{10.3.17}$$

This is the operating line of the **rectifying section**.

A mole balance around the control volume between the bottom stage and the i-th stage of the stripping section gives:

$$(L + F)x_{i+1} = Bx_1 + Vy_i$$

or

$$y_i = \left(\frac{L + F}{V}\right)x_{i+1} - \left(\frac{B}{V}x_1\right) \tag{10.3.18}$$

We define the **boilup ratio** as follows:

$$br = \frac{V}{B} \tag{10.3.19}$$

and it follows that

$$V = br\, B$$

$$L + F = B + V = (1 + br)B$$

Eq. (10.3.15) can be written in the compact form

$$y_{i-1} = \left(\frac{1 + br}{br}\right)x_i - \left(\frac{x_1}{br}\right) \tag{10.3.20}$$

This is the operating line of the **stripping section**.

The complete picture is shown in Fig. 10.22. In this figure the following lines are presented:

- the operating line for the rectifying section (Eq. (10.3.17)),
- the operating line for the stripping section (Eq. (10.3.20)),

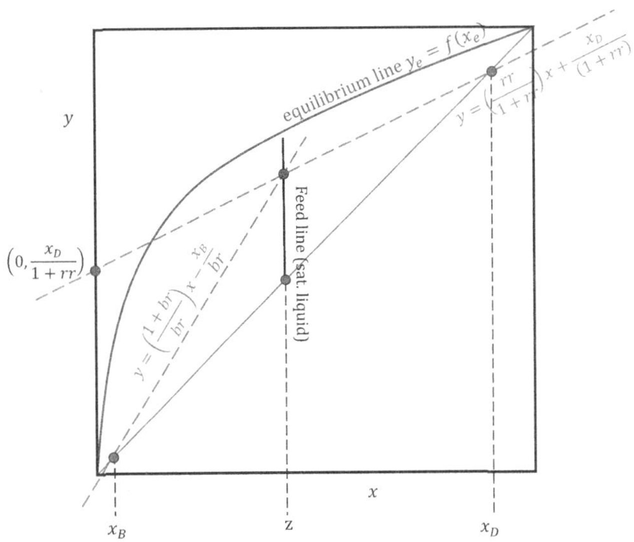

Figure 10.22 Operating line of an equilibrium flash distillation unit for a binary mixture.

- the feed line for saturated liquid feed,
- the equilibrium line.

The data necessary to complete this figure are:

- the feed flowrate, composition and vapor fraction ($\psi = 0$ in Fig. 10.22, see Fig. 10.12 for other cases),
- the top product composition $x_D = x_{NT+1}$,
- the bottom product composition $x_B = x_1$,
- the reflux ratio.

The distillate and bottom product molar flowrates are determined from the molar balances of the total process:

$$F = D + B \tag{10.3.21}$$

$$zF = x_B B + x_D D \tag{10.3.22}$$

The graphical design of a binary fractional distillation column follows the steps:

- plot the experimental data on the $x - y$ diagram,
- for the given top product composition and reflux ratio plot the operating line of the rectifying section,
- for the given bottom product composition and reflux ratio plot the operating line of the stripping section,
- start from either the top or the bottom and make vertical and horizontal movements between the operating lines and the equilibrium lines until you cover the interval $[x_B, x_D]$,

- count the number of steps necessary, which is equal to the number of equilibrium stages.

Example 10.4. Design of a binary distillation column from experimental data

An equimolar mixture of methanol and water is fed as a saturated liquid to a distillation column (feed molar flowrate $F=100$ kmol/h). The top product is 95 mol% methanol while the bottom product is 5 mol% methanol ($x_B = x_1 = 0.05, x_D = x_{NT+1} = 0.95$). Calculate the number of equilibrium stages including the exact position of the feed when the reflux ratio is $rr = 1.5$. Use the experimental data of Table 8.3.

Solution

We use the data of Table 8.3 to plot the experimental equilibrium line.

We calculate the two points that define the operating line for the rectifying section:

$$(x_D, x_D) = (0.95, 0.95)$$

$$\left(0, \frac{x_D}{1+rr}\right) = \left(0, \frac{0.95}{2.5}\right) = (0, 0.38)$$

We connect the two points and extend the line.

We calculate the two points that define the operating line for the rectifying section (show using the overall material balances that $D=B=50$ kmol/h and then $V=(1+rr)D=125$ kmol/h and finally $br=2.5$):

$$\left(\frac{x_B}{1+br}, 0\right) = (0.0143, 0)$$

$$\left(\frac{br + x_B}{1+br}, 1\right) = (0.7286, 1)$$

We connect the two points and extend the line.

We have now created the following figure:

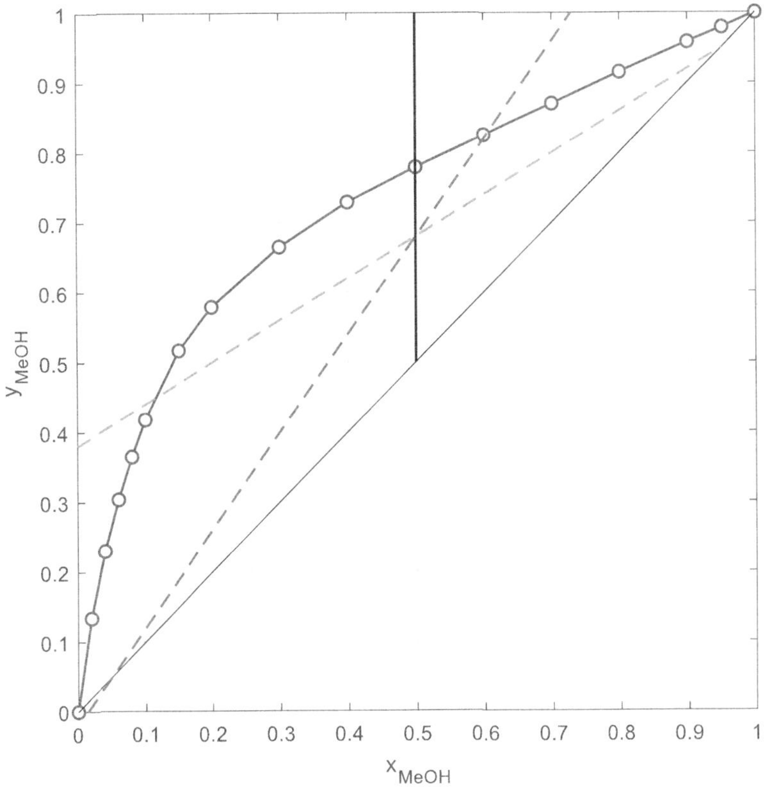

To calculate the number of equilibrium stages we start from the (x_B, x_B) point, move vertically to the equilibrium line, then horizontally to the operating line and repeat the process until the liquid composition becomes greater than x_D. The result is shown in Fig. 10.23. Six stages are necessary to achieve the requested separation. It is important to note that when the composition of the vapor becomes greater than the composition of the feed, we move from the operating line of the stripping section to the operating line of the rectifying section.

Before closing this section, it is important to discuss further the selection of the reflux ratio for a particular case study. Note that the point $(0,\ x_D/(1+rr))$, used to plot the operating line of the rectifying section, depends on the reflux ratio selected. The other point (x_D, x_D) is on the diagonal and is a fixed point independent of the reflux ratio selected. It is important to realize that:

- when the reflux ratio is increased then $x_D/(1+rr) \rightarrow 0$ and the operating line is moving away from the equilibrium line,
- when the reflux ratio is decreased then $x_D/(1+rr) \rightarrow x_D$ and the operating line is moving closer to the equilibrium line.

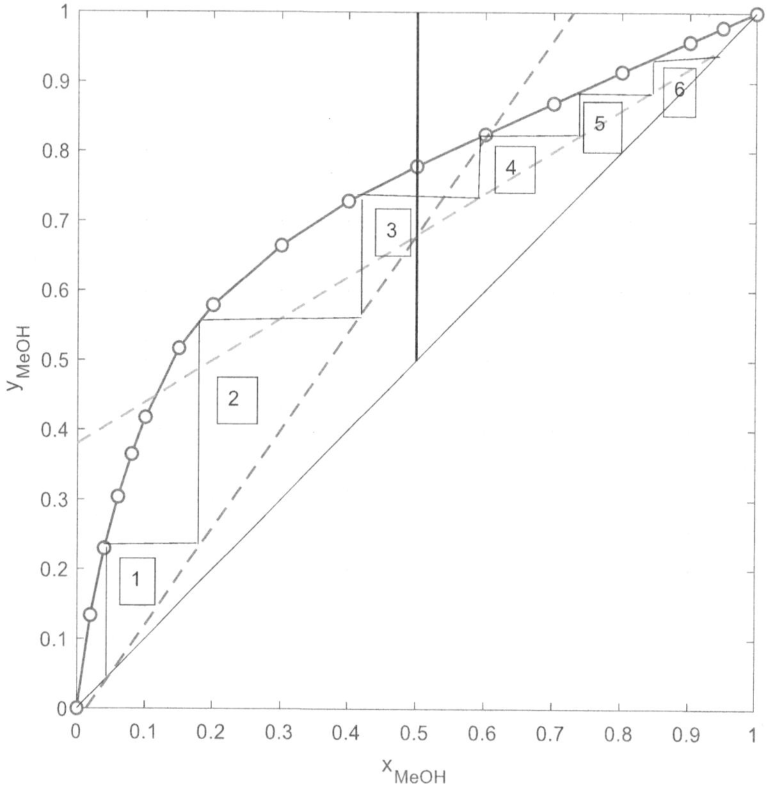

Figure 10.23 Estimating the number of equilibrium stages for the separation of methanol-water mixture using the graphical method.

An increase in the reflux ratio in a distillation column leads to a shift in the operating line away from the equilibrium line on the $x - y$ diagram. This movement enhances the efficiency of each theoretical stage, allowing the desired separation to be achieved with fewer equilibrium stages. In contrast, when the reflux ratio is reduced, the operating line moves closer to the equilibrium line. As a result, each stage becomes less efficient, and more stages are required to accomplish the same degree of separation.

This relationship highlights a fundamental trade-off in distillation design. Operating at a higher reflux ratio reduces the number of required stages and potentially the height of the column, but it increases the energy consumption due to greater reboiler and condenser duties. On the other hand, a lower reflux ratio minimizes energy demand but necessitates a taller column, leading to higher capital costs.

Thus, the design and operation of a distillation column involve a classical optimization problem: finding the optimal reflux ratio that balances operating costs (primarily energy) and capital costs (equipment and construction). By applying principles of process optimization, engineers aim to identify the most cost-effective operating point that satisfies both economic and separation performance criteria. This is the subject of the chapter that follows.

A widely used rule of thumb for selecting the reflux ratio is to choose a value between 1.2 and 1.4 times the minimum reflux ratio. To apply this guideline effectively, it is essential to understand what the minimum reflux ratio represents and how it is determined.

As previously discussed, decreasing the reflux ratio causes the operating line to move closer to the equilibrium curve. For a distillation process to be feasible, however, the operating line must remain below the equilibrium curve throughout the composition range between the feed composition (z) and the distillate composition (x_D). This requirement stems from the fundamental principle that the vapor composition in equilibrium with the liquid cannot exceed the equilibrium limit (Fig. 10.24).

The **minimum reflux ratio** corresponds to the critical condition where the operating line just touches or intersects the equilibrium curve at $x = z$. At this point, the number of required theoretical stages becomes infinite, and any further reduction in reflux ratio renders the separation thermodynamically infeasible. Thus, operating above the minimum reflux ratio is necessary to ensure that mass transfer between phases can proceed throughout the column. By choosing a reflux ratio modestly above the minimum—typically 1.2 to 1.4 times higher—we achieve a practical balance between energy efficiency and column size, avoiding both excessive energy consumption and unreasonably large equipment.

To calculate the minimum reflux ratio for separating an equimolar mixture of methanol and water, use a ruler to extend the straight line that passes through the points (0.95, 0.95) and (0.5, 0.779). The second point is obtained from Table 8.3. Extend this line until it intersects the $y = x$ diagonal at the $x = 0$ axis. The x-intercept should be approximately 0.589. This value will be used in determining the minimum reflux ratio. Next, we observe that

$$0.589 = \frac{x_D}{1 + rr_{\min}} = \frac{0.95}{1 + rr_{\min}} \Rightarrow rr_{\min} = 0.613$$

Make sure that you can reproduce and fully understand this calculation.

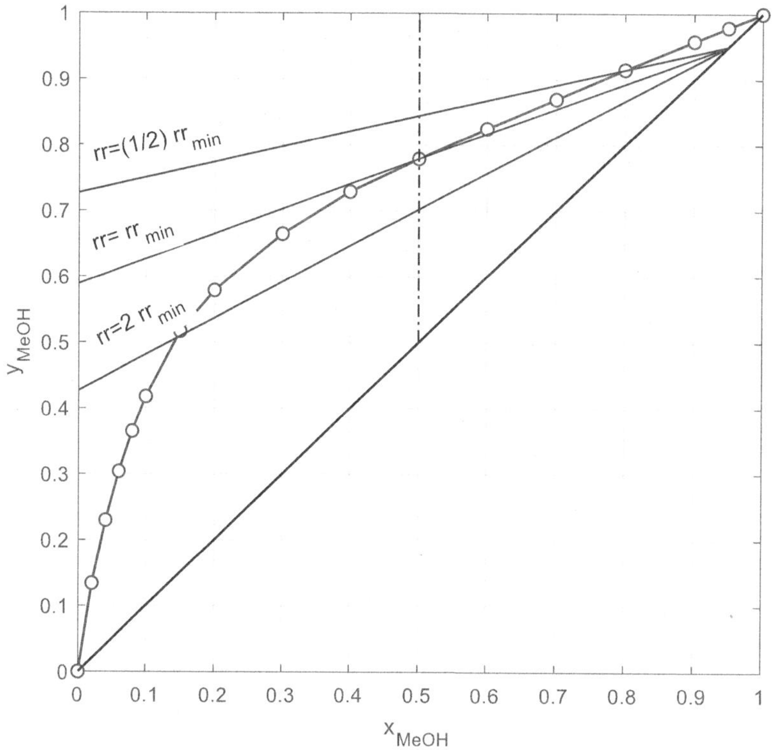

Figure 10.24 Operating lines for the rectifying section in the methanol–water separation discussed in Example 10.4. At the minimum reflux ratio, the operating line intersects the equilibrium curve at $x = z$, indicating the limit of feasible separation. When the reflux ratio is below the minimum, part of the operating line lies above the equilibrium curve, making separation thermodynamically infeasible. As the reflux ratio increases above the minimum, the operating line moves progressively farther from the equilibrium curve, improving stage efficiency and ensuring feasible operation.

Summary

In this chapter, we introduced several of the most well-known and widely used graphical methods for the design of ideal reactors and equilibrium-stage cascades. These techniques play a crucial role in the preliminary stages of process design, where engineers must manage a large number of design specifications efficiently. By using graphical methods, it is possible to obtain reasonable and insightful solutions without the need for repeatedly solving complex systems of nonlinear equations.

Such methods not only provide a visual understanding of system behavior but also allow for rapid screening of design alternatives, estimation of key parameters, and identification of operating constraints. Once a preliminary design has been established using these simplified approaches, more advanced and rigorous computational methods—such as numerical simulations or optimization techniques—can be employed to refine the design, improve accuracy, and ensure robustness under varying operating conditions.

Problems

10.1 Consider the liquid-phase reaction R→P, which is second order reaction:

$$\dot{R}''' = kC_R^2$$

with $k = 1\ \mathrm{m^3/(min{\cdot}kmol)}$. The feed stream flowrate to a PFR has a volumetric flowrate of $\dot{V} = 1\ \mathrm{m^3/min}$ and reactant concentration $C_{in,R} = 4\ \mathrm{kmol/m^3}$. Use Eq. (10.2.13) to derive an analytic expression for the volume of the PFR. For 80% conversion calculate the necessary volume of the PFR.

10.2 Assume that the exact form of the reaction rate is not known but the following experimental data are available:

Conversion	0.0	0.1	0.2	0.3	0.4	0.5	0.6	0.7	0.8
Reaction rate (kmol/(m^3 min))	16.00	12.96	10.24	7.84	5.76	4.00	2.56	1.44	0.64

Use the trapezoidal rule to determine numerically the PFR volume with $\Delta X = 0.4$, $\Delta X = 0.2$, and $\Delta X = 0.1$, based on Eq. (10.2.13). Compare your results with the exact result obtained in Problem 10.1. What is your observation?

Derive a general equation for calculating numerically the definite integral $\int_a^b f(x)dx$ using the trapezoid rule and find from the literature how the error depends on the integration interval h.

10.3 Assume that the exact form of the reaction rate is not known but the experimental data given in Problem 10.2 are available. Use Simpson's rule to determine numerically the PFR volume with $\Delta X = 0.4$, 0.2, and 0.1, based on Eq. (10.2.13). Compare your results with the exact result obtained in Problem 10.1. What is your observation?

Derive a general equation for calculating numerically the definite integral $\int_a^b f(x)dx$ using Simpson's rule and find from the literature how the error depends on the integration interval h.

Simpson's rule is derived by fitting a quadratic polynomial for any three consecutive points $(x_{i-1}, f(x_{i-1}))$, $(x_i, f(x_i))$, and $(x_{i+1}, f(x_{i+1}))$ and then integrating analytically the second order polynomial approximation derived.

10.4 Bioreactors are usually employed to develop extracellular products, such as ethanol and propanol, but also biomass (cell mass) by consuming substrates such as glucose. The rate at which biomass grows is described by the Monod kinetics:

$$\dot{R}_X''' = \mu_{max}\frac{S}{K_s + S}X$$

where μ_{max}, is the is the maximum growth rate (1/h), S is the concentration of the limiting substrate for growth (kg/m^3), K_s is the "half-velocity constant"—the value of S when $\dot{R}_X'''/\mu_{max} = 0.5$ (kg/m^3) and X is the concentration of the biomass (kg/m^3).

An important consideration in commercial cultures is how much product you are getting from the expensive substrate you are adding. This is the yield coefficient (Y), which is the ratio of the biomass formed (X, measured in kg) to the mass of substrate (e.g., glucose) consumed (S, also in kg):

$$Y = \frac{X - X_0}{S_0 - S}$$

where X_0 and S_0 are the initial biomass and substrate concentrations. For a particular microorganism and substrate we have the following parameters:

$$X_0 = 0.01\frac{kg}{m^3}, S_0 = 100\frac{kg}{m^3}, \mu_{max} = 0.4\frac{1}{h}, K_S = 40\frac{kg}{m^3}, Y = 0.1$$

The fresh feed has flowrate of $\dot{V} = 1\frac{m^3}{h}$.

a. Use MATLAB or Python to plot the reaction rate and the inverse of the reaction rate as a function of conversion.

b. What is the volume of a CSTR that achieves 80% conversion of the biomass?

c. What is the volume of a PFR that achieves 80% conversion of the biomass?

d. What is the minimum volume configuration?

10.5 A fresh liquid feed stream of $V = 100$ m^3/h with reactant concentration $C_{in,R} = 1$ kmol/m^3 is to be used to produce product P. The reaction is a second order irreversible reaction

$$\dot{R}''' = kC_R = kC_{in,R}^2(1-X)^2$$

with reaction rate constant $k = 75\ \mathrm{m^3/(kmol \cdot h)}$. If the desired conversion is 75%, answer the following questions:

a. What is the volume of a single CSTR that achieves the desired conversion?

b. What is the volume of a single PFR that achieves the desired conversion?

c. What is the minimum volume combination of two CSTRs in series?

Perform the calculations analytically and numerically using $h = 0.25$.

10.6 A fresh liquid feed stream of $\dot{V} = 100\ \mathrm{m^3/h}$ with reactant concentration $C_{in,R} = 10\ \mathrm{kmol/m^3}$ is to be used to produce product P. The reaction is an autocatalytic reaction

$$R + P \rightarrow P + P$$

with reaction rate

$$\dot{R}''' = k\left(1 + k' C_P\right) C_R$$

with reaction rate constant $k = 1\ \mathrm{h^{-1}}$ and $k' = \frac{1}{5}\ \mathrm{m^3/kmol}$.

If the desired conversion is 75%, answer the following question:

What is the minimum volume ideal reactor combination?

Table P.10.7 Experimental VLE data for methanol and water binary mixture at 1 atm.

T (°C)	x_{MeOH}	y_{MeOH}	T (°C)	x_{MeOH}	y_{MeOH}
100.0	0.00	0.000	75.2	0.40	0.729
96.4	0.02	0.134	73.1	0.50	0.779
93.5	0.04	0.230	71.2	0.60	0.825
91.2	0.06	0.304	69.3	0.70	0.870
89.3	0.08	0.365	67.5	0.80	0.915
87.7	0.10	0.418	66.0	0.90	0.958
84.4	0.15	0.517	65.0	0.95	0.979
81.7	0.20	0.579	64.5	1.00	1.000
78.0	0.30	0.665			

Data from Perry's Handbook.

10.7 In Table P10.7 experimental VLE data for the binary mixture methanol-water are given. Use these experimental VLE data to determine the molar flowrates and the compositions of all streams of the process shown in Fig. P10.7.

10.8 In Table P10.7 experimental VLE data for the binary mixture methanol-water are given. Use these experimental VLE data to determine the compositions of all streams of the process shown in Fig. P10.8.

10.9 A process stream is a binary mixture of acetone (A) and water (W) that is 20 wt% acetone, with mass flowrate of $L = 1000$ kg/h. Water is insoluble with trichloroethane (TCE), which can be used to extract acetone (acetone is more soluble in trichloroethane than in water). The equilibrium relationship of acetone in the two immiscible phases of water and TCE is as follows

$$Y = 1.75X$$

where

$$Y = \frac{\text{mass of acetone}}{\text{mass of (free of acetone) trichloroethane}}$$

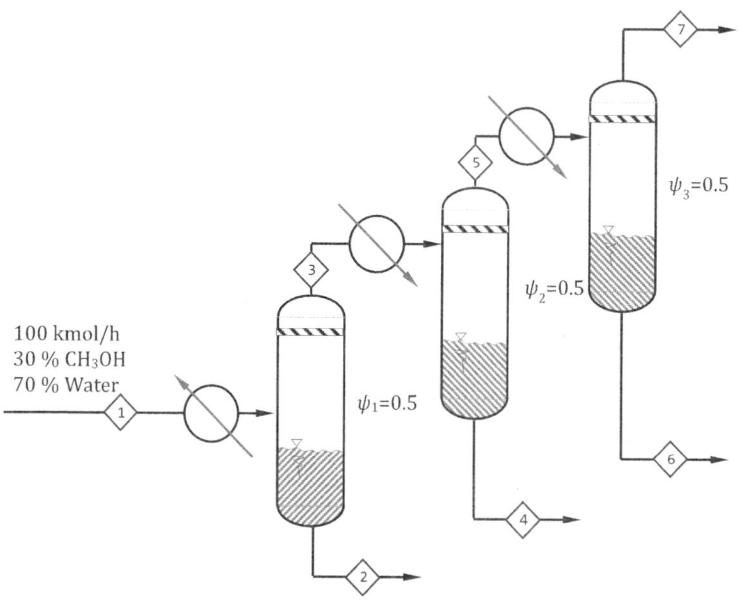

Figure P10.7 A process of three flash distillation units in series.

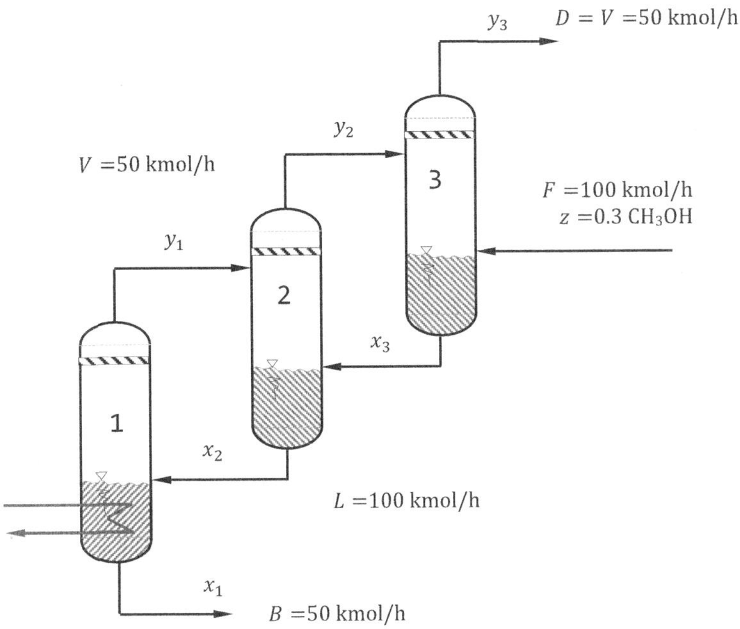

Figure P10.8 A process of three interconnected flash distillation units.

$$X = \frac{\text{mass of acetone}}{\text{mass of (free of acetone) water}}$$

How many equilibrium stages are required in order to recover 90% of acetone in a stream of TCE that is initially free of

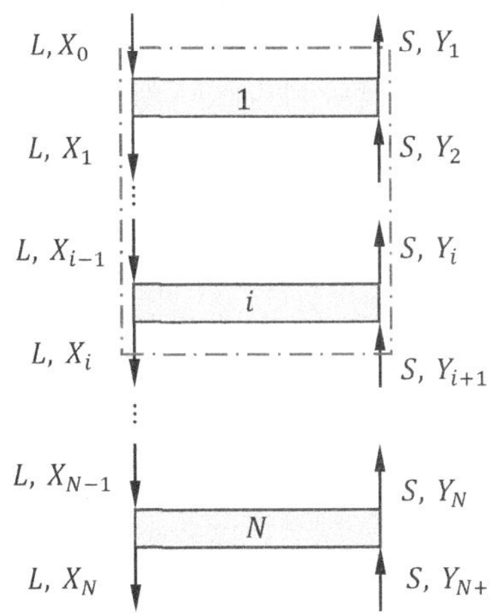

Figure P10.9 A cascade of equilibrium stages.

acetone and has a mass flowrate of $S = 500$ kg/h? See Fig. P10.9 for the notation to be used.

10.10 The relative volatility in a binary mixture, denoted by a, is defined as the equilibrium ratio of the more volatile component to the equilibrium ratio of the less volatile component. Use the definition of relative volatility to show that the mole fraction of the more volatile component in the vapor phase y is related to the mole fraction of the more volatile component in the liquid phase x through the following equation

$$y = \frac{a}{1 + (a - 1)x}x$$

In Table P10.10 experimental VLE data for the mixture of toluene and benzene are presented (data taken from: Computer Aided Data Book of Vapor-Liquid Equilibria. Von M. Hirata, S. Ohe, K. Nagahama. Elsevier Scientific Publishing Comp., Tokio, 1975). Show that this is a constant relative volatility binary mixture and find a "good" value for their relative volatility using the definition.

Table P10.10 Experimental VLE data for benzene and toluene binary mixture at 1 atm.

T (°C)	$x_{Benzene}$	$y_{Benzene}$
108.75	0.042	0.089
108.42	0.047	0.102
104.87	0.132	0.257
103.00	0.183	0.334
101.52	0.219	0.395
97.76	0.325	0.530
95.01	0.407	0.619
92.76	0.483	0.688
90.76	0.551	0.742
88.63	0.628	0.800
86.41	0.712	0.853
84.10	0.810	0.911
81.99	0.900	0.958
81.18	0.941	0.973

10.11 Consider the case of ideal binary mixtures with constant relative volatility for which the following is true:

$$y = \frac{a}{1 + (a - 1)x}x$$

The operating line for the rectifying section of a distillation column is given by

$$y = \left(\frac{rr}{1 + rr}\right)x + \left(\frac{x_D}{1 + rr}\right)$$

a. Show that for the minimum reflux ratio $rr_{\min}$ the following equation applies when the feed is a saturated liquid (and the feed line is vertical):

$$\frac{rr_{\min}}{1 + rr_{\min}} = \frac{x_D - y_F}{x_D - x_F}$$

where y_F is the composition of vapor in equilibrium with the composition of the liquid feed Fx.

b. Eliminate y_F from the equation given in a to show that

$$rr_{\min} = \left(\frac{1}{a - 1}\right)\left[\frac{x_D}{z} - a\left(\frac{1 - x_D}{1 - z}\right)\right]$$

10.12 Consider the case of ideal binary mixtures with constant relative volatility for which the feed is a saturated vapor. Follow the steps of Problem 10.11 to show that:

$$rr_{\min} = \left(\frac{1}{a - 1}\right)\left[a\frac{x_D}{z} - \left(\frac{1 - x_D}{1 - z}\right)\right]$$

10.13 Consider the binary mixture of benzene and toluene studied in Problem 10.10, which is a representative case of an ideal binary mixture. If the feed is 40% benzene, the distillate 90% benzene, and the bottom product 10% benzene.

a. Find the minimum reflux ratio if the feed is a saturated liquid.

b. Find the number of equilibrium stages necessary to achieve the separation when $rr = 1.5rr_{\min}$.

c. Find the minimum reflux ratio if the feed is a saturated vapor.

d. Find the number of equilibrium stages necessary to achieve the separation when $rr = 1.5rr_{\min}$.

10.14 A binary mixture with 25% acetone and 75% methanol is fed to a distillation column as a saturated liquid. If the reflux ratio is 1.5 times the minimum reflux ratio, find the number of equilibrium stages necessary to obtain a

Table P10.14 Experimental VLE data for acetone and methanol binary mixture at 1 atm.

T (°C)	$x_{Acetone}$	$y_{Acetone}$
64.65	0.000	0.000
61.78	0.091	0.177
59.60	0.190	0.312
58.14	0.288	0.412
56.96	0.401	0.505
56.22	0.501	0.578
55.78	0.579	0.631
55.41	0.687	0.707
55.29	0.756	0.760
55.37	0.840	0.829
55.54	0.895	0.880
55.92	0.954	0.946
56.21	1.000	1.000

distillate product containing 75% acetone and a bottom product that is 5% acetone. VLE data are given in Table P10.14.

References

Andersen, L.B., Wenzel, L.A., 1961. Introduction to Chemical Engineering. McGraw Hill, NY.

Brian, P.L.T., 1972. Staged Cascades in Chemical Processing. Prentice Hall, Inc, Englewood Cliffs, NJ.

Foust, A.S., Wenzel, L.A., Clump, C.W., Maus, L., Andersen, L.B., 1980. Principles of Unit Operations. John Wiley and Sons, NY.

Henley, E.J., Bieber, H., 1959. Chemical Engineering Calculations. McGraw Hill, NY.

Henley, E.J., Rosen, E.M., 1969. Material and Energy Balance Calculations. John Wiley & Sons, NY.

Hill, C., 1977. Chemical Engineering Kinetics and Reactor Design. John Wiley and Sons, NY.

Hougen, O.A., Watson, k.m., Ragatz, R.A., 1943. Chemical Process Principles, Part I, Material and Energy Balances. John Wiley and Sons, NY.

Kirkbride, C.G., 1947. Chemical Engineering Fundamentals. McGraw Hill, NY.

Levenspiel, O., 1972. Chemical Reaction Engineering, second ed. John Wiley and Sons, NY.

Mickley, H.S., Sherwood, T.K., Reed, C.E., 1957. Applied Mathematics in Chemical Engineering. McGraw Hill, NY.
Myers, A.L., Seider, W.D., 1976. Introduction to Chemical Engineering and Computer Calculations. Prentice Hall, Inc., Englewood Cliffs, NJ.
Reklaitis, G.V., 1983. Introduction to Material and Energy Balances. John Wiley and Sons, NY.
Rudd, D.F., Watson, C.C., 1968. Strategy of Process Engineering. John Wiley & Sons, NY.
Treybal, R.E, Mass-Transfer Operations, third ed., McGraw Hill Book Company, NY.

11

Psychrometry

11.1 Introduction

Psychrometry, originating from the Greek terms psychron (ψυχρό = cold) and metron (μετρώ = to measure), refers to a field within engineering that investigates the physical and thermodynamic behavior of gas-vapor systems. While these principles can be applied to any combination of gases and vapors, the air–water vapor mixture is by far the most commonly analyzed due to its widespread relevance in meteorology, HVAC (heating, ventilation, and air conditioning), and environmental control systems.

In terms of human comfort, temperature alone does not determine how warm or cool we feel. Since our bodies rely on perspiration to dissipate heat, the level of moisture in the surrounding air directly influences this process. When humidity is high, sweat evaporates less efficiently, making it harder for the body to regulate its temperature and maintain comfort.

Numerous substances possess hygroscopic qualities, which means they tend to absorb water from their environment, often proportionally to the surrounding relative humidity or once a critical threshold is surpassed. Materials such as cotton, paper, cellulose, sugar, quicklime (calcium oxide), various chemicals, and fertilizers exhibit this behavior. Industries that deal with these materials frequently monitor and manage humidity to prevent degradation or spoilage during both production and storage phases.

Maintaining suitable relative humidity levels also plays a vital role in ensuring workplace safety, especially in areas where combustible materials are used. Excessively dry air can promote static charge buildup, which might discharge unexpectedly and ignite fires. To reduce this risk, humidity control is often integrated into operational procedures in such environments.

In industrial drying operations—such as those involved in paper production—manufacturers aim to strike a balance between accelerating drying through low humidity and conserving energy, which typically improves as the humidity in the exhaust

Introduction to Process Systems Engineering. https://doi.org/10.1016/B978-0-443-45444-8.00011-7

air increases. Furthermore, preventing condensation is crucial in many industrial contexts, as moisture accumulation can damage products or cause corrosion on metal surfaces and equipment.

In this chapter, a comprehensive graphical design methodology will be developed, grounded in the fundamental principles of psychrometry. The chapter begins by outlining the core theoretical concepts that underpin the behavior of gas-vapor mixtures, with a particular focus on the thermodynamic interactions between air and water vapor. Following this theoretical foundation, a systematic approach for constructing psychrometric charts specific to the water–air system will be introduced.

The chapter will then explore the practical application of these diagrams in detail, illustrating how psychrometric tools can be effectively utilized to analyze and design systems involving heat and mass transfer processes. Through worked examples and case studies, it will be demonstrated that problems involving combined energy and mass balances, which are often mathematically complex when approached analytically, can be greatly simplified through graphical techniques using the psychrometric diagram.

11.2 Basic definitions and the psychrometric chart

The processes examined in this chapter focus on the interphase transfer of mass and energy that occurs when a gas comes into contact with a pure liquid in which it is largely insoluble. In such cases, the mass that transfers across the phase boundary is the substance that constitutes the liquid phase itself, which either undergoes vaporization into the gas phase or condensation back into the liquid. The situation remains complex due to the coupled nature of heat transfer and mass transfer in the gas phase, both of which must be analyzed simultaneously since they are mutually influential. These types of operations are broadly categorized under the term **humidification** operations. Despite the name, their objectives extend beyond merely increasing the moisture content of a gas. These processes can also involve **dehumidification** (removal of vapor from the gas), cooling of the gas stream, determining the vapor content of the gas, and even cooling the liquid itself. Among all humidification processes, air-based water cooling stands out as the most critical industrial application. In this scenario, water that has been heated—typically by passing through heat exchangers, condensers, or other equipment—is brought into contact with ambient air to reduce its temperature before being recirculated. A small fraction of the water evaporates due to a vapor concentration gradient in the air. The energy required for this phase change, known as the latent heat of vaporization, is drawn from the sensible heat of the liquid water, thereby cooling it. Because the latent heat of vaporization for water is quite high, even a minimal amount of evaporation can lead to significant cooling.

11.2.1 Properties of liquid water, water vapor, and dry air

Psychrometry is the study of the thermodynamic properties of **moist air**, which is a mixture of dry air and water vapor. It involves understanding and analyzing how air and water vapor behave together under different conditions of temperature, pressure, and humidity. Calculating with accuracy the properties of water and air and of their mixtures is of paramount importance. The first property of interest is the saturation pressure of water. An accurate equation for calculating the saturation pressure of water in the range 0–200 °C has been proposed by ASHRAE:

$$\ln\left(P_w^s\right) = \frac{C_{-1}}{T} + C_0 + C_1 T + C_2 T^2 + C_3 T^3 + C_4 \ln T \qquad (11.2.1)$$

where P_w^s is the saturation pressure of water (Pa), T is the absolute temperature (K), and C_{-1} to C_4 constants that are given in Table 11.1. A simpler Antoine equation is the following:

$$\ln\left(P_w^s\right) = 23.349 - \frac{3915.4}{T - 42} \qquad (11.2.2)$$

Table 11.1 Constants for Eq. (11.2.1).

Constant	Value
C_{-1}	$-5.800220\,6 \times 10^{3}$
C_0	$+1.391499\,3 \times 10^{0}$
C_1	$-4.864023\,9 \times 10^{-2}$
C_2	$+4.176476\,8 \times 10^{-5}$
C_3	$-1.445209\,3 \times 10^{-8}$
C_4	$+6.545967\,3 \times 10^{0}$

Taken from ASHRAE Fundamentals Handbook.

The maximum error of Eq. (11.2.1) is approximately 0.01% and of Eq. (11.2.2) is 0.5%. Eq. (11.2.1) will be used in computer calculations and the simpler one Eq. (11.2.2) will be used in manual calculations.

As no chemical reactions are taking place, the specific enthalpy of liquid water can be approximated by the equation

$$h_{w,L} = c_{p_L,w} t \tag{11.2.3}$$

where t is the temperature in °C and $h_{w,L}$ is the specific enthalpy of liquid water. The specific heat capacity under constant pressure for liquid water is approximately constant in the range 0 °C–200 °C and a mean value is

$$c_{p_L,w} \approx 4.186 \ \frac{\text{kJ}}{\text{kg °C}} \tag{11.2.4}$$

Note that according to Eq. (11.2.3) the enthalpy of liquid water is zero at 0 °C. Actually, in the literature, the correct basis for enthalpy calculation of liquid water is zero specific enthalpy of liquid water at the triple point of water (0.01 °C). As 0 °C and 0.01 °C make no difference when used in Eq. (11.2.3), the former is used for simplicity.

More important for the enthalpy calculation is the calculation of the enthalpy of water in the vapor phase. Again, the specific enthalpy of water is set equal to zero in the liquid state at 0 °C. When water evaporates at 0 °C its specific enthalpy becomes equal to the specific enthalpy of evaporation (latent heat) at 0 °C, which is equal to $\Delta h_{\text{evap},W} = 2501$ kJ/kg. The specific heat capacity under constant pressure for water vapor is approximately

$$c_{p_V,w} \approx 1.805 \ \frac{\text{kJ}}{\text{kg °C}} \tag{11.2.5}$$

The specific enthalpy of vapor water is therefore

$$h_{w,V} = \Delta h_{\text{evap},w} + c_{p_V,w} t \tag{11.2.6}$$

Air is the noncondensable component and we will refer to using the term dry air to make a clear distinction between **dry** (moisture-free) **air** and humid air, i.e., mixture of dry air and water vapor. The only property of dry air that we will make extensive use of is that of specific enthalpy. Again, the reference point is 0 °C where the specific enthalpy is assumed to be zero:

$$h_{\text{air}} = c_{p,\text{air}} t \tag{11.2.7}$$

where the specific heat capacity of dry air $c_{p,\mathrm{air}}$ is

$$c_{p,\mathrm{air}} \approx 1.006 \ \frac{\mathrm{kJ}}{\mathrm{kg}\ ^\circ\mathrm{C}} \tag{11.2.8}$$

Finally, the molar mass of dry air is $M_{\mathrm{air}} = 28.9645$ and the molar mass of water is $M_W = 18.02$.

11.2.2 Humidity ratio and relative humidity

Humid air is a binary mixture of dry air and water vapor. If the total pressure is P then the following is true

$$P = P_w + P_{\mathrm{air}} \tag{11.2.9}$$

where P_w is the partial pressure of water and P_{air} is the partial pressure of dry air. As ideal gas law applies to both components in a binary mixture at absolute temperature T, absolute pressure P, and volume V, we have that

$$P_w V = n_w RT \tag{11.2.10}$$

$$P_{\mathrm{air}} V = n_{\mathrm{air}} RT \tag{11.2.11}$$

We divide both sides of the two equations we obtain

$$\frac{n_w}{n_{\mathrm{air}}} = \frac{P_w}{P_{\mathrm{air}}} \tag{11.2.12}$$

We use Eq. (11.2.9) to eliminate P_{air}.

$$\frac{n_w}{n_{\mathrm{air}}} = \frac{P_w}{P - P_w} \tag{11.2.13}$$

Note that $n_w = m_w/M_w$ and $n_{\mathrm{air}} = m_{\mathrm{air}}/M_{\mathrm{air}}$ and from Eq. (11.2.12) we obtain

$$\frac{m_w}{m_{\mathrm{air}}} = \left(\frac{M_w}{M_{\mathrm{air}}}\right) \frac{P_w}{P - P_w} \tag{11.2.14}$$

Humidity ratio, also known as specific humidity or mixing ratio, is defined as the mass of water vapor per unit mass of dry air in a given air-water vapor mixture:

$$\mathscr{W} = \frac{m_w}{m_{\mathrm{air}}} \tag{11.2.15}$$

From Eqs. (11.2.14) and (11.2.15) and using the fact that $M_w/M_{\mathrm{air}} = 0.62198$ we obtain the following extremely important equation:

$$\mathscr{W} = 0.62198\left(\frac{P_w}{P - P_w}\right) \tag{11.2.16}$$

Another definition of importance is that of the **relative humidity** ($\mathscr{RH}$) which is defined as:

$$\mathscr{RH}\% = 100\left(\frac{P_w}{P_w^s(T)}\right) \tag{11.2.17}$$

where P_w is the actual partial pressure of water in the moist air and $P_w^s(T)$ is the saturation vapor pressure of water at the same temperature. If $\mathscr{RH}\%$ and temperature are known then we can determine the partial pressure of water vapor in the mixture:

$$P_w = \left(\frac{\mathscr{RH}\%}{100}\right) P_w^s(T) \tag{11.2.18}$$

Eq. (11.2.18) is replaced into Eq. (11.2.16) to obtain:

$$\mathscr{W} = 0.62198\left[\frac{\left(\frac{\mathscr{RH}\%}{100}\right) P_w^s(T)}{P - \left(\frac{\mathscr{RH}\%}{100}\right) P_w^s(T)}\right] \tag{11.2.19}$$

Eq. (11.2.16) is used to generate a basic **psychrometric diagram**. A psychrometric diagram (or psychrometric chart) is a graphical representation of the thermodynamic properties of moist air. It's a crucial tool in HVAC, meteorology, and thermal engineering for analyzing air-conditioning processes, drying, ventilation, and more. The chart plots multiple air properties together, allowing you to determine the state of air and visualize changes such as heating, cooling, humidification, and dehumidification. More specifically.

- The horizontal axis represents the dry-bulb temperature (in °C)
- The vertical axis (sometimes slanted) represents the humidity ratio (kg of water vapor per kg of dry air)

To construct a skeleton psychrometric chart at a given overall pressure, the following steps are applied:

1. Select a value for the relative humidity (i.e., $\mathscr{RH}\% = 10\%$, 20%, 30%, ..., 100%).
2. For a given value of the relative humidity vary systematically the temperature (usually in the range 0–50 °C)
3. Apply Eq. (11.2.19) to calculate and plot the humidity ratio.
4. Change the relative humidity and continue in 2.

These steps are implemented in MATLAB as shown in Table 11.2. The results are presented in Fig. 11.1. Note that in this diagram three variables are involved.

- The temperature, which from now on we will call the **dry-bulb temperature** (the reason for that will be explained shortly) in °C,
- the relative humidity (%),
- the humidity ratio in g water/kg dry air.

It is important to realize that knowledge of any two among these three parameters allows us to completely define the state of the system and the third parameter. Use Fig. 11.1

a. determine the humidity ratio if the dry-bulb temperature is 35 °C and the relative humidity is 70% (ans.: 25 g/kg),

Table 11.2 MATLAB implementation for generating the skeleton psychrometric chart.

```
clear all
% select values for relative humidity %
RH = linspace(10,100,10);
% select values for temperature in °C
t  = linspace(0,50,51);
P  = 101325;
for i=1:length(RH)
    for j=1:length(t)
        term = (RH(i)/100)*SatPrWater(t(j)+273.15);
        W(j,i) = 0.62198*term/(P-term);
    end
    plot(t,W(:,i)*1000,'LineWidth',1.2,'Color',[0.00,0.45,0.74])
    hold on
end
grid on
xlabel('Temperature (°C)')
ylabel('Humidity Ratio (g W/kg dry air)')
axis([0 50 0 30])
text(18,19,'RH=100%')
text(29,14,'50%')
text(40,6,'10%')

function Psw=SatPrWater(T)
% T   : absolute temperature in K
% Psw : saturation press. of water in Pa
Psw = exp(-5.8002206E3./T     + 1.3914993 ...
          -4.8640239E-2*T     + 4.1764768E-5*T.^2 ...
          -1.4452093E-8*T.^3 + 6.5459673*log(T));
end
```

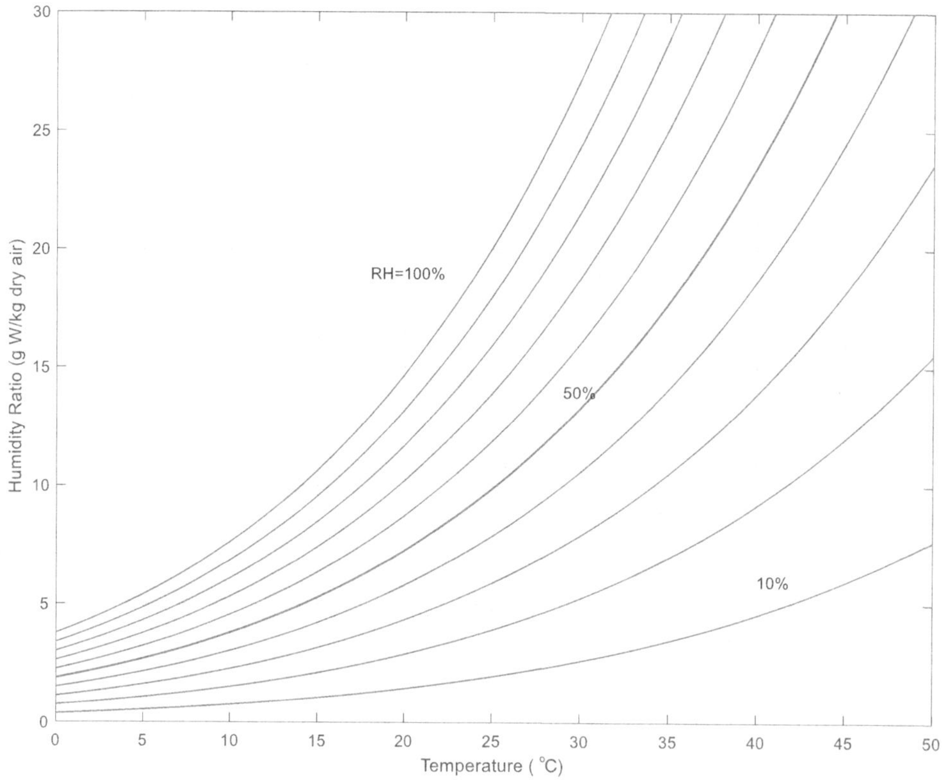

Figure 11.1 Skeleton psychrometric chart.

b. determine the dry-bulb temperature if the relative humidity is 20% and the humidity ratio is 5 g/kg (ans.: 29 °C),

c. determine the relative humidity if the humidity ratio is 10 g/kg and the dry-bulb temperature is 25 °C (ans.: 50%).

Example 11.1. Use of the skeleton psychrometric diagram

Humid air at 35 °C and 70% relative humidity is cooled at 5 °C. What is the final humidity ratio and how much water is condensed in this process? The pressure is 101.325 kPa.

Solution.

The process is shown in Fig. 11.2. The initial conditions (35 °C and 70% relative humidity) define point A in Fig. 11.2. We note from the figure that the initial humidity ratio is $\mathscr{H}_A \approx 25$ g moisture/kg dry air.

When the mixture is cooled, the line A to B is followed initially at constant humidity ratio but decreasing dry-bulb temperature until point B is reached where the mixture becomes saturated in water.

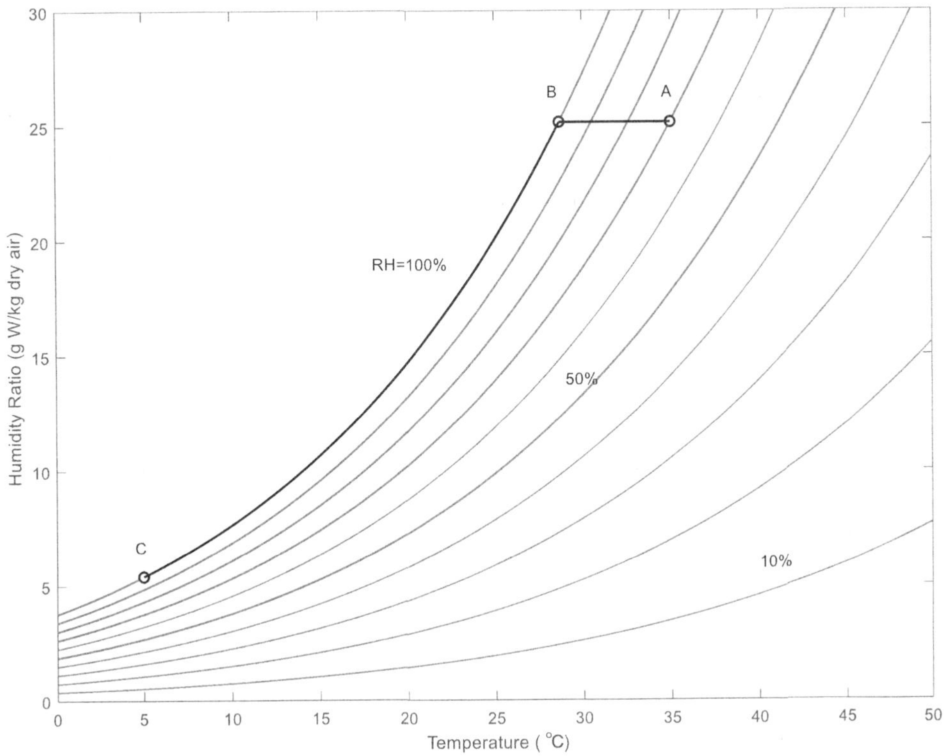

Figure 11.2 A humid-air cooling process on the skeleton psychrometric chart.

Further cooling occurs as the mixture transitions from point B to point C along the 100% relative humidity line. This process takes place after the air-water vapor mixture has reached saturation—meaning it holds the maximum amount of water vapor possible at that specific temperature. At this point, the air is fully saturated (100% relative humidity), and any additional decrease in temperature cannot be accommodated by more water vapor in the air. As the mixture continues to cool beyond this saturation point, it can no longer retain the same amount of water vapor. Consequently, the excess moisture begins to condense into liquid water. The temperature at which this condensation begins is known as the dew point, which we have analyzed in a previous chapter.

The movement from point B to point C, therefore, not only represents a drop in temperature but also a removal of moisture from the air through condensation. This is a key principle in processes such as dehumidification, air conditioning, and psychrometric analysis.

At the final point C, which corresponds to dry bulb temperature of 5 °C and 100% relative humidity, the humidity ratio is approximately $\mathscr{W}_C \approx 5.5$ g moisture/kg dry air. The amount of water removed due to condensation is therefore

$$m_w = \mathscr{W}_A - \mathscr{W}_C = (25 - 5.5) = 19.5 \frac{\text{g water}}{\text{kg dry air}}$$

11.2.3 Specific enthalpy and specific volume of humid air

We have already discussed valid approximations for calculating the specific enthalpy of water vapor and dry air. Under the assumption that water vapor and dry air form an ideal mixture, it follows that the enthalpy of the mixture is the sum of the enthalpies of the water vapor and dry air. Using 1 kg of dry air as the basis for calculating the specific enthalpy of humid air we have that

$$h(t, \mathscr{W}) = h_{\text{air}}(t) + \mathscr{W} h_{w,V} = c_{p,\text{air}} t + \mathscr{W}\left(\Delta h_{\text{evap},w} + c_{p_V,w} t\right) \tag{11.2.20}$$

where we have used Eqs. (11.2.6) and (11.2.7). Note that the specific enthalpy of humid air has units of kJ/kg dry air and that t is used to denote temperature in °C.

The question that we will answer is how we can assign the corresponding value for the specific enthalpy of humid air on the psychrometric chart. Let us assume that we are looking for all points on the psychrometric chart that share a common value for specific enthalpy, say h^*. To develop the equation that gives the coordinates of all points with the same value for the specific enthalpy we use Eq. (11.2.20), which is solved for $\mathscr{W}$

$$\mathscr{W}(t; h^*) = \frac{h^* - c_{p,\text{air}} t}{\Delta h_{\text{evap},w} + c_{p_V,w} t} \tag{11.2.21}$$

Based on Eq. (11.2.21) for each value of the specific enthalpy h^* we can vary the dry bulb temperature, calculate the humidity ratio, and finally plot the results on the skeleton psychrometric diagram.

The last property of interest is that of specific volume of humid air. The specific volume of humid air v is defined here as the volume of humid air per unit mass of dry air. Ideal gas assumption is valid for moist air. Using again 1 kg of dry air as the basis we can calculate the number of moles that correspond to the basis:

$$n = \frac{1}{M_{air}} + \frac{\mathscr{W}}{M_w} \tag{11.2.22}$$

Application of the ideal gas law gives

$$Pv = nRT$$

or

$$v = \left(\frac{1}{M_{air}} + \frac{\mathscr{W}}{M_w}\right) \frac{R(t + 273.15)}{P} \tag{11.2.23}$$

Let us assume that we are looking for all points on the psychrometric chart that share a common value for specific volume, say v^*. To develop the equation that gives the coordinates of all points with the same value for the specific volume, Eq. (11.2.23) is solved for $\mathscr{W}$

$$\mathscr{W}(t; v^*) = M_w \frac{v^* P}{R(t + 273.15)} - \frac{M_w}{M_{air}} \tag{11.2.24}$$

Based on Eq. (11.2.24) for each value of the specific volume v^* we can vary the dry bulb temperature t, calculate the humidity ratio, and finally plot the results on the psychrometric chart.

By adding the lines of constant specific enthalpy and the lines of constant specific volume of moist air we obtain the complete psychrometric diagram shown in Fig. 11.3. The MATLAB program for generating Fig. 11.3 is shown in Table 11.3. The corresponding python program is given in Table 11.4.

Example 11.2. Use of the psychrometric diagram to perform energy calculation

For the process considered in Example 11.1 calculate the cooling load necessary and the change in the volume of the humid air.

Solution.

The process is shown in Fig. 11.4. At the initial conditions (35 °C and 70% relative humidity) the specific enthalpy is $h_A \approx$ 100 kJ/kg dry air, while at the final point (5 °C and 100% relative humidity) the specific enthalpy is $h_C \approx$ 18.5 kJ/kg dry air. The necessary cooling is

$$Q = h_C - h_A = (18.5 - 100) = -81.5 \frac{\text{kJ}}{\text{kg dry air}}$$

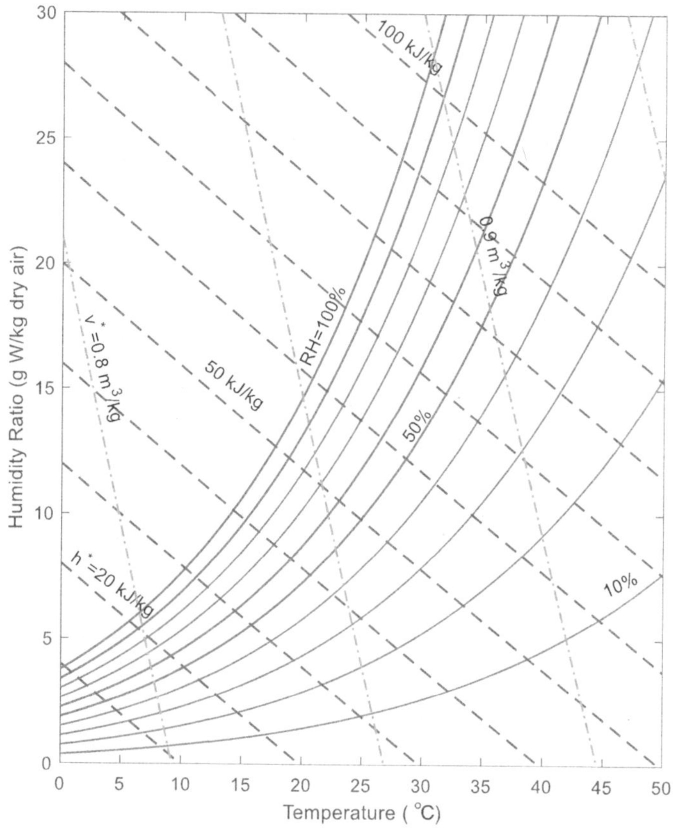

Figure 11.3 Psychrometric chart for humid air (valid for 101.325 kPa).

For the change in the volume of the mixture we note from Fig. 11.4 that

$$v_A \approx 0.910 \text{ m}^3/\text{kg dry air}$$

$$v_C \approx 0.795 \text{ m}^3/\text{kg dry air}$$

and

$$\Delta v = h_C - h_A = (0.795 - 0.910) = -0.115 \frac{\text{m}^3}{\text{kg dry air}}$$

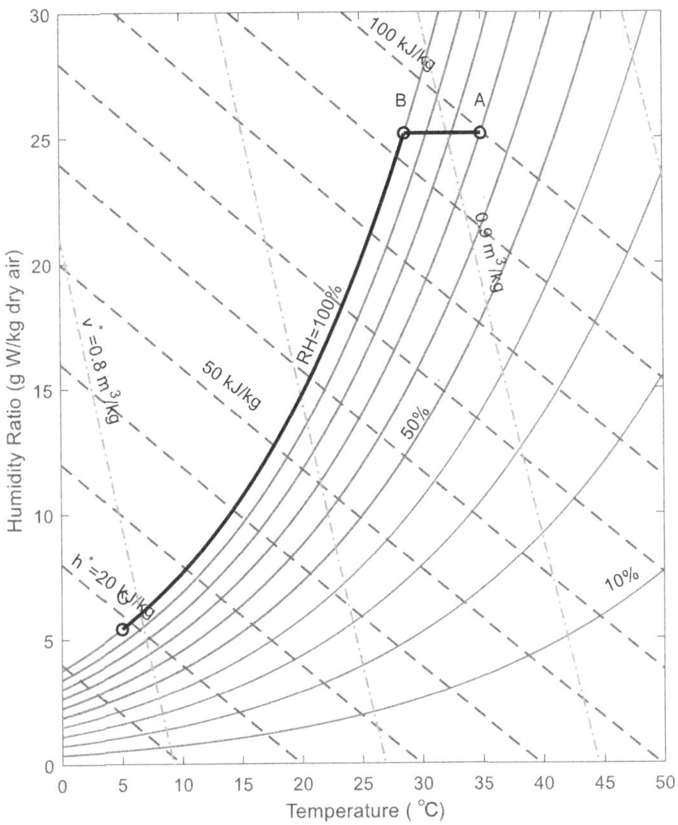

Figure 11.4 Representation of process examined in Examples 11.1 and 11.2 on the psychrometric chart.

11.2.4 Dew point temperature and wet bulb temperature

The temperature of the humid air as measured by a conventional thermometer is referred to as the dry-bulb temperature and is the ordinate used in the psychrometric chart. The temperature at which condensation of water begins when humid air is cooled at constant pressure is the **dew-point temperature**. The temperature at this point is denoted by $T_{dp}(\mathscr{W})$, and if the temperature drops any further, some water vapor condenses out, the amount of vapor in the air decreases, which results in a decrease in the humidity ratio. The air remains saturated during the condensation process and thus follows a path of 100% relative humidity (the saturated vapor line), as we have seen in Example 11.1. This happens, for instance, when you buy a cold canned drink on a hot and humid day, as dew forms on the can. The

Table 11.3 MATLAB implementation for generating the psychrometric chart.

```
clear all
%% skeleton psychrometric chart
RH = linspace(10,100,10);
t  = linspace(0,50,51);
P  = 101325;
figure(1); clf; hold on
for i=1:length(RH)
    for j=1:length(t)
        term = (RH(i)/100)*SatPrWater(t(j)+273.15);
        W(j,i) = 0.62198*term/(P-term);
    end
    plot(t,W(:,i)*1000,'LineWidth',1.2,'Color',[0.00,0.45,0.74])
end

grid on
xlabel('Temperature (°C)')
ylabel('Humidity Ratio (g W/kg dry air)')
axis([0 50 0 30])
tx=text(20,16,'RH=100%'); tx.Rotation = +65;
tx=text(28.5,13,'50%'); tx.Rotation = +55;
tx=text(45,7,'10%'); tx.Rotation = +25;

%% constant specific enthalpy lines
hstar = linspace(10,100,10); % kJ/kg dry air
Cpair = 1.006; % kJ/(kg °C)
Dhvapw= 2501;  % kJ/kg
Cpvw  = 1.805; % kJ/(kg °C)
for i=1:length(hstar)
    W = (hstar(i)-Cpair*t)./(Dhvapw+Cpvw*t)
    plot(t,W*1000,'--','LineWidth',1.2,'Color',[0.64,0.08,0.18])
end
tx=text( 1.2, 8.4,'h^{*}=20 kJ/kg'); tx.Rotation = -35;
tx=text(12.0,16.0,'50 kJ/kg'); tx.Rotation = -35;
tx=text(26.0,29.7,'100 kJ/kg'); tx.Rotation = -35;

%% constant specific volume lines
vstar= linspace(0.80,0.95,4); % kJ/kg dry air
R    = 0.0831446 ;          % m^3 bar / K kmol
Mair = 28.9645; Mw   = 18.01528;
for i=1:length(vstar)
    W = ( vstar(i)*(P/1E5)./(R*(273.15+t)) - 1/Mair)*Mw;
    plot(t,W*1000,'-.','LineWidth',1.2,'Color',[0.466 0.674 0.188])
end
tx=text( 2.5,17.9,'v^{*}=0.8 m^{3}/kg'); tx.Rotation = -75;
tx=text(35.3,22.0,'0.9 m^{3}/kg'); tx.Rotation = -75;
set(gcf,'position',[60,100,560,650])
function Psw=SatPrWater(T)
% T   : absolute temperature in K
% Psw : saturation press. of water in Pa
Psw = exp(-5.8002206E3./T    + 1.3914993 ...
          -4.8640239E-2*T    + 4.1764768E-5*T.^2 ...
          -1.4452093E-8*T.^3 + 6.5459673*log(T));
end
```

Table 11.4 Python implementation for generating the psychrometric chart.

```
import numpy as np
import matplotlib.pyplot as plt

def SatPrWater(TC):
    # saturation pressure of water in Pa
    TK = TC + 273.15 # TC in °C and TK in Kelvin
    Psw = np.exp(-5.8002206E+3/TK + 1.3914993 \
                -4.8640239E-2*TK + 4.1764768E-5*TK**2 \
                -1.4452093E-8*TK**3 + 6.5459673*np.log(TK))
    return Psw

# format the figure
fig, ax = plt.subplots(figsize=(3,4), dpi=200)
plt.title("SKELETON PSYCHROMETRIC CHART")
plt.xlabel("Temperature (°C)")
plt.ylabel("Humidity ratio (g W/g dry air)")
plt.grid(color = 'grey', linestyle = '-', linewidth = 0.5)
plt.axis([0, 50, 0, 30])

# RELATIVE SATURATION LINES
# select representative values for the relative humidity
RH = np.linspace(10,100,10) # relative humidity %

# select reasoble values for the dry bulp temperature
t  = np.linspace(0,50,51)   # temperature in °C

# select the pressure
P  = 101325.0               # pressure in Pa

W=np.zeros((51,10))
for i in range(10):
    for j in range(51):
        term = (RH[i]/100.0)*SatPrWater(t[j])
        W[j,i] = 0.62198*term/(P-term)
    plt.plot(t,1000*W[:,i], linestyle='-', linewidth=1)

# indicate the values
plt.text(19, 16.0, 'RH=100%', dict(size=4), rotation=65.)
plt.text(30, 11.6, 'RH= 50%', dict(size=4), rotation=55.)
plt.text(44,  6.2, 'RH= 10%', dict(size=4), rotation=34.)

# CONSTANT SPECIFIC ENTHALPY LINES
# select reasonalbe values for specific enthalpy
hstar = np.linspace(10,100,10) # kJ/ kg dry air

# air heat capacity
Cpair = 1.006               # kJ / (kg °C)

# enthalpy of vaporization of water at 1 atm
Dhvapw= 2501.0              # kJ / kg
```

continued

Table 11.4 Python implementation for generating the psychrometric chart.—*continued*

```
# heat capacity of water vapor
Cpvw  = 1.805                          # kJ / (kg °C)
W=np.zeros((51,10))
for i in range(10):
    for j in range(51):
        W[j,i] = (hstar[i]-Cpair*t[j])/(Dhvapw+Cpvw*t[j])
    plt.plot(t,1000*W[:,i], linestyle='--', linewidth=0.75)

# indicate the values
plt.text( 0.0, 6.4,'h$^{*}$=20 kJ/kg',dict(size=4),rotation=-40.)
plt.text(10.0,14.0,'h$^{*}$=50 kJ/kg',dict(size=4),rotation=-40.)
plt.text(25.0,27.2,'h$^{*}$=100 kJ/kg',dict(size=4),rotation=-40.)

# CONSTANT SPECIFIC VOLUME LINES
# select reasonable values for specific volume
Vstar = np.linspace(0.8,0.95,4) # kJ/ kg dry air

# ideal gas constant
R     = 0.0831446                      # m^3 bar / K kmol

# molar mass of air
Mair  = 28.9645                        # kg / kmol

# molar mass of water
Mw    = 18.01528                       # kg / kmol

W=np.zeros((51,4))
for i in range(4):
    for j in range(51):
        W[j,i] = (Vstar[i]*(P/1E5) /(R*(273.15+t[j])) - 1/Mair)*Mw
    plt.plot(t,1000*W[:,i], linestyle='-.', linewidth=0.75)

# indicate the values
plt.text(1.0,15.4,'v$^{*}$=0.8m$^{3}$/kg',dict(size=4),rotation=-80.)
plt.text(36.0,15.4,'v$^{*}$=0.9m$^{3}$/kg',dict(size=4),rotation=-80.)
```

formation of dew on the can indicates that the temperature of the drink is below the dew-point temperature of the surrounding air. The dew-point temperature can easily be determined from the psychrometric chart by moving horizontally, increasing the relative humidity until we cross the 100% relative humidity line.

The dew-point temperature can be calculated using Eq. (11.2.19) and setting $\mathscr{RH}\% = 100$:

$$\mathscr{W} = 0.62198\left(\frac{P_w^s(T_{dp})}{P - P_w^s(T_{dp})}\right) \tag{11.2.25}$$

Thus, for a given humidity ratio $\mathscr{W}$ and total pressure P we first solve (11.2.25) for $P_w^s(T_{dp})$

$$P_w^s(T_{dp}) = \frac{\mathscr{W}}{0.62198 + \mathscr{W}} P \tag{11.2.26}$$

Note that the humidity ratio $\mathscr{W}$ must be expressed in kg water vapor/kg dry air. Then, the dew-point temperature is calculated by solving numerically Eq. (11.2.1) or by solving directly the approximating Eq. (11.2.2).

In Fig. 11.5 an **adiabatic saturation process** is shown. The system consists of a long insulated channel that contains a pool of water. A steady stream of unsaturated air that has a relative humidity $\mathscr{RH}_{\text{in}}$ and a temperature of T_{in} is passed through this channel. As the air flows over the water, some water evaporates causing the moisture content of air to increase during this process, and its temperature decreases as part of the latent heat of vaporization of the water that evaporates comes from the air. If the channel is long enough, the airstream exits as saturated humid air at temperature T_s, which is called the adiabatic saturation temperature. The

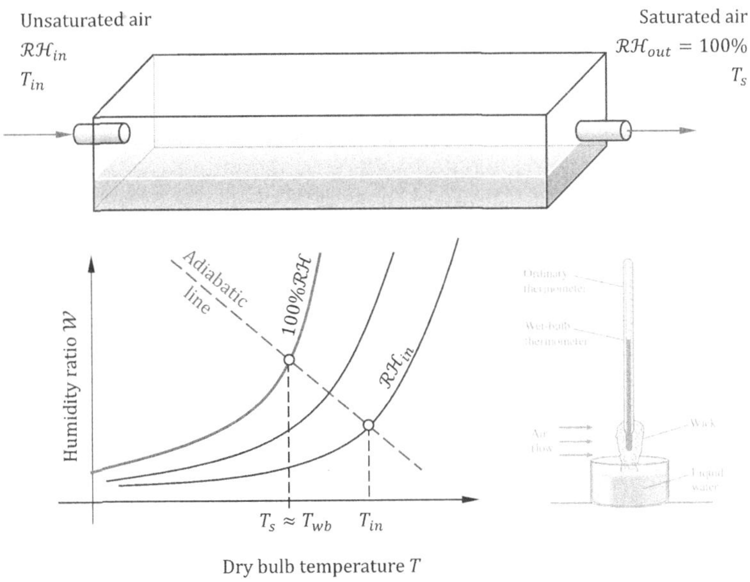

Figure 11.5 The adiabatic saturation process of humid air.

process involves no heat or work interactions, and the kinetic and potential energy changes can be neglected.

The adiabatic saturation process discussed above requires a long channel or a spray mechanism to achieve saturation conditions at the exit. A more practical approach is to use a thermometer whose bulb is covered with a cotton wick saturated with water and to blow air over the wick. The temperature measured in this manner is called the wet-bulb temperature T_{wb}, and it is commonly used in air-conditioning applications.

The basic principle involved is similar to that in adiabatic saturation. When unsaturated air passes over the wet wick, water in the wick evaporates, the temperature of the water drops, creating a temperature difference between the air and the water, resulting in heat transfer. At steady conditions the heat loss from the water by evaporation equals the heat gain from the air, and the water temperature stabilizes. The thermometer reading at this point is the wet-bulb temperature. In general, the adiabatic saturation temperature and the wet-bulb temperature are not the same. However, for air–water vapor mixtures at atmospheric pressure, the wet-bulb temperature happens to be approximately equal to the adiabatic saturation temperature. Therefore, the **wet-bulb temperature** T_{wb} can, for any practical application, be considered equal to saturation temperature T_s.

Example 11.3. Use of the psychrometric diagram

A closed space has volume $V = 1000\ m^3$ and the (dry bulb) temperature is 25 °C and the relative humidity is 50%. The pressure is 1 atm (101.325 kPa). Calculate the following:

a. the humidity ratio,
b. the saturation temperature,
c. the wet-bulb temperature,
d. the mass of water and dry air in the closed room.

The closed space is cooled to 8 °C. Calculate.

e. the mass of water removed (condensed),
f. the heat that needs to be removed,

Cooling is terminated and the closed space returns slowly to the initial temperature of 25 °C.

g. what is the final relative humidity?

Solution.

The four points that correspond to the process are pictured in Fig. 11.6. The initial state corresponds to point A (25 °C, 50% relative humidity). As the air in the closed space is cooled, the state of the humid air moves from point A horizontally (constant humidity ratio) to point B (saturation temperature) and then continues on the saturation line until the temperature becomes 8 °C (point

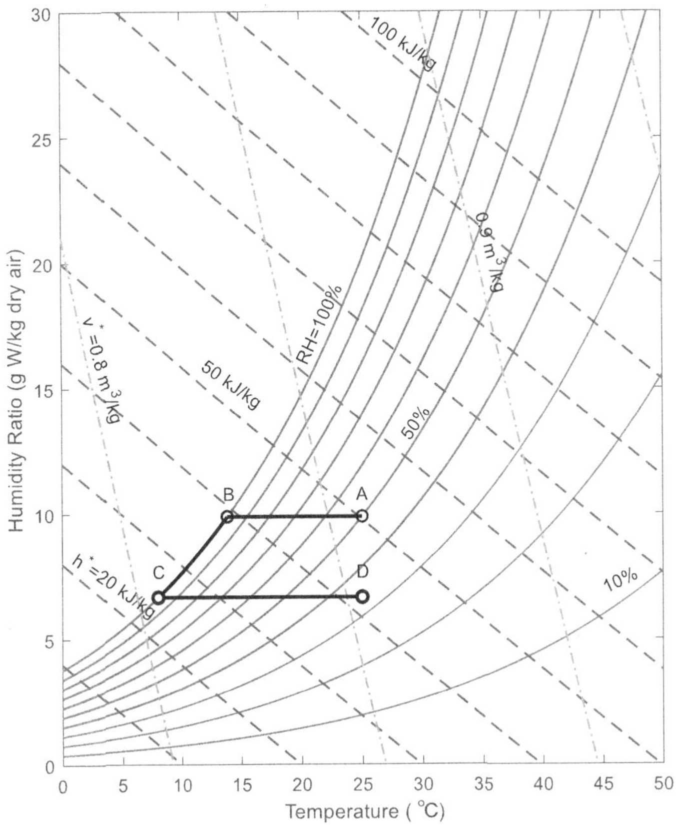

Figure 11.6 Representation of process examined in Example 11.3 on the psychrometric chart.

C). Finally, as the humid air is reheated under constant humidity ratio it reaches point D (dry-bulb temperature of 25 °C).

From the psychrometric chart we collect all data necessary for the calculations involved (the numbers in bold denote the information used to locate the point on the psychrometric chart).

Point	Dry-bulb Temperature (°C)	Relative humidity (%)	Humidity Ratio (g/kg d.a.)	Specific Enthalpy (kJ/kg d.a.)	Specific Volume (m^3/kg d.a.)	Wet-bulb Temperature (°C)
A	**25.0**	**50**	9.9	50	0.860	18.0
B	13.9	**100**	**9.9**	39	0.825	13.9
C	**8.0**	**100**	6.7	25	0.805	8.0
D	**25.0**	34	**6.7**	42	0.850	15.2

Questions a to c have been answered. To answer question d we find the mass of dry air in the closed space:

$$m_{da} = \frac{V}{v} = \frac{1000 \text{ m}^3}{0.86 \dfrac{\text{m}^3}{\text{kg dry air}}} = 1162.8 \text{ kg dry air}$$

The mass of water at point A is

$$m_{w,A} = \mathscr{W}_A m_{da} = 9.9 \frac{\text{g moisture}}{\text{kg dry air}} \times 1162.8 \text{ kg dry air}$$
$$= 11511.6 \text{ g moisture} = 11.512 \text{ kg water}$$

To calculate the amount of water that condenses between points A and C (question e) we calculate the mass of water at point C:

$$m_{w,C} = \mathscr{W}_C m_{da} = 6.7 \frac{\text{g moisture}}{\text{kg dry air}} \times 1162.8 \text{ kg dry air}$$
$$= 7790.8 \text{ g moisture} = 7.791 \text{ kg water}$$

The amount of condensate is

$$m_{\text{condensate}} = m_{w,A} - m_{w,C} = (11.512 - 7.791) \text{ kg water}$$
$$= 3.721 \text{ kg water}$$

The amount of heat that is removed between points A and C (question f) is:

$$Q_{A \to C} = m_{da}(h_C - h_A) = 1162.8 \text{ kg dry air} \times (25 - 50) \frac{\text{kJ}}{\text{kg dry air}}$$
$$= 29070 \text{ kJ}$$

Question g has already been answered.

It is interesting to repeat the calculation computationally so as to realize the amount of calculations involved and the practical usefulness of the psychrometric chart.

Example 11.4. Use of the psychrometric diagram and comparison with computations

Humid air has a dry-bulb temperature of 40 °C and wet-bulb temperature of 30 °C. Find, using the psychrometric chart, the following.

a. the humidity ratio,
b. the relative humidity,
c. the specific volume, and
d. the specific enthalpy.

Repeat using the corresponding equations. Comment on the complexity of the two approaches.

Solution.

The method that can be used to find the point on the psychrometric chart that corresponds to a pair of wet-bulb temperature T_{wb} and dry-bulb temperature T_{db} is described in Fig. 11.7. We move vertically on the psychrometric chart at T_{wb} until we cross the saturation line (100% relative humidity). Then we follow the constant enthalpy (adiabatic saturation) line until we cross the vertical line that corresponds to T_{db}. Use Fig. 11.3 and move vertically on the 30 °C: the saturation line is crossed when the humidity ratio is 27.3. Note that the constant enthalpy line at 100 kJ/kg dry air also passes very close to this point. Follow the constant enthalpy line at 100 kJ/kg dry air until you meet the vertical line of 40 °C. The final point has the following properties:

Dry-bulb Temperature (°C)	Relative humidity (%)	Humidity Ratio (g/kg d.a.)	Specific Enthalpy (kJ/kg d.a.)	Specific Volume (m^3/kg d.a.)	Wet-bulb Temperature (°C)
40.0	49	23	100	0.92	30.0

To calculate the properties is significantly more complex. Note that whet $T = T_{wb}$, $\mathscr{RH}\% = 100\%$ and from Eq. (11.2.19) we have that

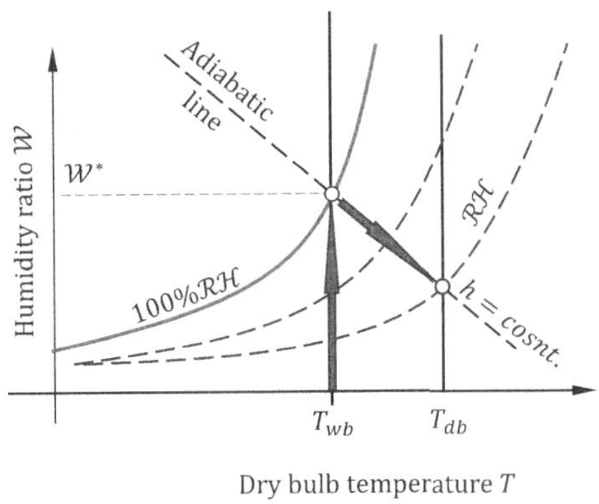

Figure 11.7 Locating a point on psychrometric chart defined by the dry-bulb temperature and the wet-bulb temperature (Example 11.4).

$$\mathscr{W}^* = 0.62198\left[\frac{P_w^s(T_{wb})}{P - P_w^s(T_{wb})}\right]$$

We use $T_{wb} = 273.15 + 30 = 303.15$ K and we replace into Eq. (11.2.1) to obtain $P_w^s(T_{wb}) = 4246$ Pa and

$$\mathscr{W}^* = 0.62198\left[\frac{P_w^s(T_{wb})}{P - P_w^s(T_{wb})}\right] = 0.62198\left(\frac{4246}{101325 - 4246}\right)$$
$$= 0.0272\,\frac{\text{kg}}{\text{kg}} = 27.2\,\frac{\text{g moisture}}{\text{kg dry air}}$$

We then calculate the specific enthalpy at the saturation point from Eq. (11.2.20)

$$h(t_{wb}, \mathscr{W}^*) = c_{p,\text{air}}t_{wb} + \mathscr{W}^*(\Delta h_{\text{evap},w} + c_{pV,w}t_{wb})$$
$$= 1.006 \times 30 + 0.0272\,\frac{\text{kg}}{\text{kg}}(2501 + 1.805 \times 30)$$
$$= 99.69\,\frac{\text{kJ}}{\text{kg dry air}}$$

Using Eq. (11.2.20) for a constant specific enthalpy we can calculate the humidity ratio for the dry-bulb temperature

$$\mathscr{W} = \frac{h(t_{wb}, \mathscr{W}^*) - c_{p,\text{air}}t_{db}}{\Delta h_{\text{evap},w} + c_{pV,w}t_{db}} = \frac{(99.69 - 1.006 \times 40)\,\dfrac{\text{kJ}}{\text{kg dry air}}}{(2501 + 1.805 \times 40)\,\dfrac{\text{kJ}}{\text{kg moisture}}}$$
$$= 0.0231\,\frac{\text{kg}}{\text{kg}} = 23.1\,\frac{\text{g moisture}}{\text{kg dry air}}$$

Finally, we solve Eq. (11.2.19) for the relative humidity:

$$\mathscr{W} = 0.62198\left[\frac{\left(\dfrac{\mathscr{RH}\%}{100}\right)P_w^s(T_{db})}{P - \left(\dfrac{\mathscr{RH}\%}{100}\right)P_w^s(T_{db})}\right] \Rightarrow \frac{\mathscr{RH}\%}{100}$$
$$= \left(\frac{\mathscr{W}}{0.62198 + \mathscr{W}}\right)\frac{P}{P_w^s(T_{db})}$$

We use $T_{db} = 273.15 + 40 = 313.15$ K and we replace into Eq. (11.2.1) to obtain $P_w^s(T_{wb}) = 7383.5$ Pa and

$$\mathscr{RH}\% = 100\left(\frac{0.0231}{0.62198 + 0.0231}\right)\frac{101320}{7383.5} = 49.15\ \%$$

The calculations can also be performed in MATLAB.

```
clear all

R      = 0.0831446261815324;      % ideal gas constant m^3 bar / K kmol
Mair   = 28.9645;                 % molecular mass of air
Mw     = 18.01528;                % molecular mas of water
P      = 101325;                  % system pressure in Pa
Cpair  = 1.006;                   % sp.heat cap. of dry air kJ/(kg °C)
Dhvapw= 2501;                     % sp.enthalpy of evaporation kJ/kg
Cpvw   = 1.805;                   % sp. heat cap. of water kJ/(kg °C)

t_db   = 40;T_db = t_db + 273.15;% dry bulb temperature °C/K
t_wb   = 30;T_wb = t_wb + 273.15;
Psw_wb= SatPrWater(T_wb);
Wstar  = (Mw/Mair)*Psw_wb/(P-Psw_wb)   % kg moisture/kg dry air
h      = Cpair*t_wb + Wstar*(Dhvapw+Cpvw*t_wb) % kJ/kg dry air
W      = (h - Cpair*t_db)/(Dhvapw+Cpvw*t_db)   % kg moisture/kg dry air
Psw_db= SatPrWater(T_db)
RH     = 100 * (W / ((Mw/Mair)+W) ) * (P/Psw_db)

function Psw=SatPrWater(T)
% T   : absolute temperature in K
% Psw : saturation press. of water in Pa
Psw = exp(-5.8002206E3./T     + 1.3914993 ...
          -4.8640239E-2*T     + 4.1764768E-5*T.^2 ...
          -1.4452093E-8*T.^3 + 6.5459673*log(T));
end
```

The computational approach involves significant workload and the improvement in the accuracy is only marginal. It is recommended that initial solutions and preliminary designs are obtained using graphical means, and once the final solution has been obtained then it can be validated using computer calculations.

11.3 Cooling towers

Cooling towers serve as an efficient and economical method for dissipating unwanted heat from water-cooled systems and various industrial processes. They achieve this by facilitating the transfer of heat from warm process or condenser water to the atmosphere. The fundamental mechanism involves passing this heated water over a specially designed heat exchange surface, commonly known as "fill." As air is either drawn or forced through the tower, a portion of the water evaporates. This phase change—liquid to vapor—absorbs a significant amount of heat, resulting in a reduction of the water temperature. The cooled water is then collected in a basin and recirculated back into the system to continue the cooling cycle.

Although all cooling towers operate on this principle of evaporative cooling, their specific arrangements and design features may vary slightly depending on application requirements and manufacturer design. Each model is tailored to meet different capacity demands, spatial constraints, and operational preferences.

There are two predominant types of factory-assembled cooling tower designs, distinguished by the direction of airflow relative to water flow: **crossflow** and **counterflow**. In a crossflow cooling tower, the water flows downward through the fill media while the air moves horizontally across it. This orthogonal movement of air and water allows for easier maintenance access and lower pump head requirements but may result in slightly larger tower footprints. In a counterflow cooling tower, both air and water move vertically, but in opposite directions—air travels upward as water descends. This configuration typically enables more compact designs and improved thermal performance due to better contact between air and water streams, although it may require more fan energy and result in more complex maintenance. By leveraging evaporative cooling, these systems play a critical role in temperature regulation for HVAC systems, power plants, and manufacturing operations, ensuring reliable and energy-efficient performance.

Natural draft cooling towers (see Fig. 11.8) represent an engineering marvel—vast, hyperbolic-shelled structures whose functionality derives entirely from natural forces rather than mechanical aid. Towering often over 200 m, they gently harness the buoyancy of heated air rising through the structure's waist to sustain continuous airflow, eliminating the need for fans or motors. The process begins as hot condenser water is pumped upward and dispersed via **spray nozzles** across a specialized fill matrix near the top of the tower. As this water trickles downward, ambient air is drawn upward from the base, contacting the descending water. Evaporation occurs—extracting heat from the water—while the now-cooled liquid collects in a basin before recirculating to the plant.

The hyperboloid profile of these towers enhances structural integrity and strengthens airflow via a venturi effect. As the airflow passes through the narrowed waist, it accelerates—naturally boosting draft performance without requiring power. Because there are no fans, electrical consumption is limited to water pumps. Operational costs remain low and noise is nearly nonexistent. However, constructing a natural draft tower demands considerable investment due to the structure's sheer

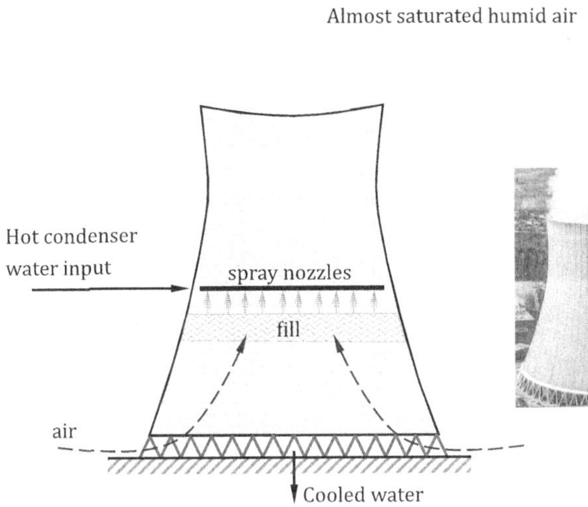

Figure 11.8 Main parts of a natural draft cooling tower.

size, depth of engineering, and reinforced concrete and land requirements. For that reason, natural draft towers are predominantly employed at large-scale facilities—coal, gas, nuclear power plants, petrochemical complexes—where their passive operation, low running costs, and high throughput (hundreds of thousands to millions of cubic meters per minute) justify the initial investment.

Cooling towers can be designed based on psychrometric charts and simple material and energy balances. The following examples demonstrate the general methodology. It is important to note that these gigantic constructions, whose operation is of paramount importance to most large-scale chemical plants, refineries, and electricity production plants, can be designed based on simple fundamental principles and understanding of the simple psychrometric chart.

Example 11.5. Use of the psychrometric diagram to analyze a cooling tower

Water stream from the condenser of a power plant has a mass flowrate of $m_{w,\text{in}} = 10\ \text{t/min}$ and temperature $T_{w,\text{in}} = 40\ ^\circ\text{C}$. A natural draft cooling tower (Fig. 11.8) is used to cool the water to $T_{w,\text{out}} = 25\ ^\circ\text{C}$. Atmospheric air at $T_{a,\text{in}} = 20\ ^\circ\text{C}$ and 50% relative humidity is used to cool down the cooling water. The air becomes saturated in water and its temperature is $T_{a,\text{out}} = 30\ °\text{C}$. Calculate:

a. the mass and volumetric flowrate of air, and
b. the amount of water evaporated in the cooling tower.

Solution.

We formulate the energy balance of the cooling tower by ignoring any energy losses:

$$\text{Energy in} = \text{Energy out}$$

or

$$\text{Energy in the incoming air} + \text{Energy in the incoming water}$$
$$= \text{Energy in the outgoing air} + \text{Energy in the outgoint water}$$

Using Fig. 11.9 we have that

$$\dot{m}_{da} h_a(T_3, \mathscr{RH}_3) + \dot{m}_{w,\text{in}} h_w(T_1) = \dot{m}_{da} h_a(T_4, \mathscr{RH}_4) + \dot{m}_{w,\text{out}} h_w(T_2)$$

From the psychrometric chart we find the following

$$h_a(T_3, \mathscr{RH}_3) = 38.5\ \frac{\text{kJ}}{\text{kg dry air}}$$

$$h_a(T_4, \mathscr{RH}_4) = 99.8\ \frac{\text{kJ}}{\text{kg dry air}}$$

Figure 11.9 Data for Example 11.5.

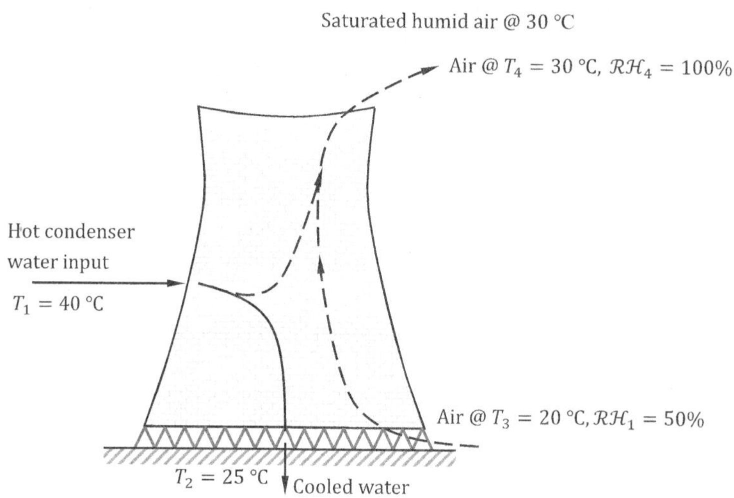

We also have that the humidity ratio at the input and output of the air stream are

$$\mathscr{W}_3 = 0.007\,\frac{\text{kg moisture}}{\text{kg dry air}}$$

$$\mathscr{W}_4 = 0.027\,\frac{\text{kg moisture}}{\text{kg dry air}}$$

Using Eq. (11.2.3)

$$h_w(T_1) = c_{p_L,w}t_1 = 4.186\frac{\text{kJ}}{\text{kg}\,^\circ\text{C}} \times 40\,^\circ\text{C} = 167.44\,\frac{\text{kJ}}{\text{kg}}$$

$$h_w(T_2) = c_{p_L,w}t_1 = 4.186\frac{\text{kJ}}{\text{kg}\,^\circ\text{C}} \times 25\,^\circ\text{C} = 104.65\,\frac{\text{kJ}}{\text{kg}}$$

We use the mass balance of water

$$\dot{m}_{w,\text{in}} + \dot{m}_{da}\mathscr{W}_3 = \dot{m}_{w,\text{out}} + \dot{m}_{da}\mathscr{W}_4$$

We replace the known numerical values

$$10000\,\frac{\text{kg}}{\text{min}} + \dot{m}_{da}0.007 = \dot{m}_{w,\text{out}} + \dot{m}_{da}0.027$$

or

$$\dot{m}_{w,\text{out}} = 10000 - 0.02\dot{m}_{da}$$

We replace the mass balance of water and all calculated values into the energy balance to obtain:

$$38.5\,\dot{m}_{da} + 167.44 \times 10^4 = 99.8\,\dot{m}_{da} + 104.65 \times (10000 - 0.02\dot{m}_{da})$$

or

$$99.8\,\dot{m}_{da} - 2.093\dot{m}_{da} - 38.5\,\dot{m}_{da} = (167.44 - 104.65) \times 10^4$$

and finally

$$\dot{m}_{da} = 10605\,\frac{\text{kg dry air}}{\text{min}}$$

The water "lost" in the air stream is calculated based on the mass flowrate of water at the exit point:

$$\dot{m}_{w,\text{out}} = 10000 - 0.02\dot{m}_{da} = 10000 - 0.02 \times 10605 = 9787.9\frac{\text{kg}}{\text{min}}$$

and it follows that (10,000−9787.9 =) 212.1 kg water/min. This needs to be replaced by make-up water. Make-up water in a cooling tower refers to the fresh water added to the system to replace losses that occur during the cooling process. Water quality must be monitored because impurities (e.g., calcium, magnesium, silica) can cause scaling, corrosion, or fouling. Pretreatment, such as softening, filtration, or chemical dosing, may be necessary, and this is the reason why minimizing the water losses is important.

11.4 Psychrometric charts at different pressures and different mixtures

The psychrometric chart, although mainly applied to humid air, and every point is defined (for a given pressure) by the combination of any two properties among the following (shown also for reference in Fig. 11.10).

- dry-bulb temperature (t_{db})
- wet-bulb temperature (t_{wb})
- humidity ratio ($\mathscr{W}$)
- dew point temperature (t_{dp})
- relative humidity ($\mathscr{RH}$)
- specific enthalpy (h)
- specific volume (v)

Psychrometric charts are typically constructed for a specific reference pressure, most commonly at standard atmospheric pressure (101.325 kPa or 1 atm). These charts illustrate the thermodynamic properties of moist air and are essential tools in

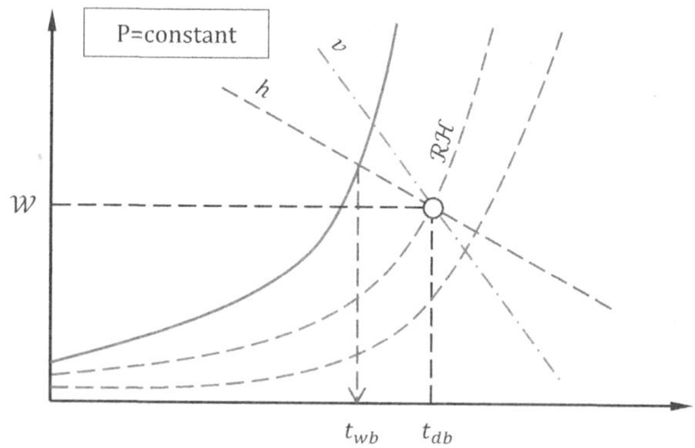

Figure 11.10 Summary of property isolines on a psychrometric chart.

HVAC design, meteorology, and process engineering. However, when the system pressure deviates from this standard value—whether lower or higher—it becomes challenging to find a readily available psychrometric chart that accurately reflects the conditions at the desired pressure. In such cases, standard charts become unreliable, and adjustments must be made to account for the pressure variation. Fortunately, MATLAB provides a solution through the program presented in Table 11.3, which enables users to generate custom psychrometric charts tailored to any specified pressure. This program operates under the assumption that the ideal gas law is valid, which holds true for pressures below atmospheric levels and remains reasonably accurate for pressures up to approximately 10 atm. By using this MATLAB or Python tool, engineers and researchers can analyze air-vapor mixtures under nonstandard conditions with confidence, ensuring precise control and optimization of thermodynamic processes across a wide range of pressures based on graphical means. It is stressed again that any final design needs to be validated using more accurate, commercial simulation packages.

While psychrometric charts are most commonly associated with water–air mixtures, the same principles can be extended to other binary gas–vapor systems. In general, psychrometric charts can be constructed for any mixture consisting of a condensable component and a second component that is effectively noncondensable under the conditions of interest. The fundamental thermodynamic relationships used to analyze humid air remain valid for these alternative systems, making it possible to model and visualize their behavior using similar techniques. Classical examples of such mixtures include water vapor with methane, nitrogen, or noble gases, as well as light hydrocarbons (such as ethane or propane) or nitrogen mixed with aromatic hydrocarbons. These systems are of practical importance in various fields such as natural gas processing, petrochemical refining, environmental control, and cryogenics. However, psychrometric charts for these alternative mixtures are rarely found in standard references or the open literature, making preliminary design and analysis more challenging.

To address this gap, the MATLAB program presented in Table 11.3 can be employed to generate psychrometric charts for these nonstandard mixtures. Provided the ideal gas approximation holds, the program allows users to create customized charts for a wide range of binary systems. This capability is particularly valuable during the conceptual or preliminary design stages, where a quick and reasonably accurate visualization of mixture behavior is needed before resorting to more complex models or detailed simulation tools.

Let us consider a concrete case study: the mixture of nitrogen and toluene at atmospheric pressure and temperatures in the range 0 and 60 °C. Nitrogen has a normal boiling point of 77 K (−196.15 °C) and is practically noncondensable at usual temperatures. The heat capacity of nitrogen is practically constant from 100 K up to 500 K with a mean value of 29.11 kJ/(kmol °C) or 1.03915 kJ/(kg °C).

Toluene, on the other hand, has a normal boiling point of 110.7 °C and has significant vapor pressure at the temperatures of interest and coexists in both vapor and liquid phases. Its enthalpy of vaporization can be calculated using the following equation and data obtained from NIST:

$$\Delta h_{\mathrm{vap,Tol}} = \frac{53090}{92.1384}\exp\left(-0.2774\times\frac{T}{591.7}\right)\left(1-\frac{T}{591.7}\right)^{0.2774} \tag{11.4.1}$$

where T is the absolute temperature in K and $\Delta h_{\mathrm{vap,Tol}}$ is the enthalpy of vaporization in kJ/kg. If we use 0 °C as the basis for enthalpy calculations we then have that the enthalpy of vaporization of toluene at 0 °C is

$$\Delta h_{\mathrm{vap,Tol}} = \frac{53090}{92.1384}\exp\left(-0.2774\times\frac{273.15}{591.7}\right)\left(1-\frac{273.15}{591.7}\right)^{0.2774} = 426.93$$

From NIST we also obtain the heat capacity of vapor toluene at the temperature of 300 K (27 °C), which is 104.4 kJ/(kmol °C), or

$$c_{p,V.\mathrm{Tol}} = 1.133\ \frac{\mathrm{kJ}}{\mathrm{kg\ ^\circ C}}$$

We can therefore propose the following model for calculating the specific enthalpy of vapor mixtures of toluene and nitrogen:

$$h(t, \mathscr{W}) = 1.03915t + \mathscr{W}(426.93 + 1.133t),\ \mathrm{in}\ \frac{\mathrm{kJ}}{\mathrm{kg\ dry\ nitrogen}} \tag{11.4.2}$$

where $\mathscr{W}$ is the humidity ratio in kg of toluene/kg of dry nitrogen and t is the temperature in °C.

Finally, we have the following Antoine's equation for toluene with constants obtained from the NIST Chemistry WebBook and adapted to units of degrees Celcius for temperature and pressure in bar (valid in the range 0–50 °C):

$$\ln P_{\text{Tol}}^{s} = 9.5363 - \frac{3172}{t + 222.643} \tag{11.4.3}$$

where P_{Tol}^{s} is the saturation pressure of toluene in bar and t is the temperature in °C. The MATLAB program of Table 11.3 for generating psychrometric charts is adapted to the case of nitrogen-toluene and is given in Table 11.5. The psychrometric chart generated for 1 bar is shown in Fig. 11.11.

Example 11.6. Use of the psychrometric chart for toluene-nitrogen mixture

A toluene-nitrogen stream with mass flowrate of 1000 kg dry nitrogen/h at 42 °C and 50% relative humidity is cooled at 1 bar down to a final temperature of 10 °C. Calculate:

a. the cooling load, and
b. the amount of toluene condensing.

Solution.

From the psychrometric chart of Fig. 11.11 we collect the following properties for the initial and the final points.

	t_{db} (°C)	$\mathcal{RH}$ (%)	$\mathcal{W}$ (g/kg)	h (kJ/kg)	v (m^3/kg)
Initial point	42	50	150	115	0.97
Final point	10	100	55	35	0.842

Using these data we can calculate the cooling load (rate of heat transfer)

$$\dot{Q} = 1000 \frac{\text{kg dry nitrogen}}{\text{h}} \times (35 - 115) \frac{\text{kJ}}{\text{kg dry nitrogen}}$$

$$= -80000 \frac{\text{kJ}}{\text{h}} \times \frac{\text{h}}{3600\ \text{s}} = -22.2\ \text{kW}$$

and the rate at which toluene is condensing

$$\dot{m}_{\text{Tol}} = 1000 \frac{\text{kg dry nitrogen}}{\text{h}} \times (55 - 150) \frac{\text{g toluene}}{\text{kg dry nitrogen}}$$

$$= -95000 \frac{\text{g toluene}}{\text{h}} \times \frac{1\ \text{kg}}{1000\ \text{g}} = -95 \frac{\text{kg toluene}}{\text{h}}$$

Table 11.5 MATLAB implementation for generating the psychrometric chart for toluene-nitrogen.

```
clear all

R      = 0.0831446261815324;      % ideal gas constant m^3 bar / K kmol
MN2     = 28.0134;                 % molecular mass of nitrogen
MTol  = 92.1384;                  % molecular mas of toluene
CpN2  = 1.03915;                  % sp. heat cap. of dry N2 kJ/(kg °C)
DhvapTol= 426.93;                   % sp. enthalpy of evaporation in kJ/kg
CpTol = 1.133;                    % sp. heat cap. of toluene kJ/(kg °C)

%% skeleton psychrometric chart
% select values for relative humidity %
RH = linspace(10,100,10);
% select values for temperature in °C
t  = linspace(0,60,61);
P  = 1.01325;
figure(1); clf;
for i=1:length(RH)
    for j=1:length(t)
        PsTol = exp( 9.5363-3172 / (222.643+t(j)) )
        term = (RH(i)/100)*PsTol;
        W(j,i) = (MTol/MN2)*term/(P-term);
    end
    plot(t,W(:,i)*1000,'LineWidth',1.2,'Color',[0.00,0.45,0.74])
    hold on
end

grid on
xlabel('Temperature (°C)')
ylabel('Humidity Ratio (g toluene/kg dry nitrogen)')
axis([0 60 0 260])
tx=text(20,110,'RH=100%'); tx.Rotation = +60;
tx=text(28.5,80,'50%'); tx.Rotation = +55;
tx=text(40,30,'10%'); tx.Rotation = +25;

%% constant specific enthalpy lines
hstar = linspace(10,150,15); % kJ/kg dry air
for i=1:length(hstar)
    W = (hstar(i)-CpN2*t)./(DhvapTol+CpTol*t)
    plot(t,W*1000,'--','LineWidth',1.2,'Color',[0.64,0.08,0.18])
    hold on;
end
tx=text( 2, 75,'h^{*}=30 kJ/kg'); tx.Rotation = -35;
tx=text( 2,120,'50 kJ/kg'); tx.Rotation = -35;
tx=text( 2,235,'100 kJ/kg'); tx.Rotation = -35;
%% constant specific volume lines
vstar= linspace(0.82,1.00,10); % kJ/kg dry air
for i=1:length(vstar)
    W = ( vstar(i)*P./(R*(273.15+t)) - 1/MN2)*MTol;
    plot(t,W*1000,'-.','LineWidth',1.2,'Color',[0.4660 0.6740 0.1880])
end
tx=text(3.6,60,'v^{*}=0.82 m^{3}/kg','FontSize',12);tx.Rotation=-75;
tx=text(30,65,'0.9 m^{3}/kg','FontSize',12); tx.Rotation = -75;
tx=text(50,210,'1.0 m^{3}/kg','FontSize',12); tx.Rotation = -75;
set(gcf,'position',[560,100,560,650])
```

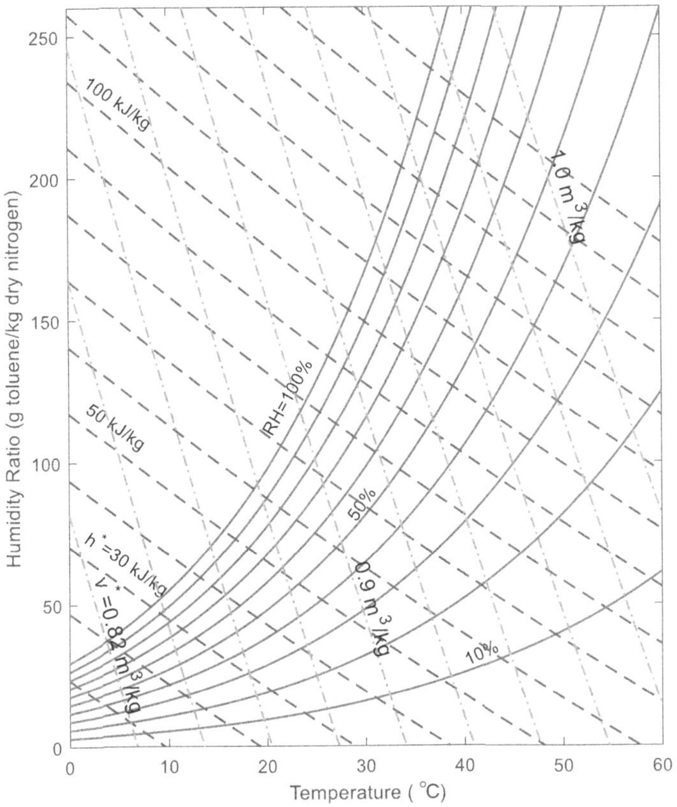

Figure 11.11 Psychrometric chart for toluene-nitrogen (valid for 101.325 kPa).

Summary

In this chapter, we introduced a systematic methodology for the design and analysis of systems involving binary mixtures composed of a noncondensable gas—such as air or nitrogen—and a condensable species, such as water vapor or heavier hydrocarbons. The fundamental principles underlying the construction of psychrometric charts for these mixtures were presented, enabling clear visualization of thermodynamic behavior across a range of operating conditions.

A basic MATLAB program was developed to facilitate the generation of custom psychrometric charts. With minor modifications, this tool can be adapted to virtually any binary gas–vapor mixture and a wide range of pressures, provided the assumption of ideal gas behavior remains valid. This flexibility makes it a powerful resource for preliminary design and analysis, particularly in cases where such charts are not readily available in the literature.

This approach exemplifies the continued relevance of graphical analysis in engineering, offering intuitive insight into complex processes. It has been applied successfully in the design of large-scale, high-cost systems—including natural draft cooling towers commonly found at thermal and nuclear power plants—where efficient thermal management is critical and early design accuracy is essential.

Problems

11.1 Find the humidity ratio and the relative humidity of air at 30 °C dry-bulb temperature and 30 °C wet-bulb temperature.

11.2 Find the specific volume and wet-bulb temperature of humid air at 30 °C dry-bulb temperature and 50% relative humidity.

11.3 Find the specific volume, percentage saturation, and moisture content of air at 30 °C dry-bulb temperature and 20 °C wet-bulb temperature.

11.4 Find the specific volume, wet-bulb temperature, moisture content, and specific enthalpy of air at 30 °C dry-bulb temperature and 30% relative humidity.

11.5 Moist air, saturated at 5 °C, enters a heating exchanger at a rate of 1 m^3/s. Air leaves the coil at 40 °C. Find the rate of heat addition.

11.6 Moist air at 30 °C dry-bulb temperature and 50% relative humidity is fed to a heat exchanger at 10 m^3/s and is cooled to a final temperature of 10 °C. Find the kW of refrigeration required.

11.7 A stream of atmospheric humid air with volumetric flowrate of 10 m^3/s at 5 °C dry-bulb temperature and 2 °C wet-bulb temperature is adiabatically mixed with 30 m^3/s of recirculated air at 25 °C dry-bulb temperature and 50% relative humidity. Find the dry-bulb temperature and wet bulb temperature of the resulting mixture.

11.8 One kilogram per s of moist air at 10 °C and 50% relative humidity is heated in a furnace to 50 °C. What is the rate of heat flow to the moist air stream? What is the final dew point of the air?

11.9 A water-cooling tower that has a blower with a capacity of 65 m^3/s of moist air. The moist air enters at 25 °C and a wet bulb temperature of 20 °C. The exit air is to leave at 35 °C and 32 °C wet-bulb. How much water can be cooled in t/h if the water to be cooled enters the tower at 50 °C, leaves the tower at 32 °C?

11.10 Atmospheric air is fed to a dryer and has a dry-bulb temperature of 32 °C and a wet-bulb temperature of 25.5 °C. It is heated to 90 °C and is fed into a dryer. In the dryer, it cools along an adiabatic cooling line as moisture from the dehydrating material is taken up and leaves the dryer fully saturated.

a. What is the dew point of the atmospheric air?
b. What is its humidity ratio?
c. What is its relative humidity?
d. How much heat is needed to heat 1 m^3 of initial air to 90 °C?
e. How much water will be evaporated per 1 m^3 of air entering the dryer?
f. At what temperature does the air leave the dryer?

11.11 A stream of preheated air with a dry-bulb temperature of 40 °C and a wet bulb temperature of 32 °C is mixed adiabatically with a stream of saturated cool air at 18 °C. The dry air mass flow rates of the warm and cool airstreams are 8 and 6 kg/s, respectively. Assuming a total pressure of 1 atm, determine (a) the temperature, (b) the specific humidity, and (c) the relative humidity of the mixture.

11.12 Use MATLAB or python to generate a psychrometric diagram of the air-water system at a pressure of 0.5 atm.

11.13 Use MATLAB or python to generate a psychrometric diagram of the nitrogen-toluene system at a pressure of 5 atm.

11.14 Use MATLAB or python to generate a psychrometric diagram of the water-methane system at a pressure of 10 atm.

Further reading

Gatley, D.P., 2013. Understanding Psychrometrics, third ed. ASHRAE, Atlanta, GA.

Mc Cabe, W., Smith, J., Harriott, P., 2004. Unit Operations in Chemical Engineering, seventh ed. McGraw Hill Book Company, NY.

Shallcross, D.C., 1997. Handbook of Psychrometric Charts: Humidity Diagrams for Engineers. Chapman & Hall, London.

Treybal, R.E, Mass-Transfer Operations, third ed., McGraw Hill Book Company, NY.

12

Process optimization

12.1 Introduction

Up to this point, our analysis of process systems has primarily focused on scenarios where the number of degrees of freedom was either zero or extremely limited. In many cases, when a degree of freedom did exist, our approach involved arbitrarily assigning an equal number of specifications to fully define the system. While such methods are suitable for ensuring feasibility and operational stability, they do not necessarily lead to optimal performance.

In this chapter, we introduce a more systematic and powerful methodology for selecting values of the available process variables—a methodology grounded in the principles of optimization. The presence of degrees of freedom in a process model indicates that there is flexibility in how the system can be configured, designed, or operated. Rather than resolving this flexibility through arbitrary choices, we aim to leverage it strategically to improve system performance.

At the heart of any **optimization**-based approach lies the concept of the **objective function**. This function quantitatively expresses the goal we wish to achieve—whether it is minimizing energy consumption, maximizing product yield, reducing emissions, or, most commonly, enhancing economic performance. Selecting an appropriate objective function is a critical decision, as it directly determines the direction and outcome of the optimization process.

In practical engineering applications, process economics often dominates as the primary concern. Accordingly, economic criteria will serve as our principal objective when utilizing the available degrees of freedom. That is, whenever we face decisions regarding the values of certain process variables—such as temperatures, pressures, flow rates, or compositions—our overarching goal will be to choose values that either minimize total cost (e.g., operating or capital costs) or maximize profitability.

This chapter will explore the formulation and solution of such optimization problems. We will define the mathematical

Introduction to Process Systems Engineering. https://doi.org/10.1016/B978-0-443-45444-8.00010-5

structure of optimization models, discuss how to appropriately select objective functions and constraints, and examine various techniques for solving these models. Ultimately, by transitioning from arbitrary specification to informed decision-making, we lay the foundation for designing and operating more efficient and cost-effective process systems.

12.2 A motivating example

Let us consider a Continuous Stirred Tank Reactor (CSTR) used to carry out a first-order, irreversible reaction of the form $R \rightarrow P$, where R is the reactant and P is the desired product. Our objective is to determine the CSTR configuration that minimizes the total cost, or equivalently, minimizes the overall reactor volume required to achieve a specified level of conversion.

We assume the system is fed with a fresh inlet stream of volumetric flow rate F, containing the reactant at an initial concentration $C_{R,0}$. The desired conversion corresponds to a final reactant concentration $C_{R,f}$ in the outlet stream, as dictated by the well-mixed nature of the CSTR. This ensures that the concentration of the reactant in the exit stream is equal to that within the reactor. For simplicity, the reactor operates at a constant temperature, resulting in a constant rate constant k for the first-order reaction. The schematic representation of the process is shown in Fig. 12.1.

To determine the required reactor volume V, we begin by applying a steady-state mole balance on the reactant:

$$F\left(C_{R,0} - C_{R,f}\right) - kC_{R,f}V = 0 \qquad (12.2.1)$$

which can be solved for the necessary reactor volume

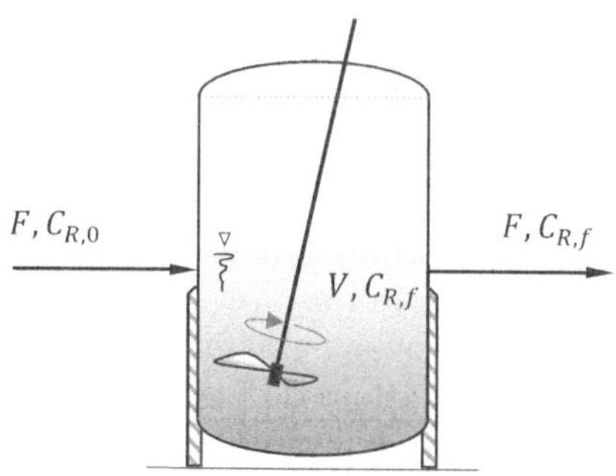

Figure 12.1 A CSTR with first-order reaction.

$$V = \frac{F}{k}\left(\frac{C_{R,0} - C_{R,f}}{C_{R,f}}\right) \tag{12.2.2}$$

Our task has been completed as there is no degree of freedom available (all variables appearing in the RHS of Eq. (12.2.2) have known and fixed values).

To make our example more concrete and specific, we assume that $k = 1\,\text{h}^{-1}$, $F = 1\frac{\text{m}^3}{\text{h}}$, $C_{R,0} = 10\,\text{kmol/m}^3$ and $C_{R,f} = 1\,\text{kmol/m}^3$ (90% conversion). We replace the numerical values into Eq. (12.2.2) to obtain

$$V = \frac{F}{k}\left(\frac{C_{R,0} - C_{R,f}}{C_{R,f}}\right) = \frac{1}{1}\left(\frac{10-1}{1}\right) = 9\,\text{m}^3 \tag{12.2.3}$$

We know from Chapter 10 that we can decrease the overall reacting volume by placing two or more CSTRs in series that operate at the same temperature. Let us start with two CSTRs in series as shown in Fig. 12.2. We formulate the mole balances for the reactant in the two CSTRs:

$$F(C_{R,0} - C_{R,1}) - kC_{R,1}V_1 = 0 \tag{12.2.4}$$

$$F\left(C_{R,1} - C_{R,f}\right) - kC_{R,f}V_2 = 0 \tag{12.2.5}$$

which can be solved for the corresponding reactor volume

$$V_1 = \frac{F}{k}\left(\frac{C_{R,0} - C_{R,1}}{C_{R,1}}\right) \tag{12.2.6}$$

$$V_2 = \frac{F}{k}\left(\frac{C_{R,1} - C_{R,f}}{C_{R,f}}\right) \tag{12.2.7}$$

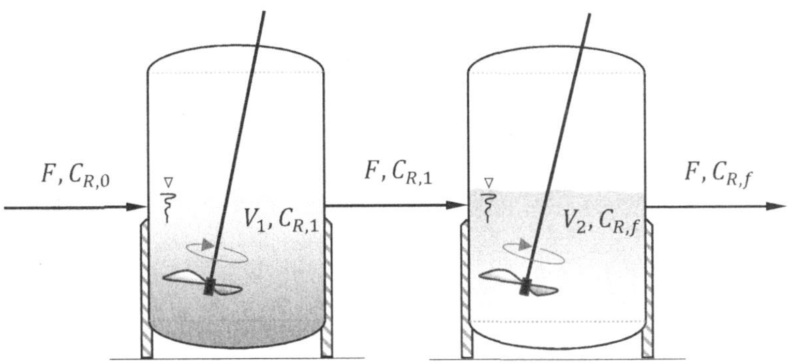

Figure 12.2 Two CSTRs in series with first-order reaction.

The overall reaction volume is

$$V(C_{R,1}) = V_1 + V_2 = \frac{F}{k}\left[\left(\frac{C_{R,0} - C_{R,1}}{C_{R,1}}\right) + \left(\frac{C_{R,1} - C_{R,f}}{C_{R,f}}\right)\right] \quad (12.2.8)$$

The bad news is that we now have a degree of freedom: we are free to select the concentration of the reactant in the first reactor. The concentration of the reactant in the second reactor is fixed by the required conversion. How do we select the value of the reactant concentration in the first reactor?

One practical approach to addressing this important design question is to systematically explore possible values for the intermediate concentration $C_{R,1}$ within the interval $[C_{R,f}, C_{R,0}]$. For each candidate value, we can perform the necessary calculations using Eq. (12.2.8) to determine the corresponding reactor volume. After evaluating all options, we can select the optimal one based on a well-defined performance criterion.

But what constitutes a realistic and meaningful **performance criterion**? From an engineering and economic standpoint, one of the most relevant objectives is to minimize the Fixed Capital Investment (FCI). FCI encompasses the total upfront cost required to bring the plant to an operational state. This includes the cost of purchasing and installing all essential, nonconsumable components such as process equipment, piping, instrumentation, and electrical systems.

However, FCI as a function of design variables can be complex and nonlinear, often involving detailed cost correlations, equipment sizing models, and empirical factors. This complexity can make direct mathematical optimization of FCI challenging, especially in early design stages where simplified analysis is preferred. Since reactor volume directly impacts equipment size and, by extension, capital cost, minimizing FCI is strongly aligned with efficient and cost-effective process design.

To overcome this, we often seek surrogate objective functions—simpler, analytically tractable quantities that are highly correlated with FCI. For example, minimizing the total reactor volume or the sum of individual reactor volumes in a series configuration can serve as a useful proxy. This approach offers a more intuitive understanding of the design trade-offs while remaining closely tied to economic performance.

In this context, our goal becomes clearer: we aim to choose the intermediate concentration $C_{R,1}$ such that the resulting total reactor volume—and by implication, the FCI—is minimized, using a mathematically manageable objective function.

Let us select values of $C_{R,1}$ in the range [1, 10] and calculate the overall reaction volume using Eq. (12.2.8). In Table 12.1 the results of the calculation are shown. Indicative calculation for $C_{R,1} = 5\ \text{kmol/m}^3$ is as follows:

$$V\left(C_{R,1} = 5\ \frac{\text{kmol}}{\text{m}^3}\right) = \frac{1}{1}\left[\left(\frac{10-5}{5}\right) + \left(\frac{5-1}{1}\right)\right] = \frac{5}{5} + \frac{4}{1}$$
$$= 1\ \text{m}^3 + 4\ \text{m}^3$$
$$= 5\ \text{m}^3$$

The volumes of the individual reactors, as well as the total combined volume, are plotted in Fig. 12.3 as functions of the reactant concentration in the first CSTR, denoted as $C_{R,1}$. As expected, the volume of the first reactor decreases with increasing $C_{R,1}$, since a higher concentration corresponds to a lower required conversion in that reactor. Conversely, the second reactor must achieve more of the overall conversion when $C_{R,1}$ is higher, causing its volume to increase.

Interestingly, the total reactor volume—the sum of the volumes of both CSTRs—exhibits a nonmonotonic behavior: it initially decreases, reaches a minimum, and then begins to increase again. This trend reflects a trade-off between the two stages, with an optimal intermediate concentration that minimizes the overall reactor volume. According to the plot, this optimal point occurs at a $C_{R,1}$ value around 3.1 kmol/m^3. The minimum overall volume is $V = 4.3\ \text{m}^3$, which is an impressive reduction in the overall reactor volume compared to the volume of a single CSTR (100(9–4.3)/9 = 52.2% reduction).

Table 12.1 Result of the analytic calculations of the total reaction volume for two CSTRs in series.

$C_{R,1}$ $\left(\frac{\text{kmol}}{\text{m}^3}\right)$	V_1 (m^3)	V_2 (m^3)	V (m^3)
1	9.000	0.000	9.000
2	4.000	1.000	5.000
3	2.333	2.000	4.333
4	1.500	3.000	4.500
5	1.000	4.000	5.000
6	0.667	5.000	5.667
7	0.429	6.000	6.429
8	0.250	7.000	7.250
9	0.111	8.000	8.111
10	0.000	9.000	9.000

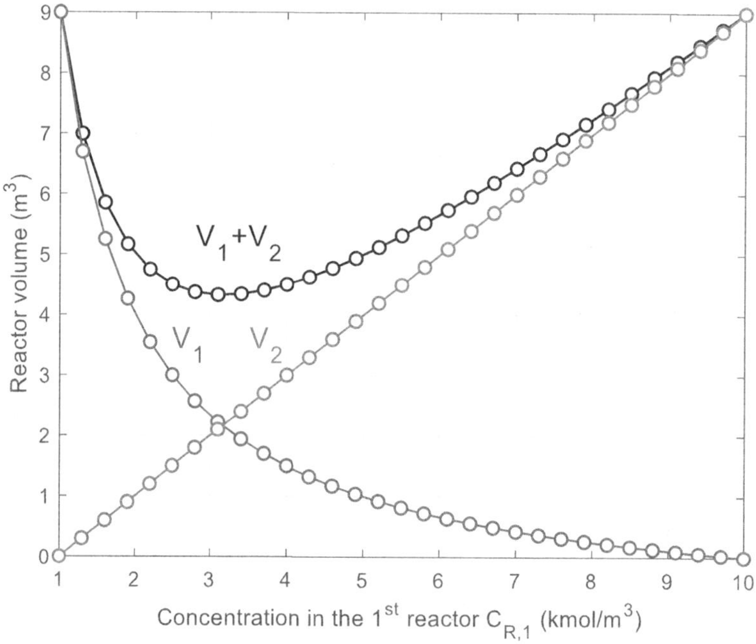

Figure 12.3 Volume of two CSTRs in series as a function of the reactant concentration in the first reactor.

The MATLAB script used to generate Fig. 12.3 is shown below.

```
clear all
F = 1;    % feed volumetric flowrate       [=] m^3/min
k = 1;    % reaction rate constant         [=] 1/min
CR0 =10;  % reactant feed concentration    [=] kmol/m^3
CR2 = 1;  % reactant final concentration   [=] kmol/m^3
% select values for the intermediate reactant concentration
CR1 = linspace(CR2,CR0,31);
% total reaction volume
V1 = (F/k)*(CR0-CR1)./CR1; % volume of first reactor in m^3
V2 = (F/k)*(CR1-CR2)./CR2; % volume of first reactor in m^3
V  = V1+V2;                % overall volume of reactors in m^3
figure(1); clf; hold on; grid on;
plot(CR1,V ,'-o','linewidth',1.2,'Color','k')
plot(CR1,V1,'-o','linewidth',1.2,'Color',[0.64,0.08,0.184])
plot(CR1,V2,'-o','linewidth',1.2,'Color',[0.00,0.45,0.740])
xlabel('Concentration in the 1^{st} reactor C_{R,1}
(kmol/m^{3})')
ylabel('Reactor volume (m^{3})')
```

One interesting aspect of Fig. 12.3 is the appearance of a trade-off. A trade-off refers to a situation where achieving more of one desirable outcome requires accepting less of another. In other words, it's the balance you strike when you cannot maximize all objectives at once—you gain something, but at the cost of something else: increasing the concentration of the reactant in the first CSTR decreases the volume of the first CSTR but increases the volume of the second CSTR. Trade-offs are central to optimization, design, and decision-making. Recognizing them helps engineers, economists, and designers choose the best compromise given constraints, priorities, or limited resources.

Encouraged by the reduction in the overall volume when two CSTRs are used instead of one, we are tempted to increase further the number of reactors. When three CSTRs are used the overall volume is (see Fig. 12.4 for notation used):

$$V(C_{R,1}, C_{R,2}) = V_1 + V_2 + V_3$$
$$= \frac{F}{k}\left[\left(\frac{C_{R,0} - C_{R,1}}{C_{R,1}}\right) + \left(\frac{C_{R,1} - C_{R,2}}{C_{R,2}}\right) + \left(\frac{C_{R,2} - C_{R,f}}{C_{R,f}}\right)\right] \tag{12.2.9}$$

It is important to note that our degrees of freedom have increased to two as we can choose two reactant concentrations $C_{R,1}$ and $C_{R,2}$ in the range $\left[C_{R,f}, C_{R,0}\right]$ while satisfying the constraint $C_{R,1} > C_{R,2}$. The method to investigate the dependency of the total volume on the two intermediate concentrations is a minor extension of the method followed in the case of a single degree of freedom. This investigation can be performed easily in MATLAB as follows:

```
clear all
F    = 1; % feed volumetric flowrate         [=] m^3/min
k    = 1; % reaction rate constant           [=] 1/min
CR0 =10; % reactant feed concentration      [=] kmol/m^3
CR3 = 1; % reactant final concentration     [=] kmol/m^3
% select values for the intermediate reactant concentrations
CR1 = linspace(CR3,CR0,31); CR2 = linspace(CR3,CR0,31);
%
for i=1:length(CR1)
   for j=1:length(CR2)
      V(j,i)= (F/k)*( ( CR0    - CR1(i) )/CR1(i) + ...
                      ( CR1(i) - CR2(j) )/CR2(j) + ...
                      ( CR2(j) - CR3    )/CR3    );
   end
end
[CR11,CR22]=meshgrid(CR1,CR2);
meshc(CR11,CR22,V,'linewidth',1.2)
axis([1 10 1 10 3 9])
xlabel('C_{R,1} (kmol/m^{3})')
ylabel('C_{R,2} (kmol/m^{3})')
zlabel('Volume (m^{3})')
grid on
```

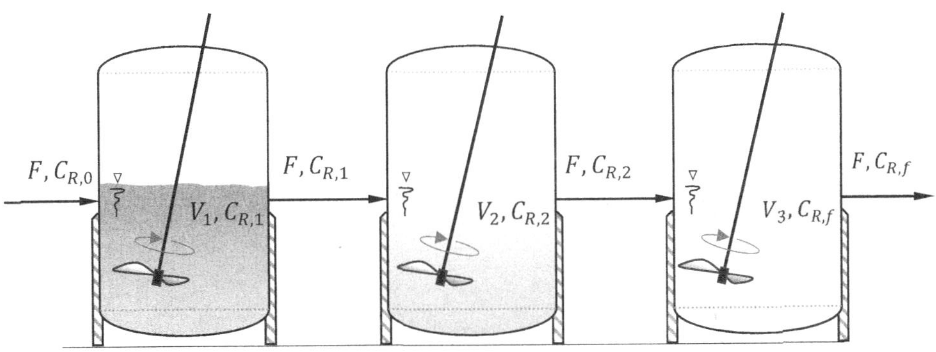

Figure 12.4 Three CSTRs in series with first-order reaction.

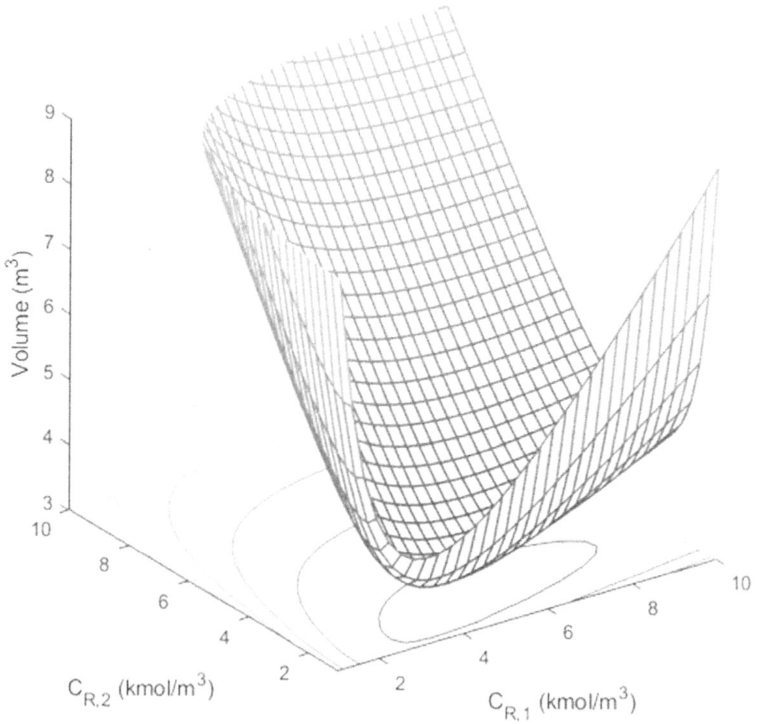

Figure 12.5 Total volume of three CSTRs in series as a function of the intermediate reactant concentrations.

The results are shown in Fig. 12.5, where the total volume of the three CSTRs is shown as a function of the two intermediate concentrations. There is again a clear minimum at the point $(C_{R,1}, C_{R,2}) \approx (4.64, 2.15)$.

Example 12.1. A two-stage air compression system with intercooling

In Fig. 12.6, a two-stage gas compression system with intercooling is shown. A gas with molar flowrate N and an initial

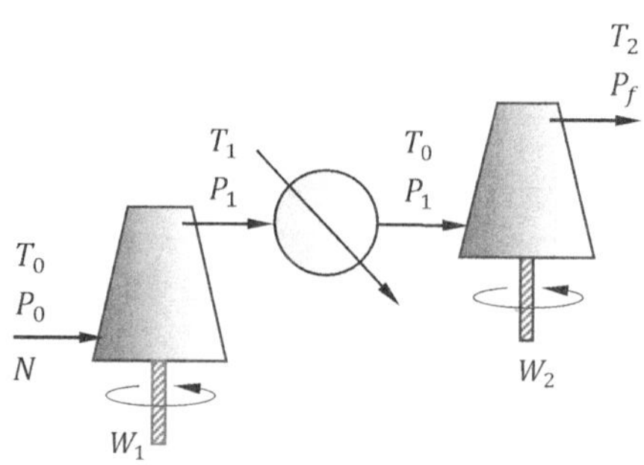

Figure 12.6 A two-stage compression system.

pressure P_0 is compressed to a final pressure of P_f. The first compressor achieves an intermediate pressure of P_1. The temperature at the suction of both compressors is T_0. The compression power necessary is given by the following equation:

$$W = N\frac{RT_0}{ea}\left\{\left[\left(\frac{P_1}{P_0}\right)^a - 1\right] + \left[\left(\frac{P_f}{P_1}\right)^a - 1\right]\right\}$$

where e is the efficiency, and

$$a = \frac{\gamma - 1}{\gamma} = \frac{C_p - C_v}{C_p}, \gamma = \frac{C_p}{C_v}$$

At a particular case, air with molar flowrate of $N = 10\ \frac{\text{mol}}{\text{s}}$, an initial absolute temperature $T_0 = 298.15$ K and a pressure of $P_0 = 1$ bar is to be compressed at a final pressure of $P_f = 10$ bar. The efficiency of the compressor is $e = 0.75$ and for air $a = 0.29$. Use MATLAB or Python to calculate the dependency of the compression power on the intermediate pressure P_1.

Solution.

The objective function consists of the power of the first compressor:

$$W_1(P_1) = N\frac{RT_0}{ea}\left[\left(\frac{P_1}{P_0}\right)^a - 1\right]$$

and of the power of the second compressor:

$$W_2(P_1) = N\frac{RT_0}{ea}\left[\left(\frac{P_f}{1}\right)^a - 1\right]$$

The problem can be solved in MATLAB using the following steps:

We assign values to model parameters.

```
clear all
N   = 0.010; % air molar flowrate kmol/s
T0  = 298.15;% feed temperature   K
a   = 0.29;  % (γ-1)/γ            -
e   = 0.75;  % efficiency         -
R   = 8.314; % ideal gas constant kJ/(kmol K)
P0  = 1;     % initial pressure in bar
Pf  = 10;    % final pressure in bar
```

Select values for the intermediate pressure and calculate the rate of work done.

```
P1  = linspace(P0,Pf,31);
% power calculations
W1 = N*R*T0*((P1 /P0).^a-1)/(e*a); % work in the ferst com-
pressor kW
W2 = N*R*T0*((Pf./P1).^a-1)/(e*a); % work in the ferst com-
pressor kW
W  = W1+W2;                          % total
```

Visualize the results (Fig. 12.7).

```
figure(1); clf;
plot(P1,W ,'-
o','linewidth',1.2,'Color','k','MarkerFaceColor','w')
hold on;
plot(P1,W1,'-
o','linewidth',1.2,'Color',[0.64,0.08,0.184],'MarkerFaceColor'
,'w')
plot(P1,W2,'-
o','linewidth',1.2,'Color',[0.00,0.45,0.740],'MarkerFaceColor'
,'w')
xlabel('Discharge pressure of first compressor P_{1} (bar)')
ylabel('W (kW)')
grid on
tx=text(6,70,'W_{1}','FontSize',14,'Color',[0.64,0.08,0.184]);
tx=text(6,24,'W_{2}','FontSize',14,'Color',[0.00,0.45,0.74]);
tx=text(3.5,100,'W_{1}+W_{2}','FontSize',14,'Color','K');
```

Fig. 12.7 is obtained.

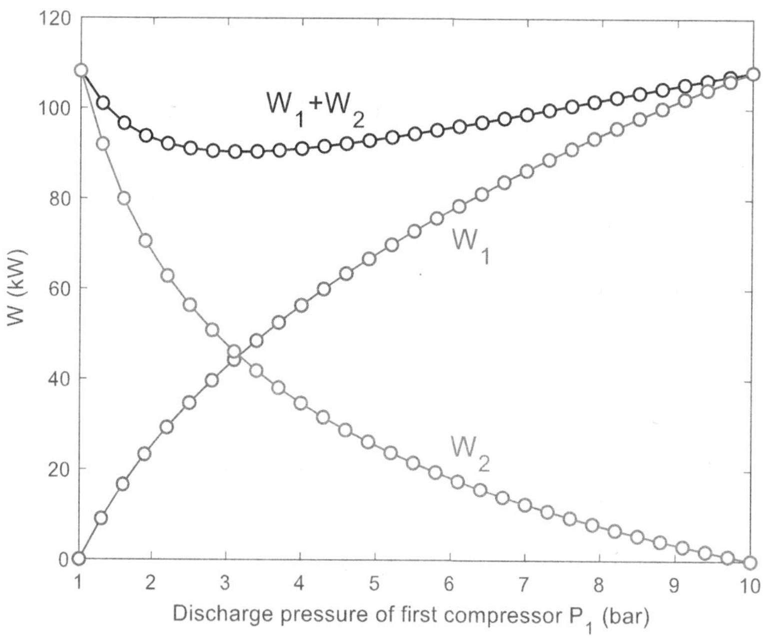

Figure 12.7 Rate of compression work as a function of the intermediate pressure.

12.3 Optimality conditions for unconstrained problems

In this section the conditions that are satisfied by solutions to the problem

$$\min_{\boldsymbol{x}} f(\boldsymbol{x}) \tag{12.3.1}$$

are discussed, where $\boldsymbol{x} \in R^n$ is an n-dimensional vector and $f : R^n \to R$ is a real valued function. We will first present the results for the case where $x \in R$ (a real variable) and then extend the results to the more general case.

12.3.1 Single-variable, unconstrained optimization

A function of one variable $f(x)$ is said to have a relative or **local minimum** at $x = x^*$ if $f(x^*) \leq f(x^* + \delta x)$ for all sufficiently small positive and negative values of δx. A single-variable optimization problem is one in which the value of $x = x^*$ is to be found in the interval $[a, b]$ such that x^* minimizes $f(x)$. The following two theorems provide the necessary and sufficient conditions for the relative minimum of a function of a single variable.

The following is only a necessary condition for $f(x)$ to have a minimum at x^*: If a function $f(x)$ is defined in the interval $a \leq x \leq b$ and has a relative minimum at x^*, where $a < x^* < b$, and if the first derivative exists as a finite number at x^*, then

$$f'(x^*) = \left.\frac{df(x)}{dx}\right|_{x=x^*} = 0 \tag{12.3.2}$$

In the one-dimensional case, there is a geometric interpretation for this condition. If f is increasing at a point x, then $f'(x) > 0$. Similarly, if f is decreasing, then $f'(x) < 0$. A point where f is increasing or decreasing cannot correspond to a minimizer. At a minimizer the function will be flat or stationary, and hence $f'(x^*) = 0$ follows.

The condition $f'(x^*) = 0$ is referred to as the first-order necessary condition for a minimizer. The term "first-order" refers to the presence of the first derivatives of f, and it is a "necessary" condition since if x^* is a local minimizer, then it "necessarily" satisfies this condition. The condition is, however, not "sufficient" to determine a local minimizer since a point satisfying $f'(x^*) = 0$ could be a local minimizer, a local maximizer, or a saddle point (a stationary point that is neither a minimizer nor a maximizer). Local minimizers can be distinguished from other stationary points by examining second derivatives. Consider again the Taylor series expansion at $x = x^*$ (see Chapter 9, Eq. (9.3.12))

$$f(x^* + \delta x) \approx f(x^*) + \left.\frac{df(x)}{dx}\right|_{x=x^*} \delta x + \frac{1}{2}\left.\frac{d^2 f(x)}{dx^2}\right|_{x=x^*} (\delta x)^2 \qquad (12.3.3)$$

At a stationary point the first derivative is zero and we then have that

$$f(x^* + \delta x) \approx f(x^*) + \frac{1}{2}\left.\frac{d^2 f(x)}{dx^2}\right|_{x=x^*} (\delta x)^2$$

or

$$f(x^* + \delta x) - f(x^*) \approx \frac{1}{2}\left.\frac{d^2 f(x)}{dx^2}\right|_{x=x^*} (\delta x)^2 \qquad (12.3.4)$$

The meaning of the last equations is clear: close to x^*, i.e., for small δx, the sign of $f(x^* + \delta x) - f'(x^*)$ is the same as the sign of the second derivative, as $(\delta x)^2$ is always positive. If x^* is a local minimizer then $f(x^* + \delta x) - f(x^*) > 0$ and the condition that needs to be satisfied by the second derivative is

$$f''(x^*) = \left.\frac{d^2 f(x)}{dx^2}\right|_{x=x^*} > 0 \qquad (12.3.5)$$

This is referred to as the second-order sufficient condition for a minimizer, with the "second-order" referring to the use of second derivatives or the second-order term in the Taylor series.

When we are looking for the minimizer of a real-valued function of a real variable, we calculate the derivative of the function and then solve the, possibly, nonlinear equation that equates the derivative to zero. Let x^* denote the solution of $f'(x^*) = 0$. Once we have found the solution, then to prove that this solution corresponds to a minimum we only need to evaluate the second derivative, which needs to be strictly positive for a minimum.

To summarize: when seeking the minimizer of a real-valued function of a real variable, we begin by computing the first derivative and solving the (potentially nonlinear) equation $f'(x) = 0$. Let x^* denote a solution to this equation. To determine whether x^* is indeed a local minimum, we evaluate the second derivative at that point. If $f''(x^*) > 0$, then x^* satisfies the second-order condition for a local minimum. Note that these criteria do not apply when $x^* = a$ or $x^* = b$.

Let us return to the two CSTRs in series example for which $x = C_{R,1}$ and $f(x) = V_1(C_{R,1}) + V_2(C_{R,2})$, which is given by Eq. (12.2.):

$$V(C_{R,1}) = \frac{F}{k}\left[\left(\frac{C_{R,0} - C_{R,1}}{C_{R,1}}\right) + \left(\frac{C_{R,1} - C_{Rf}}{C_{Rf}}\right)\right]$$

It is reminded that F, k, $C_{R,0}$ and C_{Rf} are constants. To find the minimum total volume solution we first calculate the first-order derivative of the total volume with respect to the reactant concentration in the first CSTR:

$$\frac{dV(C_{R,1})}{dC_{R,1}} = \frac{F}{k}\left[\frac{d}{dC_{R,1}}\left(\frac{C_{R,0} - C_{R,1}}{C_{R,1}}\right) + \frac{d}{dC_{R,1}}\left(\frac{C_{R,1} - C_{Rf}}{C_{Rf}}\right)\right]$$

$$= \frac{F}{k}\left[\frac{-C_{R,1} - (C_{R,0} - C_{R,1})}{C_{R,1}^2} + \frac{1}{C_{Rf}}\right] = \frac{F}{k}\left(-\frac{C_{R,0}}{C_{R,1}^2} + \frac{1}{C_{Rf}}\right)$$

We then set the derivative equal to zero:

$$\frac{F}{k}\left(-\frac{C_{R,0}}{C_{R,1}^2} + \frac{1}{C_{Rf}}\right) = 0$$

and we finally solve for the reactant concentration in the first CSTR

$$C_{R,1} = \sqrt{C_{R,0}\, C_{Rf}} \qquad (12.3.6)$$

i.e., the optimal value of $C_{R,1}$ is the geometric mean of $C_{R,0}$ and C_{Rf}. To prove that this is the minimum volume solution, we calculate the second derivative

$$\frac{d^2V(C_{R,1})}{dC_{R,1}^2} = \frac{d}{dC_{R,1}}\left[\frac{F}{k}\left(-\frac{C_{R,0}}{C_{R,1}^2} + \frac{1}{C_{Rf}}\right)\right]$$

$$= -C_{R,0}\frac{F}{k}\frac{d}{dC_{R,1}}\left(\frac{1}{C_{R,1}^2}\right) = 2\frac{F}{k}\left(\frac{C_{R,0}}{C_{R,1}^3}\right) > 0$$

which proves that the solution given by Eq. (12.3.4) corresponds to a minimum. This is also obvious from Fig. 12.3. The exact numerical value of the minimum volume solution corresponds to the following reactant concentration in the first CSTR

$$C_{R,1} = \sqrt{C_{R,0}\, C_{Rf}} = \sqrt{1 \times 10} = 3.1623\ \frac{\text{kmol}}{\text{m}^3}$$

An interesting result is obtained if Eq. (12.3.6) is substituted into Eqs. (12.2.6) and (12.2.7) to eliminate $C_{R,1}$:

$$V_1 = \frac{F}{k}\left(\frac{C_{R,0} - \sqrt{C_{R,0}\,C_{R,f}}}{\sqrt{C_{R,0}\,C_{R,f}}}\right) = \frac{F}{k}\left(\sqrt{\frac{C_{R,0}}{C_{R,f}}} - 1\right)$$

$$V_2 = \frac{F}{k}\left(\frac{\sqrt{C_{R,0}\,C_{R,f}} - C_{R,f}}{C_{R,f}}\right) = \frac{F}{k}\left(\sqrt{\frac{C_{R,0}}{C_{R,f}}} - 1\right)$$

i.e.,

$$V_1 = V_2 = \frac{F}{k}\left(\sqrt{\frac{C_{R,0}}{C_{R,f}}} - 1\right)$$

This is an amazing result as the volumes of the two reactors turn out to be equal. This is also a very desirable result in terms of industrial applications.

Example 12.2. Multistage compression system

Let us return to Example 12.1, in which for the two-stage air compression system shown in Fig. 12.6, the compression power necessary is given by the following equation:

$$W = N\frac{RT_0}{ea}\left\{\left[\left(\frac{P_1}{P_0}\right)^a - 1\right] + \left[\left(\frac{P_f}{P_1}\right)^a - 1\right]\right\}$$

where the notation and the numerical values of the parameters are given in Example 12.1. Find the optimal discharge pressure of the first compressor analytically.

Solution.

To find the optimal discharge pressure of the first compressor we calculate the first-order derivative:

$$\frac{dW(P_1)}{dP_1} = N\frac{RT_0}{ea}\left\{\frac{d}{dP_1}\left[\left(\frac{P_1}{P_0}\right)^a - 1\right] + \frac{d}{dP_1}\left[\left(\frac{P_f}{P_1}\right)^a - 1\right]\right\}$$

$$= N\frac{RT_0}{ea}\left(\frac{a}{P_0^a}P_1^{a-1} + P_f^a\frac{dP_1^{-a}}{dP_1}\right) = N\frac{RT_0}{ea}\left[a\left(\frac{P_1}{P_0}\right)^a\frac{1}{P_1} - aP_f^aP_1^{-a-1}\right]$$

$$= N\frac{RT_0}{ea}a\left[\left(\frac{P_1}{P_0}\right)^a - \left(\frac{P_f}{P_1}\right)^a\right]\frac{1}{P_1}$$

We set the derivative equal to zero:

$$N\frac{RT_0}{ea}a\left[\left(\frac{P_1}{P_0}\right)^a - \left(\frac{P_f}{P_1}\right)^a\right]\frac{1}{P_1} = 0$$

from which it follows that

$$\left(\frac{P_1}{P_0}\right)^a = \left(\frac{P_f}{P_1}\right)^a$$

or

$$P_1 = \sqrt{P_0 P_f}$$

We also note that the two compressors have the same compression ratio λ:

$$\lambda = \frac{\text{discharge pressure}}{\text{suction pressure}} = \frac{P_1}{P_0} = \frac{P_f}{P_1}$$

The following result follows immediately:

$$W_1 = N\frac{RT_0}{ea}\left[\left(\frac{P_1}{P_0}\right)^a - 1\right] = N\frac{RT_0}{ea}[\lambda^a - 1] = N\frac{RT_0}{ea}\left[\left(\frac{P_f}{P_1}\right)^a - 1\right] = W_2$$

Again, the use of identical compressors corresponds to the optimal solution.

12.3.2 Multivariable, unconstrained optimization

We now extend the results of the previous section to the case

$$\min_{x} f(\boldsymbol{x}) \tag{12.3.7}$$

where $\boldsymbol{x} \in R^n$ is an n-dimensional vector and $f : R^n \rightarrow R$ is a real valued function. Let us consider the case of $\boldsymbol{x} \in R^2$. Following the steps of the previous section we may write the following:

$$f(x_1^* + \delta x_1, x_2^* + \delta x_2) \approx f(x_1^*, x_2^*)$$

$$+ \left.\frac{\partial f(x)}{\partial x_1}\right|_{\boldsymbol{x}^*=(x_1^*,x_2^*)} \delta x_1 + \left.\frac{\partial f(x)}{\partial x_2}\right|_{\boldsymbol{x}^*=(x_1^*,x_2^*)} \delta x_2 + \frac{1}{2}\left.\frac{\partial^2 f(x)}{\partial x_1^2}\right|_{\boldsymbol{x}^*=(x_1^*,x_2^*)} (\delta x_1)^2$$

$$+ \frac{1}{2}\left.\frac{\partial^2 f(x)}{\partial x_2^2}\right|_{\boldsymbol{x}^*=(x_1^*,x_2^*)} (\delta x_2)^2 + \frac{1}{2}\left.\frac{\partial^2 f(x)}{\partial x_1 \partial x_2}\right|_{\boldsymbol{x}^*=(x_1^*,x_2^*)} \delta x_1 \delta x_2 + \frac{1}{2}\left.\frac{\partial^2 f(x)}{\partial x_2 \partial x_1}\right|_{\boldsymbol{x}^*=(x_1^*,x_2^*)} \delta x_2 \delta x_1 \tag{12.3.8}$$

We can use matrix notation to make the structure of this result clear:

$$f(x_1^*+\delta x_1, x_2^*+\delta x_2) \approx f(x_1^*, x_2^*) + \begin{bmatrix} \frac{\partial f(x)}{\partial x_1} & \frac{\partial f(x)}{\partial x_2} \end{bmatrix}_{x^*=(x_1^*,x_2^*)} \begin{bmatrix} \delta x_1 \\ \delta x_2 \end{bmatrix}$$

$$+\frac{1}{2}[\delta x_1 \quad \delta x_2] \begin{bmatrix} \frac{\partial^2 f(x)}{\partial x_1^2} & \frac{\partial^2 f(x)}{\partial x_1 \partial x_2} \\ \frac{\partial^2 f(x)}{\partial x_2 \partial x_1} & \frac{\partial^2 f(x)}{\partial x_2^2} \end{bmatrix}_{x^*=(x_1^*,x_2^*)} \begin{bmatrix} \delta x_1 \\ \delta x_2 \end{bmatrix}$$

We then define the gradient vector

$$\nabla f = \begin{bmatrix} \frac{\partial f(x)}{\partial x_1} \\ \frac{\partial f(x)}{\partial x_2} \end{bmatrix}$$

and the Hessian matrix of the second derivatives:

$$\mathbf{H} = \begin{bmatrix} \frac{\partial^2 f(x)}{\partial x_1^2} & \frac{\partial^2 f(x)}{\partial x_1 \partial x_2} \\ \frac{\partial^2 f(x)}{\partial x_2 \partial x_1} & \frac{\partial^2 f(x)}{\partial x_2^2} \end{bmatrix}$$

and Eq. (12.3.8) can be written as

$$f(\boldsymbol{x}^* + \boldsymbol{\delta x}) \approx f(\boldsymbol{x}^*) + \nabla f^T \delta \boldsymbol{x} + \frac{1}{2}\delta \boldsymbol{x}^T \mathbf{H}(\boldsymbol{x}^*)\delta \boldsymbol{x} \qquad (12.3.9)$$

Note the similarity between the multivariable case and the single-variable case:

$$f(x^* + \delta x) \approx f(x^*) + f'(x^*)\delta \boldsymbol{x} + \frac{1}{2}\delta x f''(x^*)\delta x \qquad (12.3.10)$$

Note that when we generalize from the single-variable case to the multivariable case, we have that:

$$x \leftarrow \boldsymbol{x} = \begin{bmatrix} x_1 \\ x_2 \\ \vdots \\ x_n \end{bmatrix}, \quad f' = \frac{df}{dx} \leftarrow \nabla f = \begin{bmatrix} \frac{\partial f}{\partial x_1} \\ \frac{\partial f}{\partial x_2} \\ \vdots \\ \frac{\partial f}{\partial x_n} \end{bmatrix}$$

$$f'' = \frac{d^2f}{dx^2} \leftarrow \mathbf{H} = \begin{bmatrix} \frac{\partial^2 f}{\partial x_1 \partial x_1} & \frac{\partial^2 f}{\partial x_1 \partial x_2} & \cdots & \frac{\partial^2 f}{\partial x_1 \partial x_n} \\ \frac{\partial^2 f}{\partial x_2 \partial x_1} & \frac{\partial^2 f}{\partial x_2 \partial x_2} & \cdots & \frac{\partial^2 f}{\partial x_2 \partial x_n} \\ \vdots & \vdots & \ddots & \vdots \\ \frac{\partial^2 f}{\partial x_n \partial x_1} & \frac{\partial^2 f}{\partial x_n \partial x_2} & \cdots & \frac{\partial^2 f}{\partial x_n \partial x_n} \end{bmatrix}$$

It comes as no surprise that the optimality conditions for the multivariable case are:

$$\begin{bmatrix} \frac{\partial f}{\partial x_1} \\ \frac{\partial f}{\partial x_2} \\ \vdots \\ \frac{\partial f}{\partial x_n} \end{bmatrix}_{x^*} = \begin{bmatrix} 0 \\ 0 \\ \vdots \\ 0 \end{bmatrix} \tag{12.3.11}$$

and

$$\delta \mathbf{x}^T \mathbf{H}(\mathbf{x}^*) \delta \mathbf{x}^T = [\,\delta x_1 \quad \cdots \quad \delta x_n\,] \begin{bmatrix} \frac{\partial^2 f}{\partial x_1 \partial x_1} & \frac{\partial^2 f}{\partial x_1 \partial x_2} & \cdots & \frac{\partial^2 f}{\partial x_1 \partial x_n} \\ \frac{\partial^2 f}{\partial x_2 \partial x_1} & \frac{\partial^2 f}{\partial x_2 \partial x_2} & \cdots & \frac{\partial^2 f}{\partial x_2 \partial x_n} \\ \vdots & \vdots & \ddots & \vdots \\ \frac{\partial^2 f}{\partial x_n \partial x_1} & \frac{\partial^2 f}{\partial x_n \partial x_2} & \cdots & \frac{\partial^2 f}{\partial x_n \partial x_n} \end{bmatrix}_{x^*} \begin{bmatrix} \delta x_1 \\ \delta x_2 \\ \vdots \\ \delta x_n \end{bmatrix} > 0 \tag{12.3.12}$$

or, more compactly

$$\nabla f(\mathbf{x}^*) = \mathbf{0} \tag{12.3.13}$$

$$\delta \mathbf{x}^T \mathbf{H}(\mathbf{x}^*) \delta \mathbf{x}^T > 0 \tag{12.3.14}$$

We conclude that the optimization problem given by Eq. (12.3.7) has been transformed to a problem of solving a square nonlinear (in the general case) set of algebraic equations (Eq. (12.3.13)). In the case where $\mathbf{x}$ is a single variable then conditions (12.3.13) and (12.3.14) simplify to conditions (12.3.2) and (12.3.5).

Let us consider the case of the three CSTRs in series considered in Section 12.2 and presented in Fig. 12.4. The overall reactor volume is given by Eq. (12.2.9):

$$V(C_{R,1}, C_{R,2}) = V_1 + V_2 + V_3$$
$$= \frac{F}{k}\left[\left(\frac{C_{R,0} - C_{R,1}}{C_{R,1}}\right) + \left(\frac{C_{R,1} - C_{R,2}}{C_{R,2}}\right) + \left(\frac{C_{R,2} - C_{R,f}}{C_{R,f}}\right)\right]$$

We calculate the two partial derivatives

$$\frac{\partial V(C_{R,1}, C_{R,2})}{\partial C_{R,1}} = \frac{F}{k}\frac{\partial}{\partial C_{R,1}}\left[\left(\frac{C_{R,0} - C_{R,1}}{C_{R,1}}\right) + \left(\frac{C_{R,1} - C_{R,2}}{C_{R,2}}\right) + \left(\frac{C_{R,2} - C_{R,f}}{C_{R,f}}\right)\right]$$
$$= \frac{F}{k}\left[\frac{\partial}{\partial C_{R,1}}\left(\frac{C_{R,0} - C_{R,1}}{C_{R,1}}\right) + \frac{\partial}{\partial C_{R,1}}\left(\frac{C_{R,1} - C_{R,2}}{C_{R,2}}\right)\right]$$
$$= \frac{F}{k}\left(-\frac{C_{R,0}}{C_{R,1}^2} + \frac{1}{C_{R,2}}\right)$$

$$\frac{\partial V(C_{R,1}, C_{R,2})}{\partial C_{R,2}} = \frac{F}{k}\frac{\partial}{\partial C_{R,2}}\left[\left(\frac{C_{R,0} - C_{R,1}}{C_{R,1}}\right) + \left(\frac{C_{R,1} - C_{R,2}}{C_{R,2}}\right) + \left(\frac{C_{R,2} - C_{R,f}}{C_{R,f}}\right)\right]$$
$$= \frac{F}{k}\left[\frac{\partial}{\partial C_{R,2}}\left(\frac{C_{R,1} - C_{R,2}}{C_{R,2}}\right) + \frac{\partial}{\partial C_{R,2}}\left(\frac{C_{R,2} - C_{R,f}}{C_{R,f}}\right)\right]$$
$$= \frac{F}{k}\left(-\frac{C_{R,1}}{C_{R,2}^2} + \frac{1}{C_{R,f}}\right)$$

By setting both partial derivatives equal to zero we obtain:

$$C_{R,1}^2 = C_{R,0}C_{R,2} \tag{12.3.15}$$

$$C_{R,2}^2 = C_{R,1}C_{R,f} \tag{12.3.16}$$

After some trivial algebraic manipulations, we obtain the following interesting result:

$$\frac{C_{R,0}}{C_{R,1}} = \frac{C_{R,1}}{C_{R,2}} = \frac{C_{R,2}}{C_{R,f}} = \left(\frac{C_{R,0}}{C_{R,f}}\right)^{1/3} \tag{12.3.17}$$

and

$$V_1 = V_2 = V_3 = \frac{F}{k}\left[\left(\frac{C_{R,0}}{C_{R,f}}\right)^{1/3} - 1\right] \tag{12.3.18}$$

These are in agreement with the results presented in Fig. 12.5, as

$$\left(\frac{C_{R,0}}{C_{R,f}}\right)^{1/3} = \left(\frac{10}{1}\right)^{1/3} = 2.1544$$

and

$$C_{R,2} = C_{R,f}\left(\frac{C_{R,0}}{C_{R,f}}\right)^{1/3} = 1 \times 2.1544 = 2.1544\ \frac{\text{kmol}}{\text{m}^3}$$

$$C_{R,1} = C_{R,2}\left(\frac{C_{R,0}}{C_{R,f}}\right)^{1/3} = C_{R,f}\left(\frac{C_{R,0}}{C_{R,f}}\right)^{2/3} = 4.6416\ \frac{\text{kmol}}{\text{m}^3}$$

$$V_1 = V_2 = V_3 = \frac{F}{k}\left[\left(\frac{C_{R,0}}{C_{R,f}}\right)^{1/3} - 1\right] = 1.154\ \text{m}^3$$

$$V = V_1 + V_2 + V_3 = 3.463\ \text{m}^3$$

Example 12.3. A three-stage air compression system with intercooling

In Fig. 12.8, a three-stage gas compression system with intercooling is shown. A gas with molar flowrate N and an initial pressure P_0 is compressed to a final pressure of P_f. The first compressor achieves a discharge pressure P_1 and the second compressor a discharge pressure P_2. The temperature at the suction of all compressors is T_0. The rate of compression work necessary is given by the following equation:

$$W(P_1, P_2) = N\frac{RT_0}{ea}\left\{\left[\left(\frac{P_1}{P_0}\right)^a - 1\right] + \left[\left(\frac{P_2}{P_1}\right)^a - 1\right] + \left[\left(\frac{P_f}{P_2}\right)^a - 1\right]\right\}$$

where the notation is explained in Example 12.1.

Calculate analytically the optimal intermediate pressures P_1 and P_2.

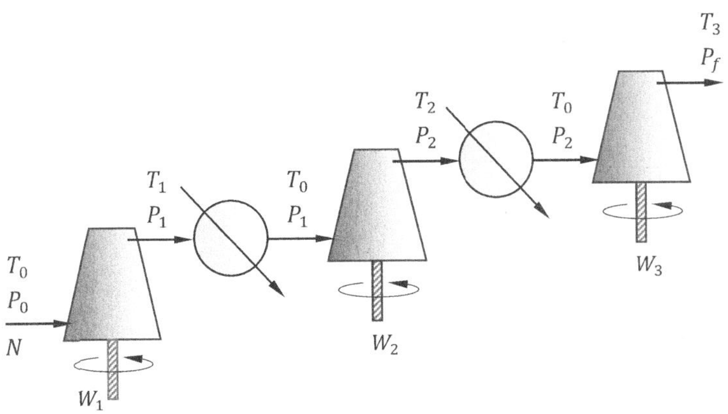

Figure 12.8 A three-stage compression system with intercooling.

Solution.

The objective function depends on two variables P_1 and P_2. We calculate the two partial derivatives:

$$\frac{\partial W(P_1,P_2)}{\partial P_1} = N\frac{RT_0}{ea}\frac{\partial}{\partial P_1}\left\{\left[\left(\frac{P_1}{P_0}\right)^a - 1\right] + \left[\left(\frac{P_2}{P_1}\right)^a - 1\right] + \left[\left(\frac{P_f}{P_2}\right)^a - 1\right]\right\}$$

$$= N\frac{RT_0}{ea}\left[\frac{\partial}{\partial P_1}\left(\frac{P_1}{P_0}\right)^a + \frac{\partial}{\partial P_1}\left(\frac{P_2}{P_1}\right)^a\right]$$

$$= N\frac{RT_0}{e}\left[\left(\frac{P_1}{P_0}\right)^a - \left(\frac{P_2}{P_1}\right)^a\right]\frac{1}{P_1}$$

$$\frac{\partial W(P_1,P_2)}{\partial P_2} = N\frac{RT_0}{ea}\frac{\partial}{\partial P_2}\left\{\left[\left(\frac{P_1}{P_0}\right)^a - 1\right] + \left[\left(\frac{P_2}{P_1}\right)^a - 1\right] + \left[\left(\frac{P_f}{P_2}\right)^a - 1\right]\right\}$$

$$= N\frac{RT_0}{ea}\left[\frac{\partial}{\partial P_1}\left(\frac{P_2}{P_1}\right)^a + \frac{\partial}{\partial P_2}\left(\frac{P_f}{P_2}\right)^a\right] = N\frac{RT_0}{e}\left[\left(\frac{P_2}{P_1}\right)^a - \left(\frac{P_f}{P_2}\right)^a\right]\frac{1}{P_2}$$

By setting the two partial derivatives equal to zero we obtain the following important result:

$$\lambda = \frac{P_1}{P_0} = \frac{P_2}{P_1} = \frac{P_f}{P_2} = \text{constant}$$

i.e., the compression ratio is constant and all compressors are identical. A direct generalization for n compressors in series with intercooling follows by the fact that

$$W(P_1, P_2, \ldots, P_{n-1}) = N\frac{RT_0}{ea}\sum_{k=1}^{n}\left[\left(\frac{P_k}{P_{k-1}}\right)^a - 1\right]$$

and

$$\frac{\partial W(P_1, P_2, \ldots, P_{n-1})}{\partial P_i} = N\frac{RT_0}{e}\left[\left(\frac{P_i}{P_{i-1}}\right)^a - \left(\frac{P_{i+1}}{P_i}\right)^a\right]\frac{1}{P_i}$$

from which it follows that

$$\frac{P_i}{P_{i-1}} = \frac{P_{i+1}}{P_i}, \quad \forall i = \{1, 2, .., n-1\}$$

Note that

$$\frac{P_f}{P_0} = \frac{P_1}{P_0}\cdot\frac{P_2}{P_1}\cdot\frac{P_3}{P_2}\cdots\frac{P_f}{P_{n-1}} = \lambda^n$$

or

$$\lambda = \frac{P_{i+1}}{P_i} = \left(\frac{P_f}{P_0}\right)^{1/n}$$

12.3.3 Numerical solution of optimization problems – Newton's method

To prove the most celebrated method for solving unconstrained, multivariable optimization problems:

$$\min_{\boldsymbol{x}} f(\boldsymbol{x}) \tag{12.3.19}$$

we consider the quadratic approximation of $f(\boldsymbol{x})$ using Taylor's approximation (see Eq. (12.3.9)):

$$f(\boldsymbol{x}+\delta\boldsymbol{x}) = f(\boldsymbol{x}) + \nabla f^T \delta\boldsymbol{x} + \frac{1}{2}\delta\boldsymbol{x}^T \mathbf{H}(\boldsymbol{x})\delta\boldsymbol{x} \tag{12.3.20}$$

To determine the step $\delta\boldsymbol{x}$ that will improve our estimation of the optimal solution we apply the optimality conditions given by Eq. (12.3.13) to the quadratic approximation:

$$\frac{df(\boldsymbol{x}+\delta\boldsymbol{x})}{d\delta\boldsymbol{x}} = \frac{d}{d\delta\boldsymbol{x}}\left(f(\boldsymbol{x}) + \nabla f^T \delta\boldsymbol{x} + \frac{1}{2}\delta\boldsymbol{x}^T \mathbf{H}(\boldsymbol{x})\delta\boldsymbol{x}\right) = \nabla f + \mathbf{H}(\boldsymbol{x})\delta\boldsymbol{x} = \mathbf{0}$$

where we have used the identities

$$\frac{\partial}{\partial \boldsymbol{x}}(\boldsymbol{c}^T\boldsymbol{x}) = \begin{bmatrix} \dfrac{\partial}{\partial x_1} \\ \dfrac{\partial}{\partial x_2} \end{bmatrix}\left(\begin{bmatrix} c_1 & c_2 \end{bmatrix}\begin{bmatrix} x_1 \\ x_2 \end{bmatrix}\right) = \begin{bmatrix} \dfrac{\partial(c_1x_1 + c_2x_2)}{\partial x_1} \\ \dfrac{\partial(c_1x_1 + c_2x_2)}{\partial x_2} \end{bmatrix} = \begin{bmatrix} c_1 \\ c_2 \end{bmatrix} = \boldsymbol{c}$$

$$\begin{aligned} \frac{\partial}{\partial \boldsymbol{x}}(\boldsymbol{x}^T\mathbf{A}\boldsymbol{x}) &= \begin{bmatrix} \dfrac{\partial}{\partial x_1} \\ \dfrac{\partial}{\partial x_2} \end{bmatrix}\left(\begin{bmatrix} x_1 & x_2 \end{bmatrix}\begin{bmatrix} a_{11} & a_{12} \\ a_{21} & a_{22} \end{bmatrix}\begin{bmatrix} x_1 \\ x_2 \end{bmatrix}\right) \\ &= \begin{bmatrix} \dfrac{\partial\left(a_{11}x_1^2 + a_{12}x_1x_2 + a_{21}x_2x_1 + a_{22}x_2^2\right)}{\partial x_1} \\ \dfrac{\partial\left(a_{11}x_1^2 + a_{12}x_1x_2 + a_{21}x_2x_1 + a_{22}x_2^2\right)}{\partial x_2} \end{bmatrix} \\ &= \begin{bmatrix} (a_{11}x_1 + a_{12}x_2) + (a_{11}x_1 + a_{21}x_2) \\ (a_{21}x_1 + a_{22}x_2) + (a_{12}x_1 + a_{22}x_2) \end{bmatrix} = \mathbf{A}\boldsymbol{x} + \mathbf{A}^T\boldsymbol{x} = 2\mathbf{A}\boldsymbol{x} \end{aligned}$$

where a symmetric matrix has been considered, i.e., $\mathbf{A} = \mathbf{A}^T$, as the Hessian matrix is a symmetric matrix. It then follows that the step δx can be calculated as the solution of the system of linear equations

$$\mathbf{H}(\boldsymbol{x})\delta\boldsymbol{x} = -\nabla f \tag{12.3.21}$$

The steps of Newton's method for solving numerically an unconstrained, multivariable optimization problem is based on Eq. (12.3.21) and has the following steps.

STEP 0: Select an initial point $\boldsymbol{x}^0$. Set $k = 0$.

STEP 1: If $\|\nabla f(\boldsymbol{x}^k)\| < \varepsilon$ then stop. $\boldsymbol{x}^k$ is a solution.

STEP 2: Solve $\mathbf{H}(\boldsymbol{x}^k)\delta\boldsymbol{x}^k = -\nabla f(\boldsymbol{x}^k)$.

STEP 3: Calculate the new estimate of the solution $\boldsymbol{x}^{k+1} = \boldsymbol{x}^k + a\delta\boldsymbol{x}^k$, with appropriate selection of a. Set $k = k + 1$ and return to STEP 1.

Let us solve a simple numerical example to get some experience of the intermediate steps. Consider the following function

$$f(x_1, x_2) = x_1^3 + x_2^3 + 2x_1^2 + 4x_2^2 + 6$$

The gradient and the Hessian matrix are

$$\nabla f(x_1, x_2) = \begin{bmatrix} \dfrac{\partial f}{\partial x_1} \\ \dfrac{\partial f}{\partial x_2} \end{bmatrix} = \begin{bmatrix} 3x_1^2 + 4x_1 \\ 3x_2^2 + 8x_2 \end{bmatrix}$$

$$\mathbf{H}(x_1, x_2) = \begin{bmatrix} \dfrac{\partial}{\partial x_1}\left(\dfrac{\partial f}{\partial x_1}\right) & \dfrac{\partial}{\partial x_1}\left(\dfrac{\partial f}{\partial x_2}\right) \\ \dfrac{\partial}{\partial x_2}\left(\dfrac{\partial f}{\partial x_1}\right) & \dfrac{\partial}{\partial x_2}\left(\dfrac{\partial f}{\partial x_2}\right) \end{bmatrix} = \begin{bmatrix} 6x_1 + 4 & 0 \\ 0 & 6x_2 + 8 \end{bmatrix}$$

Select $\boldsymbol{x}^0 = \begin{bmatrix} 1 \\ 0 \end{bmatrix}$, then $f(\boldsymbol{x}^0) = 9$

$$\nabla f(\boldsymbol{x}^0) = \begin{bmatrix} 7 \\ 0 \end{bmatrix}, \mathbf{H}(\boldsymbol{x}^0) = \begin{bmatrix} 10 & 0 \\ 0 & 8 \end{bmatrix}$$

We then solve the system of linear equations:

$$\begin{bmatrix} 10 & 0 \\ 0 & 8 \end{bmatrix} \begin{bmatrix} \delta x_1 \\ \delta x_2 \end{bmatrix}^0 = -\begin{bmatrix} 7 \\ 0 \end{bmatrix} \Rightarrow \begin{bmatrix} \delta x_1 \\ \delta x_2 \end{bmatrix}^0 = -\begin{bmatrix} 0.7 \\ 0 \end{bmatrix}$$

We find the new estimate of the solution

$$\boldsymbol{x}^1 = \boldsymbol{x}^0 + \delta\boldsymbol{x}^0 = \begin{bmatrix} 1 \\ 0 \end{bmatrix} - \begin{bmatrix} 0.7 \\ 0 \end{bmatrix} = \begin{bmatrix} 0.3 \\ 0 \end{bmatrix}$$

then $f(\boldsymbol{x}^1) = 0.3^3 + 2 \times 0.3^2 + 6 = 6.207.$

$$\nabla f(\boldsymbol{x}^1) = \begin{bmatrix} 1.47 \\ 0 \end{bmatrix}, \mathbf{H}(\boldsymbol{x}^1) = \begin{bmatrix} 5.8 & 0 \\ 0 & 8 \end{bmatrix}$$

We then solve the system of linear equations:

$$\begin{bmatrix} 5.8 & 0 \\ 0 & 8 \end{bmatrix} \begin{bmatrix} \delta x_1 \\ \delta x_2 \end{bmatrix}^1 = -\begin{bmatrix} 1.47 \\ 0 \end{bmatrix} \Rightarrow \begin{bmatrix} \delta x_1 \\ \delta x_2 \end{bmatrix}^1 = -\begin{bmatrix} 0.2534 \\ 0 \end{bmatrix}$$

We find the new estimate of the solution

$$\boldsymbol{x}^2 = \boldsymbol{x}^1 + \delta\boldsymbol{x}^1 = \begin{bmatrix} 0.3 \\ 0 \end{bmatrix} - \begin{bmatrix} 0.25345 \\ 0 \end{bmatrix} = \begin{bmatrix} 0.0466 \\ 0 \end{bmatrix}$$

then $f(\boldsymbol{x}^2) = 6.0044$. It takes a few more iterations to discover the solution, which is $\boldsymbol{x} = \mathbf{0}$.

MATLAB offers the command **fminunc** that performs unconstrained, multivariable optimization.

```
x = fminunc(fun,x0,OPTIONS)
```

fminunc starts at **x0** and attempts to find a local minimizer **x** of the function **fun**. **fun** accepts input **x** and returns a scalar function value **f** evaluated at **x** (**f = fun(x)**). For very simple functions **fminunc** can be used as follows

```
>> clear all
>> f=@(x) x(1)^3+x(2)^3+2*x(1)^2+4*x(2)^2+6

f =
  function_handle with value:
    @(x)x(1)^3+x(2)^3+2*x(1)^2+4*x(2)^2+6

>> x = fminunc(f,[1;0],optimset('display','iter'))
                                                        First-order
 Iteration  Func-count       f(x)        Step-size       optimality
     0           3              9                             7
     1           6              6         0.142857         5.96e-08

Local minimum found.

Optimization completed because the size of the gradient is
less than
the value of the optimality tolerance.

x =
     0
     0
```

When the function is more complex, then an m-file can be used, as is shown below for the case of the three CSTRs in series.

```
function V = ThreeCSTRs(C)
F   = 1; % feed volumetric flowrate      [=] m^3/min
k   = 1; % reaction rate constant        [=] 1/min
CR0 =10; % reactant feed concentration   [=] kmol/m^3
CR3 = 1; % reactant final concentration  [=] kmol/m^3
% calculate the total volume of the 3 CSTRs
V= (F/k)*( (CR0   - C(1) )/C(1) + ...
           (C(1) - C(2) )/C(2) + ...
           (C(2) - CR3  )/CR3  );
return
```

```
>> x = fminunc(@ThreeCSTRs,[5;3],optimset('display','iter'))
                                                        First-order
 Iteration  Func-count       f(x)        Step-size       optimality
     0           3         3.66667                         0.444
     1           6         3.51185             1           0.224
     2           9         3.48349             1           0.152
     3          12         3.46958             1          0.0344
     4          15         3.46651             1           0.026
     5          18         3.46334             1         0.00309
     6          21          3.4633             1        0.000254
     7          24          3.4633             1        3.64e-06
     8          27          3.4633             1         1.8e-07

Local minimum found.
Optimization completed because the size of the gradient is
less than
the value of the optimality tolerance.

x =
    4.6416
    2.1544
```

12.4 Optimality conditions for equality-constrained problems

12.4.1 Motivating example

We consider the case of producing a product P through the reaction $R \rightarrow P$, which is first order with respect to the reactant concentration $\dot{R}''' = kC_R$, with reaction rate constant $k = 0.10\ \text{s}^{-1}$. Our aim is to produce $\Pi = 0.05$ kmol/s of product P using a fresh feed of unknown volumetric flowrate F (m^3/s) with reactant concentration $C_{R,0} = 1\ \text{kmol}/\text{m}^3$ (see Fig. 12.9). The cost of reactant is $\alpha = 810\ \$/\text{kmol A}$ while the cost of the CSTR reactor is $\beta = 1\ \$/(\text{m}^3\ \text{s})$. What is the minimum-cost solution in terms of

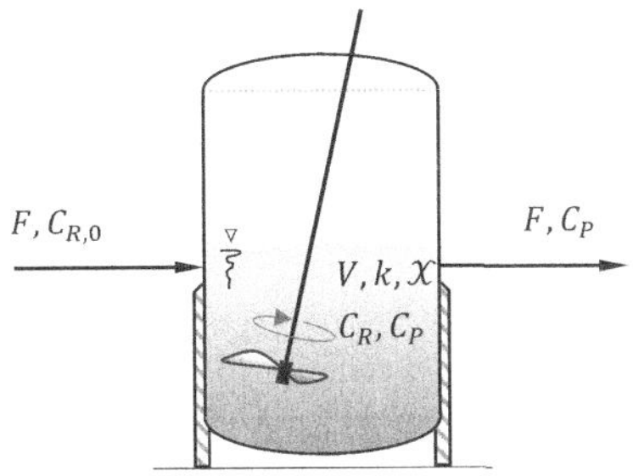

Figure 12.9 A CSTR with first-order reaction.

fresh feed volumetric flowrate (F), CSTR reactor volume (V), and reactant conversion (X)?

The first equation is the mole balance of the reactant in the CSTR

$$FC_{R,0}X - VkC_{R,0}(1 - X) = 0 \qquad (12.4.1)$$

where the definition of conversion has been used $X = \left(C_{R,0} - C_R\right)/C_{R,0}$ There is a specification on the amount of product P produced (which, due to the stoichiometry of the reaction, equals the amount of reactant consumed)

$$FC_{R,0}X = \Pi \qquad (12.4.2)$$

The operating cost consists of the cost of reactant in the feed and the cost of the CSTR:

$$\text{Cost} = \alpha FC_{R,0} + \beta V \qquad (12.4.3)$$

The final problem can be summarized as follows:

$$\min_{X,F,V} f = \alpha FC_{R,0} + \beta V \quad \text{s.t.} \qquad (12.4.4)$$

$$\begin{aligned} h_1 &= FC_{R,0}X - VkC_{R,0}(1 - X) = 0 \\ h_2 &= FC_{R,0}X - \Pi = 0 \end{aligned}$$

This is a problem of constrained optimization as the objective function (f) is augmented by two equality constraints (h_1 and h_2).

One method that can be used to solve equality-constrained problems with n variables and m equalities (with $n > m$), is to select $n - m$ variables to which we can assign values and then use the m equalities to calculate the remaining m variables. Finally, with all n variables known the objective function can be evaluated. In the cases where $n - m = 1$ or 2 we can visualize the results and locate the optimal solution.

For the case study considered above we have $n = 3$ and more specifically

$$\boldsymbol{x} = \begin{bmatrix} x_1 \\ x_2 \\ x_3 \end{bmatrix} = \begin{bmatrix} X \\ F \\ V \end{bmatrix}$$

We also have two equalities and $m = 2$. We therefore have that $n - m = 1$ and we may select one variable to assign values and then use two equalities to calculate the other two. We may select, for example, to assign values to the conversion X and then solve Eq. (12.4.2) to calculate F:

$$F = \frac{\Pi}{C_{R,0}X}$$

and Eq. (12.4.1) to calculate the reactor volume:

$$V = \frac{\Pi}{kC_{R,0}(1 - X)}$$

We finally calculate the cost from Eq. (12.4.3) and visualize the result to locate the optimal solution. This can easily be implemented in MATLAB.

```
clear all
k    = 0.10; % reaction rate constant 1/s
PI   = 0.05; % production rate in kmol/s
C0   = 1.00; % reactant concentration kmol/m^3
alfa =  810; % $/kmol A
beta =    1; % $ / (m^3 s)

X = linspace(0.001,0.999,999);
F = PI./(C0*X);
V = PI./(k*C0*(1-X));
Cost = alfa*F*C0 + beta*V;
plot(X,Cost,'linewidth',1.2,'color',[0.00,0.45,0.740])
grid on
xlabel('Fraction conversion')
ylabel('Operating cost ($/s)')
axis([0 1 40 140])
```

The results are shown in Fig. 12.10.

We could have also used Eq. (12.4.1) and (12.4.2) to eliminate F and V from the objective function (Eq. (12.4.4)) to obtain:

$$\text{Cost} = \alpha\Pi\frac{1}{X} + \frac{\beta\Pi}{kC_{R,0}}\frac{1}{(1 - X)} \tag{12.4.5}$$

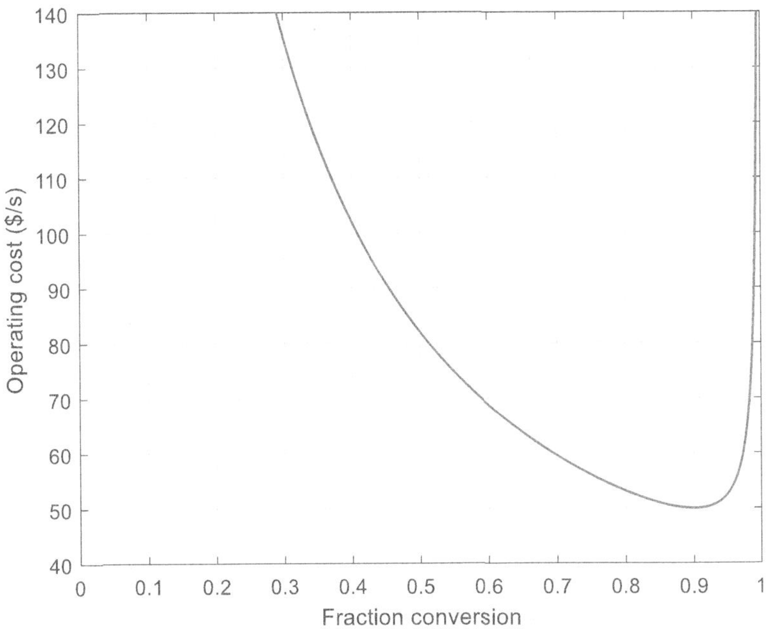

Figure 12.10 The operating cost as a function of fraction conversion for the CSTR case study with specified rate of production.

which is a real-valued function of a real variable (a single-variable unconstrained optimization problem). We then calculate the derivative

$$\frac{d\text{Cost}}{dX} = -\alpha\Pi\frac{1}{X^2} + \frac{\beta\Pi}{kC_{R,0}}\frac{1}{(1-X)^2}$$

and then set the derivative equal to zero and finally solve for X to obtain

$$X = \frac{\sqrt{\frac{a}{b}kC_{R,0}}}{1+\sqrt{\frac{a}{b}kC_{R,0}}} \tag{12.4.6}$$

We can replace the numerical values to validate the optimum shown in Fig. 12.10:

$$X = \frac{\sqrt{81}}{1+\sqrt{81}} = 0.9$$

and also $V = 5\,\text{m}^3, F = 1/18\,\frac{\text{m}^3}{\text{s}}$.

12.4.2 The Lagrange multiplier method for equality-constrained problems

The Lagrange multiplier method for equality-constrained optimization problems is a method for solving problems of the general form

$$\begin{aligned} &\min_{x} f(\boldsymbol{x}) \\ &\text{s.t.} \\ &\boldsymbol{h}(\boldsymbol{x}) = 0 \end{aligned} \tag{12.4.7}$$

where $\boldsymbol{x} \in R^n, f : R^n \to R, \boldsymbol{h} : R^n \to R^m$. Lagrange proposed to transform the problem to an unconstrained optimization problem using Lagrange multipliers $\lambda_j, j = 1, 2, \ldots, m$:

$$\mathcal{L}(\boldsymbol{x}, \boldsymbol{\lambda}) = f(\boldsymbol{x}) + \sum_{j=1}^{m} \lambda_j h_j(\boldsymbol{x}) \tag{12.4.8}$$

This is an unconstrained problem with $n+m$ variables $(x_1, x_2, \ldots, x_n, \lambda_1, \lambda_2, \ldots, \lambda_m)$ and we apply the optimality conditions to obtain:

$$\frac{\partial \mathcal{L}(\boldsymbol{x}, \boldsymbol{\lambda})}{\partial x_i} = \frac{\partial f(\boldsymbol{x})}{\partial x_i} + \sum_{i=1}^{m} \lambda_i \frac{\partial h_i(\boldsymbol{x})}{\partial x_i} = 0, \quad i = 1, 2, \ldots, n \tag{12.4.9}$$

$$\frac{\partial \mathcal{L}(\boldsymbol{x}, \boldsymbol{\lambda})}{\partial \lambda_j} = h_j(\boldsymbol{x}) = 0, \quad j = 1, 2, \ldots, m \tag{12.4.10}$$

For example, consider the problem given by (12.4.4). The Lagrange function is given by

$$\mathcal{L}(X, F, V, \lambda_1, \lambda_2) = \underbrace{\alpha F C_{R,0} + \beta V}_{f(x)} + \underbrace{\lambda_1 (F C_{R,0} X - V k C_{R,0}(1 - X))}_{h_1} + \underbrace{\lambda_2 (F C_{R,0} X - \Pi)}_{h_2}$$

The first order optimality conditions are:

$$\frac{\partial \mathcal{L}(X, F, V, \lambda_1, \lambda_2)}{\partial X} = \lambda_1 (F C_{R,0} + V k C_{R,0}) + \lambda_2 F C_{R,0} = 0$$

$$\frac{\partial \mathcal{L}(X, F, V, \lambda_1, \lambda_2)}{\partial F} = \alpha C_{R,0} + \lambda_1 C_{R,0} X + \lambda_2 C_{R,0} X = 0$$

$$\frac{\partial \mathcal{L}(X, F, V, \lambda_1, \lambda_2)}{\partial V} = \beta - \lambda_1 k C_{R,0}(1 - X) = 0$$

$$\frac{\partial \mathscr{L}(X,F,V,\lambda_1,\lambda_2)}{\partial \lambda_1} = FC_{R,0}X - VkC_{R,0}(1-X) = 0$$

$$\frac{\partial \mathscr{L}(X,F,V,\lambda_1,\lambda_2)}{\partial \lambda_2} = FC_{R,0}X - \Pi = 0$$

This needs to be solved as a system of nonlinear algebraic equations. Constrained problems are solved using a family of numerical algorithms that have become known as successive quadratic programming techniques. As they are outside the scope of this introductory text, the reader is referred to the references at the end of this chapter for further study.

Example 12.4. An application of the Lagrange multiplier method

A rectangular cuboid storage tank is to be constructed as shown in Fig. 12.11. The tank will be used to hold a raw material and its overall volume must be equal to V. The raw material will be kept refrigerated and, in order to minimize losses to the environment, it must have the minimum external surface S. The dimensions of the rectangular cuboid tank will be x, y, and z, as shown in Fig. 12.11. Find what will be the sizes of x, y, and z so as to minimize the external surface for a given volume.

Solution.

The volume of the tank is

$$V = xyz$$

while the external area is

$$S = 2(xy + xz + yz)$$

The optimization problem is

$$\min_{x,y,z} f(x,y,z) = 2(xy + xz + yz)$$

$$h(x,y,z) = xyz - V = 0$$

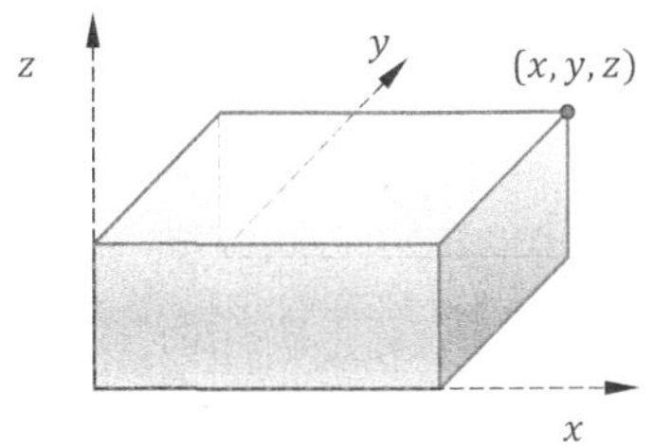

Figure 12.11 A rectangular cuboid storage tank.

We introduce the Lagrange multiplier λ and we form the Lagrange function

$$\mathscr{L}(x,y,z,\lambda) = 2(xy + xz + yz) + \lambda(xyz - V)$$

We apply the first order conditions

$$\frac{\partial \mathscr{L}(x,y,z,\lambda)}{\partial x} = 2(y+z) + \lambda yz = 0$$

$$\frac{\partial \mathscr{L}(x,y,z,\lambda)}{\partial y} = 2(x+z) + \lambda xz = 0$$

$$\frac{\partial \mathscr{L}(x,y,z,\lambda)}{\partial z} = 2(x+y) + \lambda xy = 0$$

$$\frac{\partial \mathscr{L}(x,y,z,\lambda)}{\partial \lambda} = xyz - V = 0$$

We multiply the first condition by x, the second by $-y$ and we add them to obtain

$$2x(y+z) + \lambda xyz - 2y(x+z) - \lambda xyz = 0$$

or:

$$(x-y)z = 0$$

or $x = y$. In the same way we can prove that $y = z$ and finally $x = y = z = \sqrt[3]{V}$, the optimal solution corresponds to a cube!

12.4.3 Solving optimization problems with constraints in MATLAB

We will show how Problem (12.4.4), which is an equality-constrained optimization problem, can be solved with MATLAB. The command for solving nonlinearly (equality and inequality) constrained optimization problems in MATLAB is fmincon.

```
x = fmincon(objfun,x0,A,b,Aeq,beq,lb,ub,nonlincons,OPTIONS)
```

fmincon starts at **x0** and attempts to find a local minimizer **x** of the function **objfun** subject to several linear and/or nonlinear inequality and/or equality constraints. obj**fun** accepts input **x** and returns a scalar function value **f** evaluated at **x** (**f = objfun (x)**). The linear inequality and equality constraints are of the form

$$\boldsymbol{Ax} \leq \boldsymbol{b}$$

$$\boldsymbol{A}_{\text{eq}}\boldsymbol{x} = \boldsymbol{b}_{\text{eq}}$$

In addition to that, linear lower and upper bounds on the optimization variables can be given explicitly

$$\boldsymbol{lb} \leq \boldsymbol{x} \leq \boldsymbol{ub}$$

The **nonlincons** is a function that returns a vector with the nonlinear inequalities followed by the nonlinear equalities for any given vector $\boldsymbol{x}$: **[g, h] = nonlincons(x)**.

Let us use **fmincon** to solve the problem given by (12.4.4). To achieve that, we first need to create the m-files for the **objfun** and **nonlincons**.

```
function f =objfun(x)
CR0   = 1.00; % reactant concentration kmol/m^3
alfa =  810; % $/kmol A
beta =    1; % $ / (m^3 s)
f     = alfa*x(2)*CR0+beta*x(3);
end
```

```
function [g,h] =nonlincons(x)
k     = 0.10; % reaction rate constant 1/s
PI    = 0.05; % production rate in kmol/s
CR0   = 1.00; % reactant concentration kmol/m^3
alfa =  810; % $/kmol A
beta =    1; % $ / (m^3 s)
g     = [];   % there are no inequality constraints
h(1) = x(2)*CR0*x(1) - x(3)*k*CR0*(1-x(1));
h(2) = x(2)*CR0*x(1) - PI;
end
```

With these functions available we can use **fmincon** to solve the problem (observe that there are no linear equality or inequality constraints).

```
clear all
x0 = [0.5;0.1;10]
lb = zeros(size(x0))
ub = [1;1;10];
options = optimset('Display','iter')
[xsol,fsol,flag]=fmincon(@objfun,x0,[],[],[],[],lb,ub,@nonlinc
ons,options)
```

```
>> chapter_12_demo_fmincon
                                                First-order      Norm of
 Iter F-count            f(x)  Feasibility    optimality         step
    0       4    9.001000e+01   4.005e-01     3.902e+00
    1       8    3.589727e+01   1.329e-01     5.465e+02    1.044e+00
    2      12    5.480717e+01   2.561e-03     2.244e+02    1.133e-01
    3      16    5.249675e+01   1.073e-05     1.117e+02    1.543e-01
    4      20    5.122513e+01   1.419e-03     8.348e+01    1.656e+00
    5      24    5.037621e+01   1.659e-03     3.397e+01    1.398e+00
    6      28    4.922879e+01   7.252e-03     4.243e+00    2.517e+00
    7      32    4.998303e+01   1.831e-04     4.683e+00    8.216e-01
    8      36    5.000066e+01   9.012e-07     1.519e+00    1.447e-02
    9      40    4.999943e+01   5.243e-06     4.164e-01    5.209e-02
   10      44    4.999999e+01   1.222e-07     4.395e-03    7.620e-03
   11      48    5.000000e+01   2.673e-11     1.429e-04    1.226e-04

Local minimum found that satisfies the constraints.

xsol =
      0.9000
      0.0556
      5.0000

fsol =
    50.0000

flag =
      1
```

12.4.4 Solving optimization problems with constraints in Python

We can perform nonlinearly constrained optimization in Python using **scipy.optimize.minimize**—one of the most powerful and flexible tools in SciPy. The general structure of the **minimize** command is.

```
sol = minimize(fun, x0, method='SLSQP', constraints=cons,
               bounds=bnds)
```

Arguments:

fun: the objective function to minimize

x0: initial guess

method: optimization algorithm (for nonlinear constraints, typically **"SLSQP"** or **"trust-constr"**)

constraints: a list of constraint dictionaries

bounds: optional variable bounds

The following Python program solves the optimization problem given by (12.4.4).

```
import numpy as np
from scipy.optimize import minimize

k    = 0.10 # reaction rate constant 1/s
PI   = 0.05 # production rate in kmol/s
CR0  = 1.00 # reactant concentration kmol/m^3
alfa =  810 # $/kmol A
beta =    1 # $ / (m^3 s)

def objfun(x):
    X, F, V = x
    f    = alfa*F*CR0+beta*V # calculate the cost in $
    return f

nonlincons =[ {'type':'eq',  'fun': \
               lambda x: x[1]*CR0*x[0]-x[2]*k*CR0*(1-x[0]) },
               # : F C_R0 X - V k C_R0 (1-X) = 0
              {'type':'ineq','fun': \
               lambda x: x[1]*CR0*x[0]-PI                  }]
               # : 0 ≤ F C_R0 X - Π
# Print callback function
def print_progress(x):
    global Nfeval
    print('Iteration  {0:3d},  X=  {1:3.4f},  F=  {2:3.4f},  V=
{3:3.4f}, f= {4:3.4f}'\
          .format(Nfeval,x[0],x[1],x[2],objfun(x)) )
    Nfeval += 1
Nfeval = 0
# ----------------------

x0 = np.array([0.5,0.1,5])
bnds = ((1E-6, 1.0), # bounds on X: 1e-6 ≤ X ≤ 1
        (1E-6, 1.0), # bounds on F: 1e-6 ≤ F ≤ 1
        (1E-6, 10.0))# bounds on V: 1e-6 ≤ V ≤ 10

print_progress(x0)
sol = minimize(objfun,
               x0,
               method='SLSQP',
               constraints=nonlincons,
               bounds=bnds,
               callback=print_progress
               )

print('Optimization status = ',sol.message)
print(f"Optimal conversion     X = {sol.x[0]:4.4f}")
print(f"Optimal vol. flowrate  F = {sol.x[1]:4.4f} m\u00b3/s")
print(f"Optimal reactor volume V = {sol.x[2]:4.4f} m\u00b3")
print(f"Minimum cost achieved  C = {sol.fun:4.3f} $/s")
```

```
[in] runfile('K:/....')

Iteration  0, X= 0.5000, F= 0.1000, V= 5.0000, f= 86.0000
Iteration  1, X= 1.0000, F= 0.0000, V= 5.9999, f= 6.0008
Iteration  2, X= 0.9167, F= 0.0500, V= 4.7770, f= 45.2773
Iteration  3, X= 0.8713, F= 0.0547, V= 1.0805, f= 45.3480
Iteration  4, X= 0.9013, F= 0.0553, V= 5.2606, f= 50.0902
Iteration  5, X= 0.9007, F= 0.0555, V= 5.0327, f= 49.9986
Iteration  6, X= 0.9004, F= 0.0555, V= 5.0186, f= 50.0000
Iteration  7, X= 0.9000, F= 0.0556, V= 5.0001, f= 49.9999
Iteration  8, X= 0.9000, F= 0.0556, V= 5.0000, f= 50.0000
Iteration  9, X= 0.9000, F= 0.0556, V= 5.0000, f= 50.0000

Optimization status =  Optimization terminated successfully

Optimal conversion     X = 0.9000
Optimal vol. flowrate  F = 0.0556 m³/s
Optimal reactor volume V = 5.0000 m³

Minimum cost achieved  C = 50.000 $/s
```

One characteristic that needs further elaboration is the use of the callback function **print_progress**. In SciPy's optimization routines, the function **callback(x)** is called automatically after each iteration, where **x** represents the current parameter vector as a NumPy array. Depending on the optimization method used, the callback function may also receive additional information about the current iteration. This feature allows users to monitor and control the optimization process in real time—for example, by printing or logging the current iteration, tracking the objective function values, saving intermediate results for plotting convergence, or even stopping the optimization early if a specific condition is met. It should also be noted that we have replaced the second equality in (12.4.4) with an inequality that forces the productivity to be greater than 0.05 kmol/s, as the result is the same. This is done in order to show how nonlinear inequality constraint can be used in **scipy.optimize.minimize**.

12.5 Parameter estimation using least squares

Arguably, the most important application of multivariable unconstrained optimization is that of parameter estimation, which is usually achieved using a least-squares objective function. To introduce the method we consider the data given in Table 12.2 for the saturation or vapor pressure of water at different temperatures. A model for calculating the saturation pressure that we have used in this book is that of Antoine's equation:

Table 12.2 Data for the saturation pressure of water at different temperatures.

i	T (°C)	P^{sat} (kPa)
1	0	0.6113
2	5	0.8726
3	10	1.2281
4	15	1.7056
5	20	2.3388
6	25	3.1690
7	30	4.2455
8	35	5.6267
9	40	7.3814
10	45	9.5898
11	50	12.3440
12	55	15.7520
13	60	19.9320
14	65	25.0220
15	70	31.1760
16	75	38.5630
17	80	47.3730
18	85	57.8150
19	90	70.1170
20	95	84.5290
$21 = N$	100	101.3200

$$\ln P^{sat} = A - \frac{B}{C + T} \tag{12.5.1}$$

where P^{sat} is the vapor pressure of a substance, T is the absolute temperature of the substance, and A, B, and C are substance-specific coefficients. A simpler form of the equation with only two coefficients is sometimes used:

$$\ln P^{sat} = A - \frac{B}{T} \tag{12.5.2}$$

The question that we answer in this section is the following: given experimental observations and a corresponding mathematical model, how can we accurately estimate the model's parameters? The importance of this question cannot be overstated.

Let us first consider the simple model given by Eq. (12.5.2). We will use the symbol $\mathscr{Y}^{\text{exp}}$ to denote the measured variable and the symbol $\mathscr{Y}^{\text{model}}$ to denote the model predictions. The deviation between the model prediction and the experimentally measured variable is called the **error**, **deviation**, or **residual**, and the symbol e will be used:

$$e(\mathscr{X}_i) = \mathscr{Y}^{\text{exp}}(\mathscr{X}_i) - \mathscr{Y}^{\text{model}}(\mathscr{X}_i) \tag{12.5.3}$$

where $\mathscr{X}$ denotes the independent variable in the experimental data (which is the temperature in the vapor pressure data of Table 12.2) and i denotes a specific experiment ($i = 1,2, \ldots, N$). For the case of the experimental data of Table 12.2 and the mathematical model given by Eq. (12.5.2) we have $\mathscr{Y}^{\text{exp}} = \ln P^{\text{sat}}$ and $\mathscr{X} = 1/T$ and

$$e(\mathscr{X}_i) = \underbrace{\ln P^{\text{sat}}(T_i)}_{\mathscr{Y}^{\text{exp}}(\mathscr{X}_i)} - \underbrace{\left(A - B\frac{1}{T_i}\right)}_{\mathscr{Y}^{\text{model}}(\mathscr{X}_i)} \tag{12.5.4}$$

or

$$e(\mathscr{X}_i) = \ln P^{\text{sat}}(T_i) - \begin{bmatrix} 1 & -\frac{1}{T_i} \end{bmatrix} \begin{bmatrix} A \\ B \end{bmatrix} \tag{12.5.5}$$

Eq. (12.5.5) is written for all 21 experimental points of Table 12.2 in matrix format:

$$\begin{bmatrix} e(\mathscr{X}_1) \\ e(\mathscr{X}_2) \\ \vdots \\ e(\mathscr{X}_i) \\ \vdots \\ e(\mathscr{X}_{21}) \end{bmatrix} = \begin{bmatrix} \ln P^{\text{sat}}(T_1) \\ \ln P^{\text{sat}}(T_2) \\ \vdots \\ \ln P^{\text{sat}}(T_i) \\ \vdots \\ \ln P^{\text{sat}}(T_{21}) \end{bmatrix} - \begin{bmatrix} 1 & -1/T_1 \\ 1 & -1/T_2 \\ \vdots & \vdots \\ 1 & -1/T_i \\ \vdots & \vdots \\ 1 & -1/T_{21} \end{bmatrix} \begin{bmatrix} A \\ B \end{bmatrix} \tag{12.5.6}$$

To make the last equation more compact we define the following general notation and its application in this case study

$$\boldsymbol{e} = \begin{bmatrix} e(\mathscr{X}_1;\boldsymbol{\theta}) \\ e(\mathscr{X}_2;\boldsymbol{\theta}) \\ \vdots \\ e(\mathscr{X}_i;\boldsymbol{\theta}) \\ \vdots \\ e(\mathscr{X}_N;\boldsymbol{\theta}) \end{bmatrix} \tag{12.5.7}$$

$$
\mathcal{Y}^{\exp} = \begin{bmatrix} \mathscr{Y}_1^{\exp} \\ \mathscr{Y}_2^{\exp} \\ \vdots \\ \mathscr{Y}_i^{\exp} \\ \vdots \\ \mathscr{Y}_N^{\exp} \end{bmatrix} = \begin{bmatrix} \ln P^{\text{sat}}(T_1) \\ \ln P^{\text{sat}}(T_2) \\ \vdots \\ \ln P^{\text{sat}}(T_i) \\ \vdots \\ \ln P^{\text{sat}}(T_N) \end{bmatrix}, \tag{12.5.8}
$$

$$
\mathcal{X} = \begin{bmatrix} \mathscr{X}_1 \\ \mathscr{X}_2 \\ \vdots \\ \mathscr{X}_i \\ \vdots \\ \mathscr{X}_N \end{bmatrix} = \begin{bmatrix} x_{1,1} & x_{1,2} & \cdots & x_{1,K} \\ x_{2,1} & x_{2,2} & \cdots & x_{2,K} \\ \vdots & \vdots & \ddots & \vdots \\ x_{i,1} & x_{i,2} & \cdots & x_{i,K} \\ \vdots & \vdots & \ddots & \vdots \\ x_{N,1} & x_{N,2} & \cdots & x_{N,K} \end{bmatrix} = \begin{bmatrix} 1 & -1/T_1 \\ 1 & -1/T_2 \\ \vdots & \vdots \\ 1 & -1/T_i \\ \vdots & \vdots \\ 1 & -1/T_{21} \end{bmatrix} \tag{12.5.9}
$$

$$
\boldsymbol{\theta} = \begin{bmatrix} \theta_1 \\ \theta_2 \\ \vdots \\ \theta_K \end{bmatrix} = \begin{bmatrix} A \\ B \end{bmatrix} \tag{12.5.10}
$$

where $\boldsymbol{\theta}$ is the vector of the K unknown parameters. Take time to familiarize yourself with this notation—it is essential for understanding some of the more challenging material ahead. Eq. (12.5.6) can be written as

$$
\boldsymbol{e} = \mathcal{Y}^{\exp} - \mathcal{X}\boldsymbol{\theta} \tag{12.5.11}
$$

To estimate the model parameters, we need an appropriate objective function. The most used criterion in parameter estimation problems is that of the sum of squares of errors:

$$
f(\boldsymbol{\theta}) = \frac{1}{2}\sum_{i=1}^{N} e(\mathscr{X}_i)^2 = \frac{1}{2}\left(e(\mathscr{X}_1;\boldsymbol{\theta})^2 + e(\mathscr{X}_2;\boldsymbol{\theta})^2 + \ldots + e(\mathscr{X}_N;\boldsymbol{\theta})^2\right) \tag{12.5.12}
$$

To apply the first order optimality conditions for multivariable, unconstrained problems we first determine all partial derivatives with respect to the model parameters $\theta_k, k = 1, 2, \ldots, K$:

$$
\begin{aligned}
\frac{\partial f(\boldsymbol{\theta})}{\partial \theta_k} &= \frac{1}{2}\frac{\partial}{\partial \theta_k}\left(e(\mathscr{X}_1;\boldsymbol{\theta})^2 + e(\mathscr{X}_2;\boldsymbol{\theta})^2 + \ldots + e(\mathscr{X}_N;\boldsymbol{\theta})^2\right) \\
&= \frac{1}{2}\left(2e(\mathscr{X}_1;\boldsymbol{\theta})\frac{\partial e(\mathscr{X}_1;\boldsymbol{\theta})}{\partial \theta_k} + 2e(\mathscr{X}_2;\boldsymbol{\theta})\frac{\partial e(\mathscr{X}_2;\boldsymbol{\theta})}{\partial \theta_k} + \ldots + 2e(\mathscr{X}_N;\boldsymbol{\theta})\frac{\partial e(\mathscr{X}_N;\boldsymbol{\theta})}{\partial \theta_k}\right) \\
&= e(\mathscr{X}_1;\boldsymbol{\theta})\frac{\partial(\mathscr{Y}^{\text{exp}}(\mathscr{X}_1) - \mathscr{X}_1\boldsymbol{\theta})}{\partial \theta_k} + e(\mathscr{X}_2;\boldsymbol{\theta})\frac{\partial(\mathscr{Y}^{\text{exp}}(\mathscr{X}_2) - \mathscr{X}_2\boldsymbol{\theta})}{\partial \theta_k} \\
&\quad + \ldots + e(\mathscr{X}_N;\boldsymbol{\theta})\frac{\partial(\mathscr{Y}^{\text{exp}}(\mathscr{X}_N) - \mathscr{X}_N\boldsymbol{\theta})}{\partial \theta_k} \\
&= -e(\mathscr{X}_1;\boldsymbol{\theta})\frac{\partial \mathscr{X}_1\boldsymbol{\theta}}{\partial \theta_k} - e(\mathscr{X}_2;\boldsymbol{\theta})\frac{\partial \mathscr{X}_2\boldsymbol{\theta}}{\partial \theta_k} - \ldots + e(\mathscr{X}_N;\boldsymbol{\theta})\frac{\partial \mathscr{X}_N\boldsymbol{\theta}}{\partial \theta_k} \\
&= -\begin{bmatrix} x_{1,k} & x_{2,k} & \cdots & x_{N,k} \end{bmatrix}\begin{bmatrix} e(\mathscr{X}_1;\boldsymbol{\theta}) \\ e(\mathscr{X}_2;\boldsymbol{\theta}) \\ \vdots \\ e(\mathscr{X}_N;\boldsymbol{\theta}) \end{bmatrix}
\end{aligned}
$$

It then follows that the gradient of $f(\boldsymbol{\theta})$ with respect to $\boldsymbol{\theta}$ is given by:

$$
\nabla_{\boldsymbol{\theta}} f = \begin{bmatrix} \frac{\partial f(\boldsymbol{\theta})}{\partial \theta_1} \\ \frac{\partial f(\boldsymbol{\theta})}{\partial \theta_2} \\ \vdots \\ \frac{\partial f(\boldsymbol{\theta})}{\partial \theta_K} \end{bmatrix} = -\begin{bmatrix} x_{1,1} & x_{2,1} & \cdots & x_{i,1} & \cdots & x_{N,1} \\ x_{1,2} & x_{2,2} & \cdots & x_{i,2} & \cdots & x_{N,2} \\ \vdots & \vdots & \ddots & \vdots & \ddots & \vdots \\ x_{1,K} & x_{2,K} & \cdots & x_{i,K} & \cdots & x_{N,K} \end{bmatrix}\begin{bmatrix} e(\mathscr{X}_1;\boldsymbol{\theta}) \\ e(\mathscr{X}_2;\boldsymbol{\theta}) \\ \vdots \\ e(\mathscr{X}_i;\boldsymbol{\theta}) \\ \vdots \\ e(\mathscr{X}_N;\boldsymbol{\theta}) \end{bmatrix}
$$

$$
= -\begin{bmatrix} x_{1,1} & x_{1,2} & \cdots & x_{1,K} \\ x_{2,1} & x_{2,2} & \cdots & x_{2,K} \\ \vdots & \vdots & \ddots & \vdots \\ x_{i,1} & x_{i,2} & \cdots & x_{i,K} \\ \vdots & \vdots & \ddots & \vdots \\ x_{N,1} & x_{N,2} & \cdots & x_{N,K} \end{bmatrix}^T \begin{bmatrix} e(\mathscr{X}_1;\boldsymbol{\theta}) \\ e(\mathscr{X}_2;\boldsymbol{\theta}) \\ \vdots \\ e(\mathscr{X}_i;\boldsymbol{\theta}) \\ \vdots \\ e(\mathscr{X}_N;\boldsymbol{\theta}) \end{bmatrix}
$$

or

$$
\nabla_{\boldsymbol{\theta}} f = -\mathcal{X}^T(\mathcal{Y}^{\text{exp}} - \mathcal{X}\boldsymbol{\theta}) = \mathbf{0} \qquad (12.5.13)
$$

The last equation can be written as

$$
-\mathcal{X}^T\mathcal{Y}^{\text{exp}} + \mathcal{X}^T\mathcal{X}\boldsymbol{\theta} = \mathbf{0}
$$

We have reached the following result, which is arguably one of the most important equations in optimization, linear algebra, and machine learning applications:

$$\boldsymbol{\theta} = \left(\mathcal{X}^T\mathcal{X}\right)^{-1}\mathcal{X}^T\mathcal{Y}^{\exp} \tag{12.5.14}$$

Eq. (12.5.14) is the solution of the more general problem of the linear-least squares (LLS) problem, which has a large number of important applications. It is noted that this problem is called linear as the model we use (see Eq. (12.5.2)) is linear in the vector of parameters, as can be also seen from Eq. (12.5.5).

There is a more elegant and compact way to prove the same result by first noting that

$$f(\boldsymbol{\theta}) = \frac{1}{2}\sum_{i=1}^{N} e(\mathscr{X}_i)^2 = \frac{1}{2}\boldsymbol{e}^T(\boldsymbol{\theta})\boldsymbol{e}(\boldsymbol{\theta}) = \frac{1}{2}(\mathcal{Y}^{\exp} - \mathcal{X}\boldsymbol{\theta})^T(\mathcal{Y}^{\exp} - \mathcal{X}\boldsymbol{\theta}) \tag{12.5.15}$$

We expand the last expression

$$\begin{aligned}(\mathcal{Y}^{\exp} - \mathcal{X}\boldsymbol{\theta})^T(\mathcal{Y}^{\exp} - \mathcal{X}\boldsymbol{\theta}) &= \left(\mathcal{Y}^{\exp T} - \boldsymbol{\theta}^T\mathcal{X}^T\right)(\mathcal{Y}^{\exp} - \mathcal{X}\boldsymbol{\theta}) \\ &= \mathcal{Y}^{\exp T}\mathcal{Y}^{\exp} - \mathcal{Y}^{\exp T}\mathcal{X}\boldsymbol{\theta} - \boldsymbol{\theta}^T\mathcal{X}^T\mathcal{Y}^{\exp} + \boldsymbol{\theta}^T\mathcal{X}^T\mathcal{X}\boldsymbol{\theta}\end{aligned}$$

where we have used the property $(\boldsymbol{AB})^T = \boldsymbol{B}^T\boldsymbol{A}^T$. We also note that $\mathcal{Y}^{\exp T}\mathcal{X}\boldsymbol{\theta}$ and $\boldsymbol{\theta}^T\mathcal{X}^T\mathcal{Y}^{\exp}$, as they are scalars, they are equal. We therefore have the result

$$f(\boldsymbol{\theta}) = \frac{1}{2}\left(\mathcal{Y}^{\exp T}\mathcal{Y}^{\exp} - 2\mathcal{Y}^{\exp T}\mathcal{X}\boldsymbol{\theta} + \boldsymbol{\theta}^T\mathcal{X}^T\mathcal{X}\boldsymbol{\theta}\right)$$

We now calculate the gradient by taking the derivative with respect to the vector $\boldsymbol{\theta}$ and using the identities

$$\frac{\partial}{\partial \boldsymbol{x}}\left(\boldsymbol{c}^T\boldsymbol{x}\right) = \boldsymbol{c}$$

$$\frac{\partial}{\partial \boldsymbol{x}}\left(\boldsymbol{x}^T\mathbf{A}\boldsymbol{x}\right) = \mathbf{A}^T\boldsymbol{x} = 2\mathbf{A}\boldsymbol{x}$$

we obtain Eq. (12.5.14).

The following MATLAB program applies Eq. (12.5.14) to the experimental data of Table 12.2 and estimates the parameters A and B of model (12.5.2):

```
clear all
data=[  0 0.6113
        5 0.8726
        ...
        95      84.5290
        100     101.3200];
Yexp = log(data(:,2));                          % equation (12.5.4)
N    = length(Yexp);
X    = [ones(N,1) -1./(data(:,1)+273.15)]; % equation (12.5.5)
theta = (X'*X)\X'*Yexp

figure(1); clf;
plot(data(:,1),data(:,2),'o','MarkerFaceColor','w')
grid on; hold on;
xlabel('Temperature (°C)')
ylabel('Water vapor pressure (kPa)')
Ymodel = exp(theta(1)-theta(2)./(data(:,1)+273.15) )
plot(data(:,1),Ymodel,'-
','linewidth',1.5,'Color',[0.64,0.08,0.184])
axis([0 100 0 110])

SSE=(Ymodel-exp(Yexp))'*(Ymodel-exp(Yexp))

figure(2); clf;
plot(data(:,1),data(:,2)-Ymodel,'-o','LineWidth',1.4)
grid on; hold on;
xlabel('Temperature (°C)')
ylabel('Model error in water vapor pressure (kPa)')
```

The results are presented in Fig. 12.12 ($A = 18.59{,}867$, $B = 5204.46{,}165$ K). The error becomes significant as the temperature increases. We therefore need to investigate model (12.5.1), which has an additional parameter and offers the potential of improvement.

The parameter estimation in Python follows the same lines as in MATLAB.

```
import numpy as np

Pexp = np.array( [0.6113,   0.8726, 1.2281, 1.7056, 2.3388, \
                  3.1690,   4.2455, 5.6267, 7.3814, 9.5898, \
                  12.3440,  15.7520, 19.9320, 25.0220, 31.1760,
\
                  38.5630,  47.3730, 57.8150, 70.1170, 84.5290,
\
                  101.3200] )
Y=np.log(Pexp.T)
Texp = 273.15+np.linspace(0,100,21)
Np = len(Pexp)
X = np.array( [np.ones(Np), -1/Texp] ).T
A = (X.T)@(X)
b = (X.T)@Y
theta = np.linalg.solve(A,b)
print(f"Antoine eq. constant A = {theta[0]:4.6f}")
print(f"Antoine eq. constant B = {theta[1]:4.6f}")
```

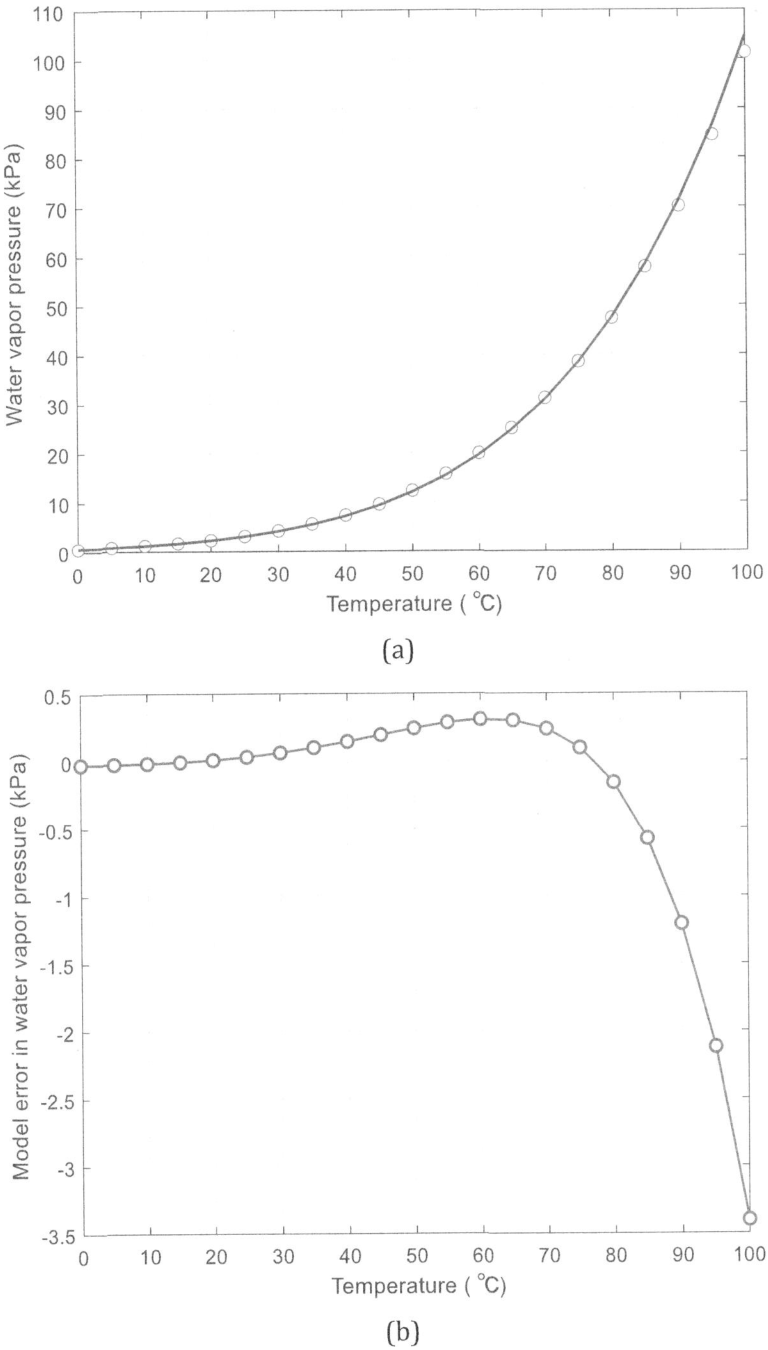

Figure 12.12 Comparison of experimental data and model predictions (model (12.5.2)) for the vapor pressure of water (a), and modeling error (b) as a function of the temperature.

```
In [81]: runfile('...')
Antoine eq. constant A = 18.598672
Antoine eq. constant B = 5204.461649
```

When we use Antoine's equation (Eq. 12.5.1) to model vapor pressure data, we lose the advantage of working with a model that is linear in its parameters. This is primarily because the parameter C appears in a nonlinear fashion. However, the key benefit of using Antoine's equation lies in its ability to directly utilize the raw experimental measurements of vapor pressure, without requiring a nonlinear transformation—such as taking the natural logarithm of the data.

This distinction becomes especially important when we consider the impact of measurement error. Suppose our experimental vapor pressure measurements are subject to random error. For the least-squares regression methodology to yield statistically sound parameter estimates, certain assumptions about the error must be satisfied: the errors should be random, uncorrelated, have zero mean, and possess constant variance (i.e., homoscedasticity). When we apply a nonlinear transformation—such as a logarithmic transformation—to the dependent variable, these assumptions may no longer hold, even if they were initially valid in the original measurement space. Specifically, a transformation can distort the distribution of errors, introducing correlation, heteroscedasticity, or bias in the transformed model. As a result, the statistical basis for using least squares becomes questionable in the transformed domain.

Therefore, although Antoine's equation is nonlinear in its parameters, its ability to preserve the statistical integrity of the original measurement errors makes it a preferable choice in certain cases, particularly when the error characteristics of the raw data are well understood and meet the required assumptions for regression analysis.

We now must solve the following unconstrained, multivariable optimization problem:

$$\min_{A,B,C} f(A,B,C) = \frac{1}{2}\sum_{i=1}^{N}\left(P^{sat}(T_i) - \exp\left(A - \frac{B}{C+T_i}\right)\right)^2 \tag{12.5.16}$$

We first create an m-file which returns $f(A,B,C)$ for given values of A, B, and C and then use fminunc to perform the minimization in MATLAB:

```
clear all
% experimental data
data=[  0 0.6113
        5 0.8726
        ...
        95      84.5290
        100     101.3200];
Psat = data(:,2); t     = data(:,1); T     = t+273.15;
theta0 = [15;3000;-50];
options= optimset('display','iter');
theta  = fminsearch(@SSE_fun,theta0,options,Psat,T)

figure(1);
clf;
plot(t,Psat,'o','MarkerFaceColor','w')
grid on; hold on;
xlabel('Temperature (°C)')
ylabel('Water vapor pressure (kPa)')
Ymodel = exp(theta(1)-theta(2)./(theta(3)+T) );
plot(t,Ymodel,'-','linewidth',1.5,'Color',[0.64,0.08,0.184])
axis([0 100 0 110])

figure(2);
clf;
plot(t,Psat-Ymodel,'-o','LineWidth',1.4)
grid on; hold on;
xlabel('Temperature (°C)')
ylabel('Model error in water vapor pressure (kPa)')

function f=SSE_fun(theta,Psat,T)
e    = Psat(:) - exp(theta(1) - theta(2)./(theta(3) +T(:) ) );
f    = (1/2)*e'*e;
end
```

The results are presented in Fig. 12.13. The improvement over the simpler two-parameter model is impressive (compare Figs. 12.12b and 12.13b). The proposed model for calculating the saturation pressure is

$$\ln P^{sat} = 16.40311 - \frac{3888.48047}{T - 43.190864} \tag{12.5.16}$$

where the temperature is in K and the pressure in kPa.

Note that, in order to perform the optimization in MATLAB, the command **fminsearch** has been used as it is extremely popular and efficient in solving nonlinear parameter estimation problems based on the Nelder-Mead method. This is a direct search method that does not use gradients of the cost function, and so continuity and differentiability of the function is not required to find a local minimum of the function. The basic idea is that of a simplex, which is a geometric figure formed by a set of n + 1 points in the n-dimensional space, and in two dimensions, the simplex is just a triangle. The basic idea of the Nelder–Mead method is to evaluate the objective function at the vertices of the simplex and move this simplex toward the minimum point. At each iteration, the vertex with the worst (largest)

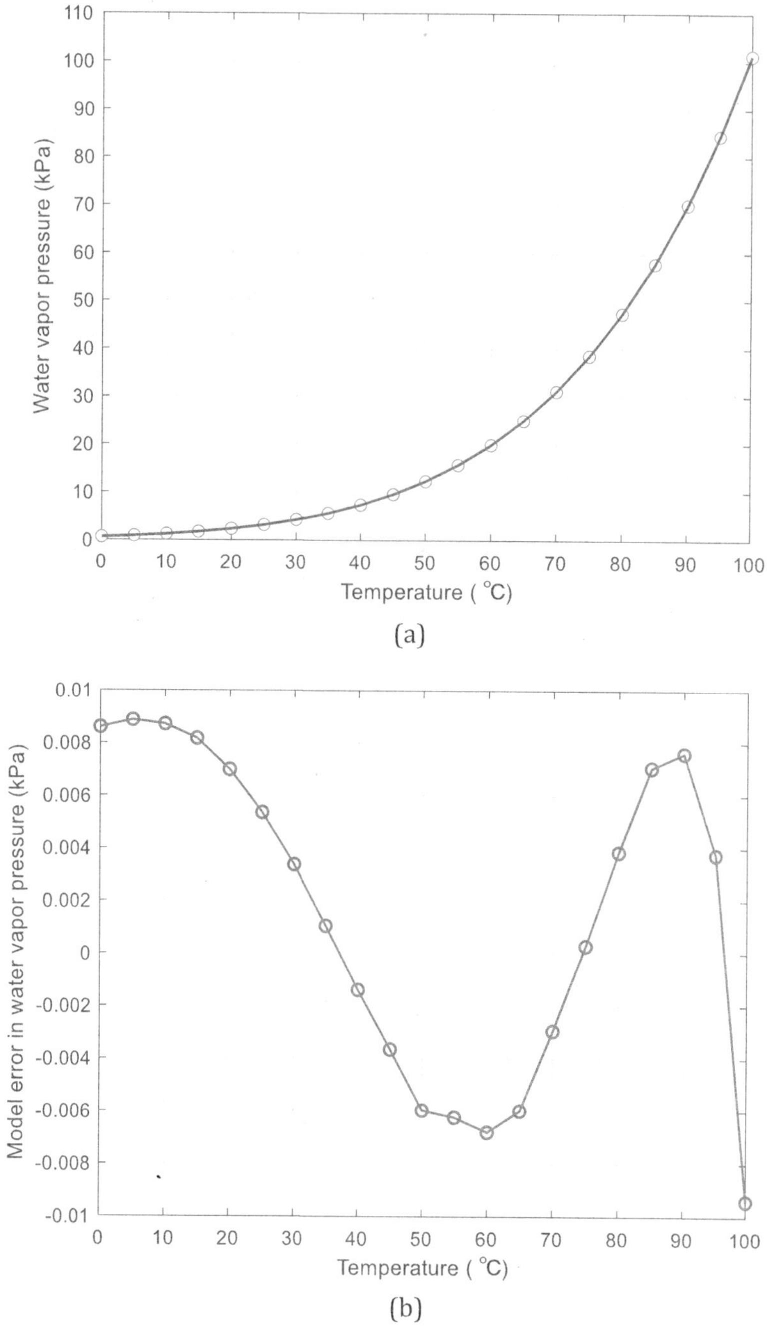

Figure 12.13 Comparison of experimental data and model predictions (model (12.5.1)) for the vapor pressure of water (a), and modeling error (b) as a function of the temperature.

objective function value is replaced with another vertex with an improved objective function value. The algorithm has been proven extremely efficient for solving nonlinear parameter estimation problems.

The corresponding implementation in Python is the following.

```
import numpy as np
import matplotlib.pyplot as plt
from scipy.optimize import least_squares

Pexp  =  np.array(  [0.6113,   0.8726,   1.2281,   1.7056,   2.3388,
3.1690, \
                  4.2455,    5.6267,    7.3814,    9.5898,    12.3440,
15.7520,\
                 19.9320,   25.0220,   31.1760,   38.5630,   47.3730,
57.8150,\
                 70.1170, 84.5290, 101.3200] )
Texp = 273.15+np.linspace(0,100,21)

def error(X,xexp,yexp):
    A, B, C = X
    e = np.exp(A-B/(C+xexp)) - yexp
    return e

X0 = np.array([15,3000,-50])

sol  = least_squares(error,
                     X0,
                     args=(Texp,Pexp),
                     bounds   = ([10, 1000, -100], [20, 5000,
0]),
                     max_nfev = 1e4,
                     verbose  = 2 )
A, B, C = sol.x
print('The solution ..................')
print(f"A = {A:>12.6f}")
print(f"B = {B:>12.6f}")
print(f"C = {C:>12.6f}")

plt.plot(Texp,Pexp,'ro')
plt.plot(Texp,np.exp(A-B/(C+Texp)),'b-')
plt.xlabel('Temperature (K)')
plt.ylabel('Saturation pressure (kPa)')
plt.axis([270, 380, 0, 110])
```

Note that we are not using a general-purpose unconstrained optimization function but the function specific to nonlinear least-squares optimization problems, **least_squares**. A similar function named **lsqnonlin** is also available in MATLAB.

Summary

In this chapter the fundamental principles of process optimization have been presented. More specifically, the case of multivariable unconstrained optimization has been discussed. Emphasis has been placed on one of the most important applications of unconstrained optimization, which is the problem of model parameter estimation. The equality-constrained optimization problem has also been discussed, and the optimality conditions have been derived and applied to representative case studies.

Problems

12.1 Find the minimum (if it exists) of the following functions.

(a) $f(x) = x^2$

(b) $g(x) = x^3$

(c) $h(x) = -x^4$

12.2 Find analytically and numerically the minimum (if it exists) of the following.

(a) $f(x) = x^3 - x^2 - 4x - 4, x^0 = 0$

(b) $f(x_1, x_2) = x_1^2 + 2x_1x_2 + 2x_2^2 - 2x_1 + x_2 + 8,$

$$\left(x_1^0, x_2^0\right) = (0, 0)$$

12.3 Consider the design of a cylindrical tank of a given volume with torispherical heads. Find the diameter (d) and the height (h) of the tank that result in the minimum cost of the tank. The cost of the tank is given by

$$c = c_0\left(\frac{w}{w_0}\right)^n$$

where c is the cost of a tank with mass w and c_0 is the known cost of a tank with mass w_0 and $n = 0.6$. The thickness of the cylindrical part of the tank is t and that of the torispherical heads is 1.77 t, while the area of the torispherical heads is 1.124 times the cross-sectional area of the cylindrical part of the tank. Assume, for simplicity, that $V = 4/\pi$. What is the optimal h/d ratio?

12.4 Consider the operation of a continuous bioreactor with volume V and a feed volumetric flowrate F. The feed contains a substrate at a concentration S_0. In the bioreactor cells are produced at a rate equal to $\dot{r}_X''' = \dot{\mu}X$. There is also an extracellular product produced at a rate $\dot{r}_P''' = a\,\dot{r}_X'''$, i.e., the extracellular product is associated with cell growth. The specific growth rate is given by a Monod equation with product inhibition

$$\dot{\mu} = \frac{\dot{r}_X'''}{X} = \mu_{\max}\frac{S}{K_s + S}\exp\left(-\frac{P}{P_{\max}}\right)$$

where X is the biomass concentration (in g/L), S is the substrate concentration, P is the product concentration and $\mu_{\max}$, K_s and a are constants. We finally define the substrate-to-biomass yield coefficient by

$$Y_{X/S} = \frac{\text{change in biomass concentration}}{\text{change in substrate concentration}} = \frac{\dot{r}_X'''}{\dot{r}_S'''}$$

Find the optimal reactor volume for a given substrate concentration in the fresh feed and a given volumetric flowrate of the fresh feed that maximizes the productivity of the bioreactor defined as:

$$\Pi = \frac{FP}{V}$$

Use the following numerical values: $\mu_{\max} = 0.41\frac{1}{h}$, $K_s = 0.4\frac{g}{L}$, $a = 2.44$, $Y_{X/S} = 0.1$, $S_0 = 100\frac{g}{L}$, $F = 1\frac{m^3}{h}$,
$P_{max} = 25\frac{g}{L}$.

12.5 Consider the case study of Problem 12.4. Investigate how the optimal operating point varies as a function of $P_{\max}$ in the range [10 g/L, 90 g/L].

12.6 Consider the case study of Problem 12.4. Investigate what the optimal operating point is when there is no product inhibition, i.e., $P_{\max} \to +\infty$, and

$$\dot{\mu} = \frac{\dot{r}_X'''}{X} = \mu_{\max}\frac{S}{K_s + S}$$

12.7 Consider the following reactions that are first order with respect to the concentration of the reactant in each reaction.

$$A \to B \to C.$$

Show that for a batch reactor and only A initially present with concentration C_{A0} the concentration of the main product B is

$$\frac{C_B(t)}{C_{A,0}} = \frac{k_1}{k_2 - k_1}\left(e^{-k_1 t} - e^{-k_2 t}\right)$$

What is the time t_{opt} at which the concentration of B is maximum?

Use the numerical values $k_1 = 1\frac{1}{h}, k_2 = 0.1\frac{1}{h}, C_{A,0} = 1\frac{\text{kmol}}{\text{m}^3}$

12.8 The potential energy of an ion pair A^+B^- is given by

$$E(r) = -\frac{a}{r} + \frac{b}{r^8}$$

where $a = 4 \times 10^{-28}$ and $b = 7 \times 10^{-96}$, r is the center-to-center distance in nm and E is the binding energy of the system in eV. If at the stable condition the bonding energy is minimum, calculate the center to center distance of the ion pair.

12.9 Fig. P12.9 shows a system consisting of a heat exchanger (HE) and a CSTR in series. The fresh feed is heated up in the HE and is then fed into the CSTR. The fresh feed has a volumetric flow rate of $F = 0.01$ m^3/s, a reactant concentration $C_0 = 10$ kmol/m^3 and temperature $T_0 = 25°C$. The feed is heated in the HE using low-pressure steam at $T_s = 100°C$ ($\Delta h_s = 2087$ kJ/kg steam). The overall heat transfer coefficient in the HE is $U = 0.2$ kJ/(m^2 °C) while the area available for heat transfer is A (m^2). The CSTR has volume $\mathscr{V}$ and the reaction Reactant→Product takes place, which is first order, i.e., $R = kC$. The reaction has almost negligible heat of reaction, resulting in isothermal reactor operation. The reacting mixture has a constant density $\rho = 1000$ kg/m^3 and a constant specific heat capacity $c_p = 4$ kJ/(kg°C).

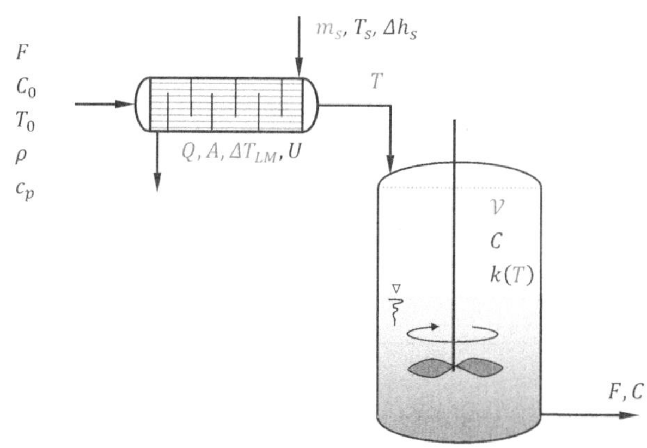

Figure P12.9 The process studied in Problem 12.9.

The reaction rate constant is given by

$$k\left([=]\frac{1}{s}\right) = k_0 \exp\left(-\frac{E}{RT}\right) = 1 \times \exp\left(-\frac{30000}{R_g(273.15+T)}\right),$$

$$R_g = 8.314 \frac{\text{kJ}}{\text{kmol K}}, T\,[=]\,^\circ C$$

Find the optimal design of the system for achieving 90% conversion ($C = 1$ kmol/m^3) if the annual cost of the reactor CSTR and the HE are given by:

$$C_{\text{CSTR}} = 100000\ \mathscr{V}^{0.85}\,[=]\,\frac{\$}{y}, \quad \mathscr{V}\,[=]\,\text{m}^3$$

$$C_{\text{HE}} = 2500\ A^{0.6}\,[=]\,\frac{\$}{y}, \quad A[=]\,\text{m}^2$$

and the unit cost of the low-pressure steam is $c_s = 0.015$ \$/kg. The operating time is 8300 h/y.

Show that the mathematical model has seven variables $(T, Q, m_s, \Delta T_{lm}, A, V, \text{TAC})$ and six equations:

$$Q = F\rho c_p(T - T_0) \tag{12.1}$$

$$m_s = \frac{Q}{\Delta h_s} \tag{12.2}$$

$$\Delta T_{LM} = \frac{(T_s - T_0) - (T_s - T)}{\ln\left(\frac{T_s - T_0}{T_s - T}\right)} \tag{12.3}$$

$$A = \frac{Q}{U \Delta T_{LM}} \tag{12.4}$$

$$\mathscr{V} = \frac{F}{k(T)} \frac{(C_0 - C)}{C} \tag{12.5}$$

$$\text{TAC} = 10000\ \mathscr{V}^{0.6} + 2500\ A^{0.6} + 8300 \times 3600 m_s c_s \tag{12.6}$$

Select temperatures in the range [40°C, 90°C] and calculate the total annual operating cost (TAC) to find the minimum-cost solution.

12.10 In Table P12.10 data for the heat capacity of water vapor are presented (taken from NIST Chemistry WebBook). The following equation is proposed for calculating the heat capacity of water vapor:

$$c_p = A + Bt + Ct^2 + Dt^3 + Et^{-2}$$

Table P12.10 Heat capacity of water vapor as a function of temperature.

T (K)	c_p (kJ/kmol K)
500	35.22
600	36.32
700	37.50
800	38.74
900	40.00
1000	41.27
1100	42.52
1200	43.75
1300	44.94
1400	46.06
1500	47.11
1600	48.07
1700	48.92

where $t = T[K]/1000$.

Estimate the constants A, B, C, D, and E using LLS (closed-form solution) and compare your results with the values given by NIST:

$$A = 30.092, \quad B = 6.832514, \quad C = 6.793435, \quad D = -2.53448, \quad E = 0.082139$$

Table P12.11 Heat of evaporation of water vapor as a function of temperature.

T (°C)	Δh_v (kJ/kg)
0.01	**2501.3**
50	2382.7
T_{nb} = **100**	**2257.0**
150	2144.7
200	1940.7
250	1716.2
300	1404.9
340	1027.9
T_c = **374.1**	**0.0**

12.11 In Table P12.11 data for the enthalpy of evaporation of water are presented (taken from several sources). The following equation, known as Watson's equation, is proposed for calculating the enthalpy of evaporation of water:

$$\Delta h_v = \Delta h_v(T_{nb})\left(\frac{T - T_c}{T_{nb} - T_c}\right)^{a+b\left(\frac{T}{T_c}\right)}$$

where $T_c = 374.1°C$ is the critical temperature of water and $T_{nb} = 100°C$ is the normal boiling point of water. Determine the constants a and b using LLS.

12.12 The success of a medical treatment does not only depend on the pharmacodynamic activity of the drug, but also on the availability of the active agent at the site of action in the human body. To improve the efficiency of a pharmacotherapy, time-controlled delivery systems can be used. The drug can be embedded within a polymeric matrix restricting its release. Various types of polymers and dosage-form designs (compositions, sizes, and shapes) have been studied. Poly(lactic-co-glycolic acid) (PLGA)-based microparticles (spherical beads, generally 1–100 μm in diameter) offer various important advantages over other controlled drug delivery systems. Siepman et al., (Biomacromolecules, 6, p. 2312, 2005) have presented the experimental results shown on Table P12.12 for the release of lidocaine (also known as lignocaine and sold under the brand name

Table P12.12 Percent amount of lidocaine as a function of time.

Time t (d)	% Released
1	27
3	38
7	59
10	69
12	75
14	78
20	85
21	88
28	95

Table P12.13 Data on voltage as a function of current density for a solid oxide fuel cell.

i (A/cm^2)	E (V)
0.02	0.835
0.11	0.837
0.6	0.764
1.09	0.69
0.59	0.609
2.09	0.533
2.58	0.447
3.08	0.346
3.46	0.242
3.84	0.062

Xylocaine among others, is a local anesthetic) from $d = 7.2\ \mu m$ in diameter spherical PLGA particles.

A mathematical model that can be used to describe the release of the active agent from the polymeric microparticles is the following

$$\frac{n(t)}{n_\infty} = 1 - \frac{6}{\pi^2}\exp\left[-\left(\frac{\pi}{r_p}\right)^2 D\cdot t\right]$$

where $n(t)$ is the amount of active agent that has been released up to time t, n_∞ is the total amount of active agent that is released in infinite time, r_p is the radius of the spherical particles, and D is the diffusion coefficient. The aim of the experiments is to determine the diffusion coefficient. Estimate D using the experimental data and compare your results with the value $D = 1.2335 \times 10^{-14}\ cm^2/s$.

12.13 A solid oxide fuel cell (or SOFC) is an electrochemical conversion device that produces electricity directly from oxidizing a fuel such as hydrogen. In Table P12.13 experimental data for the current density i in A/cm^2 and voltage E in V for a SOFC are given (data taken from Huang C. et al., Journal of Power Sources, 195, p. 5176, 2010). An appropriate model that can be used to describe the observed power curve of the SOFC is the following

$$E(i;\theta_1,\theta_2,\theta_3) = \theta_1 - \theta_2 i + \alpha \ln\left(1 - \frac{i}{\theta_3}\right)$$

where $a = RT/2F = 0.0376\ V$. Use the experimental data to determine the parameters of the model by minimizing

(a) the sum of the errors squared,

(b) the sum of the absolute values of the errors,

(c) the maximum absolute error.

12.14 In Table P12.14 data for the power curve of a specific wind turbine are given (data for Enercon, 2050 E82 wind turbine). The data give the power produced by the wind turbine $P(u)$ as a function of the wind velocity u at the rotor. An equation that can be used to describe the power curve is the following:

$$\frac{P(u)}{P_R} = \frac{1}{1 + \exp\left(\sum_{k=0}^{n} c_k u^k\right)}$$

Table P12.14 Data for the power curve of a wind turbine.

u (m/s)	*P* (kW)
1	0
2	0
3	25
4	82
5	174
6	321
7	532
8	815
9	1180
10	1580
11	1810
12	1980
13	2050
14	2050
15	2050
16	2050
17	2050
18	2050
19	2050
20	2050
25	2050

where P_R is the rated power of the wind turbine, i.e., the maximum power that can be produced by the turbine and c_k are constants. Determine the constants $c_k, k = 0, 1, \ldots, 4$ using the data of Table P12.14.

12.15 Determine the rectangle of maximum area that can be inscribed within the unit circle.

12.16 Consider a system of n states each one having its own energy e_i. If the probability of each state is π_i and the average energy is $\mathscr{E}$ find an expression of the probability that maximizes the entropy $\mathscr{S}$ of the system:

$$\mathscr{S} = -\mathscr{k} \sum_{i=1}^{n} \pi_i \ln \pi_i$$

Further reading

Arora, J.S., 2017. Introduction to Optimum Design, fourth ed. Elsevier, London, UK.

Bazaraa, M.S., Sherali, H.D., Shetty, C.M., 2006. Nonlinear Programming, third ed. John Wiley and Sons, NJ.

Books on chemical process and Engineering Optimization.

Edgar, T.F., Himmelblau, D.M., Lasdon, L.S., 2001. Optimization of Chemical Processes, second ed. McGraw Hill, NY.

Griva, I., Nash, S.G., Sofer, A., 2009. Linear and Nonlinear Optimization, second ed. SIAM, Philadelphia.

Grossmann, I.E., 2021. Advanced Optimization for Process Systems Engineering. Cambridge University Press, Cambridge, UK.

Kookos, I.K., 2022. Practical Chemical Process Optimization. Springer Nature Switzerland, AG.

Luenberger, D.G., Ye, Y., 2016. Linear and Nonlinear Programming. Springer, NY.

Maranas, C.D., Zomorrodi, A.R., 2016. Optimization Methods in Metabolic Networks. John Wiley & Sons, NJ.

Rao, S.S., 2009. Engineering Optimization, Theory and Practice. John Wiley and, Sons, NJ.

Ravindran, A., Ragsdell, K.M., Reklaitis, G.V., 2006. Engineering Optimization, Methods and Applications, second ed. John Wiley and Sons, NJ.

Taha, H.A., 2007. Operations Research, an Introduction, eighth ed. Prentice Hall, NJ.

Vassiliadis, V.S., Kahn, W., Chanona, E.A., Yuan, Y., 2020. Optimization for Chemical and Biochemical Engineering. Cambridge University Press, Cambridge, UK.

13

Dimensional analysis and scale-up

13.1 Introduction

Up to this point, we have analyzed systems using classical mathematical tools grounded in the fundamental laws of chemistry and physics—such as conservation of mass, energy, and momentum. These well-established principles have allowed us to describe and design chemical process systems using both analytical and graphical methods. These approaches serve us well when operating within the conventional boundaries of engineering practice, where assumptions are clear and the methodologies for predicting system performance are mature and reliable.

However, we now stand at the edge of deeper exploration. It is time to take a step beyond the familiar terrain of traditional analysis and into a more abstract—but no less essential—dimension of engineering thought. We are about to pose two questions that, at first glance, may seem philosophical or even irrelevant to the day-to-day concerns of an engineer. Yet they strike at the very core of what it means to innovate, to design, and to understand the complexity of large-scale industrial systems. These questions challenge not only how we design systems, but why we make the choices we do—and what principles truly guide us when established tools begin to falter.

We now pose a set of foundational questions that will serve as a conceptual anchor for this chapter. Upon completion of the material, the reader is expected not only to answer these questions but also to recognize their critical relevance to the principles of process systems engineering and their implications for large-scale industrial practice. The questions are as follows:

First question: Can you develop a plausible scientific relationship that describes the walking speed of humans based on relevant physical or biomechanical factors (such as leg length,

Introduction to Process Systems Engineering. https://doi.org/10.1016/B978-0-443-45444-8.00008-7

stride frequency, or body mass)? Using this relationship or a generalized version of it, can you then extrapolate to estimate the potential running speed of *Tyrannosaurus rex*—the iconic dinosaur (another bipedal animal) made famous in Steven Spielberg's film *Jurassic Park*?

Second question: Fig. 13.1 shows a poster for the fictional horror film *Attack of the 50 Foot Woman*. The movie tells the story of a wealthy heiress who, after a close encounter with a giant alien and its spherical spacecraft, undergoes a dramatic transformation—growing to enormous proportions. This raises an intriguing question: Can the human body be scaled up uniformly by a constant factor and remain physically and biologically functional? In other words, does biology allow for simple geometric scaling of organisms? To explore this idea, consider the difference between a mouse and an elephant—animals with a size ratio on the order of 60:1 and a weight ratio of 200000:1. If you were to scale up a mouse uniformly by a factor of 60, would you end up with an elephant-like creature? Similarly, if you enlarged a lizard by a similar factor, would the result be physically and biologically comparable to an alligator or a dinosaur?

As we will see, nature does not scale animals this way. There are fundamental physical and physiological constraints—rooted in geometry, biomechanics, and metabolism—that prevent such direct scaling. These constraints explain why simply "supersizing" an organism leads to significant changes in structure and function. This fact turns out to have significant implications when we scale up or down engineering systems.

To address these questions, we must build upon the foundational concepts introduced in Chapter 2, particularly dimensional homogeneity and systems of units. We begin by introducing the concepts of nondimensionalization of equations and variable scaling, which are essential tools for analyzing complex systems. We then provide a concise overview of dimensional analysis and the principle of similarity, both of which are fundamental to understanding physical modeling and system behavior. The chapter concludes with a discussion of the critical challenge of scale-up—specifically, the risks and limitations involved when extrapolating experimental results from small-scale pilot plant equipment to full-scale industrial processes.

Figure 13.1 Film poster by Reynold Brown for the film Attack of the 50 Foot Woman (1958) starring Allison Hayes. Attack of the 50 Foot Woman is a 1958 independently made American science fiction horror film directed by Nathan H. Juran (credited as Nathan Hertz) and starring Allison Hayes, William Hudson, and Yvette Vickers. It was produced by Bernard Woolner. The screenplay was written by Mark Hanna, and the original music score was composed by Ronald Stein. The film's storyline concerns the plight of a wealthy heiress whose close encounter with an enormous alien in its round spacecraft causes her to grow into a giantess, complicating her marriage, which is already troubled by a philandering husband. The film was remade in 1993 by HBO under the same title, Attack of the 50 Ft. Woman. It was directed by Christopher Guest, with a script by Thirtysomething writer Joseph Dougherty. Daryl Hannah produced the film and starred in the title role. (from Wikipedia, this poster art entered the public domain in 1987).

13.2 Nondimensionalization of equations

The principle of dimensional homogeneity ensures that all additive terms in an equation share identical dimensions. Therefore, if each term is divided by a combination of variables and constants whose product possesses the same dimensions, the resulting equation becomes **nondimensional**. In any nondimensional equation, all terms are inherently dimensionless. During the nondimensionalization of physical equations it is common for a few key **dimensionless parameters** (such as the **Reynolds** and **Froude numbers**) to emerge.

As an introductory example consider the mass balance of the reactant in a constant-temperature and volume CSTR where a first-order reaction ($R \rightarrow P$) is taking place

$$V\frac{dC}{dt} = F(C_0 - C) - Vk(T)C \tag{13.2.1}$$

where V is the volume of the reacting mixture, F is the volumetric flowrate of the feed, C_0 is the concentration of the reactant R in the feed, C is the concentration of the reactant in the reactor, and $k(T)$ is the reaction rate constant:

$$k(T) = Ae^{-\frac{E}{RT}} \tag{13.2.2}$$

where A is the pre-exponential constant, E is the activation energy, R is the ideal gas constant, and T is the absolute temperature.

In the context of physical modeling, **dimensional variables** are quantities that both vary throughout the problem and carry physical units. For instance, in the differential equation presented in Eq. (13.2.1) the reactant concentration $C(t)$ (with units of mol/L) and time t (with units of time) are considered dimensional variables because their values change and they possess measurable dimensions. In contrast, **dimensionless variables** (or **nondimensional** variables) also change over the course of the problem but do not have associated physical units. A common example is an angle of rotation, which—when expressed in degrees or radians—has no inherent dimension.

Some quantities, like the ideal gas constant, do not vary within the scope of the problem but still have units. These are referred to as **dimensional constants**. In the case of the example at hand, other dimensional constants include the reaction rate constant $k(T)$, the reactor volume V, or the volumetric flowrate of the fresh feed F. Although these constants may differ between different case studies, they remain fixed within a given scenario, setting them apart from variables. The constant e is a constant called **pure constant**. Another common example of a pure constant is π.

To nondimensionalize Eq. (13.2.1) we must note that all terms of the equation have the same units. Following Chapter 2 we note that:

$$\left[V\frac{dC}{dt}\right] = [FC] = [VkC] = \mathrm{N\,T^{-1}} \tag{13.2.3}$$

i.e., units of mol/s. To nondimensionalize Eq. (13.2.1), we need to select **scaling parameters**, based on the primary dimensions contained in the original equation. Let us divide all the terms of the differential Eq. (13.2.1) by FC_0.

$$\left(\frac{V}{F}\right)\frac{1}{C_0}\frac{dC}{dt} = 1 - \frac{C}{C_0} - \left(\frac{V}{F}\right)k(T)\frac{C}{C_0}$$

We note at this point that V/F has units of time and can be used to scale time, i.e., if t^* is the dimensionless time, then

$$t^* = \left(\frac{F}{V}\right)t \tag{13.2.4}$$

We also can use the concentration of the reactant in the feed to scale concentration

$$C^* = \frac{C}{C_0} \tag{13.2.5}$$

The differential equation becomes

$$\frac{dC^*}{dt^*} = 1 - C^* - \left(\frac{V}{F}\right)k(T)C^* \tag{13.2.6}$$

We finally note that the following dimensionless number, known as the **Damköhler number** (Da), has been formed:

$$\mathscr{D}a = \left(\frac{V}{F}\right)k \tag{13.2.7}$$

The final form of the differential equation is the following:

$$\frac{dC^*}{dt^*} + (1 + \mathscr{D}a)C^* = 1 \tag{13.2.8}$$

An immediate conclusion is that the solution of this differential equation will have the general form

$$C^* = f(\mathscr{D}a, t^*) \tag{13.2.9}$$

In the initial form of the differential Eq. (13.2.8) six variables and parameters appear: t, C, C_0, F, V, k. However, in the nondimensional form only three variables and parameters appear: $t^*, C^*, \mathscr{D}a$. It seems that all information conveyed by four parameters C_0, F, V, k has been condensed into a single dimensionless parameter $\mathscr{D}a$. Is this important?

Imagine that you did not have a mathematical model available for the isothermal CSTR and that you have been asked to perform experiments through which you can predict the dynamic behavior of any CSTR with a first-order reaction. Furthermore, you have been asked to present your experimental results graphically for any combination of C_0, F, V, k. How many experiments do you need to perform? If you vary each variable using n_p values within the range of its minimum and maximum value then: for each value of C_0 you need to examine n_p values of F and for each combination of C_0, F values you need to consider n_p values of V, etc. Overall, you need to perform n_p^4 experiments. This number grows very fast with the number of values selected for each variable. If you select 10 levels then you need to perform 10,000 experiments. This is not feasible, not even for the simplest experimental systems. Using the nondimensional model instead, requires performing n_p experiments only!

Consider now the case of a constant-volume mixing tank which has the following mathematical description

$$V\frac{dC(t)}{dt} = F(C_0(t) - C(t)) \tag{13.2.10}$$

We divide again by FC_0

$$\left(\frac{V}{F}\right)\frac{1}{C_0}\frac{dC}{dt} = \frac{C_0(t)}{C_0} - \frac{C}{C_0} \tag{13.2.11}$$

Using the same scaling factors as in the case of the CSTR we obtain

$$\frac{dC^*(t^*)}{dt^*} + C^*(t^*) = C_0^*(t^*) \tag{13.2.12}$$

Let us consider the case of starting up the operation of the mixing tank. Initially $C_0^*(t^*) = 0, t^* < 0$ and then $C_0^*(t^*) = 1, t^* \geq 0$. The solution to this linear first order differential equation is

$$C^*(t^*) = 1 - \exp(-t^*) \tag{13.2.13}$$

This result is also shown in Fig. 13.2.

Consider now a tank with constant volume V and a constant volumetric flowrate F. The temperature of the feed is $T_0(t)$ and the temperature in the tank is $T(t)$. If the density and the heat capacity are constant then the temperature dynamics is described by the following energy balance:

$$V\rho C_p\frac{dT(t)}{dt} = F\rho C_p(T_0(t) - T(t)) \tag{13.2.14}$$

or, dividing by $F\rho C_p$

$$\left(\frac{V}{F}\right)\frac{dT(t)}{dt} + T(t) = T_0(t) \tag{13.2.15}$$

Eq. (13.2.15) that describes the temperature dynamics has exactly the same structure as Eq. (13.2.10) that describes the concentration dynamics, and they have exactly the same solution given by Eq. (13.2.13).

A fundamentally different system is shown in Fig. 13.3. This is a classical RL electrical circuit where $v(t)$ is the voltage of the source, $i(t)$ is the electrical current, R is a resistor, which opposes the flow of current and dissipates energy as heat, and L is an inductor, which opposes changes in current by storing energy in a magnetic field.

The differential equation that describes the system dynamics is the following

$$L\frac{di(t)}{dt} + Ri(t) = v(t) \tag{13.2.16}$$

or

$$\left(\frac{L}{R}\right)\frac{di(t)}{dt} + i(t) = \frac{1}{R}v(t) \tag{13.2.17}$$

The ratio of inductance over resistance has units of time, which are also the units of volume divided by the volumetric flowrate. As a result, Eq. (13.2.17) has exactly the same form as

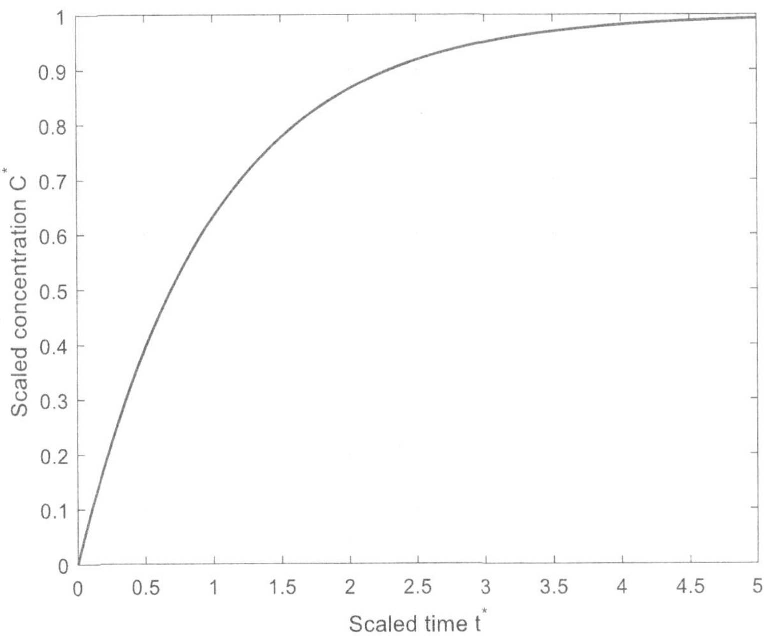

Figure 13.2 Starting up of a mixing tank in scaled variables.

Eqs. (13.2.10) and (13.2.15). They all belong to the general family of first order dynamical systems

$$\tau\frac{dy(t)}{dt}+y(t) = ku(t) \tag{13.2.18}$$

There are significant analogies between electrical, mechanical, thermal, and chemical systems, particularly in the way they can be modeled, analyzed, and understood. These similarities allow researchers and engineers to study one type of system, often the one that is simpler, more cost-effective, or easier to build and operate, in order to gain insights into another, more complex system. By developing dimensionless representations of these systems (e.g., through nondimensionalization or the use of similarity parameters), experimental results obtained from the more accessible system can be transferred or scaled to others. This approach leverages the underlying analogies between system behaviors such as mass and energy storage, dissipation, and transfer across different physical domains. As a result, analog modeling becomes a powerful tool in system analysis, enabling predictions and optimizations without the need to directly experiment with every type of system.

Example 13.1. Nondimensionalizing an equation from classical mechanics

Consider the equation of motion describing the elevation h of a body falling by gravity (in a uniform gravitational field) without air resistance (vacuum), as shown in Fig. 13.4. The initial elevation is h_0 and the initial velocity $u_0 > 0$. From high school physics

$$\frac{d^2h(t)}{dt^2} = -g, \quad h(0) = h_0, \frac{dh}{dt}(0) = u(0) = u_0 > 0 \tag{a}$$

The solution of this differential equation is.

$$h(t; h_0, u_0, g) = h_0 + u_0 t - \frac{1}{2}gt^2 \tag{b}$$

Propose a nondimensional form of this equation.

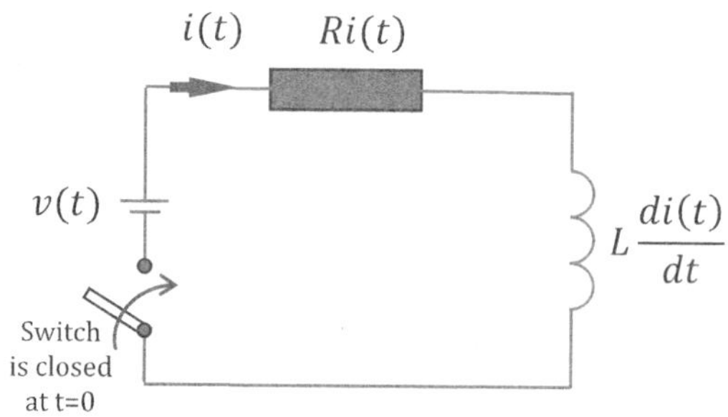

Figure 13.3 An RL electrical circuit.

As the solution depends on three parameters you need to execute n_p^3 experiments to represent graphically the evolution of the position of an object that falls in vacuum in a uniform gravitational field. Can you propose a more satisfactory approach?

Solution.

To nondimensionalize the mathematical model, we need to find scaling parameters for the elevation and the time. However, the problem does not give us any specific time scale or elevation scale (apart from the initial elevation, which, however, may be selected to be zero and is inappropriate for scaling as division by zero is possible). One interesting point is the point of maximum elevation h_{max}, which is observed at time t_{max}. To determine this point, we set the derivative of Eq. (b) equal to zero and solve for the time:

$$\frac{dh(t;h_0,u_0,g)}{dt} = u_0 - gt = 0 \quad \Rightarrow \quad t_{max} = \frac{u_0}{g} \tag{c}$$

The maximum elevation is obtained by replacing (c) into (b):

$$\Delta h_{max} = h_{max} - h_0 = u_0 t_{max} - \frac{1}{2} g t_{max}^2 = \frac{1}{2}\left(\frac{u_0^2}{g}\right) \tag{d}$$

We therefore propose using the scaled variables

$$t^* = \frac{t}{t_{max}} = \frac{t}{(u_0/g)} \tag{e}$$

$$h^* = \frac{h - h_0}{\Delta h_{max}} = \frac{h - h_0}{(u_0^2/2g)} \tag{f}$$

All terms in Eq. (b) have units of length, and we therefore divide by Δh_{max}.

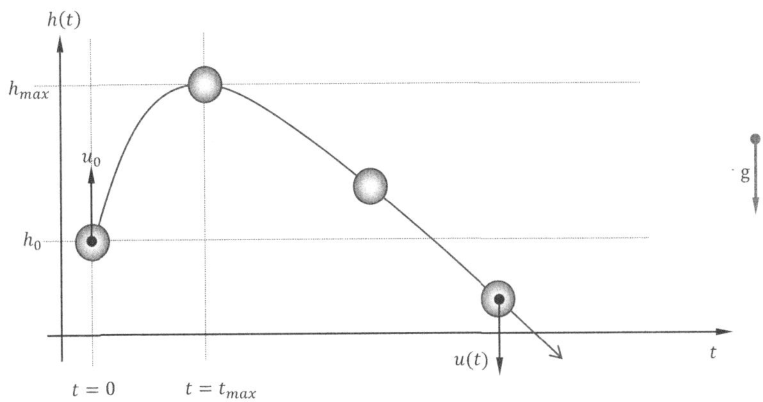

Figure 13.4 Object falling in a vacuum.

$$\frac{h - h_0}{\Delta h_{\max}} = \frac{u_0 t}{(u_0^2/2g)} - \frac{1}{2}\frac{gt^2}{(u_0^2/2g)}$$

or

$$h^* = 2\left(\frac{g}{u_0}\right)t - \frac{1}{2}2\left[\left(\frac{g}{u_0}\right)t\right]^2 = t^*(2 - t^*)$$

This result is shown in Fig. 13.5. It is impressive that this figure can be used to describe any experiment of free falling of an object in vacuum in a uniform gravitational field.

13.3 Dimensional similarity and the Buckingham π theorem

13.3.1 Similarity

The method that we applied in the previous section for the nondimensionalization is known as nondimensionalization by inspection. Its applicability is restricted to cases where the governing equations are already known, and this turns out to be a rather restrictive prerequisite. In many real-world engineering problems, the equations may be unknown or too complex to solve analytically. In such situations, experimentation often becomes the most practical approach to obtain reliable data. To

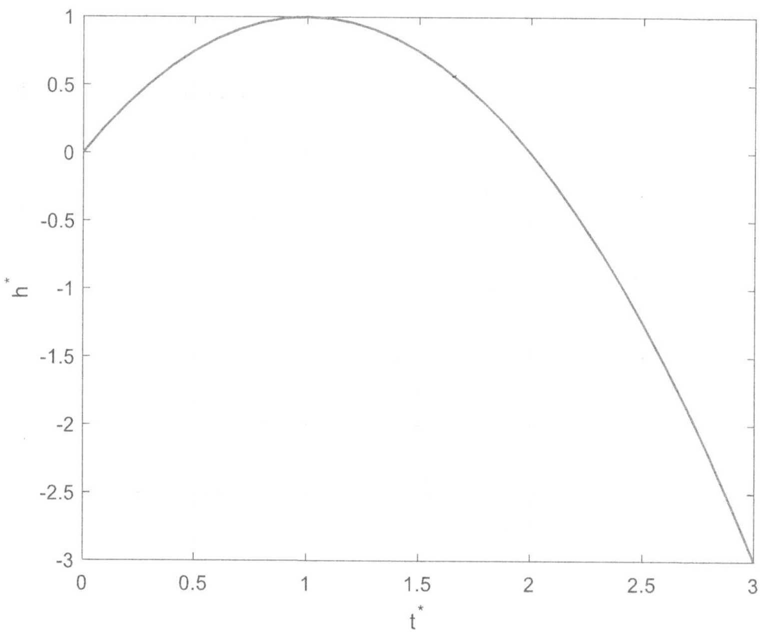

Figure 13.5 Object falling in a vacuum in dimensional variables.

reduce cost and time, experiments are typically conducted on geometrically scaled models rather than full-scale systems. When using models, it is essential to ensure the results are properly scaled to apply to the real system.

In many engineering experiments, conducting tests on the full-scale prototype can be impractical due to high costs, time constraints, or logistical challenges. To address this, engineers often use geometrically scaled models—smaller versions that preserve the shape and proportions of the original system. These models allow for more manageable and cost-effective testing while still capturing the essential behavior of the system under study.

When using scaled models, it is crucial to ensure that the results can be accurately translated—or scaled up—to predict the performance of the full-scale system. This process requires a systematic understanding of how different physical quantities relate across scales. To achieve this, we introduce a powerful analytical tool known as **dimensional analysis**. This technique allows engineers and scientists to identify the key dimensionless parameters that govern a system's behavior, ensuring that the experimental results from the model remain valid and applicable to the real-world application. This is where dimensional analysis becomes a valuable tool across all areas of engineering and science—particularly when designing and interpreting experiments. You are encouraged to apply this versatile method, as it can offer powerful insights in many disciplines.

Before introducing the technique of dimensional analysis itself, it is essential to understand the foundational concept on which it is based: the **principle of similarity**. This principle governs the conditions under which experimental results obtained from a scaled model (a bioreactor of 5 L in volume) can be reliably applied to a real-world system (a bioreactor of 100 m^3), known as the prototype. For complete similarity between a model and its prototype to exist, three distinct types of similarity must be satisfied: geometric, kinematic, and dynamic similarity.

Geometric Similarity. Geometric similarity requires that the model and the prototype have identical shapes, although they may differ in size. This means that all corresponding dimensions (lengths, widths, heights) of the model and prototype must relate through a constant length scale factor. For example, if a bioreactor is scaled up by a factor of 100:1, then every linear dimension in the model is 1:100 of the corresponding dimension in the prototype. Geometric similarity is the most basic of the three and serves as a prerequisite for the other types of similarity.

Kinematic Similarity. Kinematic similarity extends beyond shape and requires that the motion of fluid particles (or solid components) in the model corresponds proportionally to that in the prototype. Specifically, the velocities at corresponding points in the model and prototype must: be scaled by a constant velocity scale factor and point in the same relative directions. In this sense, kinematic similarity implies time-scale equivalence between model and prototype motion. For example, if the model is one-10th the size of the prototype (geometric scale <1), but the flow velocity around it is higher (velocity scale >1), the model flow may still be kinematically similar, provided the velocity vectors maintain proportional magnitudes and directions.

Dynamic Similarity. The most restrictive condition is dynamic similarity, which requires that the forces acting on the fluid or structure in the model scale proportionally with those in the prototype. This means that all relevant forces—such as pressure, viscous, inertial, and gravitational forces—must maintain the same relative magnitudes in both systems. This is often expressed in terms of dimensionless numbers (e.g., Reynolds, Froude, or Power numbers, that will be introduced shortly), which must match between the model and the prototype to achieve dynamic similarity.

13.3.2 The Buckingham π theorem

The **Buckingham π Theorem** is a fundamental theorem in dimensional analysis that provides a systematic way to reduce the number of variables in a physical problem by expressing them in terms of **dimensionless groups** (called **π-groups**). The theorem says:

If a physical problem involves n variables and these variables involve *k fundamental dimensions* (e.g., mass, length, time), then the problem can be reduced to a relationship among $(n - k)$ independent dimensionless groups (called π terms).

It is noted that:

- A dimensionless group is a product of variables raised to powers such that the combination has no units.
- These groups are often denoted $\pi_1, \pi_2, \ldots, \pi_{n-k}$, hence the name "$\pi$-theorem."
- The theorem helps simplify complex problems and is essential when similarity and scaling are important (e.g., model testing, equipment scale-up or down, heat or mass transfer, etc.).

Suppose a physical phenomenon is described by a relationship among n variables:

$$f(x_1, x_2, \cdots, x_n) = 0 \tag{13.3.1}$$

For the case of the free falling of an object in vacuum

$$f(h - h_0, t, u_0, g) = 0$$

which can also be expressed as

$$h - h_0 = f(t, u_0, g) \tag{13.3.2}$$

In this formulation, elevation $\Delta h = h - h_0$ is identified as the dependent variable, while time t serves as the independent variable. The parameters u_0 and g are specified by the underlying mathematical model, which is a distinct advantage when analyzing well-defined systems—an advantage often lacking in more complex or poorly understood systems.

Each variable has physical units and can be expressed in terms of k fundamental dimensions (such as mass M, length L, time T, temperature Θ, etc.). The Buckingham π theorem states that the equation can be recast into a dimensionless form:

$$f(\pi_1, \pi_2, \cdots, \pi_{n-k}) = 0 \tag{13.3.3}$$

where each π_i is a dimensionless product of the original variables. There are $n - k$ such independent dimensionless groups that capture the essential physics of the problem and are critical for modeling and scaling.

The following are the steps to apply the Buckingham π Theorem

Step 1: List all variables involved; include both dependent and independent variables. Do not include dependent variables (i.e., if you include the radius of a sphere, you cannot include its external area or volume).

Step 2: Express dimensions of each variable and write each variable in terms of basic dimensions (M, L, T, etc.).

Step 3: Count variables and fundamental dimensions. Let n = number of variables, k = number of independent fundamental dimensions involved.

Step 4: Determine the number of π groups: you will form $n - k$ independent π groups.

Step 5: Choose repeating variables: select k variables that a) are dimensionally independent, b) together contain all fundamental dimensions, and c) do not include the dependent variable.

Step 6: Form π groups. Each π group is formed by multiplying the dependent variable with powers of the repeating variables such that the result is dimensionless.

Step 7: Simplify and interpret.

To fully understand these steps, consider again the case of the free-falling object for which the elevation is the dependent variable and time is the independent variable. There are also two parameters involved, as shown in Eq. (13.3.2) (Step 1). The dimensions of each one of these variables are (Step 2).

$$[\Delta h] = \mathrm{L}, [u_0] = \mathrm{L\,T^{-1}}, [\mathrm{g}] = \mathrm{L\,T^{-2}}, [t] = \mathrm{T}$$

We therefore conclude that $n = 4$, $k = 2$ (Step 3) and $n - k = 2$ (Step 4), i.e., there are two π groups (Step 5). We choose u_0 and g to be the repeating groups as they satisfy all requirements of Step 5.

The first group will be determined

$$\pi_1 = \Delta h \cdot u_0^a \, g^b$$

where a and b are constant exponents to be determined. For π_1 to be dimensionless, the following needs to be satisfied

$$L^0T^0 = L \cdot \left(L\,T^{-1}\right)^a \left(L\,T^{-2}\right)^b$$

The following linear algebraic equations can be used to determine a and b

$$L: 1 + a + b = 0$$

$$T: -a - 2b = 0$$

From the second we have that $a = -2b$ and we replace in the first to obtain

$$b = 1, a = -2$$

The first π group follows immediately

$$\pi_1 = \Delta h \frac{g}{u_0^2} \tag{13.3.4}$$

For the second π group we have that

$$\pi_2 = t \cdot u_0^a \, g^b$$

or

$$L^0T^0 = T \cdot \left(L\,T^{-1}\right)^a \left(L\,T^{-2}\right)^b$$

The following linear equations ensure that π_2 is dimensionless

$$L: a + b = 0$$

$$T: 1 - a - 2b = 0$$

From the first one we have that $-a = b$ and we replace in the second to obtain

$$b = 1, a = -1$$

The second π group follows immediately

$$\pi_2 = t\frac{g}{u_0} \tag{13.3.5}$$

Finally, we write our results in the form of Eq. (13.3.3)

$$f\left(\Delta h\frac{g}{u_0^2}, t\frac{g}{u_0}\right) = 0 \tag{13.3.6}$$

This is Step 7 and the application of the Buckingham π theorem has been completed.

Let us now consider the CSTR with first-order reaction studied in the previous section (see Eq. (13.2.1)). The concentration of the reactant in the CSTR is the dependent variable and time is the independent variable. There are also four parameters involved (C_0, F, V, k). The dimensions of each one of these variables and parameters are (Step 2).

$$[C] = [C_0] = \mathrm{N\,L^{-3}}, \quad [t] = \mathrm{T}, \quad [F] = \mathrm{L^3\,T^{-1}}, \quad [V] = \mathrm{L^3},$$
$$[k] = \mathrm{T^{-1}}$$

We therefore conclude that $n = 6$, $k = 3$ (Step 3) and $n - k = 3$ (Step 4), i.e., there are three π groups (Step 5). We choose C_0, F and k to be the repeating variables as they satisfy all requirements of Step 5.

The first group will be determined

$$\pi_1 = C{\cdot}C_0^a\, F^b k^c$$

where a, b and c are constant exponents to be determined. For π_1 to be dimensionless, the following needs to be satisfied

$$\mathrm{N^0L^0T^0} = \mathrm{N\,L^{-3}}{\cdot}\left(\mathrm{N\,L^{-3}}\right)^a\left(\mathrm{L^3\,T^{-1}}\right)^b\left(\mathrm{T^{-1}}\right)^c$$

The following linear algebraic equations can be used to determine a and b

$$\mathrm{N}:\ 1 + a = 0$$

$$\mathrm{L}:\ -3 - 3a + 3b = 0$$

$$\mathrm{T}:\ -b - c = 0$$

From the first one it follows that $a = -1$, we replace in the second to obtain $b = 0$. From the third one $c = -b = 0$. The first π group follows immediately

$$\pi_1 = \frac{C}{C_0} = C^* \tag{13.3.7}$$

For the second π group we have that

$$\pi_2 = t \cdot C_0^a F^b k^c$$

or

$$\mathrm{L}^0\mathrm{T}^0 = \mathrm{T}\cdot\left(\mathrm{N}\,\mathrm{L}^{-3}\right)^a\left(\mathrm{L}^3\,\mathrm{T}^{-1}\right)^b\left(\mathrm{T}^{-1}\right)^c$$

The following linear equations ensure that π_2 is dimensionless

$$\mathrm{N}: a = 0$$

$$\mathrm{L}: -3a + 3b = 0$$

$$\mathrm{T}: 1 - b - c = 0$$

From the first one we have that $a = 0$ and we replace in the second to obtain $b = 0$. From the third equation we obtain $c = 1$ and the second π group is

$$\pi_2 = tk \tag{13.3.8}$$

For the third π group we have that

$$\pi_3 = V \cdot C_0^a F^b k^c$$

or

$$\mathrm{L}^0\mathrm{T}^0 = \mathrm{L}^3\cdot\left(\mathrm{N}\,\mathrm{L}^{-3}\right)^a\left(\mathrm{L}^3\,\mathrm{T}^{-1}\right)^b\left(\mathrm{T}^{-1}\right)^c$$

The following linear equations ensure that π_2 is dimensionless

$$\mathrm{N}: a = 0$$

$$\mathrm{L}: 3 - 3a + 3b = 0$$

$$\mathrm{T}: -b - c = 0$$

from which we obtain $a = 0$, $b = -1$, $c = 1$, i.e.,

$$\pi_3 = k\frac{V}{F} = \mathscr{D}a \tag{13.3.9}$$

Finally, we write our results in the form of Eq. (13.3.3)

$$f\left(\frac{C}{C_0}, kt, \mathscr{D}a\right) = 0 \tag{13.3.10}$$

This is Step 7 and the application of the Buckingham π theorem has been completed. Note that π_2 which corresponds to scaled time is different from the scaled time defined in Eq. (13.2.4). However, we may write:

$$kt\frac{1}{\mathscr{D}a} = kt\frac{F}{kV} = \frac{F}{V}t = t^*$$

and we may also conclude that

$$f(C^*, t^*, \mathscr{D}a) = 0 \qquad (13.3.11)$$

Example 13.2. Law of corresponding states

Consider the case of a real gas. In describing the P–V-T behavior of a real gas, the following eight variables are considered important:

1. the absolute pressure P.
2. the volume occupied by the gas V.
3. the number of moles of the gas N.
4. the absolute temperature T.
5. the critical pressure P_C.
6. the critical temperature T_C.
7. the critical volume V_C.
8. the ideal gas constant R.

Derive the general form of the real gas equation that describes its P–V-T behavior.

Solution.

In Table 13.1, the dimensions of many basic physical quantities are summarized for easy reference. From this table we obtain the dimensions of the eight variables appearing in the model under study:

$$[P] = [P_c] = \mathrm{L^{-1}M^1T^{-2}}, \quad [T] = [T_c] = \Theta, \quad [\mathrm{V}] = [\mathrm{V_c}] = \mathrm{L}^3,$$
$$[N] = \mathrm{M}, \quad [R] = \mathrm{L^2M^0T^{-2}}\Theta^{-1}$$

There are four fundamental dimensions involved and $n = 8$, $k = 4$ (Step 3) and $n - k = 4$ (Step 4), i.e., there are four π groups (Step 5). We choose P, V, T, and R to be the repeating variables as they satisfy all requirements of Step 5.

The first group will be determined

$$\pi_1 = N{\cdot}P^aV^bT^cR^d$$

where a, b, c, and d are constant exponents to be determined. For π_1 to be dimensionless, the following needs to be satisfied

$$L^0M^0T^0\Theta^0 = M\cdot(L^{-1}M^1T^{-2})^a(L^3)^b(\Theta)^c(L^2T^{-2}\Theta^{-1})^d$$

$$L:\ -a+3b+2d = 0$$

$$M:\ 1+a = 0$$

$$T:\ -2a-2d = 0$$

$$\Theta:\ c-d = 0$$

It then follows that $a = b = -1, c = d = 1$ or

$$\pi_1 = N\cdot\frac{RT}{PV} = Z$$

i.e., π_1 is the compressibility factor. We continue with π_2:

Table 13.1 Dimensions of basic physical quantities.

Physical quantity	Symbol	L	M	T	Θ	LMTΘ
Length	l	1				$L^1M^0T^0\Theta^0$
Mass	m,n		1			$L^0M^1T^0\Theta^0$
Time	t			1		$L^0M^0T^1\Theta^0$
Temperature	T				1	$L^0M^0T^0\Theta^1$
Velocity	u	1		−1		$L^1M^0T^{-1}\Theta^0$
Acceleration	a = l	1		−2		$L^1M^0T^{-2}\Theta^0$
Area	A	2				$L^2M^0T^0\Theta^0$
Volume	V	3				$L^3M^0T^0\Theta^0$
Volumetric flowrate	F	3		−1		$L^3M^0T^{-1}\Theta^0$
Density	ρ = m/V	−3	1			$L^{-3}M^1T^0\Theta^0$
Force	f = ma	1	1	−2		$L^1M^1T^{-2}\Theta^0$
Pressure	P = f/A	−1	1	−2		$L^{-1}M^1T^{-2}\Theta^0$
Energy (heat, work)	E = f l	2	1	−2		$L^2M^1T^{-2}\Theta^0$
Power	P = E/t	2	1	−3		$L^2M^1T^{-3}\Theta^0$
Ideal gas constant	R	2		−2	−1	$L^2M^0T^{-2}\Theta^{-1}$
Viscosity	μ	−1	1	−1		$L^{-1}M^1T^{-1}\Theta^0$
Specific heat capacity	c_p	2		−2	−1	$L^2M^0T^{-2}\Theta^{-1}$
Heat transfer coefficient	h		1	−3	−1	$L^0M^1T^{-3}\Theta^{-1}$
Thermal conductivity	k	1	1	−3	−1	$L^1M^1T^{-3}\Theta^{-1}$
Diffusivity	D, a	2		−1		$L^2M^0T^{-1}\Theta^0$

$$\pi_2 = P_c \cdot P^a V^b T^c R^d$$

where a, b, c, and d are constant exponents to be determined. For π_2 to be dimensionless, the following needs to be satisfied

$$L^0 M^0 T^0 \Theta^0 = L^{-1} M^1 T^{-2} \cdot \left(L^{-1} M^1 T^{-2}\right)^a \left(L^3\right)^b (\Theta)^c \left(L^2 T^{-2} \Theta^{-1}\right)^d$$

$$L: \; -1 - a + 3b + 2d = 0$$

$$M: 1 + a = 0$$

$$T: \; -2 - 2a - 2d = 0$$

$$\Theta: c - d = 0$$

It then follows that $a = -1, b = c = d = 0$ or

$$\pi_2 = \frac{P_c}{P}$$

or equivalently

$$\pi_2 = \frac{P}{P_c} = P_r$$

i.e., π_2 is the reduced pressure. In a similar fashion we "discover" π_3.

$$\pi_3 = T_c \cdot P^a V^b T^c R^d$$

where a, b, c, and d are constant exponents to be determined. For π_3 to be dimensionless, the following needs to be satisfied

$$L^0 M^0 T^0 \Theta^0 = \Theta \cdot \left(L^{-1} M^1 T^{-2}\right)^a \left(L^3\right)^b (\Theta)^c \left(L^2 T^{-2} \Theta^{-1}\right)^d$$

$$L: \; -a + 3b + 2d = 0$$

$$M: a = 0$$

$$T: \; -2a - 2d = 0$$

$$\Theta: 1 + c - d = 0$$

It then follows that $a = b = d = 0, c = -1$, or

$$\pi_3 = \frac{T_c}{T} \quad \Rightarrow \quad \pi_3 = T_r$$

Finally, for π_4 we have that

$$\pi_4 = V_c \cdot P^a V^b T^c R^d$$

where *a*, *b*, *c*, and *d* are constant exponents to be determined. For π_4 to be dimensionless, the following needs to be satisfied

$$L^0M^0T^0\Theta^0 = L^3 \cdot (L^{-1}M^1T^{-2})^a (L^3)^b (\Theta)^c (L^2T^{-2}\Theta^{-1})^d$$

$$L: 3 - a + 3b + 2d = 0$$

$$M: a = 0$$

$$T: \; -2a - 2d = 0$$

$$\Theta: c - d = 0$$

It then follows that $a = c = d = 0, b = -1$, or

$$\pi_4 = V_r$$

These results imply that there must be a functional relationship of the following form:

$$f(Z, P_r, T_r, V_r) = 0$$

This equation can be used to prove the law of corresponding states.

Example 13.3. Period of a pendulum

A pendulum is essentially a weight (often called a "bob") suspended from a fixed pivot point, allowing it to swing freely back and forth under the influence of gravity. Its key characteristics and concepts are:

Swinging Motion: When a pendulum is displaced from its resting position (equilibrium) and released, gravity acts as a restoring force, pulling it back toward the center. This causes it to swing back and forth in an oscillatory motion.

Period T: The time it takes for a pendulum to complete one full back-and-forth swing (one complete cycle) is called its period. For small swings, the period of a simple pendulum is remarkably consistent.

One student believes that the factors affecting period are the length l (the primary factor determining a simple pendulum's period is its length, and longer pendulums have longer periods as they swing more slowly), gravity g (the local acceleration due to gravity also affects the period, and a stronger gravitational field will result in a shorter period), its mass m (for a simple pendulum), and the angular displacement of the pendulum from its equilibrium position θ.

Can you develop what are the important dimensionless groups of parameters that determine the period of oscillation of a pendulum?

Solution.

It is suggested that

$$f(T, l, g, m, \theta) = 0$$

The period T is the dependent variable and we may consider θ to be the independent variable. Note that

$$[T] = \mathrm{T}, \quad [l] = \mathrm{L}, \quad [g] = \mathrm{L\,T^{-2}}, \quad [\theta] = \text{dimensionless}, \quad [m] = \mathrm{M}$$

We note that the dimension M appears in only one variable, the mass m. However, when a fundamental dimension (like M in our case) appears in only one of the chosen variables, it becomes impossible to form a dimensionless group that includes that variable. Why? Because to make a group dimensionless, you need to combine variables in such a way that all the dimensions cancel out. If mass M only exists in m, there's no other variable to "cancel out" the M dimension. When you encounter this situation, you have two primary ways to proceed:

1. *Re-evaluate the Variable List*: the first alternative is to meticulously re-examine your initial list of variables. This involves asking: "Is there any other physical quantity relevant to the pendulum's dynamics that I've missed, and importantly, does it contain the 'missing' dimension (in this case, mass)?" For a simple pendulum, discovering an additional mass-related variable that significantly impacts its fundamental period is indeed unlikely. This is because, ideally, the simple pendulum's motion is primarily governed by length and gravity. However, in more complex scenarios, you might consider: mass of the string: If the string itself has significant mass, its distribution could affect the dynamics, but this would make it a "compound pendulum." Air density/viscosity: If air resistance were a significant factor, air density (which has mass dimensions) would be relevant. While these might be relevant for more advanced analyses, for the basic simple pendulum, it's generally accepted that mass does not play a role in the period.
2. *Discard the Problematic Variable*: the second and often correct alternative for a simple pendulum is to discard the variable that's causing the dimensional isolation problem. In this specific case, it means concluding that the mass (m) of the

pendulum bob does not affect its period of oscillation. This isn't to say mass is irrelevant to all aspects of the pendulum (e.g., its momentum or kinetic energy certainly depends on mass). However, the Buckingham π Theorem, when applied to the period of a simple pendulum, strongly suggests that mass drops out of the relationship. This aligns perfectly with experimental observations, as you may recall from elementary physics. Therefore, for a simple pendulum, the conclusion from dimensional analysis would be to remove m from the list of influencing variables when determining the period and revise our functional relationship:

$$f(T, l, \mathrm{g}, \theta) = 0$$

As we have four variables and two dimensions, $n - k = 4 - 2 = 2$. We select g, l as the repeating variables and form two dimensionless groups:

$$\pi_1 = T \cdot \mathrm{g}^{\mathrm{a}} l^b$$

or

$$\mathrm{L}^0\mathrm{T}^0 = \mathrm{T} \cdot \left(\mathrm{L\,T}^{-2}\right)^a (\mathrm{L}\,)^b$$

It then follows that

$$\mathrm{L}: a + b = 0$$

$$\mathrm{T}: 1 - 2a = 0$$

and $a = -b = \frac{1}{2}$, or

$$\pi_1 = T \cdot \sqrt{\frac{\mathrm{g}}{L}}$$

In a similar way

$$\pi_2 = \theta \cdot \mathrm{g}^{\mathrm{a}} l^b$$

or

$$\mathrm{L}^0\mathrm{T}^0 = \left(\mathrm{L\,T}^{-2}\right)^a (\mathrm{L}\,)^b$$

It then follows that

$$\mathrm{L}: a + b = 0$$

$$\mathrm{T}: -2a = 0$$

and $a = -b = 0$, or

$$\pi_2 = \theta$$

The conclusion is that there must be a functional relationship of the form

$$f\left(T\sqrt{\frac{g}{L}}, \theta\right) = 0$$

Example 13.4. Spherical object moving with constant speed in a fluid

Consider a spherical object with diameter d moving with a constant speed u in a stagnant viscous fluid with density ρ_f and viscosity μ_f, as shown in Fig. 13.6. There is a friction force, referred to as drag force, acting on the spherical object denoted by F_D. Use the Buckingham π theorem to develop a functional form of the dependence of the drag force on ρ_f, μ_f, d and u.

Solution.

The following is suggested

$$f\left(F_D, u, \rho_f, \mu_f, d\right) = 0$$

The dimensions of the relative physical quantities are (see Table 13.1)

$$[F_D] = \mathrm{L\,M\,T^{-2}}, \quad [u] = \mathrm{L\,T^{-1}}, \quad \left[\rho_f\right] = \mathrm{L^{-3}\,M}, \quad \left[\mu_f\right] = \mathrm{L^{-1}\,M\,T^{-1}}, \quad [d] = \mathrm{L}$$

There are five variables and three dimensions, i.e., $n - k = 5 - 3 = 2$. We need to select three repeating variables. Select ρ_f, d, u. There are two dimensionless groups. For the first one we have that

$$\pi_1 = F_D \cdot \rho_f^a d^b u^c$$

or

$$\mathrm{L^0 M^0 T^0} = \mathrm{L\,M\,T^{-2}} \cdot \left(\mathrm{L^{-3}\,M}\right)^a (\mathrm{L})^b \left(\mathrm{L\,T^{-1}}\right)^c$$

$$\mathrm{L}: 1 - 3a + b + c = 0$$

$$\mathrm{M}: 1 + a = 0$$

$$\mathrm{T}: -2 - c = 0$$

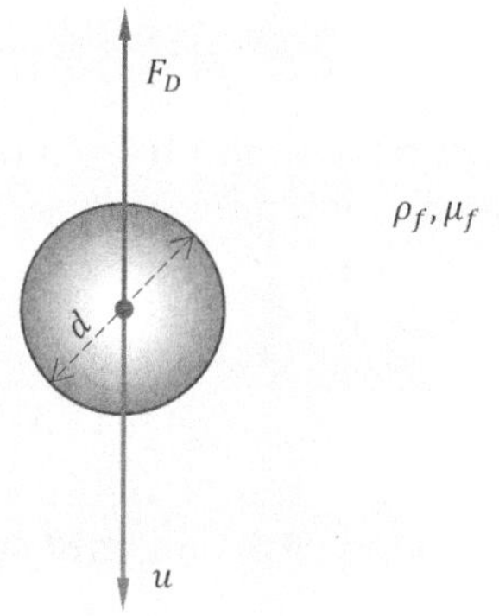

Figure 13.6 An object falling in a viscous fluid.

The solution is $a = -1, c = -2$ and $b = -2$, or

$$\pi_1 = \frac{F_D}{\rho_f d^2 u^2}$$

For the second dimensionless group

$$\pi_2 = \mu_f \cdot \rho_f^a d^b u^c$$

or

$$L^0 M^0 T^0 = L^{-1}\, M\, T^{-1} \cdot (L^{-3}\, M)^a (L)^b (L\, T^{-1})^c$$

$$L: -1 - 3a + b + c = 0$$

$$M: 1 + a = 0$$

$$T: -1 - c = 0$$

The solution is $a = b = c - 1$, or

$$\pi_2 = \frac{\mu_f}{\rho_f d u}$$

The drag coefficient, that is defined in fluid mechanics, is identical to the dimensionless group π_1 up to a numerical constant:

$$C_D = \frac{F_D}{\frac{1}{2}\rho_f u^2 \left(\frac{\pi d^2}{4}\right)} \sim \pi_1$$

π_2 is the inverse of the most well known dimensionless number known as the Reynolds number and denoted by Re:

$$\text{Re} = \frac{ud\rho}{\mu} = \frac{1}{\pi_2}$$

We therefore expect that the following holds true:

$$C_D = f(\text{Re})$$

Actual experimental data for the drag coefficient of spherical objects over a wide range of Reynolds numbers are given in Fig. 13.7. For low Reynolds numbers the **Stokes law** is valid

$$C_D = \frac{24}{\text{Re}}, \text{Re} < 2$$

while for $2 \leq \text{Re} \leq 200{,}000$

$$C_D = \frac{24}{\text{Re}}\left(1 + 0.15\text{Re}^{0.681}\right) + \frac{0.407\text{Re}}{\text{Re} + 8710}$$

The predictions of the Buckingham π theorem are validated by these experimental results for a wide range of Reynolds numbers.

13.3.3 Two interesting applications of the Buckingham π theorem

In this section, we will present two challenging applications of the Buckingham π Theorem. The first one deals with the characteristics of walking in humans and bipedal animals. We will see that the Buckingham π Theorem predicts a reasonably accurate model for human walking that can be extended to make reasonable predictions for other bipedal animals, including *T. Rex*! The second application deals with the estimation of the power of (nuclear) explosions from their visualization. This is another successful application of the Buckingham π Theorem that clearly demonstrates how careful application of dimensional analysis can provide reliable solutions for experimental modeling of complex chemical and physical phenomena, even when our theoretical understanding remains limited.

Human walking is an extraordinary process. A properly functioning locomotor system relies on a complex interplay of signals from the motor cortex, cerebellum, and basal ganglia, along with sensory feedback from vision, the vestibular system, and proprioception. These inputs are integrated to generate precise motor

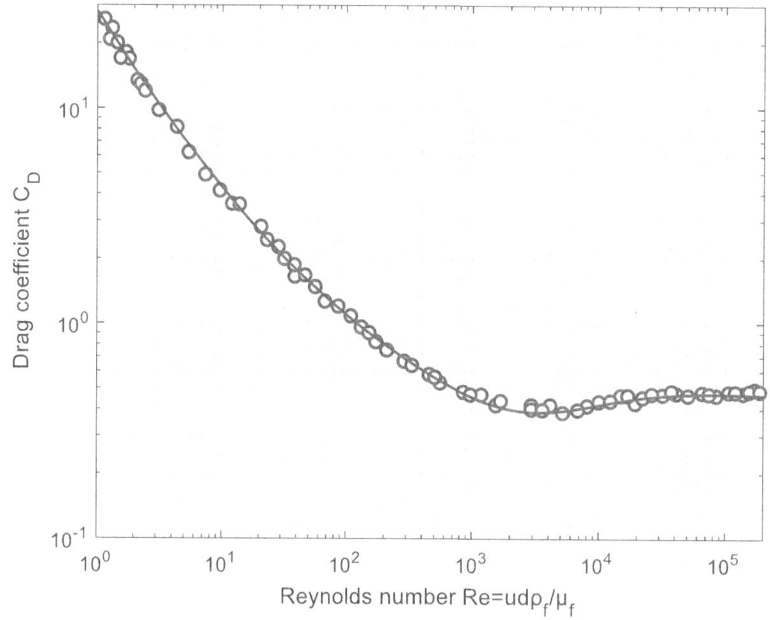

Figure 13.7 The drag coefficient for spherical objects as a function of the Reynolds number. Data taken from Brown and Lawler. March 2003. J. of Environ. Eng. 222

outputs that coordinate muscle activity and limb motion. When all components operate effectively, this hierarchical neural network maintains a stable and repeatable walking pattern. Remarkably, the forces, joint motions, and muscle actions involved in gait tend to remain consistent across successive steps, even during natural, unstructured walking. Considering that the upright human body functions like an inherently unstable inverted pendulum, achieving smooth, balanced forward movement with such regularity is quite a sophisticated accomplishment.

Each forward movement of a foot is a step. A stride consists of two steps, right and left, and represents the gait cycle. The heel is the most commonly used point of reference because the heel is—or should be—the first point of contact with the supporting surface (see Fig. 13.8). Stride length (s) and walking speed (u) can be measured easily using a stopwatch to measure the time needed to cover a known distance and also count the number of steps taken.

Neutral or balanced leg length (l) is defined as 50% of our total stature (h). That ratio helps achieve the most efficient biomechanical human gait.

We claim that there exists a functional relationship of the following form

$$f(u, s, l, g) = 0 \tag{13.3.12}$$

Note that body mass is not included in the list of important variables (see the case study of the pendulum). The dimensions are

$$[u] = \mathrm{L\,T^{-1}}, \quad [s] = [l] = \mathrm{L}, \quad [g] = \mathrm{L\,T^{-2}}$$

There are four variables and two dimensions, i.e., $n - k = 4 - 2 = 2$. We need two repeating variables. We select the repeating variables to be the leg length l and the acceleration of gravity g. The first dimensionless group is

$$\pi_1 = u{\cdot}l^a g^b$$

or

$$\mathrm{L^0M^0T^0} = \mathrm{L\,T^{-1}}{\cdot}(\mathrm{L})^a\left(\mathrm{L\,T^{-2}}\right)^b$$

$$\mathrm{L}: 1 + a + b = 0$$

$$\mathrm{T}: -1 - 2b = 0$$

The solution is $b = -1/2, a = -1/2$, or

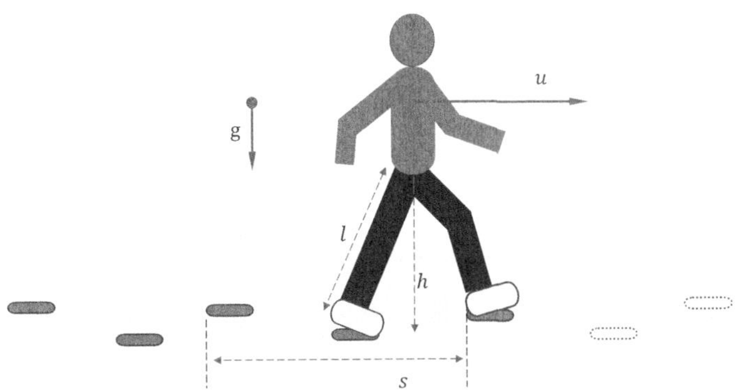

Figure 13.8 Description of main parameters appearing in human walking.

$$\pi_1 = \frac{u}{\sqrt{gl}} \; or \; \pi_1 = \frac{u^2}{gl}$$

For the second dimensionless group

$$\pi_2 = s \cdot l^a g^b$$

or

$$L^0 M^0 T^0 = L \cdot (L)^a \left(L\, T^{-2}\right)^b$$

$$L: 1 + a + b = 0$$

$$T: -2b = 0$$

The solution is $b = 0, a = -1$, or

$$\pi_2 = \frac{s}{l}$$

These two dimensionless groups imply a relationship of the following form:

$$\frac{u^2}{gl} = f\left(\frac{s}{l}\right) \tag{13.3.13}$$

The validity of Eq. (13.3.13) can be tested by using a stopwatch to count the time necessary to cover a given distance and at the same time measure the number of steps taken. The stride length is calculated as the distance divided by the number of steps multiplied by 2. The leg length is taken to be half of the height, and the walking speed is simply the distance divided by the time. This is so simple that you can try yourself. Data generated in this way are shown in Fig. 13.9, where the first dimensionless group is shown as a function of the scaled stride length. The linear least squares line that is shown in the figure is the following

$$\frac{u^2}{gl} = 0.0447\left(\frac{s}{l}\right)^{3.5323} \tag{13.3.14}$$

We will now try to answer the first question posed at the opening of this chapter: what was the running speed of *T. rex?* On the big screen, the *T. rex* is often depicted as a predator that could easily catch up to a speeding car with a few swift stomps. That's probably because paleontologists had suggested that the *T. rex* clocked a top speed of 50 km/h and a walking speed between 7 and 10 km/h. Let us use Eq. (13.3.14) to get our own estimation. We need the leg length and the stride length. The leg length of *T. rex* reported in the literature is around 3.5 m.

In the 2009 World Athletics Championships in Berlin, Usain Bolt achieved a legendary feat by breaking his own world records in the 100 m. He clocked 9.58 s in the 100 m, significantly improving his previous Olympic record of 9.69 s set in Beijing the previous year. For *T. rex* to run as fast as Usain Bolt when he achieved the world record in 2009 Olympic Games of Berlin.

$$\frac{u^2}{gl} = \frac{\left(\dfrac{100\ \text{m}}{9.58\ \text{s}}\right)^2}{9.81\dfrac{\text{m}}{\text{s}^2} \times 3.5\ \text{m}} = 3.1735$$

The relative stride length for achieving such a speed is (see Eq. (13.3.14))

$$\frac{s}{l} = \left[\frac{1}{0.0447}\left(\frac{u^2}{gl}\right)\right]^{\frac{1}{3.5323}} = 3.34$$

and

$$s = 3.34 \times l = 3.34 \times 3.5\ \text{m} = 11.7\ \text{m}$$

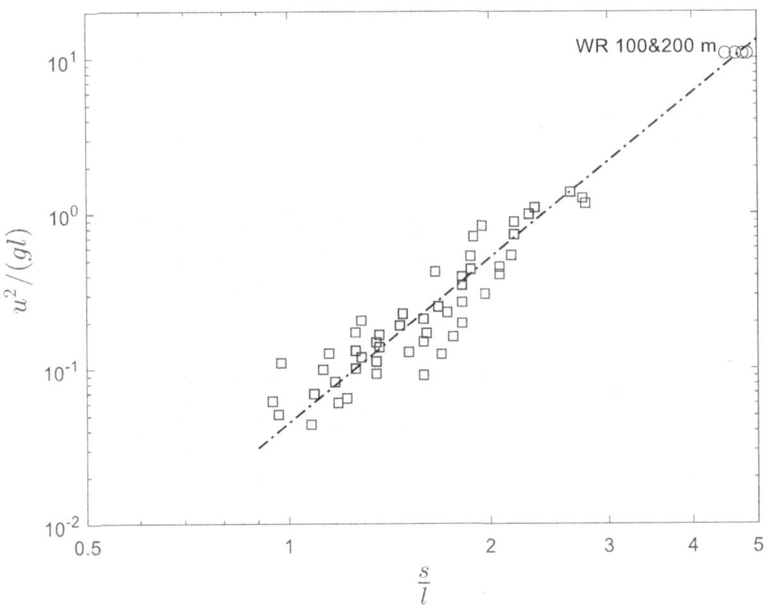

Figure 13.9 Data on the walking or running speed of humans as a function of the scaled stride length.

T. rex specimen "Sue" FMNH PR 2081 on display at the Field Museum of Natural History, Chicago, Illinois is reported as the largest, most extensive, and best-preserved *T. rex* specimen ever found, being reconstructed at more than 12 m long, which is comparable to the stride length calculated above.

A recent detailed study by Sellers et al. (2017) Sellers et al. (2017) titled "Investigating the running abilities of Tyrannosaurus rex using stress-constrained multibody dynamic analysis" concluded that *T. rex* was incapable of achieving a stride length greater than 7 m. Applying Eq. (13.3.14), with a stride-to-leg-length ratio of $s/l = 7/3.5 = 2$, yields an estimated speed $u \approx 4.2$ m/s, or roughly 15 km/h. This is less than half the top speed reached by Usain Bolt during his 100-meter world record sprint, and well within the running ability of many humans. These findings highlight the implausibility of scenes in Jurassic Park where a *T. rex* is shown outrunning both people and vehicles—such portrayals are clearly inconsistent with biomechanical evidence.

Data on the second interesting case study are presented in Table 13.2 and Fig. 13.10. Some motion picture records of the first atomic explosion in New Mexico that took place in 1945, were declassified in 1949. These pictures show not only the shape of the luminous globe which rapidly spread out from the detonation center, but also gave the time, t, of each exposure after the instant of initiation. On each series of photographs a scale is also marked so that the rate of expansion of the globe, or "ball of fire," can be found. Two series of declassified photographs are shown in an early publication (Taylor, 1950). These photographs show that the ball of fire assumes at first the form of a rough sphere, but that its surface rapidly becomes smooth. The atomic explosive was fired at a height of 100 ft (30.5 m) above the ground and the bottom of the ball of fire reached the ground in less than 1 ms. The impact on the ground does not appear to have disturbed the conditions in the upper half of the globe, which continued to expand as a nearly perfect luminous hemisphere bounded by a sharp edge which must be taken as a shock wave.

Our aim is to develop the functional form of a potential model that relates the energy of explosion E to the time t a photograph is taken, the blast wave radius R, and the density of undisturbed air ρ_0, i.e.,

$$f(E, t, R, \rho_0) = 0 \tag{13.3.15}$$

We note that there are four variables and three dimensions (see Table 13.1)

Table 13.2 Radius *R* of blast wave at time *t* after firing an atomic explosion.

Time after Firing (ms)	Ball radius (m)	Time after Firing (ms)	Ball radius (m)
Test 1		**Test 3**	
0.10	11.1	3.26	59.0
0.24	19.9	3.53	61.1
0.38	25.4	3.80	62.9
0.52	28.8	4.07	64.3
0.66	31.9	4.34	65.6
0.80	34.2	4.61	67.3
0.94	36.3		
Test 2		**Test 4**	
1.08	38.9	15	106.5
1.22	41.0	25	130.0
1.36	42.8	34	145.0
1.50	44.4	53	175.0
1.65	46.0	62	185.0
1.79	46.9		
1.93	48.7		

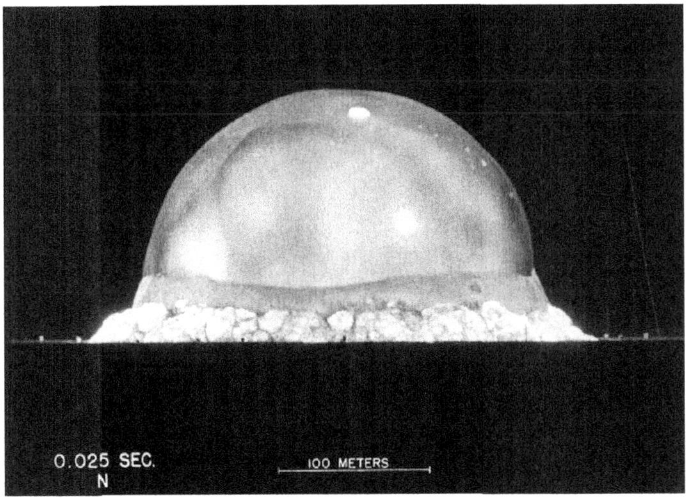

Figure 13.10 Nuclear test in New Mexico 0.025 s after firing– public domain photo from Wikipedia (July 16, 1945, 5:29:45 a.m., Trinity Site, Alamogordo Test Range, Jornada del Muerto Desert. Yield: 19 - 21 kt, U.S. Department of Defense).

$$[E] = L^2M^1T^{-2}, \quad [t] = T, \quad [R] = L, \quad [\rho_0] = M\,L^{-3}$$

We conclude that $n = 3, k = 3$ and $n - k = 1$, i.e., only one dimensionless parameter is involved which can easily be determined:

$$\pi = E^a t^b R^c \rho_0^d$$

and

$$L^0M^0T^0 = \left(L^2M^1T^{-2}\right)^a (T)^b (L)^c \left(M\,L^{-3}\right)^d$$

It then follows that

$$\pi = \rho_0 \frac{R^5}{t^2 E} = \text{constant} \tag{13.3.16}$$

The last one is usually written as follows

$$\underbrace{\frac{5}{2}\log_{10}(R(t))}_{Y(t)} = \underbrace{\frac{1}{2}\log_{10}\left(\text{constant}\cdot\frac{E}{\rho_0}\right)}_{a} + \underbrace{\log_{10}(t)}_{X(t)} \tag{13.3.17}$$

It is noted that several studies have shown that *constant* $= 1.168$ and $\rho_0 = 1.23\ \text{kg/m}^3$.

In Fig. 13.11 the data of Table 13.2 are shown after they have been transformed into the form of Eq. (13.3.17). It is shown that

$$\frac{1}{2}\log_{10}\left(\text{constant}\cdot\frac{E}{\rho_0}\right) = 6.942 \tag{13.3.18}$$

from which it follows that

$$\begin{aligned} E = \frac{\rho_0}{\text{constant}} 10^{2\times 6.942} &= \frac{1.23}{1.168} 10^{2\times 6.942} \\ &= 80.624 \times 10^{12}\ \text{J} \times \frac{1\ \text{TJ}}{10^{12}\ \text{J}} \times \frac{240\ \text{t TNT}}{1\ \text{TJ}} \\ &= 19.3\ \text{kt TNT} \end{aligned}$$

In the official documents released by the U.S. government the power of the nuclear test is reported as 15-20 kt TNT.

We can use the model for explosion power given by Eq. (13.3.17) to estimate the power of the explosion that took place in Beirut, Lebanon on August 4, 2020 (https://en.wikipedia.org/wiki/2020_Beirut_explosion). On August 4, 2020, a fire broke out in Warehouse 12 at the Port of Beirut. Warehouse 12, which was waterside and adjacent to the grain elevator, stored the ammonium nitrate that had been confiscated from MV *Rhosus* alongside a stash of fireworks. The explosion resulted in at least 218 fatalities, 7000 injuries, and approximately 300000 displaced individuals,

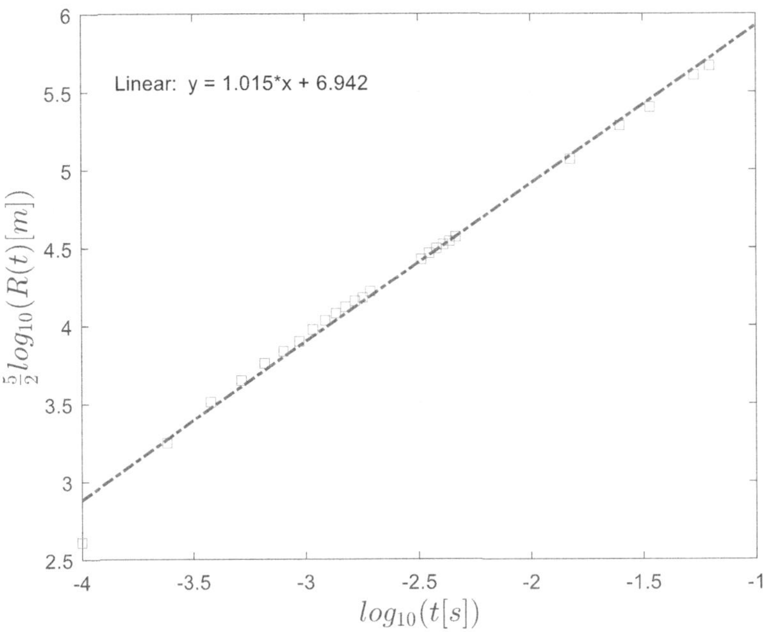

Figure 13.11 Data of Table 13.2 in the form of Eq. (13.3.17).

alongside property damage estimated at US$15 billion. The explosion was among the most powerful non-nuclear explosions ever recorded and the largest single detonation of ammonium nitrate.

From pictures reported in the press it was calculated that the radius of the blast wave increased to 100 m 0.083 s after the explosion. From this pair of data we may estimate the power of the Beirut port ammonium nitrate explosion

$$\begin{aligned} E &= \frac{\rho_0}{\text{constant}} 10^{\left[\log\left(R^5(t)\right)-\log\left(t^2\right)\right]} \\ &= \frac{1.23}{1.168} 10^{\log_{10}\left(100^5(t)\right)-\log_{10}\left(0.083^2\right)} \\ &= 1.53 \times 10^{12}\text{J} = 1.53\ \text{TJ} \approx 367\ \text{t TNT} \end{aligned}$$

The U.S. Federal Bureau of Investigation estimated that approximately 500 t of ammonium nitrate remained in the warehouse at the time of the explosion—a figure later cited by the Lebanese Prime Minister Hassan Diab and close to our findings (see also the relevant discussion on Wikipedia). Ammonium nitrate is typically assigned a TNT equivalence of 25%–50%, meaning that 1 t of ammonium nitrate is roughly equivalent to 0.25 to 0.5 t of TNT in explosive energy. Based on this range, the estimated

TNT-equivalent yield from the ammonium nitrate stored in Warehouse 12 would be between 730 t and 1470 t.

13.4 Process scale-up

13.4.1 The 50 Foot Woman—A challenging scale-up problem

Process equipment scale-up is the transition of a chemical, pharmaceutical, or industrial process from a smaller laboratory or pilot scale to a larger production or commercial scale. The goal is to maintain product quality, safety, and process efficiency while increasing the volume of production. Assume that one has performed a large number of experiments and has succeeded in optimizing a new process in a laboratory reactor that is 5 L in volume. However, achieving a commercially relevant rate of production of 1 kt per year requires a 10000 L reactor. The question is how to scale-up such a process so as to maintain optimality of operation? This is far from a trivial question to answer.

To introduce the concept of scaling up, let us begin with a more transparent example. Refer to Fig. 13.1, where the body of the actress was chosen to be scaled up linearly for commercial purposes. In this representation, no part of the body has been distorted or adjusted nonlinearly to account for biological or structural robustness reasons. The principle of uniform, proportional scaling—where a constant factor is applied uniformly to all dimensions—is further illustrated in Fig. 13.12. The actress in question, Allison Hayes, had a height of approximately 169 cm and an estimated body weight between 55 and 60 kg. Suppose we wish to estimate her body weight when she is scaled linearly to a height of 50 ft (or 1524 cm).

For simplicity, we model the human body as a composite of six cylindrical segments, as shown in Fig. 13.12. This geometric approximation allows us to relate the weight of the scaled-up figure (the prototype) to that of the original (the model). Let d_i and h_i represent the diameter and height, respectively, of the i-th cylindrical segment of the model. By analyzing how volume—and therefore mass—scales with these dimensions, we can estimate the total weight of the enlarged figure under the assumption of constant body density ρ and constant scaling factor λ:

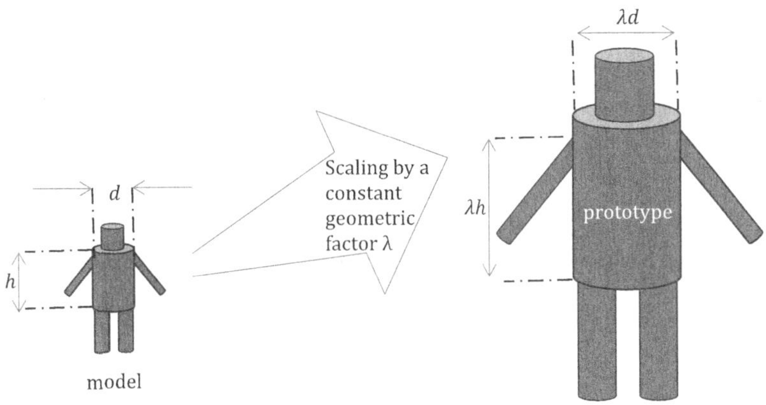

Figure 13.12 Linear scale-up of the human body approximation.

$$m_{\text{prototype}} = \rho \sum_{i=1}^{6} \frac{\pi(\lambda d_i)^2}{4} (\lambda h_i) = \lambda^3 \cdot \rho \sum_{i=1}^{6} \frac{\pi(d_i)^2}{4} (h_i) = \lambda^3 m_{\text{model}} \tag{13.4.1}$$

As the actress's height has been scaled by a constant **geometric factor** $\lambda = 1524 \text{ cm}/169 \text{ cm} = 9$ we can estimate the potential body weight of the prototype:

$$m_{\text{prototype}} = \lambda^3 m_{\text{model}} \approx 9^3 \times 60 = 43740 \text{ kg} \tag{13.4.2}$$

i.e., although the height is scaled by a factor of 9, the body weight is scaled by a factor of $9^3 = 729$! Is this a matter of concern?

The femur is the thigh bone—the longest, heaviest, and strongest bone in the human body. The axial stress received by the femur when standing on one foot at rest is (the diameter of a woman's femur is 2.5 cm):

$$(\text{Axial stress})_{model} = \frac{\text{Force}}{\text{Area}} = \frac{m_{model} g}{\frac{\pi d_{femur,model}^2}{4}} = \frac{60 \text{ kg} \times 9.81 \frac{\text{m}}{\text{s}^2}}{\pi \frac{0.025^2}{4} \text{ m}^2}$$
$$= 1.2 \times 10^6 \frac{\text{kg}}{\text{m s}^2} = 1.2 \text{ MPa} \tag{13.4.3}$$

The ultimate compressive strength of the femur is around 170–230 MPa (depends on age, health, and bone density).

We now repeat the stress calculation for the prototype:

$$(\text{Axial stress})_{\text{prototype}} = \frac{\text{Force}}{\text{Area}} = \frac{m_{\text{prototype}}g}{\frac{\pi d^2_{femur,\text{prototype}}}{4}} = \frac{\lambda^3\, m_{\text{model}}g}{\lambda^2 \frac{\pi d^2_{femur,\text{model}}}{4}}$$
$$= \lambda(\text{Axial stress})_{\text{model}} = 12 \text{ Mpa} \tag{13.4.4}$$

When running or jumping the axial stress might be 5 times that of the axial stress at rest. We therefore see that the structural integrity of the prototype is not ensured. Things get even worse when bending and torsion are also considered, as these can bring the total effective stress close to failure limits.

If we decide to scale up the actress's body by keeping the axial stress constant by increasing the diameter of the femur, then from Eq. (13.4.4) we have that

$$(\text{Axial stress})_{\text{model}} = (\text{Axial stress})_{\text{prototype}} \Rightarrow \frac{m_{\text{model}}g}{\frac{\pi d^2_{\text{femur,model}}}{4}}$$
$$= \frac{m_{\text{prototype}}g}{\frac{\pi d^2_{\text{femur,prototype}}}{4}} \Rightarrow \frac{\text{m}_{\text{model}}}{d^2_{\text{femur,model}}} = \frac{m_{\text{prototype}}}{d^2_{\text{femur,prototype}}}$$

or

$$d_{\text{femur,prototype}} = d_{\text{femur,model}}\sqrt{\frac{m_{\text{prototype}}}{m_{\text{model}}}} \approx d_{\text{femur,model}}\,\lambda^{3/2} \tag{13.4.5}$$

The diameter of the femur of the prototype that will keep the axial stress constant is approximately

$$d_{\text{femur,prototype}} \approx \lambda^{3/2} d_{\text{femur,model}} = 9^{1.5} \times 0.025 = 27 \times 0.025$$
$$= 0.675 \text{ m}$$

For linear scaling

$$d_{\text{femur,prototype}} \approx \lambda d_{\text{femur,model}} = 9 \times 0.025 = 0.225 \text{ m}$$

i.e., structural integrity consideration would result in a totally deformed body of the prototype with a thigh 3 times thicker relative to the body of a linearly scaled human.

Imagine scaling up a human body to create a prototype where the thigh relative diameter is 3 times greater than that of the original actress. At this scale, the femur alone would appear massively oversized, raising significant questions about the commercial success of the movie. Scaling up any biological system is not simply a matter of proportional enlargement—it introduces serious biomechanical and material challenges. This leads us to a critical question: How can we scale up the human body safely while preserving both its structural soundness and visual believability, especially in a context designed for commercial appeal, such as a film? To address this, we must consider how volume, mass, and strength scale differently:

- Mass scales with the cube of the size increase ($\propto L^3$)
- Cross-sectional area (and strength) scales with the square of the size increase ($\propto L^2$)

Therefore, as size increases, stress on bones and joints increases disproportionately. In other words, simply enlarging the actress's body using a constant scaling factor may result in a visually convincing model, but one that would collapse under its own weight if built in reality. This is a classic example of how engineering principles like scaling laws must be considered alongside artistic and commercial goals. In the following sections, we will explore how to apply mechanical reasoning, anatomical modeling, and scaling analysis to ensure that an enlarged figure—like a 50-foot-tall woman—can be both structurally plausible and functionally consistent.

Consider now a classical problem in chemical engineering, the scaling up of an agitated, aerated, and jacketed process tank. Such a tank is very common in the process industries for performing chemical or biochemical reactions. Experiments are normally performed in the laboratory in 1–10 L reactors while the industrially relevant reactors can be 50000 L in volume, so $\lambda = 10^3 \sim 10^4$. Before going into the details of scaling up we will introduce several concepts relevant to the operation of the equipment under consideration. We will concentrate mainly on bioreactors as their scale-up is more complex and their commercial importance more crucial when compared to classical chemical reactors.

13.4.2 The power requirements for mixing nonaerated fluids

In Fig. 13.13 the details of an agitated and **baffled bioreactor** are shown. The height of the liquid in the tank H_L is usually equal to the diameter of the tank D_T while the diameter of the impeller D_I is 1/3 of the diameter of the tank.

Baffles are internal structures commonly installed in tanks used for mixing, reaction, or storage processes to prevent the formation of a vortex during agitation, which can reduce mixing efficiency, and to promote turbulence, to facilitate more effective

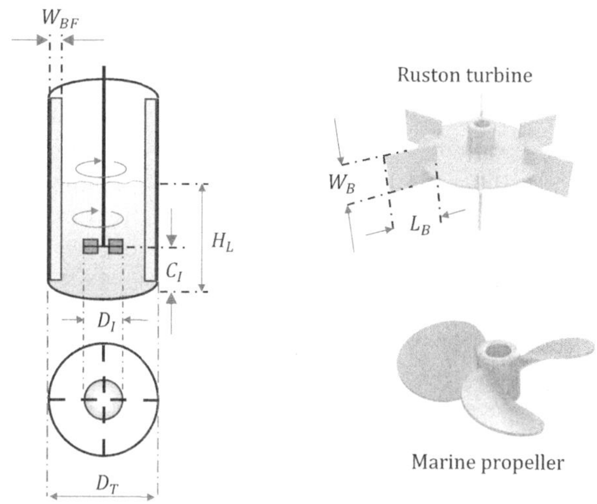

Figure 13.13 Basic geometric characteristics of nonaerated mixing tanks.

axial and radial mixing patterns by interrupting the circular flow of liquid, and also help to eliminate dead zones—areas of stagnant fluid. This results in improved heat and mass transfer, as well as a more uniform distribution of temperature, concentration, and suspended solids within the tank.

There are certain dimensionless numbers used to design and operate efficiently nonaerated mixing tanks. These are closely related to the power delivered by the impeller to the fluid. The power required to achieve a given stirrer speed depends on the magnitude of the frictional forces and form drag that resist rotation of the impeller and friction in the motor and gearbox. The **power number** N_P is an extremely useful number through which we can estimate the quality of mixing in a bioreactor. It is defined as follows

$$N_P = \frac{P}{\rho N_I^3 D_I^5} \tag{13.4.5}$$

where P is the power, ρ is the fluid density, and N_I, D_I are the stirrer speed (in revolutions per s or 1/s) and diameter (usually selected in the range 1/3 to 1/2 for Rushton turbines). To see how this dimensionless variable appears in the analysis of mixing tanks, let us assume that there is a functional relationship of the power of agitation (P), fluid density (ρ) and viscosity (μ) and the impeller rotational speed and diameter:

$$f(P, \rho, \mu, N_I, D_I) = 0 \tag{13.4.6}$$

The dimensions of the variables are (see Table 13.1):

$$[P] = L^2MT^{-3}, \quad [\rho] = L^{-3}M, \quad [\mu] = L^{-1}MT^{-1}, \quad [N_I] = T^{-1},$$
$$[D_I] = L$$

i.e., there are five parameters, three dimensions, and two dimensionless groups. We choose ρ, D_I, and N_I as the repeating variables. We then have

$$\pi_1 = P{\cdot}\rho^a D_I^b N_I^c$$

or

$$L^0M^0T^0 = L^2MT^{-3}{\cdot}(L^{-3}M)^a(L)^b(T^{-1})^c$$

It follows that $c = -3$, $a = -1$ and $b = -5$, or

$$\pi_1 = N_P = \frac{P}{\rho N_I^3 D_I^5} \tag{13.4.7}$$

We also have that

$$\pi_2 = \mu{\cdot}\rho^a D_I^b N_I^c$$

or

$$L^0M^0T^0 = L^{-1}MT^{-1}{\cdot}(L^{-3}M)^a(L)^b(T^{-1})^c$$

It follows that $c = -1$, $a = -1$ and $b = -2$, or

$$\frac{1}{\pi_2} = N_{Re} = \frac{\rho N_I D_I^2}{\mu} \tag{13.4.8}$$

i.e, π_2 is the inverse of a modified Reynolds number. Dimensional analysis predicts the potential existence of a functional relationship

$$N_P = f(N_{Re}) \tag{13.4.9}$$

Experimental data for the power number as a function of the Reynolds number is shown in Fig. 13.14. From this figure we conclude that for $N_{Re} < 10$.

$$N_P\,N_{Re} = \frac{P}{\rho N_I^3 D_I^5} \cdot \frac{\rho N_I D_I^2}{\mu} = \frac{P}{\mu N_I^2 D_I^3} \approx 70, \quad N_{Re} < 10 \tag{13.4.10}$$

while for $N_{Re} > 10^3$, which is the area of interest, it holds true that

$$N_P = \frac{P}{\rho N_I^3 D_I^5} \approx 5, \quad N_{Re} > 10^3 \tag{13.4.11}$$

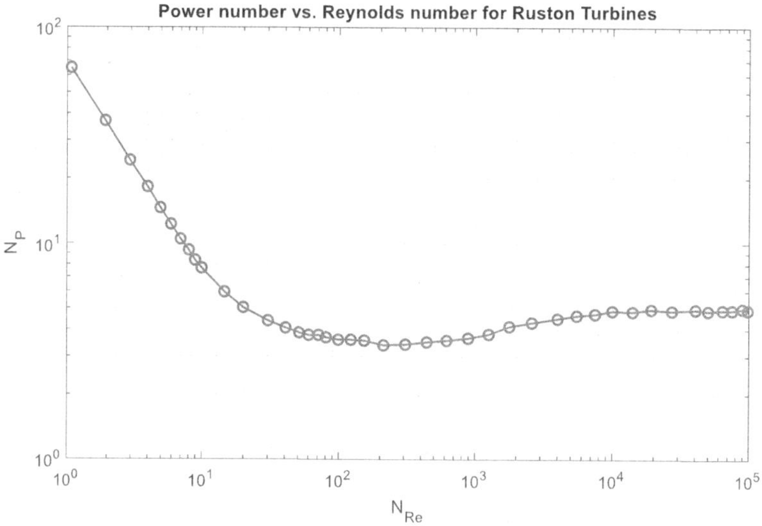

Figure 13.14 Experimental data of the power number as a function of the Reynolds number. Data taken from Bates et al. 1963. An examination of some geometric parameters of impeller power. I&EC Proc. Des. and Develop. 2 (4), p. 310.

The experimental results validate the predictions of the dimensional analysis.

Another key parameter in the design and operation of mixing and reaction vessels is the **mixing time** (θ_{mix}). This refers to the time required for the contents of the tank to reach a specified level of uniformity, starting from an initially segregated state. In practice, mixing time is often determined by introducing a tracer—a detectable substance—into the system and monitoring its concentration over time at a fixed location within the tank. Common tracers include acidic or basic solutions, or concentrated salts, which can be tracked using pH sensors or conductivity probes, respectively. An alternative approach involves adding a small volume of liquid at a different temperature and recording the thermal response using a temperature probe. This method is particularly useful in nonreactive systems or when chemical tracers might interfere with the process. Mixing time is a critical indicator of how efficiently a system distributes components and is especially important in processes where reaction kinetics or product quality depend on uniform mixing (Fig. 13.15).

It is reasonable to anticipate that the time required to achieve complete mixing in a stirred tank will be influenced by several operational and design factors, including the dimensions of the vessel and impeller, the physical characteristics of the fluid, and the speed of agitation. Through experimental studies—particularly involving common impellers such as the Rushton turbine

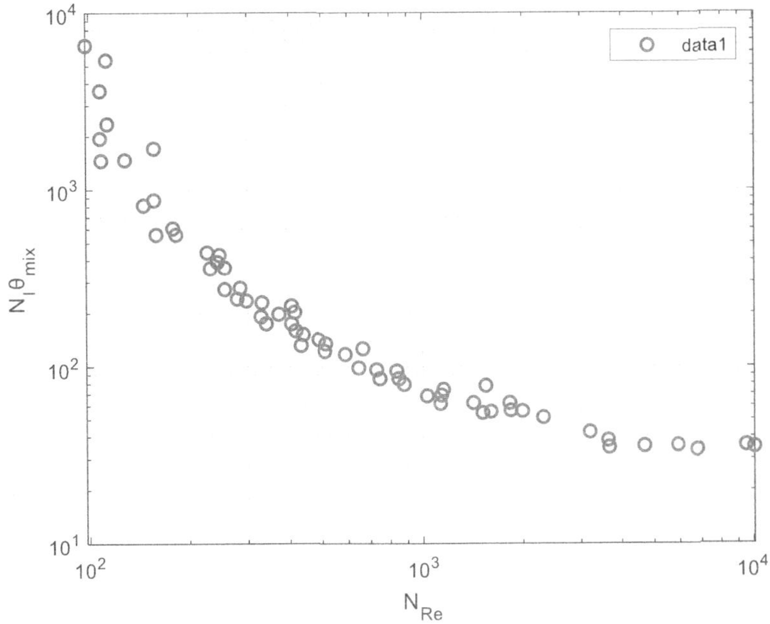

Figure 13.15 Dimensionless mixing time as a function of Reynolds number for a Rushton turbine in a baffled tank. ($D_I = C_I = D_T/3, H_L = D_T$; 4 *baffles* 10% *of* D_T). Data from Hoogendoorn, C.J., den Hartog, A.P. 1967. Model studies on mixers in the viscous flow region. Chem. Eng. Sci. 22, p. 1689.

in baffled tanks—researchers have characterized how these variables affect mixing dynamics. One widely used parameter in this analysis is the dimensionless mixing time $N_I\theta_{mix}$, where the mixing time is defined as the time needed for the system to reach within 10% of the final concentration after a tracer is introduced.

When plotted against the impeller Reynolds number (N_{Re}), $N_I\theta_{mix}$ provides insight into flow regime behavior. Such a figure is shown in Fig. 13.15. In the low to intermediate Reynolds number range $N_I\theta_{mix}$ increases sharply as flow becomes more viscous and mixing becomes less efficient. However, as the system enters the fully turbulent regime (typically beyond $N_{Re} \approx 5000$), this dimensionless mixing time approaches a plateau, indicating that additional increases in Reynolds number do not significantly reduce the number of revolutions needed to achieve homogeneity. This trend is observed across a variety of impeller designs and is considered a general characteristic of turbulent mixing. In practical terms, once turbulence is fully developed, mixing time becomes inversely proportional to the impeller speed, meaning that increasing the rotational rate directly decreases the time required to achieve uniformity. For Rushton turbines and in high N_{Re}, $N_I\theta_{\text{mix}}$ can be approximated by the following equation:

$$N_I\theta_{\text{mix}} \approx 3.16\left(\frac{D_T}{D_I}\right)^2, \quad N_{Re} > 5 \times 10^3 \qquad (13.4.12)$$

13.4.3 Scale up of agitated, baffled, and jacketed tanks

Stirred-tank reactors (STRs) remain a cornerstone technology in bioprocessing and are among the most widely adopted bioreactor configurations for both research and industrial-scale applications. Their popularity stems from several key advantages, including effective fluid circulation, efficient gas–liquid mass transfer—particularly for oxygen delivery—and their compatibility with a broad range of impeller designs. Moreover, STRs are well-aligned with regulatory expectations and good manufacturing practice (GMP) standards, making them suitable for commercial biopharmaceutical production.

Despite these strengths, STRs are not without challenges. Their operation typically involves high energy input, and the mechanical agitation can generate considerable shear forces, especially in systems with high rotational speeds or narrow clearances. These shear conditions may be tolerable or even beneficial in microbial cultures, but they pose a significant threat to the viability of more fragile cell types, such as mammalian or plant cells, which are commonly used in the production of complex biologics. Additionally, in large-scale or vertically extended STRs, mechanical design issues such as shaft stability, bearing wear, and reliable sealing become more pronounced and require careful engineering solutions.

To address the sensitivity of shear-prone cells, extensive efforts have been made to redesign impeller geometries and optimize agitation strategies. This includes the use of low-shear impellers, such as marine blades or segmented helical ribbons, and strategies that reduce tip speed while maintaining adequate mixing and oxygen transfer. The aim is to strike a balance between achieving homogeneity and preserving cell integrity under hydrodynamic stress.

Remarkably, these innovations have enabled the successful deployment of STRs at commercial production volumes ranging from 10,000 to 20,000 L for mammalian cell culture processes. These large-scale systems support the manufacturing of therapeutic proteins, monoclonal antibodies, and other biologics, highlighting the scalability and adaptability of STR technology when appropriately optimized for the biological system in use.

The primary goal in selecting, designing, and operating a bioreactor is to create an optimal environment for biological activity by controlling key parameters such as temperature, pH, oxygen transfer, mixing, and nutrient availability. Understanding the interplay between cell metabolism, fluid dynamics, and mass transfer is essential for effective bioreactor operation. Adequate mixing ensures even distribution of nutrients and prevents local accumulation of inhibitory byproducts. In suspension cultures, mixing time is a critical performance metric influenced by factors such as fluid rheology, impeller design, power input, and vessel geometry.

Reactors are central to processes in chemical technology and biotechnology, including pharmaceuticals, fermentation, food, and environmental engineering. A major challenge during scale-up is maintaining performance observed at small scale. Loss of productivity at larger volumes often results from poor control of shear, oxygen delivery, and gas-liquid interactions. To guide scale-up, engineers often aim to maintain key parameters such as

- specific power input (P/V),
- mixing time,
- gas flow per volume (vvm) or superficial gas velocity,
- the volumetric oxygen transfer coefficient (k_La),

all of which directly affect cell health and process efficiency. Scale-up can be performing by setting a single criterion for scale-up and accepting that all other criteria will be violated to a smaller or larger extent. Scaling up a reactor based on the constant specific power input criterion may result in a different mixing time. Let us investigate a few specific scale-up strategies.

Scale up at constant specific (per **unit volume) power input**. Keeping the power per unit volume (P/V) is a widely used criterion in reactor scale-up, especially in bioprocesses. In general, this scale-up strategy is used for processes that are not aerated or where the transfer of oxygen does not turn out to be so significant. We have shown that in bioreactors (cylindrical tanks with baffles and stirred by impellers) the power number is a function mainly of the Reynolds number. For large Reynolds numbers Eq. (13.4.11) offers a reasonable approximation. If we use the subscript m for the laboratory equipment (model) and the subscript p for the prototype (industrial-scale) equipment, then, assuming that the Reynolds number is high enough:

$$P_m \approx 5\rho N_{I,m}^3 D_{I,m}^5 \qquad (13.4.13)$$

$$P_p \approx 5\rho N_{I,p}^3 D_{I,p}^5 \tag{13.4.14}$$

If both the laboratory reactor and the industrial reactor are cylindrical tanks with the height of the liquid equal to the diameter of the tank, then the volume of the reacting mixture is

$$V_m = \frac{\pi D_{T,m}^2}{4} H_{L,m} = \frac{\pi}{4} D_{T,m}^3 \tag{13.4.15}$$

$$V_p = \frac{\pi D_{T,p}^2}{4} H_{L,p} = \frac{\pi}{4} D_{T,p}^3 \tag{13.4.16}$$

Keeping P/V constant results in the following condition

$$\left(\frac{P}{V}\right)_m = \frac{5\rho N_{I,m}^3 D_{I,m}^5}{\frac{\pi}{4} D_{T,m}^3} = \left(\frac{20\rho}{\pi}\right) \frac{N_{I,m}^3 D_{I,m}^5}{D_{T,m}^3}$$

$$\left(\frac{P}{V}\right)_p = \frac{5\rho N_{I,p}^3 D_{I,p}^5}{\frac{\pi}{4} D_{T,p}^3} = \left(\frac{20\rho}{\pi}\right) \frac{N_{I,p}^3 D_{I,p}^5}{D_{T,p}^3}$$

We then divide the two sides to obtain

$$N_{I,m}^3 D_{I,m}^2 \left(\frac{D_{I,m}}{D_{T,m}}\right)^3 = N_{I,p}^3 D_{I,p}^2 \left(\frac{D_{I,p}}{D_{T,p}}\right)^3$$

If a constant geometric factor is used then the ratio of the impeller diameter to the tank diameter is the same for both the model and the prototype, and the last equation simplifies to

$$N_{I,m}^3 D_{I,m}^2 = N_{I,p}^3 D_{I,p}^2 \tag{13.4.17}$$

Scaling up by a constant specific power consumption for mixing is achieved if the rotational speed for the prototype is selected according to the following equation:

$$N_{I,p} = N_{I,m} \left(\frac{D_{I,m}}{D_{I,p}}\right)^{2/3} = N_{I,m} \left(\frac{1}{\lambda}\right)^{2/3} \tag{13.4.18}$$

where λ is the constant geometric scaling factor calculated by

$$\lambda = \left(\frac{V_p}{V_m}\right)^{1/3} \tag{13.4.19}$$

How are the Reynolds numbers for the model and the prototype related? Let us start with the Reynolds number of the prototype and express it as a function of the Reynolds number of the model

$$N_{Re,p} = \frac{\rho N_{I,p} D_{I,p}^2}{\mu} = \frac{\rho}{\mu} N_{I,m} \left(\frac{1}{\lambda}\right)^{2/3} \lambda^2 D_{I,m}^2 = \frac{\rho N_{I,m} D_{I,m}^2}{\mu} \lambda^{4/3}$$
$$= \lambda^{4/3} N_{Re,m} \tag{13.4.20}$$

What about the mixing time? Let us again start with the mixing time of the prototype given by Eq. (13.4.12):

$$N_{I,p}\theta_{mix,p} \approx 3.16\left(\frac{D_{T,p}}{D_{I,p}}\right)^2 = 3.16\left(\frac{\lambda D_{T,m}}{\lambda D_{I,m}}\right)^2 = 3.16\left(\frac{D_{T,m}}{D_{I,m}}\right)^2 = N_{I,m}\theta_{mix,m}$$

or, using Eq. (13.4.18)

$$\theta_{mix,p} = \lambda^{2/3} \theta_{mix,m} \tag{13.4.21}$$

Scale up at constant mixing time. If the mixing time is kept constant when scaling up the model to obtain the prototype, then from Eq. (13.4.12) we have that the rotational speed needs to be kept constant:

$$N_{I,p} = N_{I,m} \tag{13.4.22}$$

For the Reynolds number we obtain the following result:

$$N_{Re,p} = \frac{\rho N_{I,p} D_{I,p}^2}{\mu} = \frac{\rho N_{I,m} \lambda^2 D_{I,m}^2}{\mu} = \lambda^2 N_{Re,m} \tag{13.4.23}$$

while for power consumption

$$\left(\frac{P}{V}\right)_p = \frac{5\rho N_{I,p}^3 D_{I,p}^5}{\frac{\pi}{4} D_{T,p}^3} = \left(\frac{20\rho}{\pi}\right) \frac{N_{I,m}^3 \lambda^5 D_{I,m}^5}{\lambda^3 D_{T,m}^3} = \lambda^2 \left(\frac{20\rho}{\pi}\right) \frac{N_{I,m}^3 D_{I,m}^5}{D_{T,m}^3}$$

or

$$\left(\frac{P}{V}\right)_p = \lambda^2 \left(\frac{P}{V}\right)_m \tag{13.4.24}$$

i.e., both the Reynolds number and the power consumption per unit volume increase significantly in the prototype, and this is probably a reason to avoid using the equal mixing time criterion for scaling-up purposes.

Scale up at constant Reynolds number. When the Reynolds number is kept constant during scaling up, it follows that

$$N_{Re,p} = N_{Re,m} \Rightarrow \frac{\rho N_{I,p} D_{I,p}^2}{\mu} = \frac{\rho N_{I,m} D_{I,m}^2}{\mu}$$

or

$$N_{I,p} = N_{I,m}\left(\frac{D_{I,m}^2}{D_{I,p}^2}\right) = N_{I,m}\left(\frac{1}{\lambda^2}\right) \tag{13.4.25}$$

For the mixing time we have that

$$N_{I,p}\theta_{\text{mix},p} = N_{I,m}\theta_{\text{mix},m}$$

or

$$\theta_{\text{mix},p} = \lambda^2\theta_{\text{mix},m} \tag{13.4.26}$$

For power consumption

$$\left(\frac{P}{V}\right)_p = \frac{5\rho N_{I,p}^3 D_{I,p}^5}{\frac{\pi}{4}D_{T,p}^3} = \left(\frac{20\rho}{\pi}\right)\frac{N_{I,m}^3\left(\frac{1}{\lambda^2}\right)^3\lambda^5 D_{I,m}^5}{\lambda^3 D_{T,m}^3} = \frac{1}{\lambda^4}\left(\frac{20\rho}{\pi}\right)\frac{N_{I,m}^3 D_{I,m}^5}{D_{T,m}^3}$$

or

$$\left(\frac{P}{V}\right)_p = \frac{1}{\lambda^4}\left(\frac{P}{V}\right)_m \tag{13.4.27}$$

i.e., the power consumption per unit volume decreases significantly in the prototype while the mixing time also increases significantly, and this can be a limiting factor.

Scale up at constant impeller tip speed. The impeller tip speed is calculated by the following equation

$$u_{I,tip} = \pi D_I N_I \tag{13.4.28}$$

Keeping the impeller tip speed constant when scaling up to the industrial-size tank results in the following equality:

$$D_{I,p}N_{I,p} = D_{I,m}N_{I,m} \tag{13.4.29}$$

or, using the volume geometric scaling factor λ

$$N_{I,p} = \frac{1}{\lambda}N_{I,m} \tag{13.4.30}$$

For the Reynolds number we obtain the following result:

$$N_{Re,p} = \frac{\rho N_{I,p} D_{I,p}^2}{\mu} = \frac{\rho \frac{1}{\lambda} N_{I,m}\lambda^2 D_{I,m}^2}{\mu} = \lambda\, N_{Re,m} \tag{13.4.31}$$

while for power consumption

$$\left(\frac{P}{V}\right)_p = \frac{5\rho N_{I,p}^3 D_{I,p}^5}{\frac{\pi}{4} D_{T,p}^3} = \left(\frac{20\rho}{\pi}\right) \frac{\frac{1}{\lambda^3} N_{I,m}^3 \lambda^5 D_{I,m}^5}{\lambda^3 D_{T,m}^3} = \frac{1}{\lambda}\left(\frac{20\rho}{\pi}\right) \frac{N_{I,m}^3 D_{I,m}^5}{D_{T,m}^3}$$

or

$$\left(\frac{P}{V}\right)_p = \frac{1}{\lambda}\left(\frac{P}{V}\right)_m \tag{13.4.32}$$

The mixing time of the prototype given by Eq. (13.4.12):

$$N_{I,p}\theta_{\text{mix},p} \approx 3.16\left(\frac{D_{T,p}}{D_{I,p}}\right)^2 = 3.16\left(\frac{\lambda D_{T,m}}{\lambda D_{I,m}}\right)^2 = 3.16\left(\frac{D_{T,m}}{D_{I,m}}\right)^2 = N_{I,m}\theta_{\text{mix},m}$$

or, using Eq. (13.4.18)

$$\theta_{\text{mix},p} = \lambda \theta_{\text{mix},m} \tag{13.4.21}$$

i.e., both the Reynolds number and the mixing time increase linearly with the geometric scaling factor while the power consumption decreases with increasing scaling factor.

In Table 13.3, the results of the analysis for the three alternative scale-up criteria are shown for ease of reference.

Example 13.5. Scaling up a mixing tank

A laboratory $V_{T,m} = 74$ L reactor with a diameter $D_{T,m} = 0.455$ m, is equipped with a six blade, Rushton turbine with diameter $D_{I,m} = 0.15$ m. The laboratory reactor is operated using $N_{I,m} = 15$ 1/s and the power consumed is 90 W.

It has been estimated that we need to increase the volume to $V_{T,p} = 9.25$ m^3 to achieve the requested production rate. Can you propose a reasonable scale-up strategy (see Fig. 13.16)?

The fluid properties can be assumed identical to the properties of liquid water at 60 °C, if needed.

Solution.

We first estimate the geometric scaling factor

$$\lambda = \left(\frac{V_{T,p}}{V_{T,m}}\right)^{1/3} = \left(\frac{9.25}{0.074}\right)^{1/3} = 5$$

Constant specific power consumption scale-up strategy. We calculate the specific energy consumption in the pilot-scale reactor

Table 13.3 Summary of results of different scale-up strategies.

Scale-up criterion	$\left(\frac{P}{V}\right)_p / \left(\frac{P}{V}\right)_m$	$\theta_{mix,p}/\theta_{mix,m}$	$N_{Re,p}/N_{Re,m}$	$u_{I,tip,p}/u_{I,tip,m}$
Constant $\left(\frac{P}{V}\right)$	1	$\lambda^{2/3}$	$\lambda^{4/3}$	$\lambda^{1/3}$
Constant θ_{mix}	λ^2	1	λ^2	λ
Constant N_{Re}	$\frac{1}{\lambda^4}$	λ^2	1	$\frac{1}{\lambda}$
Constant tip speed ($u_{I,tip}$)	$\frac{1}{\lambda}$	λ	λ	1

$$\left(\frac{P}{V}\right)_m = \frac{0.090\ \text{kW}}{0.074\ \text{m}^3} = 1.216\ \frac{\text{kW}}{\text{m}^3}$$

The specific power consumption for the prototype reactor is

$$\left(\frac{P}{V}\right)_p = \left(\frac{P}{V}\right)_m = 1.216\ \frac{\text{kW}}{\text{m}^3} \Rightarrow P_P$$
$$= 1.216\ \frac{\text{kW}}{\text{m}^3} \times 9.25\ \text{m}^3 = 11.25\ \text{kW}$$

We estimate the Reynolds number for the pilot-scale reactor:

Figure 13.16 Scale up of a nonaerated mixing tank.

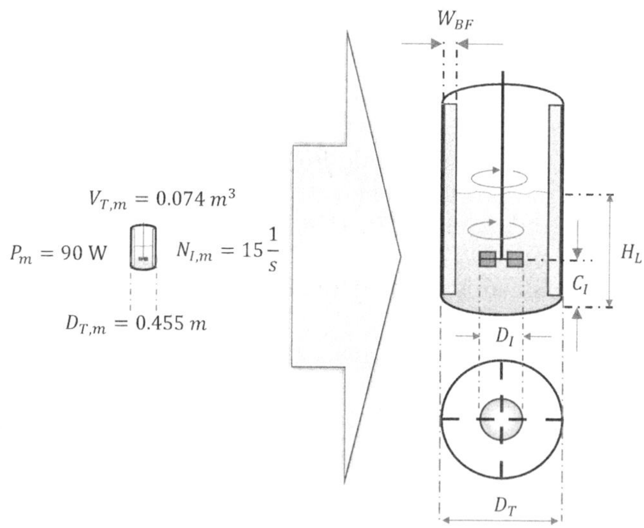

$$N_{Re} = \frac{\rho N_{I,p} D_{I,p}^2}{\mu} = \frac{983.13\ \dfrac{\text{kg}}{\text{m}^3} \times 15\ \dfrac{1}{\text{s}} \times 0.15\ \text{m}^2}{0.000467\ \dfrac{\text{kg}}{\text{s}\cdot\text{m}}} = 4.74 \times 10^6$$

We use Eq. (3.4.12) to estimate the mixing time for the pilot-scale reactor

$$\theta_{\text{mix},m} \approx \frac{3.16}{N_{I,m}} \left(\frac{D_{T,m}}{D_{I,m}} \right)^2 = \frac{3.16}{15\ \dfrac{1}{\text{s}}} \left(\frac{0.45}{0.15} \right)^2 = 1.9\ \text{s}$$

The mixing time for the industrial-scale reactor is

$$\theta_{\text{mix},p} = \lambda^{2/3} \theta_{\text{mix},m} = 5^{2/3} \times 1.9\ \text{s} = 5.56\ \text{s}$$

The calculations are summarized in Table 13.4. Based on this table a comprehensive system analysis can be performed based on the operational characteristics of the process.

Chapter summary

In this chapter we have introduced the idea of nondimensional equations and a method for nondimensionalizing algebraic or differential equations. We have also introduced a powerful and systematic methodology for developing functional relationships among dimensionless variables for complex chemical and physical phenomena that are not amenable to classical methods of model building. We have also discussed briefly the idea of process scale-up and have given a detailed analysis of the scale-up of nonaerated mixing tanks and (anaerobic bio)reactors.

Problems

13.1 Consider an isothermal CSTR where the following first-order reactions are taking place: $A \xrightarrow{k_1} B \xrightarrow{k_2} C$. The volume V of the reactor is constant and the volumetric flowrate of the feed is constant.

a Show that the material balances for A and B are the following

$$V \frac{dC_A}{dt} = F(C_{A,0} - C_A) - V k_1 C_A$$

$$V\frac{dC_B}{dt} = F\left(\underbrace{C_{B,0}}_{=0} - C_B\right) + Vk_1C_A - Vk_2C_B$$

where C_A is the concentration of A in the reactor, $C_{A,0}$ is the concentration of A in the feed, C_B is the concentration of B in the reactor, $C_{B,0}$ is the concentration of B in the feed, and k_1, k_2 are the reaction rate constants.

b Define a dimensionless time and dimensionless concentrations and derive a dimensionless model of the CSTR.

13.2 A metallic cylindrical bar with a diameter D and a length L, has a constant temperature T_0 at its base. The bar is exposed to a flowing stream of water with temperature T_∞. The bar is losing heat to the flowing stream of water, and the heat flux is described by Newton's law:

$$\dot{q}'' = h(T(x) - T_\infty)$$

The heat is transferred in the metallic bar due to conduction described by Fourier's law:

$$\dot{q}''_{x,\text{cond}} = -k\frac{dT(x)}{dx}$$

where k is the thermal diffusivity.

Table 13.4 Summary of the characteristics of the industrial-scale reactor and of the pilot-scale reactor using different scale-up criteria.

Scale-up criterion	$\left(\frac{P}{V}\right)\left(\frac{kW}{m^3}\right)$	θ_{mix} (s)	N_{Re}	$u_{l,\text{tip},p}$ (m/s)
Pilot-scale reactor	1.216	1.9	4.74×10^6	7.07
Industrial-scale reactor				
Constant $\left(\frac{P}{V}\right)$	1.216	5.56	40.53×10^6	12.09
Constant θ_{mix}	30.4	1.9	118.50×10^6	35.34
Constant N_{Re}	0.002	47.5	4.74×10^6	1.41
Constant tip speed ($u_{l,\text{tip}}$)	0.2432	9.5	23.70×10^6	7.07

a Show that the steady-state energy balance can be written as

$$k\frac{d^2T(x)}{dx^2} = \left(\frac{4h}{D}\right)(T(x) - T_\infty),\ T(0) = T_0$$

b Derive a dimensionless form of the equation.

13.3 Consider a mechanical system consisting of a spring with constant k (N/m), a damper c (N s/m) and a mass m (kg). What are the dimensionless constants that determine the dynamics of the spring-mass-damper system?

13.4 Consider the case of two infinite parallel plates at a distance h with a fluid (density ρ, viscosity μ) trapped between them. The two plates are initially stationary. The top plate starts moving with a constant speed V. Initially the velocity of the fluid changes as a function of the time t. Find the dimensionless parameters necessary to describe the velocity of the fluid as a function of the distance y from the stationary plate, $u(y)$.

13.5 When tiny particles or droplets (density ρ_p and diameter d_p) fall in a fluid (density ρ_f, viscosity μ_f) the creeping flow approximation is often valid: the velocity of the particle u_p is practically constant and depends only on d_p, μ_f, g and the density difference $\rho_p - \rho_f$. Use dimensional analysis to discover the relevant dimensionless groups and propose a functional relationship among them.

13.6 When a free moving fluid (density ρ, viscosity μ) with uniform velocity u_∞ meets a solid horizontal surface a boundary layer forms (see Fig. P13.6). The thickness of the boundary layer $\delta(x)$ is the distance from the horizontal surface at which the disruption to the flow is insignificant. Usually, the distance at which $u(y) = 0.99u_\infty$. The thickness of the boundary layer is a function of x. Derive the important dimensionless variables of the problem.

13.7 An explosion takes place in the atmosphere, producing a shock (or blast) wave that propagates outward in all directions from the point of detonation. The pressure difference across the blast wave, denoted ΔP, and its radial distance from the center, $r(t)$, depends on time t, the speed of sound c, and the total energy E released by the explosion. Develop dimensionless expressions that relate ΔP to the other parameters.

13.8 A liquid with density ρ and dynamic viscosity μ flows under the influence of gravity through a hole of diameter d located at the bottom of a tank with overall diameter D. Initially, the

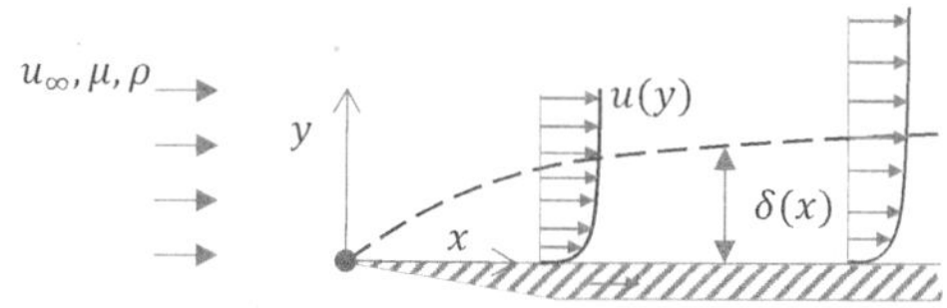

Figure P13.6 Representation of the boundary layer.

liquid surface is at a height h above the tank's bottom. The liquid exits as a downward jet with an average velocity u. Using dimensional analysis, derive a dimensionless relationship expressing u in terms of the other relevant parameters in the system. Since there are three characteristic lengths in this problem, use h as the reference length scale.

13.9 Suppose we are interested in studying the deformation of elastic balls upon impact with a rigid wall. One specific aspect of this investigation could be determining the factors that influence the size of the contact area during the collision. For instance, we may want to understand what governs the diameter d of the circular imprint left on the wall when a ball, freshly coated with dye, strikes the surface and then rebounds. This imprint provides a visual measure of the deformation that occurs at the moment of impact, and analyzing it could offer insights into the mechanics of collision.

13.10 If the (per unit volume) energy production of the human body remains constant, how can this affect the 50 Foot Woman? If, in the same spirit, the mixing tank studied in Example 13.5 has a constant volumetric heat production rate, what is the heat transfer problem that we are faced with?

13.11 A cylindrical tank, as the one shown in Fig. 13.3, has a diameter of $D_T = H_L = 3$ m and a Rushton turbine with $D_T/D_I = 3$ is used for mixing. The density of the fluid is approximately equal to that of water, $\rho = 1000$ kg/m^3. Generate a plot of the electricity cost per year for N_I in the range [1 1/s, 5 1/s] and a viscosity of $\mu = 0.001$ kg/(m s). The cost of electricity is 0.15 \$/kWh.

13.12 The baffled reactor of Fig. 13.3 has a volume $V_m = 0.1$ m^3 is scaled to $V_p = 6$ m^3. The pilot-scale reactor operates with a Rushton turbine with $N_I = 3$ 1/s. Propose a scale-up strategy. The fluid has a density of $\rho = 1000$ kg/m3 and a viscosity of $\mu = 0.005$ kg/(m s).

References

Sellers, W.I., Pond, S.B., Brassey, C.A., Manning, P.L., Bates, K.T., 2017. Investigating the running abilities of Tyrannosaurus rex using stress-constrained multibody dynamic analysis. PeerJ 5, e3420. https://doi.org/10.7717/peerj.3420.

Taylor, G.I., 1950. The formation of a blast wave by a very intense explosion I. Theoretical discussion. Proc. R. Soc. Lond. Ser. A Math. Phys. Sci. 201 (1064), 159–174.

Further reading

Books on dimensional analysis and scale-up.

Cengel, Y.A., Cimbala, J.M., 2018. Fluid Mechanics, fourth ed. McGraw Hill, NY.

Conejo, A.N., 2012. Fundamentals of Dimensional Analysis. Springer, Singapore.

Huntley, H.E., 1952. Dimensional Analysis. Macdonald and Co, London.

Ipsen, D.C., 1960. Units, Dimensions and Dimensionless Numbers. McGraw-Hill, NY.

Johnstone, R.E., Wooldridge Thring, M., 1957. Pilot Plants, Models, and Scale-Up Methods in Chemical Engineering. McGraw Hill, NY.

Lokarnik, M., 1991. Dimensional analysis and scale-up. In: Chemical Engineering. Springer Verlag, Berlin.

Sedov, L.I., 1959. Similarity and Dimensional Methods in Mechanics. Academic Press, NY.

Index

'*Note:* Page numbers followed by "f" indicate figures and "t" indicate tables.'

9780443454448